Electrolytes for Electrochemical Supercapacitors

ELECTROCHEMICAL ENERGY STORAGE AND CONVERSION

Series Editor: Jiujun Zhang
National Research Council Institute for Fuel Cell Innovation
Vancouver, British Columbia, Canada

Published Titles

Electrochemical Supercapacitors for Energy Storage and Delivery: Fundamentals and Applications
Aiping Yu, Victor Chabot, and Jiujun Zhang

Proton Exchange Membrane Fuel Cells
Zhigang Qi

Graphene: Energy Storage and Conversion Applications
Zhaoping Liu and Xufeng Zhou

Electrochemical Polymer Electrolyte Membranes
Jianhua Fang, Jinli Qiao, David P. Wilkinson, and Jiujun Zhang

Lithium-Ion Batteries: Fundamentals and Applications
Yuping Wu

Lead-Acid Battery Technologies: Fundamentals, Materials, and Applications
Joey Jung, Lei Zhang, and Jiujun Zhang

Solar Energy Conversion and Storage: Photochemical Modes
Suresh C. Ameta and Rakshit Ameta

Electrochemical Energy: Advanced Materials and Technologies
Pei Kang Shen, Chao-Yang Wang, San Ping Jiang, Xueliang Sun, and Jiujun Zhang

Electrolytes for Electrochemical Supercapacitors
Cheng Zhong, Yida Deng, Wenbin Hu, Daoming Sun, Xiaopeng Han, Jinli Qiao, and Jiujun Zhang

Electrolytes for Electrochemical Supercapacitors

Edited by

Cheng Zhong • Yida Deng • Wenbin Hu
Daoming Sun • Xiaopeng Han • Jinli Qiao
Jiujun Zhang

CRC Press
Taylor & Francis Group
Boca Raton London New York

CRC Press is an imprint of the
Taylor & Francis Group, an **informa** business

CRC Press
Taylor & Francis Group
6000 Broken Sound Parkway NW, Suite 300
Boca Raton, FL 33487-2742

First issued in paperback 2019

ISBN-13: 978-0-4987-4755-4 (hbk)
ISBN-13: 978-0-367-87536-7 (pbk)

Library of Congress Cataloging-in-Publication Data

Names: Zhong, Cheng (Electrical engineer), editor.
Title: Electrolytes for electrochemical supercapacitors / editors, Cheng Zhong, Yida Deng, Wenbin Hu, Daoming Sun, Xiaopeng Han, Jinli Qiao, Jiujun Zhang.
Description: Boca Raton : Taylor & Francis, 2016. | Series: Electrochemical energy storage and conversion | Includes bibliographical references and index.
Identifiers: LCCN 2015048071 | ISBN 9781498747554 (alk. paper)
Subjects: LCSH: Supercapacitors--Materials. | Electrolytes. | Electrolytic capacitors.
Classification: LCC TK7872.C65 E46 2016 | DDC 621.31/5--dc23
LC record available at http://lccn.loc.gov/2015048071

Visit the Taylor & Francis Web site at
http://www.taylorandfrancis.com

and the CRC Press Web site at
http://www.crcpress.com

Contents

Series Preface

The goal of the *Electrochemical Energy Storage and Conversion* book series is to provide comprehensive coverage of the field, with titles focusing on fundamentals, technologies, applications, and the latest developments, including secondary (or rechargeable) batteries, fuel cells, supercapacitors, CO_2 electroreduction to produce low-carbon fuels, electrolysis for hydrogen generation/storage, and photoelectrochemistry for water splitting to produce hydrogen, among others. Each book in this series is self-contained, written by scientists and engineers with strong academic and industrial expertise who are at the top of their fields and on the cutting edge of technology. With a broad view of various electrochemical energy conversion and storage devices, this unique book series provides essential reading for university students, scientists, and engineers, and allows them to easily locate the latest information on electrochemical technology, fundamentals, and applications.

Jiujun Zhang
National Research Council of Canada
Richmond, British Columbia

Preface

The electrochemical supercapacitor (ES), also known as a supercapacitor, ultracapacitor, or electrochemical capacitor, is a special type of capacitor that can have relatively high-energy density when compared to conventional dielectric capacitors and electrolytic capacitors. As an electrochemical energy storage/conversion device the ES has several advantages, such as fast charging–discharging capability, immense power density, wide operating temperature ranges, and long service lifetimes. As a result, ESs have been found to be readily applicable in a wide variety of important applications such as hybrid and electrical vehicles, portable electronics, backup power supplies, aircraft, and smart grids. In particular, they can play a significant role in the effective use of intermittent energies from renewable sources.

Although ESs are now available in commercial markets, they still face some challenges, such as relatively low-energy density compared to both batteries and fuel cells. This low-energy density will limit their wide applications in many high energy–demanding systems, and also cause high manufacturing cost. In recent years, there have been tremendous efforts focusing on the development of novel electrode materials and electrolytes as well as electrode/electrolyte configurations to improve the energy density for the next generation of supercapacitors.

For fundamental aspects, in Conway's excellent pioneering book (*Electrochemical Supercapacitors: Scientific Fundamentals and Technological Applications*), published in 1999, he provided a comprehensive discussion of the science of ESs in the twentieth century. In 2013, Yu et al. published *Electrochemical Supercapacitors for Energy Storage and Delivery: Fundamentals and Applications* and Béguin et al. published *Supercapacitors: Materials, Systems, and Applications*, covered comprehensive science and technology, as well as coverage of the latest developments of ESs including electrical double-layer capacitors and pseudocapacitors. The critical roles of both ES electrode materials and electrolytes in improving energy density are emphasized in these books.

On the basis of our recent review paper published in *Chemical Society Reviews* (2015, 44: 7484–7539), we have largely extended this paper's content and completed this book covering the key aspects of novel and advanced electrolyte systems and compositions. This book focuses on both the fundamental and technological aspects of all kinds of ES electrolytes. The effects of electrolytes on ES performance are highlighted. Some comparisons among different electrolytes in terms of their associated ES performance are presented. This book also discusses the compatibility issue or interactions between ES electrolytes, active components such as electrode materials, and inactive components such as current collectors, binders, additives, and separators. The book also illuminates the practical aspects of designing and optimizing electrolyte systems for improving ES performance. We have a strong desire to have this book be beneficial for the research and development activities of academic researchers, graduate/undergraduate students, industry professionals, and manufacturers of electrode/electrolyte systems, and also for other electrochemical energy devices such as batteries, and end users of these devices.

We acknowledge and express our sincere thanks to the contributions of all those who involved in preparing and developing this book. We also express our appreciation to CRC Press for giving us this great opportunity to lead this book project, particularly to Allison Shatkin, Jill Jurgensen, and Ariel Crockett for their guidance and support in the book preparation process. Finally, we are also grateful for the support of the National Natural Science Foundation for Distinguished Young Scholars of China (no. 51125016), and the National Natural Science Foundation of China (no. 51472178).

If there are any technical errors in the book, we would deeply welcome any constructive comments for further improvement.

Editors

Cheng Zhong is an associate professor in the School of Materials Science and Engineering at Tianjin University. Prior to joining the faculty at Tianjin University, he worked as an associate professor in the State Key Laboratory of Metal Matrix Composites, Department of Materials Science and Engineering at Shanghai Jiao Tong University. He earned his BSc and PhD in materials science from Fudan University in 2004 and 2009, respectively. Dr. Zhong's recent research interests focus on the development of electrochemical metallurgy methods for preparing micro/nanostructured materials for electrochemical and electrocatalysis applications.

Yida Deng is a professor in the School of Materials Science and Engineering, Tianjin University. He earned his PhD from Shanghai Jiao Tong University in 2006. Dr. Deng's research interests include metal and metal oxide nanostructures for electrochemical and energy applications.

Wenbin Hu is a professor and dean of the School of Materials Science and Engineering at Tianjin University. Prior to joining the faculty at Tianjin University, he worked as a professor in the Department of Materials Science and Engineering at Shanghai Jiao Tong University. He graduated from Central-South University with a BSc in 1988, and earned an MSc from Tianjin University in 1991. He earned a PhD from Central-South University in 1994. Dr. Hu is a member of the expert group on advanced structural and composite materials in the new materials field of China's 863 Program

(National High-Tech Research and Development Program). He received the support of the National Science Foundation for Distinguished Young Scholars of China in 2011. Dr. Hu's research interests focus on the design, synthesis, and characterization of advanced micro/nanomaterials for energy storage and conversion applications.

Daoming Sun is a chief technology officer at the Suzhou LiCeram Electronic Technology Co., Ltd (www.liceram.cn). He earned his PhD in physical electronics from Fudan University and holds a bachelor's degree in materials science and engineering from Xi'an University of Technology, China. His research focuses on the development of materials for electrochemical environmental sensors. Dr. Sun worked as a team leader at LiCeram, and he has successfully developed a novel high-sensitive sensor series that detects the time of wetness and corrosion current for the corrosion evaluation of metal.

Xiaopeng Han is currently a lecturer in the School of Materials Science and Engineering at Tianjin University. He earned his BSc in chemical engineering and technology from Tianjin University (2010) and PhD in material physics and chemistry from Nankai University (2015). Dr. Han's research interest focuses on functional materials with micro/nanostructures for oxygen electrocatalysis and metal-air batteries.

Jinli Qiao is a professor, PhD supervisor, and scientific core-competency leader at Donghua University, China. She earned her PhD in electrochemistry from Yamaguchi University, Japan, in 2004 before joining the National Institute of Advanced Industrial Science and Technology (AIST), Japan, as a research scientist. From 2004 to 2008, Dr. Qiao carried out seven fuel cell projects including two NEDO projects in Japan on the development of novel proton-conducting membranes, new binders for membrane electrode assembly (MEA)

fabrication, and nonplatinum catalysts. From 2008 to the present, she has carried out a total of 12 projects funded by the Chinese government including NNSF of China and IACEP of the Shanghai Science and Technology Committee (SSTC). Dr. Qiao has more than 20 years of scientific research experience, particularly in the areas of electrochemical material development and energy storage and conversion including polymer electrolyte membrane (PEM) fuel cells, supercapacitors, CO_2 electroreduction, and metal-air batteries.

Jiujun Zhang is a principal research officer at the National Research Council of Canada (NRC), and a fellow of the International Society of Electrochemistry (ISE). Dr. Zhang's technical expertise areas are electrochemistry, photoelectrochemistry, spectroelectrochemistry, electrocatalysis, fuel cells (PEMFC, SOFC, and DMFC), batteries, supercapacitors, and electrolysis. Dr. Zhang earned his BS and MSc in electrochemistry from Peking University in 1982 and 1985, respectively, and his PhD in electrochemistry from Wuhan University in 1988. Starting in 1990, he carried out three terms of postdoctoral research at the California Institute of Technology, York University, and the University of British Columbia. Dr. Zhang holds more than 14 adjunct professorships, including one at the University of Waterloo, one at the University of British Columbia, and one at Peking University. To date, Dr. Zhang has approximately 400 publications with more than 16,000 citations, including 230 refereed journal papers with an *H-index* of 60, 15 edited/coauthored books, 37 book chapters, 110 conference oral and keynote/invited presentations, as well as more than 10 US/EU/WO/JP/CA patents, and has produced over 90 industrial technical reports. Dr. Zhang serves as an editor/editorial board member for several international journals as well as an editor for a book series (Electrochemical Energy Storage and Conversion, CRC Press).

1 Fundamentals of Electrochemical Supercapacitors

Cheng Zhong, Yida Deng, Wenbin Hu, Jinli Qiao, and Jiujun Zhang

CONTENTS

1.1 INTRODUCTION AND APPLICATIONS OF SUPERCAPACITORS

With the increasing consumption of globally limited fossil fuels and the resultant environmental problems such as greenhouse gas emissions and other air pollutants, there is an urgent need to engage sustainable, renewable, and clean energy sources such as the sun, winds, and tides. Since most of these sustainable (renewable) energy sources are dependent on the time of day and location, it is highly desirable to develop related energy storage devices in order to effectively harness these intermittent energy sources. Electrochemical energy storage devices, such as batteries and electrochemical supercapacitors (ESs), will play an increasingly critical role in the sustainable and large-scale use of energy from renewable sources.

ESs, also referred to as supercapacitors, ultracapacitors, or electrochemical capacitors, have received a great deal of attention due to their high power density (potentially above 10 kW kg^{-1}), long lifetime compared to batteries, and much higher energy density compared to conventional dielectric and electrolytic capacitors [1]. Depending on the charge storage mechanism, ESs can be briefly grouped into electrical double-layer capacitors (EDLCs), pseudocapacitors, and hybrid capacitors, which will be discussed further in Section 1.3. To date, EDLCs make up the majority of commercial ESs, mainly due to their technical maturity. The role of ESs among various typical electrochemical energy storage devices can be clearly seen in terms of their power density and energy density (Ragone plot) as shown in

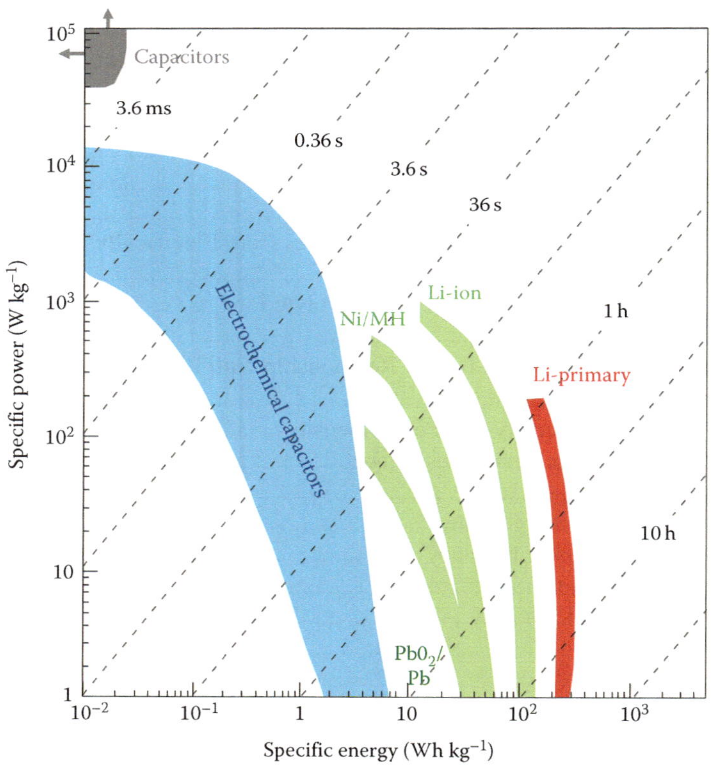

FIGURE 1.1 Ragone plots for representative energy storage devices of capacitors, batteries, and supercapacitors. (Reprinted by permission from Macmillan Publishers Ltd. *Nature Materials*, Simon, P., and Y. Gogotsi. Materials for electrochemical capacitors, 7:845–854, Copyright 2008.)

Figure 1.1 [2]. ESs currently fill the gap between conventional dielectric capacitors and batteries. Although ESs have a much lower energy density than batteries, they can store and release the energy in a very short time period (usually on the order of seconds in the case of EDLCs) and thus provide significantly higher power density. Therefore, besides the renewable energy storage applications, ESs are also very attractive for a wide range of important applications, such as [3] (1) backup energy storage in power electronics; (2) memory protection; (3) electric or hybrid electric vehicles, where ESs can provide high power density during acceleration and recover the energy during braking; (4) battery enhancement, where ESs can relieve batteries from the most severe load demands by meeting the peak power requirements and allowing the batteries to provide average loads; and (5) military and aerospace applications.

However, with the growing need of applications requiring high energy density, ESs currently suffer from their low energy densities compared to both batteries and fuel cells. For example, the energy density of commercial EDLCs is generally less than 10 Wh kg^{-1}. Even for pseudocapacitors and hybrid ESs, their energy densities are generally limited below 50 Wh kg^{-1} [1]. Therefore, a major effort has been focused on improving the energy density of ESs [4,5], in order to greatly expand ES applications in portable electronics, automobiles, and load leveling based on the use of intermittent renewable energy sources, etc. Since the energy density of ESs is proportional to the capacitance and the square of the voltage, increasing the capacitance and/or increasing the cell voltage are effective ways to improve the energy density. This can be achieved through the development of electrolytes, electrode materials, and integrated systems (i.e., a new structure of ESs with new concepts). Our recent review briefly summarized developments made so far in this area [6]. Furthermore, it is worth noting that increasing the cell voltage will contribute more to the improved energy density than enhancing the capacitance since the energy density is proportional to the square of the voltage. If the electrode materials are stable under the operating cell voltage range, the electrochemical stability of electrolytes will largely determine the working cell voltage of ESs. For example, the operating potential window of aqueous electrolyte-based ESs is generally about 1.0–1.3 V whereas those of organic electrolyte-based and ionic liquid (IL)-based ESs can have larger values of 2.5–2.7 and 3.5–4.0 V, respectively. Besides, the electrolytes also greatly affect other critical properties of ESs such as equivalent series resistance (ESR), power density, rate capability, operational temperature range, cycling stability, lifetime, and self-discharge rate (Figure 1.2). These properties are also crucial for the practical use of ESs. Furthermore, since all components (e.g., electrolyte, current collector, electrode material, related binders and additives, and the separator) of an ES are strongly coupled with each other, the interaction or compatibility between electrolytes and other components of ESs will also greatly affect the overall performance of ESs. For example, to achieve high capacitance and/or high power density, it is highly essential to match the electrolyte ion size and the pore size (including size distribution and pore structure) of electrode materials—especially carbon-based materials. It is possible that the interaction between electrolytes and current collectors could remarkably decrease the operative cell voltage of ESs and in turn reduce the energy density of ESs. It is also interesting to note that the development of some new electrolytes

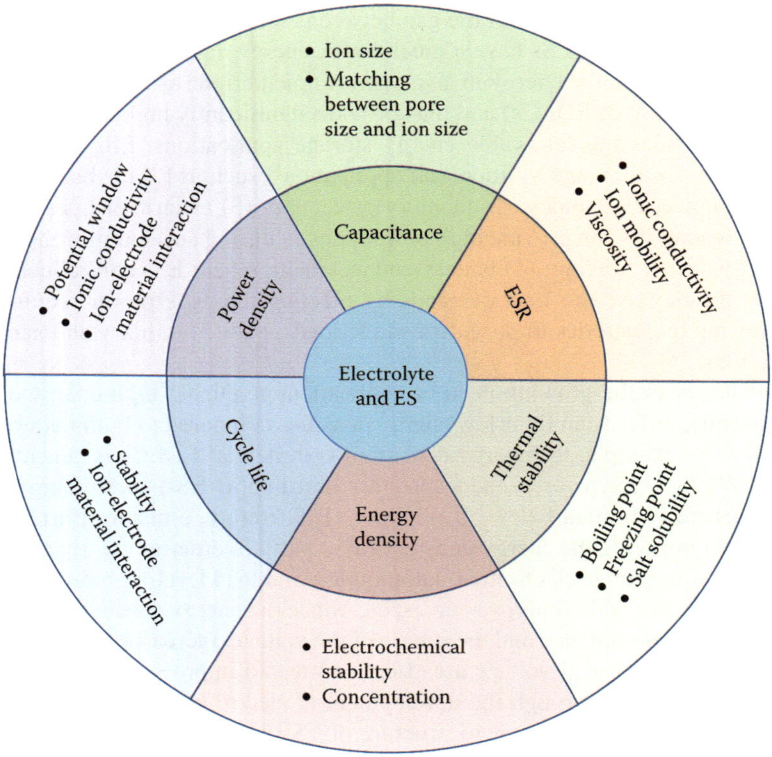

FIGURE 1.2 Effects of the electrolyte on the ES performance. (Zhong, C. et al. 2015. A review of electrolyte materials and compositions for electrochemical supercapacitors. *Chemical Society Reviews* 44:7484–7539. Reproduced by permission of The Royal Society of Chemistry.)

such as solid-state electrolytes has contributed to the invention or the development of some new types of ESs such as micro-ESs, stretchable ESs, and flexible ESs.

General requirements for an ideal electrolyte include high ionic conductivity; high chemical and electrochemical stability; chemical and electrochemical inertness to ES components such as electrodes, current collectors, and packages; good thermal stability, low volatility, and flammability; environmental friendliness; and low cost. In practical terms, it is extremely difficult for any electrolyte to meet all the above requirements, and each electrolyte has its own advantages and disadvantages. Therefore, tremendous efforts have been devoted to the exploration and development of novel electrolytes to contribute to the improvement of the overall performance of a given electrolyte and thus to the performance of the corresponding ESs. A wide variety of electrolyte types, such as aqueous, organic, ILs, solid or semisolid, and redox-active, have so far been investigated for use in ESs. To date, remarkable progress has been made in the development of specific electrolytes for specific ESs. For example, the energy density of aqueous electrolyte-based ESs has been reported to be greatly improved through the use of neutral aqueous electrolytes with a higher electrochemically stable window [7] or by the use of redox-active electrolytes that can provide extra pseudocapacitance [8]. The exploration of IL electrolytes could further increase the operating cell voltage of ESs up to 4 V. In order to address the toxicity issue and to widen the working cell voltage of commercial acetonitrile- and propylene carbonate-based ESs, some novel electrolytes with new conducting salts and/or organic solvents have been developed. The progress in research in solid or semisolid electrolytes has greatly facilitated the development of flexible and all-solid-state ESs and thereby addresses the potential leakage problem of liquid electrolytes and expands the use of ESs for various applications [9,10].

To date, many excellent reviews regarding electrode materials have been published [11–37]. However, there have been rather limited publications focusing on electrolytes for ESs [9,10,12,38–40], and most of them focused on polymer electrolytes for ESs [9,10,38,39]. In the light of the latest achievements in the field of ES electrolytes, the aim of this book is to provide an insight into the development of electrolytes for ESs. This chapter will discuss the fundamental concepts of ESs, including basic structure of different types of ESs, the development of electrical double-layer (EDL) theory, EDL formation in the geometrically confined pore, and performance metrics of various ESs and their relationship with electrolytes.

1.2 PRINCIPLES OF CAPACITORS

To understand the capacitive mechanism of ESs, it is necessary to briefly review the basic principles of conventional electrostatic capacitors. A capacitor is a device that stores energy by charge separation in an electric field and can release the stored charge (energy) in the form of electric current. Classical electrostatic capacitors consist of two parallel metal electrodes, which are separated by a thin dielectric material such as air, paper, polyester, and ceramic (Figure 1.3a). When metal electrodes are connected to an external circuit and a voltage is applied across the electrodes, electrons start flowing and the opposite charges accumulate on the electrode surface until the charges are balanced between the two electrodes. Capacitance is a measure of the stored charge for a given voltage between the electrodes. In the case of capacitors, the amount of the electric charge stored on each electrode (Q) is proportional to the voltage between the positive and negative electrodes (V) and the capacitance (C) is defined as

$$C = \frac{Q}{V} \tag{1.1}$$

The capacitance of conventional capacitors is proportional to the surface area of the electrodes (A), and inversely proportional to the distance between the electrodes (d):

$$C = \frac{\varepsilon_r \varepsilon_0 A}{d} \tag{1.2}$$

where ε_0 and ε_r are dielectric constants of the vacuum (8.854×10^{-12} F m^{-1}) and the dielectric material between the electrodes, respectively. As will be discussed below,

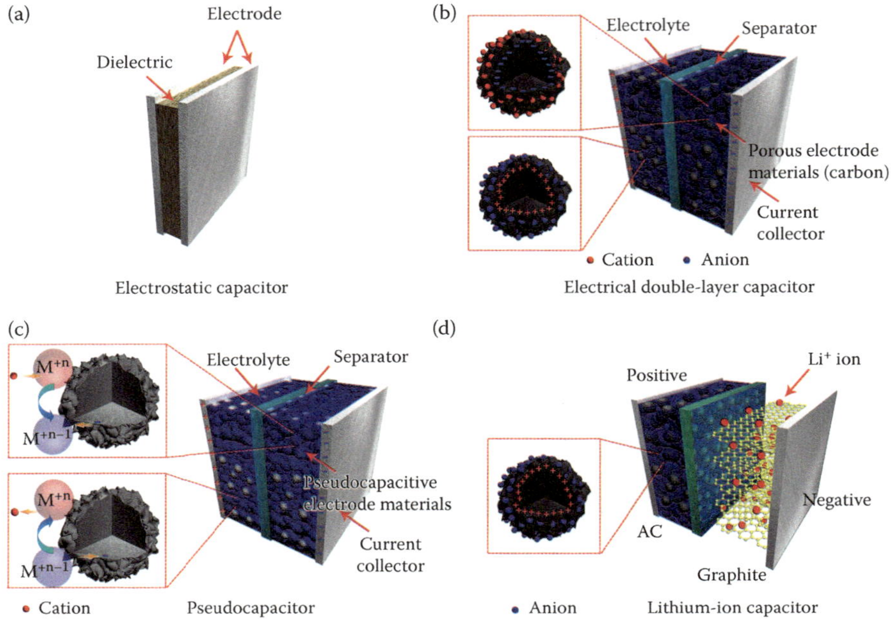

FIGURE 1.3 Schematic diagram for (a) electrostatic capacitor, (b) electric double-layer capacitor, (c) pseudocapacitor, and (d) hybrid capacitor. (Zhong, C. et al. 2015. A review of electrolyte materials and compositions for electrochemical supercapacitors. *Chemical Society Reviews* 44:7484–7539. Reproduced by permission of The Royal Society of Chemistry.)

C may not be constant with the changing V for EDL capacitance and pseudocapacitance. In this case, a differential capacitance is generally given and defined as [41]

$$C = \frac{dQ}{dV} \tag{1.3}$$

1.3 ELECTROCHEMICAL CAPACITORS

The ES is a special type of capacitor and can be distinguished by several ways such as the charge storage mechanism, the electrolyte, the cell structure, and the electrode material. Depending on the charge storage mechanism, ESs can be grouped into three categories: (1) electrical double-layer capacitors (EDLCs), where the capacitance is generated by the electrostatic charge separation at the electrode/electrolyte interface, as shown in Figure 1.3b; (2) pseudocapacitors, which are based on fast and reversible faradaic redox reactions to store the charges, as shown in Figure 1.3c; and (3) hybrid ESs. This type of ESs generally refers to ESs that use both the EDL and faradaic capacitive mechanisms to store charges. However, with the rapid developments in this area, some new battery-type hybrid devices (e.g., lithium-ion capacitor [LIC] [Figure 1.3d] and carbon//PbO_2) have been developed and are also regarded as hybrid ESs since they have one EDL or pseudocapacitive electrode although they have a rechargeable battery-type electrode. In this book, all of these types of ESs will be grouped into hybrid ESs as long as they have an EDL or pseudocapacitive electrode.

It should also be noted that although it is convenient to discuss the EDL capacitive and pseudocapacitive mechanisms separately, they usually coexist in actual ESs (where one of them plays a major role).

1.3.1 Electrical Double-Layer Capacitors

In the nineteenth century, von Helmholtz first introduced the term "double-layer" in his studies concerning colloidal particles [42]. It was only decades later, in 1957, that Becker from General Electric described the first EDLC, in which the capacitance arises from the charge stored in the EDL at the interface between the surface of porous carbon and an aqueous electrolyte [43].

1.3.1.1 Double-Layer Model

Developing high-performance ESs and related electrolytes requires a fundamental understanding of the electrochemical phenomena occurring at the interface between the electrode and the electrolyte. The understanding of the EDL formation at the electrode/electrolyte interface has developed gradually. As shown in Figure 1.4, the model of double-layer was first proposed by Helmholtz [42] and later developed by Gouy and Chapman [44,45], Stern [46], Grahame [47], and Bockris et al. [48]. In his pioneering work, Helmholtz [42] proposed a capacitor-like model, with two oppositely charged layers formed at the interface between the electrode and the electrolyte. The charged layers were separated by a very small distance depending on the size of ions in the electrolyte. He suggested that one layer is formed by the excess charge on the metal side and the other layer is formed due to the attraction of the oppositely charged ions in the solution. Thus, the term "double-layer" was coined. Figure 1.4a shows a schematic of the Helmholtz model, in which the potential decreases across the double-layer. The structure of the double-layer proposed by Helmholtz is similar to that of the conventional two-plate capacitors and the capacitance can be described by

$$C_H = \frac{\varepsilon_r \varepsilon_0 A}{d} \tag{1.4}$$

where ε_r is the dielectric constant of the electrolyte and d is the distance separating the charges (i.e., the thickness of the EDL). From Equation 1.4, it is seen that the capacitance of the EDL is not a function of the applied voltage and the ionic concentration of the electrolyte. This contradicts the experimental observations, and the capacitance values obtained by this model are much higher than the predicted values. In the early 1910s, Gouy and Chapman [44,45] independently developed a model (now termed the Gouy–Chapman theory), in which capacitance is dependent on the applied potential and the concentration of the electrolyte. As shown in Figure 1.4b, the Gouy–Chapman model considered the fact of the thermal motions of the adsorbed ions and their diffusion in the solution, and thus the surface charge of the electrode is balanced by a "diffuse layer" of oppositely charged ions instead of

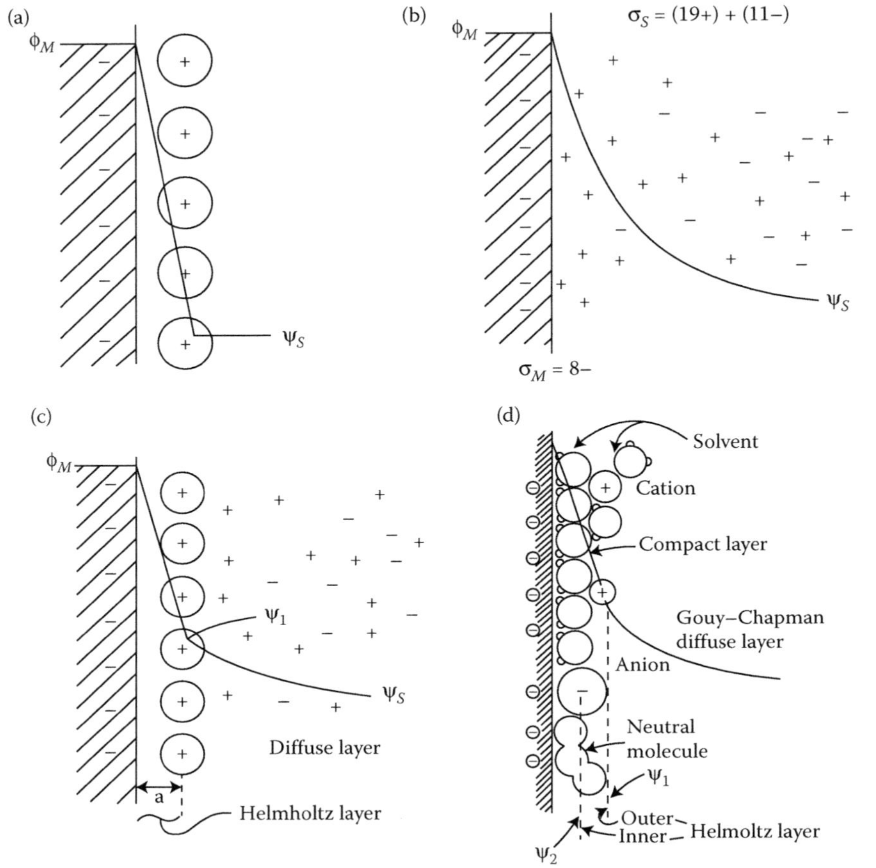

FIGURE 1.4 Double-layer models: (a) Helmholtz model, (b) Gouy–Chapman model, (c) Stern model, and (d) Grahame model. (With kind permission from Springer Science+Business Media: *Electrochemical Supercapacitors: Scientific Fundamentals and Technological Applications*, 1999, Conway, B.E. Originally published by Kluwer Academic/Plenum Publishers, New York in 1999.)

the rigid counter layer proposed by Helmholtz. This model treats the ions as point charges and used the Boltzmann equation to determine the concentration profiles of ions and the Poisson equation to relate the charge density with the potential. The resulting capacitance is described by [49]

$$C_D = \left(\frac{2\varepsilon\varepsilon_0 z^2 e_0^2 c_0}{kT} \right)^{1/2} \cosh \frac{z e_0 \psi_M}{2kT} \tag{1.5}$$

where z is the charge on the ion (coulombs), c_0 is the concentration of ion in the bulk solution, e_0 is the electron charge (1.6×10^{-19} C), k is the Boltzmann constant

(1.38×10^{-23} J K^{-1}), T is the temperature (K), and ψ_M is the potential at the metal surface (V). The Gouy–Chapman model can explain the effect of potential and temperature change on the capacitance compared to the Helmholtz model. However, since the Gouy–Chapman model is based on the assumption that the ions are point charges without considering the finite ion size, ionic charge distribution can be continuous up to the surface of the electrode [41]. This allows a large space charge at the distance very close to the electrode surface, leading to an unrealistically large calculated capacitance [41,50,51]. Later in 1924, Stern [46] modified the Gouy–Chapman model by accounting for both a compact layer of adsorbed ions with finite size very close to the electrode surface (Helmholtz layer, also referred to as the Stern layer) and a diffuse layer beyond this layer (Figure 1.4c). This is important for predicting capacitance in the electrolyte with high concentrations, which is often the case in ESs. Therefore, the total capacitance, C_T, of the double-layer consists of two capacitors in series, which is given by

$$\frac{1}{C_T} = \frac{1}{C_H} + \frac{1}{C_D} \tag{1.6}$$

where C_H is the capacitance of the Helmholtz layer and C_D is the capacitance of the diffuse layer (Gouy–Chapman). Since the smallest capacitance dominates the total capacitance, C_H determines the overall capacitance at high electrode potentials or high electrolyte concentrations while the contribution of C_D to the total capacitance becomes significant in less concentrated solutions. The Stern model was further refined by Grahame [47] in 1947, who discovered the effects of cation/anion and solvents in electrolytes on the formation of the EDL when he investigated the EDL capacitance on a mercury electrode. Figure 1.4d shows the schematic of the Grahame model and the corresponding potential distribution. Owing to the relatively small size of most cations compared to anions, cations tend to have a larger solvation shell. Therefore, cations and anions exhibit different distances to the charged electrode surface. Grahame further divided the Helmholtz layer into an inner Helmholtz layer consisting of specifically adsorbed ions and an outer Helmholtz layer consisting of solvated ions. The diffuse layer starts at the outer Helmholtz layer. Bockris et al. [48] further proposed that a water molecule layer is present in the inner Helmholtz layer and these water molecules near the electrode surface would be strongly oriented due to the charge interaction with the electrode. Other layers of water molecules would not be as oriented as the first layer. Specifically adsorbed ions would displace some of the water molecules within the inner Helmholtz layer. To date, the EDL model is still under development, as discussed by several critical reviews and books [41,52,53].

Based on the above discussion, there are a number of factors that affect the double-layer behavior and the corresponding EDL capacitance, such as the concentration and size of ions, the ion-specific adsorption, the ion–solvent interaction, and the solvent in the electrolytes. The thickness of the EDL is typically on the order of several angstroms in aqueous solution. Since the distance separating the charges in an EDL is extremely small, the specific capacitance (capacitance normalized by the effective surface area) of EDL can reach a very high value in the aqueous electrolyte. In contrast, the specific capacitance of a typical parallel-plate capacitor is quite small due to the relatively large charge separation distance (tens of nanometers to micrometers depending on the dielectric material). In addition, the electrodes of EDLCs are typically made from porous carbon materials with very large specific surface area (SSA) on the order of 2000 $m^2 g^{-1}$. This combined with the extremely small EDL thickness, giving EDLCs surprisingly high volumetric capacitances. Current commercial EDLCs with organic electrolyte and porous carbon electrode can generally achieve capacitances of above 100 F cm^{-3} [54,55], which are three orders of magnitude larger than those obtained by the conventional electrostatic capacitors (0.1 F cm^{-3}) [56–58]. Because of this, these ESs are often called supercapacitors or ultracapacitors. It should be noted that the specific capacitance of the EDLCs also depends on the electrolyte used. Specific capacitance obtained in the aqueous acid or alkaline electrolytes is usually higher than that in the organic electrolytes, as will be discussed in Chapter 2.

The double-layer models discussed above are all based on a planar electrode. However, as mentioned earlier, typical EDLC electrodes are made from porous carbon with a high SSA to maximize the electrode surface area in contact with the electrolyte. Depending on their size, the pores in the carbon material are categorized into three types by the International Union of Pure and Applied Chemistry: the macropore has a size larger than 50 nm, the size of a mesopore is between 2 and 50 nm, and micropores are less than 2 nm [59]. Since the ES generally has all these types of pores, the capacitive behavior of the ES is also dependent on the geometric features of these pores and the EDL formation in the porous electrode (i.e., geometrically confined system) is more complex than that in the planar electrode (i.e., infinite system). It is therefore essential to understand the EDL behavior in these pores in order to improve the EDLC performance through the suitable selection or design of the electrolytes and pore structures of electrode materials. It was traditionally believed that solvated ions cannot enter the subnanometer pores due to their larger size as compared to the pore size. Therefore, such subnanometer pores were considered not to contribute to the charge storage (i.e., the capacitance) of the EDLCs. However, breakthrough work by Gogotsi et al. [60] found an anomalous increase in the specific capacitance of porous carbon at pore sizes less than 1 nm. Figure 1.5 shows the relationship between the specific capacitance (normalized by the SSA) and the average pore size [60]. In region I, when pores were substantially larger than twice the size of the solvated ions, the capacitance was contributed from compact layers on ions on both adjacent pore walls (Figure 1.5b). The authors believed although the diffuse layer that exists on a planar electrode was absent or diminished in size, the capacitance was largely unaffected in this case because the compact layer contributed most of the capacitance. As shown in region II, when the pore size was further decreased to less than twice the solvated ion size (Figure 1.5c), the normalized capacitance was decreased. This was due to the reason that the compact ion layers from adjacent pore walls impinged and the effective surface area for EDL formation decreased. However, in even smaller pores with a size less than that of the solvated ion (region III), the capacitance remarkably increases with the decrease of the pore size. This result challenged the long-held presumption that pores smaller than the size of the solvated electrolyte ions do not contribute to the charge storage. Gogotsi et al. attributed such an anomalous increase in the

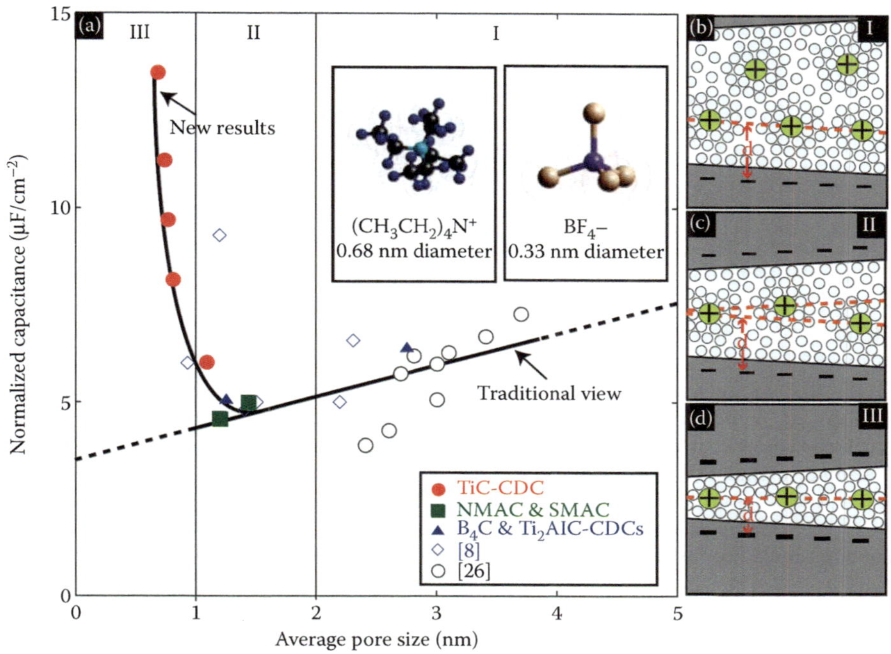

FIGURE 1.5 (a) Plot of specific capacitance normalized by Brunauer–Emmett–Teller (BET) SSA for the carbons in this study and in two other studies with identical electrolytes. The normalized capacitance decreased with decreasing pore size until a critical value was reached, unlike the traditional view that assumed that capacitance continually decreased. It would be expected that as the pore size becomes large enough to accommodate diffuse charge layers, the capacitance would approach a constant value. (b–d) Drawings of solvated ions residing in pores with distance between adjacent pore walls (b) greater than 2 nm, (c) between 1 and 2 nm, and (d) less than 1 nm illustrate this behavior schematically. (From Chmiola, J. et al. 2006. Anomalous increase in carbon capacitance at pore sizes less than 1 nanometer. *Science* 313:1760–1763. Reprinted with permission of American Association for the Advancement of Science (AAAS).)

normalized capacitance to the high distortion of the solvation shells of ions (Figure 1.5d). This allows closer approach of the ion center to the electrode surface, leading to an improved capacitance. This assumption was later confirmed by studies using solvent-free IL electrolytes, which showed that a maximum capacitance was achieved when the ion size and pore size are close [61].

To explain the anomalous enhancement of specific capacitance in the nanopores, Huang et al. [62,63] reported a heuristic theoretical model that takes the pore curvature into account. Depending on the pore size of the electrode material, different capacitive models were proposed. In the macropore regime (>50 nm), the size of pores are large enough so that pore curvature is not significant and hence the EDL model in this case can be described by the conventional model for a planer electrode (Figure 1.6a). When the pore size is in the mesopore regime (2–50 nm), solvated

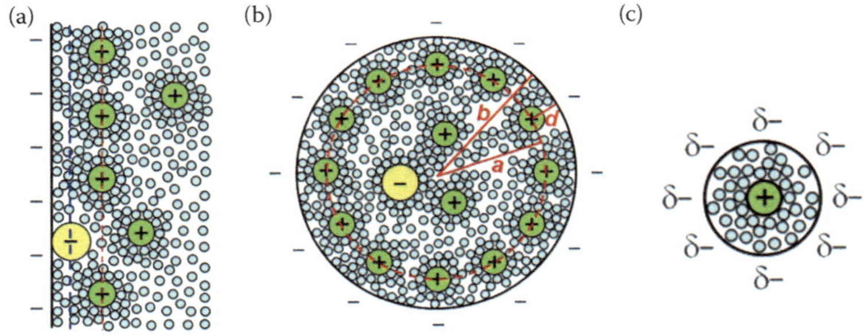

FIGURE 1.6 Schematic diagrams of (a) an electric double-layer capacitor (EDLC) formed at the carbon/electrolyte interface (anode) in which the inner and outer Helmholtz planes (IHP, OHP) are represented by blue and red dashed lines, respectively. Cations are shown to be solvated by the solvent molecules. Top views of (b) a negatively charged mesopore with solvated cations approaching the pore wall to form an EDCC with radii b and a for the outer and inner cylinders, respectively, separated by a distance d, and (c) a negatively charged micropore of radius b with solvated cations of radius a_0 lining up to form an EWCC. (Huang, J. S., B. G. Sumpter, and V. Meunier: A universal model for nanoporous carbon supercapacitors applicable to diverse pore regimes, carbon materials, and electrolytes. *Chemistry—A European Journal*. 2008. 14:6614–6626. Copyright Wiley-VCH Verlag GmbH & Co. KGaA. Reproduced with permission.)

counterions enter mesoporous carbon and approach the pore wall to form an electric double-cylinder capacitor (EDCC) (Figure 1.6b). By assuming the mesopores are cylindrical, the capacitance of EDCC is given by [62–64]

$$C = \frac{2\pi\varepsilon_r\varepsilon_0 L}{\ln(b/a)} \tag{1.7}$$

$$\frac{C}{A} = \frac{\varepsilon_r\varepsilon_0}{b\ln(b/(b-d))} \tag{1.8}$$

where L is the pore length, b and a are the radii of the outer and inner cylinders, respectively, and d is the distance between the outer and inner cylinders. In the micropore regime (<2 nm), the small pores do not allow the formation of a double cylinder. The authors assumed that the solvated (or desolvated) counterions can enter the pores and line up along the pore axis to form an electric wire-in-cylinder capacitor (EWCC) (Figure 1.6c). The capacitance of EWCC in this case is described by [62]

$$\frac{C}{A} = \frac{\varepsilon_r\varepsilon_0}{b\ln(b/a_0)} \tag{1.9}$$

where a_0 is radius of the solvated cations. The authors found that the EDCC and EWCC models fitted the experimental data quite well for EDLCs of nanoporous carbon materials with diverse pore sizes and electrolytes, including organic electrolytes, aqueous electrolytes, and an IL electrolyte. The development of the above EDL models provides a fundamental insight into the development of high-performance EDLCs and related electrolytes.

1.3.1.2 Working Principle and Structure of EDL Capacitor

Figure 1.3a shows the basic operating principles of an EDLC. The charging/discharging processes for an EDLC can be simply expressed by the following equations [65]:

On the positive electrode:

$$E_{S1} + A^- \underset{\text{discharging}}{\overset{\text{charging}}{\rightleftarrows}} E_{S1}^+ // A^- + e^- \tag{1.10}$$

On the negative electrode:

$$E_{S2} + e^- + C^+ \underset{\text{discharging}}{\overset{\text{charging}}{\rightleftarrows}} E_{S2}^- // C^+ \tag{1.11}$$

And the overall charging and discharging process is

$$E_{S1} + E_{S2} + A^- + C^+ \underset{\text{discharging}}{\overset{\text{charging}}{\rightleftarrows}} E_{S1}^+ // A^- + E_{S2}^- // C^+ \tag{1.12}$$

where E_{S1} and E_{S2} represent the two electrode surface, // is the EDL where charges are stored on either side of the interface, and A^- and C^+ are anions and cations of the electrolyte, respectively. During the charging process, electrons start to flow from the positive electrode to the negative electrode through an external power source. Within the electrolyte, cations move toward the negatively charged electrode while anions move toward the positive electrode. The excess charge on the electrode is balanced by the formation of the EDL, which has been discussed above. As shown in Figure 1.3b, there is a membrane (separator) between the two electrodes, allowing the migration of ions but preventing a short circuit. The cell voltage is therefore increased and energy is stored. During the discharging process, electrons flow from the negative electrode to the positive electrode through an external load and the stored charge (energy) is thus released. Adsorbed ions on the electrode surface are also released back into the bulk electrolyte. Based on the above discussion, some key characteristics of EDLCs can be found. For this type of ES, energy (charge) is stored in the EDL formed at the interface between the electrode surface and the electrolyte. Ideally, there are no charges transferred across the electrode/electrolyte interface (i.e., nonfaradaic) during the charging/discharging process. Owing to this electrostatic charge storage and release mechanism, there is almost no structural or volume change of the electrode materials. As a result, EDLCs can have extremely long life cycles of charging/discharging (usually between 10^5 and 10^6 cycles) [66]. Moreover, since many kinetically limited processes such as chemical reactions and diffusion processes are not involved in this charging/discharging process, the whole process is highly reversible and storage and release of charge can be very quick, resulting in a high power density of EDLCs (>500 W kg^{-1}) [66,67]. However, a major drawback of EDLCs is their much lower energy densities (<10 Wh kg^{-1}) compared to batteries (35–150 Wh kg^{-1}) because the charge storage is through a physical process and is confined at the electrode/electrolyte interface. Some progress has been made to address this issue and some recently developed ESs, such as pseudocapacitors and hybrid ESs, have shown much improved energy density. Some of them are comparable to traditional lead-acid batteries, as will be briefly discussed later.

1.3.2 Pseudocapacitors

In contrast to the EDLC, the pseudocapacitor (also known as the faradaic or redox ES) stores charges based on the fast and reversible redox reactions on the surface or near the surface of electroactive materials. Although a pseudocapacitor is similar to a battery as it requires charge transfer across the electrolyte/electrode interface that is faradaic in nature, the voltage response on the electrode to the accumulated charge exhibits capacitor-type characteristics. In this case, the change in the charge (dQ) during the faradaic reaction depends linearly, or approximately linearly, on the change in the electrode potential (dV), resulting in a measurable capacitance (i.e., $C = dQ/dV$) [41]. Since the origin of the capacitance does not arise from the traditional charge separation as in the case of classical capacitors or EDLCs, the term "pseudo" capacitance has been used. This pseudocapacitance behavior was first noted by Trasatti and Buzzanca in 1971 and they found that the cyclic voltammetry behavior of RuO_2 electrodes was similar to that of a capacitor [68]. Conway et al. [41,69,70] also did extensive research on RuO_2-based electrochemical capacitors that achieve high specific capacitance.

The nature of the electrolyte and its interaction with the electrode materials has a strong influence on the pseudocapacitive behavior and thus the pseudocapacitance. Therefore, the pseudocapacitive mechanism is briefly reviewed. As shown in Figure 1.7, there are several types of electrochemical processes that can contribute to the pseudocapacitance, including reversible adsorption of ions from the electrolyte, redox reactions involving ions from the electrolyte for some transition metal oxides, reversible electrochemical doping/dedoping in conductive polymer-based electrodes, and lattice intercalation [11,41,69,71].

1.3.2.1 Adsorption Pseudocapacitance

Adsorption pseudocapacitance can come from two-dimensional surface reactions that involve faradaic desorption and adsorption of an electroactive species from the electrolyte at a metal surface [69]. One good example is the adsorption/desorption of hydrogen (H) at Pt in the acid solution, following the reaction [41,69]

$$\begin{array}{l} H_3O^+ + \ Pt + e^- \xrightleftharpoons{K} PtH_{ads}O + H_2O \\ C_{H^+} \quad 1-\theta \quad V \end{array} \tag{1.13}$$

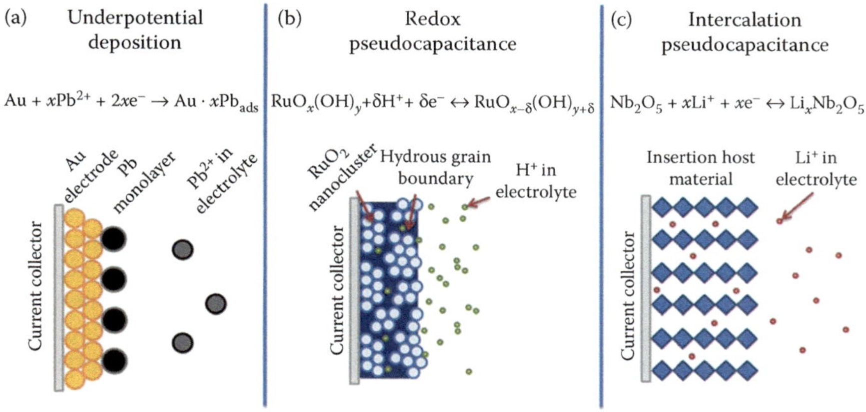

FIGURE 1.7 Different types of reversible redox mechanisms that give rise to pseudocapacitance: (a) underpotential deposition, (b) redox pseudocapacitance, and (c) intercalation pseudocapacitance. (Augustyn, V., P. Simon, and B. Dunn. 2014. Pseudocapacitive oxide materials for high-rate electrochemical energy storage. *Energy & Environmental Science* 7:1597–1614. Reproduced by permission of The Royal Society of Chemistry.)

where θ is the H coverage on the Pt surface, V is the potential, K is the equilibrium constant, and C_{H^+} is the H^+ concentration in the solution. Conway et al. [41] used a Langmuir-type electrosorption isotherm to describe the origin of this pseudocapacitance, and the surface coverage θ of H on Pt as a function of potential (V) can be expressed as

$$\frac{\theta}{1-\theta} = K_{C_{H^+}} \exp\left(\frac{VF}{RT}\right) \tag{1.14}$$

where R is ideal gas constant, F is Faraday's constant, and T is temperature. Considering the charge for a monolayer formation of H on Pt is q (210 μC cm^{-2} for polycrystalline Pt), a pseudocapacitance (C_ϕ) is derived as [41]

$$C_\phi = q\left(\frac{d\theta}{dV}\right) = \frac{qF}{RT} \cdot \frac{K_{C_{H^+}} \exp(VF/RT)}{[1 + K_{C_{H^+}} \exp(VF/RT)]^2} \tag{1.15}$$

From Equation 1.14, C_ϕ can also be expressed as

$$C_\phi = \frac{qF}{RT} \cdot \theta(1-\theta) \tag{1.16}$$

which has a maximum value at $\theta = 0.5$. By taking into account the value of q, the maximum in C_ϕ is about 2200 μF cm^{-2}, which is about 100 times higher than the EDL capacitance for Pt [41]. In addition, Conway et al. [41] further pointed out that from a thermodynamic point of view, pseudocapacitance originates whenever certain property, y, proportional to the transferred charge, has the following relationship with the potential:

$$\frac{y}{1-y} = K\exp\left(\frac{VF}{RT}\right) \tag{1.17}$$

The quantity of y can be the fraction of coverage θ on the electrode surface (as discussed here), some extent of conversion of an oxidized species to a reduced species in a redox system (e.g., RuO_2, as will be discussed later), or an extent of fractional adsorption or intercalation of electrolyte ions into the electrode material. As long as y is proportional to the charge transferred in each process, the differential of y with respect to the potential (dy/dV) is proportional to a corresponding pseudocapacitance, which is experimentally measurable. Unfortunately, the high cost of Pt and narrow range of the working potentials limit the practical use of this system for ESs, but understanding the origin of the pseudocapacitance would benefit from discussing this process.

1.3.2.2 Redox Reactions

This type of pseudocapacitance originates from the redox reactions on some electroactive materials such as metal oxides and hydroxides. RuO_2 is the first and one of the most extensively studied redox-type pseudocapacitive material for ESs because of its high specific capacitance from 200 to 1000 F g^{-1}, depending on its structure. The redox reactions of RuO_2 in the acid solution involve fast reversible electron transfer accompanied by electro-adsorption of protons on the surface or insertion of protons into the bulk RuO_2 [68,72,73]:

$$RuO_x(OH)_y + \delta H^+ + \delta e^- \leftrightarrow RuO_{x-\delta}(OH)_{y+\delta} \tag{1.18}$$

Ru changes oxidation state from Ru^{II} to Ru^{III} to Ru^{IV} as the electrode potential increases. This successive multiple redox process results in a nearly constant change in the transferred charge with the potential change in the certain voltage range. Consequently, a nearly rectangular shape (due to the overlapping of the redox peaks) of the cyclic voltammogram (CV) can be observed for RuO_2-based ESs, which is the characteristic of a capacitor. Based on the above discussion about adsorption pseudocapacitance, this process can also be regarded as the overlapping of individual Langmuir isotherms, which will enlarge the potential window with constant capacitance [41]. This type of pseudocapacitor generally has a similar structure compared to the EDLC, which consists of two electrodes with the same material separated by a separator and immersed in the electrolyte. During the charging/discharging process, electrons flow through the external circuit while the protons move from one electrode to the other through a separator (Figure 1.3b). RuO_2-based ESs offer advantages of not only high specific capacitances but also relatively long cycle life (over 10^5 cycles). However, Ru is expensive and rare, limiting the widespread applications of RuO_2-based ESs. Therefore, extensive efforts have been made to investigate other

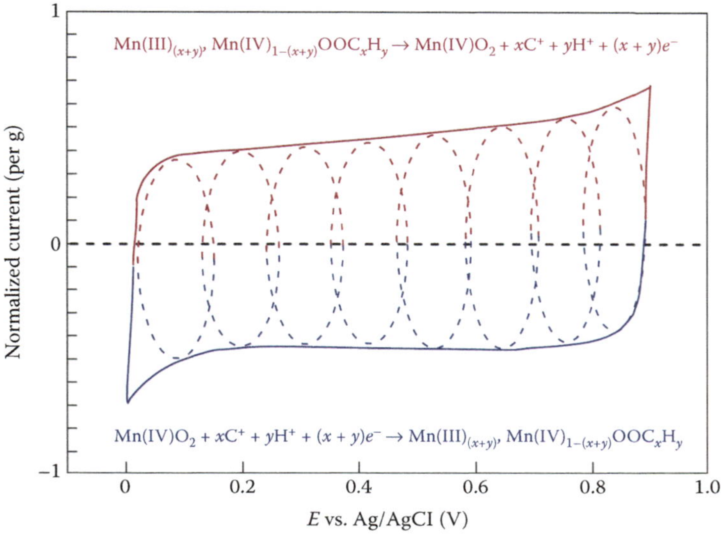

FIGURE 1.8 Cyclic voltammetry. This schematic of cyclic voltammetry for a MnO_2-electrode cell in mild aqueous electrolyte (0.1 M K_2SO_4) shows the successive multiple surface redox reactions leading to the pseudocapacitive charge storage mechanism. The red (upper) part is related to the oxidation from Mn(III) to Mn(IV) and the blue (lower) part refers to the reduction from Mn(IV) to Mn(III). (Reprinted by permission from Macmillan Publishers Ltd. *Nature Materials* (Simon, P., and Y. Gogotsi. Materials for electrochemical capacitors, 7:845–854, copyright 2008.)

less expensive oxides such as manganese oxide, nickel oxide, and cobalt oxide as possible alternatives for RuO_2, which have been discussed by many excellent reviews [11,27]. Among them, MnO_2 has been the most extensively investigated pseudocapacitive electrode material. The charge storage mechanism is based on surface adsorption of electrolyte cations such as K^+ and Na^+ as well as proton incorporation. Figure 1.8 shows a CV of a MnO_2 electrode in mild aqueous electrolyte [2]. The fast and reversible successive redox reactions result in the near-rectangular shape of the CV. Although the shape is similar to that of the EDLC, they have totally different charge storage mechanisms.

1.3.2.3 Pseudocapacitance Arises from Electrically Conducting Polymers

Some electrically conducting polymers (ECPs), such as polypyrrole (PPy), polyaniline (PANI), and poly[3,4-ethylenedioxythiophene] (PEDOT), also exhibit rapid and reversible redox processes in certain electrolytes, and thus have attracted considerable interest as pseudocapacitive materials for ES applications due to their low cost and light weight [65]. Pseudocapacitive ECPs involve the oxidation (doping) or reduction (dedoping) of electroactive functional groups of the polymer backbone and simultaneously the transfer of electrolyte ions into or from the polymer backbone simultaneously. Figure 1.9 shows the scheme of the simplified charge storage mechanism related with the p-doping/dedoping and n-doping/dedoping as the ECP

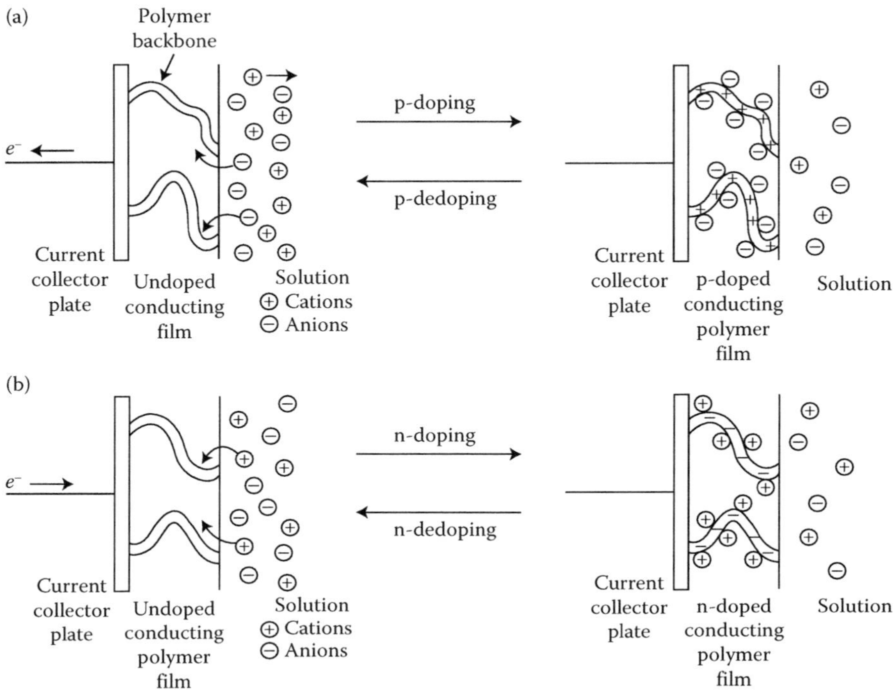

FIGURE 1.9 Schematic representation of the charging and discharging processes at ECP electrodes associated with (a) p-doping and (b) n-doping. (Reprinted from *Journal of Power Sources*, 47, Rudge, A. et al., Conducting polymers as active materials in electrochemical capacitors, 89–107, Copyright 1994, with permission from Elsevier.)

electrode undergoes the charging/discharging process [74]. This process involves the intercalation of the ions from the electrolyte into three-dimensional (3D) chain entanglements of the polymers, leading to a much higher volumetric capacitance and energy density compared to the EDLCs. Furthermore, these conductive polymers can be used in various nonaqueous electrolytes, which can greatly increase the operating cell voltage. However, the electrolyte ion intercalation/extraction during the electrochemical cycling and the chemical degradation of the conductive polymers in the electrolytes may lead to the swelling and shrinking of conductive polymers, reducing their long-term cyclic stability [75,76].

1.3.2.4 Intercalation of Electrolyte Ions

As shown in Figure 1.7, the pseudocapacitance can also arise from the intercalation of electrolyte ions (e.g., Li^+) into the tunnels, van der Waals gaps, or lattice of redox-active electrode materials (e.g., MoO_3) accompanied by a faradaic charge transfer [11,77–79]. It is worth noting that only when such intercalation process is fast enough, the electrode exhibits pseudocapacitive behavior. Otherwise, it will behave like a battery-type electrode.

As mentioned above, this pseudocapacitive behavior shares a very similar thermodynamic equation (Equation 1.17), and the resulting pseudocapacitance is strongly dependent on the nature of these processes (i.e., the electrons involved in these reactions). The first two pseudocapacitive charge storage mechanisms mainly involve surface reactions and consequently the specific capacitance strongly depends on the surface area of the electrode materials in contact with the electrolytes. The last two charge storage mechanisms involve the intercalation/deintercalation of electrolyte ions into or out of the bulk electrode materials and the resulting pseudocapacitances are not only dependent on the surface area but also related to the electrolyte ion intercalation/deintercalation process. The pseudocapacitance can be 10–100 times larger than the capacitance of EDLCs since pseudocapacitors store energy through chemical bonds whereas EDLCs mainly store energy through electrostatic process without faradaic reactions [41]. However, pseudocapacitors usually suffer from relatively lower power density compared to EDLCs because faradaic processes are normally slower than purely electrostatic processes. Besides, the repeated oxidation and reduction (or intercalation/deintercalation) processes may cause structural changes of the electrode materials, leading to the decrease of cycling stability. It should be noted, generally, that both EDL capacitance and pseudocapacitance exist for ESs. EDLCs with carbon-based electrodes have a small percentage of their capacitance as pseudocapacitance due to the faradaic reactivity of surface functional groups, and pseudocapacitors also exhibit some EDL capacitance proportional to their electrochemically effective interfacial areas [41].

1.3.3 Hybrid ESs

The ESs mentioned above consist of two electrodes with the same type of capacitive materials made from either EDL capacitive materials or pseudocapacitive materials (symmetrical configuration). In order to further increase the operating potential window, energy, and power density, a new type of ES has been developed, which is known as hybrid capacitors. With extensive achievements in this area, various types of hybrid ESs have been developed. Generally, hybrid capacitors utilize both the EDL capacitance and faradaic reaction to store charges. The hybrid capacitors reviewed in this book include (1) ESs based on composite electrodes made from both EDL capacitive materials and pseudocapacitive materials; (2) asymmetric ESs with one EDL electrode and another pseudocapacitive or battery-type electrode; and (3) asymmetric ESs with one pseudocapacitive electrode and another rechargeable battery-type electrode.

Composite electrodes integrate carbon-based materials with pseudocapacitive materials (e.g., metal oxides and ECPs) and thus combine both electrostatic and faradaic charge storage mechanisms in a single electrode [80]. The carbon-based materials not only provide EDL capacitance but also provide a high-surface-area framework or matrix that allows a uniform deposition of pseudocapacitive materials and a 3D charge distribution [80,81]. The pseudocapacitive materials further increase the total capacitance of the composite electrode materials through utilizing the pseudocapacitance. Asymmetric hybrid ESs are designed with one carbon-based or pseudocapacitive material-based electrode and the other electrode consisting of a pseudocapacitive or a rechargeable battery-type material. The hybrid ES with one battery-type electrode is also known as battery-type hybrid ES. One of the important purposes of developing hybrid ESs is to improve the operating cell voltage and thus the energy density of the symmetrical ESs with aqueous electrolyte. For example, for a symmetric carbon-based EDLC with aqueous electrolytes such as KOH or H_2SO_4, it has a limited potential window due to the gas evolution reactions and electrochemical oxidation of carbon electrodes at high potentials, as will be discussed in more detail in Chapter 2. Through the use of a faradaic positive electrode with high oxygen evolution reaction overpotential (e.g., Pb/PbO_2 and $Ni(OH)_2$) instead of the carbon-based positive electrode, the negative carbon-based electrode and the positive electrode can operate in a complementary electrochemical window [5,82]. This increases the cell voltage effectively and also improves the total capacitance compared to symmetric carbon-based ESs. Furthermore, a recently developed LIC has received particular attention due to its high operating voltage of about 3.8–4.0 V (Figure 1.10 [12]), high energy density (more than 20 Wh kg^{-1}), and increased cycle life (compared to lithium-ion batteries) [12,83]. The LIC has a carbon-based EDL positive electrode, which is the same to that of the positive electrode of the EDLC. The negative electrode consists of a properly designed material for intercalation and deintercalation of Li ions, which is similar to that of lithium-ion batteries. Naoi et al. [84,85] have prepared a lithium titanate ($Li_4Ti_5O_{12}$)–carbon nanofiber composite as negative electrode for LIC with AC positive electrode. This LIC exhibited a much higher energy density of 40 Wh L^{-1} and high power density of 7.5 kW L^{-1}. During the charging process, electrons move from the positive to negative electrodes through the external circuit and, simultaneously, the EDL is formed on the positive electrode while the negative electrode undergoes Li-ion intercalation. During the discharging process, they move the opposite way.

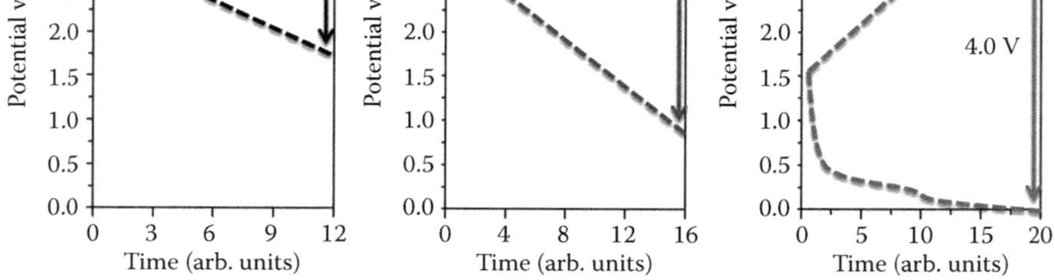

FIGURE 1.10 Comparison of the operating voltage and electrode potential profiles of EDLCs (currently available and high voltage) and LICs subjected to an identical current density. (Béguin, F. et al.: Carbons and electrolytes for advanced supercapacitors. *Advanced Materials*. 2014. 26:2219–2251. Copyright Wiley-VCH Verlag GmbH & Co. KGaA. Reproduced with permission.)

1.4 PERFORMANCE METRICS AND THEIR RELATIONSHIP WITH ELECTROLYTES

1.4.1 CAPACITANCE

The ES can be considered as two capacitors in series since both electrode/electrolyte interfaces represent a capacitor. The overall capacitance (C_T) of the ES can be expressed as

$$\frac{1}{C_T} = \frac{1}{C_1} + \frac{1}{C_2} \tag{1.19}$$

where C_1 and C_2 are the capacitances of the two electrodes, respectively. Besides the C_T, the specific capacitance is often used as an important parameter to characterize an electrode material for ES. The specific capacitance can be defined in several ways such as the gravimetric or mass specific capacitance (C_G), volumetric capacitance (C_V), and capacitance normalized by the effective surface area (C_A). C_G is the most commonly used metric to characterize the electrode materials, with a unit of Faraday per gram (F g^{-1}) and can be expressed as

$$C_G = \frac{C_i}{W} \tag{1.20}$$

where W is the weight of the electrode material and C_i is the electrode capacitance (anode or cathode). The C_G value can be used to compare the performance of different electrode materials. C_G is dependent on the C_A and the effective SSA (surface area per unit mass), which can be expressed as

$$C_T = C_A \cdot \text{SSA} \tag{1.21}$$

The C_A is an intrinsic capacitance that is related to the nature of the electrode material, electrolyte, and charge storage mechanism as discussed above. It should be noted that the total device weight of the ES is also dependent on the mass of the electrolyte. However, the C_G does not consider the weight of electrolyte in the electrodes. To increase the C_G and the total capacitance of the ES, the electrodes are typically made from nanomaterials with high SSA. Many nanomaterials, such as nanotubes or graphene, have a low packing density (<0.5 g cm^{-3}), leading to a large fraction of empty space in the electrode that will be filled in by the electrolyte and consequently increase the weight of the ES without adding to the total capacitance [86]. Therefore, the volumetric capacitance, C_V (capacitance per unit volume, F cm^{-3}), is also an important metric to evaluate the electrode material. C_V is dependent on the C_G and the density of electrode material, which is related to the pore volume (empty space) in the electrode. For some commercial applications (e.g., portable devices and microscale, on-chip applications) where the size of the ES is more important than its weight, electrode material with high C_V is required so that the volume of the ES could be minimized. Based on the above discussion, it is seen that the development of high-surface-area materials is often in contradiction with a high C_V and a balance between the C_G and C_V should be considered.

1.4.2 Energy Density and Power Density

The theoretical (or maximum) energy (E) and power densities (P) of the ES can be expressed as [35,87]

$$E = \frac{1}{2}CV^2 = \frac{QV}{2} \tag{1.22}$$

$$P = \frac{1}{4R_s}V^2 \tag{1.23}$$

where V is the voltage applied across the two electrodes, Q is the stored total charges in the ES, and R_S is the ESR of the ES. From these two equations, it can be seen the performance of ES is dependent on the value of V, C, and R_S. The energy density is generally expressed as per mass (Wh kg^{-1}) or per volume (Wh m^{-3}) of the ES [88]. In order to increase energy and power densities of an ES, it is required to increase the values of both V and C and reduce the value of R_S. Since both energy density and power density are proportional to the square of the operative voltage, it would be more effective to increase the operative voltage to enhance the energy and power densities of ES compared to increasing the capacitance or reducing ESR. The electrochemical stability of the electrolytes greatly determines the maximum working voltage of ESs. For example, when carbon is used as the electrode material for aqueous acid or alkaline electrolytes (e.g., H_2SO_4 or KOH), the operating voltage window of the ES is generally about 1 V, while for an ES with commercial organic electrolytes, the cell voltage is generally in the range of 2.6–2.7 V. Although organic electrolytes can increase the working potential window, the C_A in organic electrolytes is generally lower than that in the aqueous electrolytes as will be discussed in Chapter 2, which has a negative effect on improving the energy density. Recently, IL electrolytes have attracted great interest owing to their high decomposition potential of about 4 V; however, they usually suffer from poor conductivity, high viscosity, and high cost. Besides, the ES configuration also affects the potential window. As mentioned above, the operating potential of asymmetric ESs could be increased to 1.5–2 V compared to symmetric ESs in aqueous electrolytes due to the presence of overpotential in the electrodes [7,89]. Besides, as will be discussed in Chapter 2, the concentration of the electrolyte also plays an important role in the achieved energy density of the ES, especially in the case of the organic electrolyte [65,90].

1.4.3 Equivalent Series Resistance

Equation 1.23 shows the power density increases with the decreasing ESR value. ESR is dependent on a series of resistances associated with (a) the intrinsic resistance of the electrode material and electrolyte solution; (b) mass transfer resistance of the electrolyte ions; and (c) the contact resistance at the current collector/electrode interface [35]. Generally, when nonaqueous electrolytes such as organic, IL, and solid-state electrolytes are used in ESs, the resistance of the bulk electrolyte and the electrolyte in the pores of the electrode tends to dominate the ESR [6]. Therefore, it is beneficial to use an electrolyte with high ionic conductivity in order to improve the power density of the ES. However, there usually exists a trade-off between the ionic conductivity and other properties (e.g., electrochemical stability that determines the cell voltage) of the electrolytes. The ionic conductivities of aqueous electrolytes (e.g., H_2SO_4 and KOH) are high, but the operative voltage windows of them are low. On the contrary, nonaqueous electrolytes (e.g., organic electrolyte and IL) have high electrochemical stability and thus contribute to the high operative cell voltage of ES. However, the ionic conductivity of nonaqueous electrolytes is much lower than that of aqueous electrolytes. This results in a higher ESR, limiting the power density of the ES. Therefore, it is highly desirable to develop novel electrolytes with a large working voltage window while ensuring high conductivity in order to be used for ESs with both high energy and power densities. This will be discussed in Chapter 2.

1.4.4 Cycle Life

Cycle life is also an important issue to evaluate the stability of the ES. General test procedures for stability analysis involve the charging and discharging of the electrode for certain cycles and comparing the capacitance before and after the tests [91]. EDLCs using carbon electrodes tend to have a long cycle lifetime and can be cycled at high rates for 10^5–10^6 cycles with a relatively small degradation in the performance (10%–20% decrease in the capacitance and resistance) [66]. However, the cycling stability of pseudocapacitive electrode materials is generally lower because of the interactions between the electrode materials and electrolyte ions. The cycling life of the ES is affected by a series of factors such as the electrode configuration (type of the ES), electrode material, electrolyte, working voltage and temperature, and testing conditions such as scan rate and charging/discharging currents, which will be discussed in detail in Chapter 2.

1.4.5 Self-Discharge Rate

The self-discharge rate is associated with the potential decrease of a charged electrode over a period of time [91,92]. Large self-discharge rates may limit the use of ESs for applications requiring energy retention for a relatively long period. Charged ESs possess higher free energy than that of discharged ones, and thereby the thermodynamic driving force would cause the self-discharge of ESs at open-circuit conditions. To understand the potential variation during the self-discharge process of ES, several mechanisms have been proposed [92,93]. As will be discussed in Chapter 2, the kinetics and mechanisms of self-discharge of ESs also depend on the nature of the electrolyte and they are sensitive to residual gases and impurities in the electrolytes.

1.4.6 Coulombic Efficiency

Coulombic efficiency is defined as the ratio of the number of discharged electrons to the number of electrons needed to be recharged in order to bring an energy storage device back to its initial charge state [94]. The coulombic efficiency of an ES, η, can be measured using Equation 1.24 with the same charging and discharging current [95]

$$\eta = \frac{t_D}{t_C} \times 100\% \tag{1.24}$$

where t_D and t_C are the times for galvanostatic discharging and charging, respectively.

1.4.7 Relaxation Time

The relaxation time constant, τ_0, represents the minimum time required to discharge all the energy from an ES with an efficiency of greater than 50% [60], which can be determined from the complex impedance spectra by [96]

$$\tau_0 = \frac{1}{f_k} \tag{1.25}$$

where f_k is the knee frequency at which an ES transitions between a resistive and capacitive behavior.

1.4.8 Thermal Stability

Since most applications require an operative temperature range from −30°C to 70°C for ESs [5], extending the working temperature range can further widen the application of ESs. For example, some electronic devices used in space avionics are required to work below −55°C [5] whereas applications such as fuel cell vehicles demand a relatively high operating temperature for ESs. The operating temperature affects a series of properties of ESs such as energy and power densities, ESR, rate performance, cycle life, and self-discharge rate. The effect of temperature on the ES's performance also depends on the properties of the electrolyte such as the nature of the conducting salt (especially in the case of organic electrolytes and IL), and the properties of the solvent such as electrochemical stability, viscosity, freezing point, boiling point, and flash point. This is discussed in more detail for specific electrolytes throughout this book.

1.5 SUMMARY

This chapter describes the fundamentals of different kinds of electrochemical capacitors (ESs), including electrical double-layer capacitors (EDLCs), pseudocapacitors, and hybrid ESs. This serves as a foundation for understanding ES science and technology, and also for comprehending the fundamental role of electrolytes in ESs. Regarding EDLCs, the basic principles of double-layer theory at the electrode–electrolyte interface between electrode and electrolyte are outlined. In particular, recent studies on double-layer formation in geometrically confined porous carbons are summarized. In addition, basic principles and mechanisms behind pseudocapacitance and hybrid ESs are given. Furthermore, common performance metrics for ESs and their relationship with electrolytes are discussed. Based on the above discussion, it can be seen that the important characteristics of ESs strongly depend on electrolytes. Therefore, the development of electrode materials and electrolytes is equally important to improve the performance of ESs. To date, there have been many comprehensive reviews focusing on electrode materials [11–37]. In the following chapters, an overview of recent developments of all kinds of electrolytes for various types of ESs are given in detail.

REFERENCES

1. Burke, A. 2007. R&D considerations for the performance and application of electrochemical capacitors. *Electrochimica Acta* 53:1083–1091.
2. Simon, P., and Y. Gogotsi. 2008. Materials for electrochemical capacitors. *Nature Materials* 7:845–854.
3. Yu, A., V. Chabot, and J. Zhang. 2013. *Electrochemical Supercapacitors for Energy Storage and Delivery: Fundamentals and Applications*. CRC Press: Boca Raton, FL.
4. Naoi, K., S. Ishimoto, J. I. Miyamoto, and W. Naoi. 2012. Second generation "nanohybrid supercapacitor": Evolution of capacitive energy storage devices. *Energy & Environmental Science* 5:9363–9373.
5. Lu, M., F. Béguin, and E. Frąckowiak. 2013. *Supercapacitors: Materials, Systems and Applications*. John Wiley & Sons: Weinheim, Germany.
6. Zhong, C., Y. D. Deng, W. B. Hu, J. L. Qiao, L. Zhang, and J. J. Zhang. 2015. A review of electrolyte materials and compositions for electrochemical supercapacitors. *Chemical Society Reviews* 44:7484–7539.
7. Fic, K., G. Lota, M. Meller, and E. Frackowiak. 2012. Novel insight into neutral medium as electrolyte for high-voltage supercapacitors. *Energy & Environmental Science* 5:5842–5850.
8. Fic, K., E. Frackowiak, and F. Béguin. 2012. Unusual energy enhancement in carbon-based electrochemical capacitors. *Journal of Materials Chemistry* 22:24213–24223.
9. Choudhury, N. A., S. Sampath, and A. K. Shukla. 2009. Hydrogel-polymer electrolytes for electrochemical capacitors: An overview. *Energy & Environmental Science* 2:55–67.
10. Gao, H., and K. Lian. 2014. Proton-conducting polymer electrolytes and their applications in solid supercapacitors: A review. *RSC Advances* 4:33091–33113.
11. Augustyn, V., P. Simon, and B. Dunn. 2014. Pseudocapacitive oxide materials for high-rate electrochemical energy storage. *Energy & Environmental Science* 7:1597–1614.
12. Béguin, F., V. Presser, A. Balducci, and E. Frackowiak. 2014. Carbons and electrolytes for advanced supercapacitors. *Advanced Materials* 26:2219–2251.
13. Bose, S., T. Kuila, A. K. Mishra, R. Rajasekar, N. H. Kim, and J. H. Lee. 2012. Carbon-based nanostructured materials and their composites as supercapacitor electrodes. *Journal of Materials Chemistry* 22:767–784.
14. Cao, X. H., Z. Y. Yin, and H. Zhang. 2014. Three-dimensional graphene materials: Preparation, structures and application in supercapacitors. *Energy & Environmental Science* 7:1850–1865.

15. Chen, S., W. Xing, J. Duan, X. Hu, and S. Z. Qiao. 2013. Nanostructured morphology control for efficient supercapacitor electrodes. *Journal of Materials Chemistry A* 1:2941–2954.
16. Ghosh, A., and Y. H. Lee. 2012. Carbon-based electrochemical capacitors. *ChemSusChem* 5:480–499.
17. Huang, Y., J. J. Liang, and Y. S. Chen. 2012. An overview of the applications of graphene-based materials in supercapacitors. *Small* 8:1805–1834.
18. Jiang, H., P. S. Lee, and C. Z. Li. 2013. 3D carbon based nanostructures for advanced supercapacitors. *Energy & Environmental Science* 6:41–53.
19. Jiang, H., J. Ma, and C. Li. 2012. Mesoporous carbon incorporated metal oxide nanomaterials as supercapacitor electrodes. *Advanced Materials* 24:4197–4202.
20. Lee, S. W., B. M. Gallant, H. R. Byon, P. T. Hammond, and Y. Shao-Horn. 2011. Nanostructured carbon-based electrodes: Bridging the gap between thin-film lithium-ion batteries and electrochemical capacitors. *Energy & Environmental Science* 4:1972–1985.
21. Li, X., and B. Q. Wei. 2013. Supercapacitors based on nanostructured carbon. *Nano Energy* 2:159–173.
22. Lu, Q., J. G. G. Chen, and J. Q. Xiao. 2013. Nanostructured electrodes for high-performance pseudocapacitors. *Angewandte Chemie—International Edition* 52: 1882–1889.
23. Mahmood, N., C. Z. Zhang, H. Yin, and Y. L. Hou. 2014. Graphene-based nanocomposites for energy storage and conversion in lithium batteries, supercapacitors and fuel cells. *Journal of Materials Chemistry A* 2:15–32.
24. Peng, X., L. L. Peng, C. Z. Wu, and Y. Xie. 2014. Two dimensional nanomaterials for flexible supercapacitors. *Chemical Society Reviews* 43:3303–3323.
25. Sevilla, M., and R. Mokaya. 2014. Energy storage applications of activated carbons: Supercapacitors and hydrogen storage. *Energy & Environmental Science* 7:1250–1280.
26. Tan, Y. B., and J. M. Lee. 2013. Graphene for supercapacitor applications. *Journal of Materials Chemistry A* 1:14814–14843.
27. Wang, G., L. Zhang, and J. Zhang. 2012. A review of electrode materials for electrochemical supercapacitors. *Chemical Society Reviews* 41:797–828.
28. Wang, K., H. P. Wu, Y. N. Meng, and Z. X. Wei. 2014. Conducting polymer nanowire arrays for high performance supercapacitors. *Small* 10:14–31.
29. Wei, L., and G. Yushin. 2012. Nanostructured activated carbons from natural precursors for electrical double layer capacitors. *Nano Energy* 1:552–565.
30. Wei, W., X. Cui, W. Chen, and D. G. Ivey. 2011. Manganese oxide-based materials as electrochemical supercapacitor electrodes. *Chemical Society Reviews* 40:1697–1721.
31. Yan, J., Q. Wang, T. Wei, and Z. J. Fan. 2014. Recent advances in design and fabrication of electrochemical supercapacitors with high energy densities. *Advanced Energy Materials* 4:1300816.
32. Yu, G., X. Xie, L. Pan, Z. Bao, and Y. Cui. 2013. Hybrid nanostructured materials for high-performance electrochemical capacitors. *Nano Energy* 2:213–234.
33. Zhai, Y. P., Y. Q. Dou, D. Y. Zhao, P. F. Fulvio, R. T. Mayes, and S. Dai. 2011. Carbon materials for chemical capacitive energy storage. *Advanced Materials* 23:4828–4850.
34. Zhang, H., G. Cao, and Y. Yang. 2009. Carbon nanotube arrays and their composites for electrochemical capacitors and lithium-ion batteries. *Energy & Environmental Science* 2:932–943.
35. Zhang, L. L., and X. S. Zhao. 2009. Carbon-based materials as supercapacitor electrodes. *Chemical Society Reviews* 38:2520–2531.
36. Zhao, X., B. M. Sánchez, P. J. Dobson, and P. S. Grant. 2011. The role of nanomaterials in redox-based supercapacitors for next generation energy storage devices. *Nanoscale* 3:839–855.

37. Zhi, M. J., C. C. Xiang, J. T. Li, M. Li, and N. Q. Wu. 2013. Nanostructured carbon–metal oxide composite electrodes for supercapacitors: A review. *Nanoscale* 5:72–88.
38. Samui, A. B., and P. Sivaraman. 2010. Solid polymer electrolytes for supercapacitors. In *Polymer Electrolytes: Fundamentals and Applications*, C. Sequeira and D. Santos (Eds), pp. 431–470, Woodhead Publishing Limited: Cambridge, UK.
39. Fang, X., and D. Yao. 2013. An overview of solid-like electrolytes for supercapacitors. *Proceedings of the ASME 2013 International Mechanical Engineering Congress and Exposition*, San Diego, California.
40. Brandt, A., S. Pohlmann, A. Varzi, A. Balducci, and S. Passerini. 2013. Ionic liquids in supercapacitors. *MRS Bulletin* 38:554–559.
41. Conway, B. E. 1999. *Electrochemical Supercapacitors: Scientific Fundamentals and Technological Applications*. Kluwer Academic/Plenum: New York.
42. Helmholtz, H. V. 1853. On the conservation of force. *Annals of Physics (Leipzig)* 89:21.
43. Becker, H. I. 1957. Low voltage electrolytic capacitor. US Patent 2800616, July 23.
44. Gouy, G. 1910. Constitution of the electric charge at the surface of an electrolyte. *Journal of Physics* 9:457–467.
45. Chapman, D. L. 1913. A contribution to the theory of electrocapillarity. *The London, Edinburgh, and Dublin Philosophical Magazine and Journal of Science* 25:475–481.
46. Stern, O. Z. 1924. Theory of the electrical double layer (In German). *Electrochemistry* 30:508.
47. Grahame, D. C. 1947. The electrical double-layer and the theory of electrocapillarity. *Chemical Reviews* 41:441–501.
48. Bockris, J. O'M., M. A. V. Devanathan, and K. Muller. 1963. On the structure of charged interfaces. *Proceedings of the Royal Society of London. Series A: Mathematical and Physical Sciences* 274:55–79.
49. Bockris, J. O'M., A. K. N. Reddy, and M. E. G. Adelco. 1998. *Modern Electrochemistry 2A: Fundamentals of Electrodics*. KluwerAcademic: New York.
50. Zanello, P. 2003. *Inorganic Chemistry Theory, Practice and Application*. The Royal Society of Chemistry: Cambridge.
51. Butt, H. J., K. Graf, and M. Kappl. 2003. *Physics and Chemistry of Interfaces*. Wiley-VCH, Weinheim, Germany.
52. Bard, A. J., and L. R. Faulkner. 2001. *Electrochemical Methods: Fundamentals and Applications*, 2nd edn. John Wiley & Sons, Inc: New York.
53. Lyklema, J. 1995. *Fundamentals of Interface and Colloid Science*. Academic Press: New York.
54. Chang, Y., H. Li, L. Wu, and T. Lu. 1997. Irreversible capacity loss of graphite electrode in lithium-ion batteries. *Journal of Power Sources* 68:187–190.
55. Ning, G., B. Haran, and B. N. Popov. 2003. Capacity fade study of lithium-ion batteries cycled at high discharge rates. *Journal of Power Sources* 117:160–169.
56. Banerjee, P., I. Perez, L. Henn-Lecordier, S. B. Lee, and G. W. Rubloff. 2009. Nanotubular metal–insulator–metal capacitor arrays for energy storage. *Nature Nanotechnology* 4:292–296.
57. Emmenegger, C., P. Mauron, A. Züttel et al. 2000. Carbon nanotube synthesized on metallic substrates. *Applied Surface Science* 162:452–456.
58. Kötz, R., and M. Carlen. 2000. Principles and applications of electrochemical capacitors. *Electrochimica Acta* 45:2483–2498.
59. Sing, K. S. W. 1985. Reporting physisorption data for gas/solid systems with special reference to the determination of surface area and porosity. *Pure and Applied Chemistry* 57:603–619.

60. Chmiola, J., G. Yushin, Y. Gogotsi, C. Portet, P. Simon, and P. L. Taberna. 2006. Anomalous increase in carbon capacitance at pore sizes less than 1 nanometer. *Science* 313:1760–1763.
61. Largeot, C., C. Portet, J. Chmiola, P. L. Taberna, Y. Gogotsi, and P. Simon. 2008. Relation between the ion size and pore size for an electric double-layer capacitor. *Journal of the American Chemical Society* 130:2730–2731.
62. Huang, J. S., B. G. Sumpter, and V. Meunier. 2008. A universal model for nanoporous carbon supercapacitors applicable to diverse pore regimes, carbon materials, and electrolytes. *Chemistry—A European Journal* 14:6614–6626.
63. Huang, J. S., B. G. Sumpter, and V. Meunier. 2008. Theoretical model for nanoporous carbon supercapacitors. *Angewandte Chemie—International Edition* 47:520–524.
64. Tipler, P. A. 1976. *Physics*. Worth: New York.
65. Zheng, J. P., J. Huang, and T. R. Jow. 1997. The limitations of energy density for electrochemical capacitors. *Journal of the Electrochemical Society* 144:2026–2031.
66. Burke, A. 2000. Ultracapacitors: Why, how, and where is the technology. *Journal of Power Sources* 91:37–50.
67. Burke, A. F. 2007. Batteries and ultracapacitors for electric, hybrid, and fuel cell vehicles. *Proceedings of the IEEE* 95:806–820.
68. Trasatti, S., and G. Buzzanca. 1971. Ruthenium dioxide—A new interesting electrode material—Solid state structure and electrochemical behaviour. *Journal of Electroanalytical Chemistry and Interfacial Electrochemistry* 29:A1–A5.
69. Conway, B. E., V. Birss, and J. Wojtowicz. 1997. The role and utilization of pseudocapacitance for energy storage by supercapacitors. *Journal of Power Sources* 66:1–14.
70. Liu, T. C., W. G. Pell, and B. E. Conway. 1997. Self-discharge and potential recovery phenomena at thermally and electrochemically prepared RuO_2 supercapacitor electrodes. *Electrochimica Acta* 42:3541–3552.
71. Lu, P., D. Xue, H. Yang, and Y. Liu. 2013. Supercapacitor and nanoscale research towards electrochemical energy storage. *International Journal of Smart and Nano Materials* 4:2–26.
72. Dmowski, W., T. Egami, K. E. Swider-Lyons, C. T. Love, and D. R. Rolison. 2002. Local atomic structure and conduction mechanism of nanocrystalline hydrous RuO_2 from x-ray scattering. *The Journal of Physical Chemistry B* 106:12677–12683.
73. Trasatti, S. 1991. Physical electrochemistry of ceramic oxides. *Electrochimica Acta* 36:225–241.
74. Rudge, A., J. Davey, I. Raistrick, S. Gottesfeld, and J. P. Ferraris. 1994. Conducting polymers as active materials in electrochemical capacitors. *Journal of Power Sources* 47:89–107.
75. Niu, C. M., E. K. Sichel, R. Hoch, D. Moy, and H. Tennent. 1997. High power electrochemical capacitors based on carbon nanotube electrodes. *Applied Physics Letters* 70:1480–1482.
76. Jurewicz, K., K. Babeł, R. Pietrzak, S. Delpeux, and H. Wachowska. 2006. Capacitance properties of multi-walled carbon nanotubes modified by activation and ammoxidation. *Carbon* 44:2368–2375.
77. Brezesinski, T., J. Wang, S. H. Tolbert, and B. Dunn. 2010. Ordered mesoporous α-MoO_3 with iso-oriented nanocrystalline walls for thin-film pseudocapacitors. *Nature Materials* 9:146–151.
78. Brezesinski, K., J. Wang, J. Haetge et al. 2010. Pseudocapacitive contributions to charge storage in highly ordered mesoporous group V transition metal oxides with iso-oriented layered nanocrystalline domains. *Journal of the American Chemical Society* 132:6982–6990.
79. Rauda, I. E., V. Augustyn, B. Dunn, and S. H. Tolbert. 2013. Enhancing pseudocapacitive charge storage in polymer templated mesoporous materials. *Accounts of Chemical Research* 46:1113–1124.

80. Kiamahalleh, M. V., S. H. S. Zein, G. Najafpour, S. A. Sata, and S. Buniran. 2012. Multiwalled carbon nanotubes based nanocomposites for supercapacitors: A review of electrode materials. *Nano* 7:1230002.
81. Chukwuka, C., and K. A. Folly. 2012. Batteries and super-capacitors. *IEEE PES PowerAfrica 2012 Conference and Exposition*, Johannesburg, South Africa, July 9–13.
82. Long, J. W., D. Bélanger, T. Brousse, W. Sugimoto, M. B. Sassin, and O. Crosnier. 2011. Asymmetric electrochemical capacitors—Stretching the limits of aqueous electrolytes. *MRS Bulletin* 36:513–522.
83. Wang, D. W., H. T. Fang, F. Li et al. 2008. Aligned titania nanotubes as an intercalation anode material for hybrid electrochemical energy storage. *Advanced Functional Materials* 18:3787–3793.
84. Naoi, K., W. Naoi, S. Aoyagi, J. Miyamoto, and T. Kamino. 2013. New generation "nanohybrid supercapacitor." *Accounts of Chemical Research* 46:1075–1083.
85. Naoi, K., S. Ishimoto, Y. Isobe, and S. Aoyagi. 2010. High-rate nano-crystalline $Li_4Ti_5O_{12}$ attached on carbon nano-fibers for hybrid supercapacitors. *Journal of Power Sources* 195:6250–6254.
86. Gogotsi, Y, and P. Simon. 2011. True performance metrics in electrochemical energy storage. *Science* 334:917–918.
87. Pan, H., J. Li, and Y. P. Feng. 2010. Carbon nanotubes for supercapacitor. *Nanoscale Research Letters* 5:654–668.
88. Frackowiak, E., Q. Abbas, and F. Béguin. 2013. Carbon/carbon supercapacitors. *Journal of Energy Chemistry* 22:226–240.
89. Lu, X. H., M. H. Yu, G. M. Wang, Y. X. Tong, and Y. Li. 2014. Flexible solid-state supercapacitors: Design, fabrication and applications. *Energy & Environmental Science* 7:2160–2181.
90. Zheng, J. P., and T. R. Jow. 1997. The effect of salt concentration in electrolytes on the maximum energy storage for double layer capacitors. *Journal of the Electrochemical Society* 144:2417–2420.
91. Davies, A., and A. Yu. 2011. Material advancements in supercapacitors: From activated carbon to carbon nanotube and graphene. *Canadian Journal of Chemical Engineering* 89:1342–1357.
92. Xiong, G. P., C. Z. Meng, R. G. Reifenberger, P. P. Irazoqui, and T. S. Fisher. 2014. A review of graphene-based electrochemical microsupercapacitors. *Electroanalysis* 26:30–51.
93. Niu, J., W. G. Pell, and B. E. Conway. 2006. Requirements for performance characterization of C double-layer supercapacitors: Applications to a high specific-area C-cloth material. *Journal of Power Sources* 156:725–740.
94. Mallika, S., and K. R. Saravana. 2011. Review on ultracapacitor—Battery interface for energy management system. *International Journal of Engineering and Technology* 3:37–43.
95. Fonseca, C. P., J. E. Benedetti, and S. Neves. 2006. Poly(3-methyl thiophene)/PVDF composite as an electrode for supercapacitors. *Journal of Power Sources* 158:789–794.
96. Taberna, P. L., P. Simon, and J. F. Fauvarque. 2003. Electrochemical characteristics and impedance spectroscopy studies of carbon–carbon supercapacitors. *Journal of the Electrochemical Society* 150:A292–A300.

2 Electrolytes for Electrochemical Supercapacitors

Cheng Zhong and Wenbin Hu

CONTENTS

2.1 INTRODUCTION

The electrolyte is one of the central components of electrochemical supercapacitors (ESs) and consists of a solute dissolved in the solvent or pure salt (in the case of solvent-free ionic liquids [ILs]). The electrolyte plays an important role in the ES in several ways such as: (1) providing ionic conductivity and thus allowing for charge compensation on each electrode in the ES cell; (2) contributing to the formation of an electrical double-layer (EDL) in electrical double-layer capacitors (EDLCs); and (3) participating in the charge storage processes (reversible redox reactions) in pseudocapacitors. As a result, the electrolyte can greatly influence the performance of ESs, as shown in Figure 1.2. In addition, Table 2.1 lists some commercialized ESs and their corresponding performance properties [1]. The majority of these ESs use organic electrolytes with a cell voltage of 2.5–2.8 V (only three of them can achieve 3.3 and 3.8 V cell voltage). Most ESs with organic electrolytes use the acetonitrile (ACN) solvent while others apply the propylene carbonate (PC) solvent.

The characteristics of an electrolyte, such as ion type and size; the ion concentration and solvent; the interaction between the ion and the solvent; and the interaction between the electrolyte and the electrode materials, all have a critical influence on the EDL capacitance and pseudocapacitance. The maximum operational voltage of an ES is largely dependent on the electrochemical stability of the electrolyte, which in turn determines the energy and power densities of the ES. The equivalent series

TABLE 2.1
Summary of the Performance Characteristics of Commercial ESs

Device	V Rated	C (F)	R (mΩ)	RC (s)	Wh kg^{-1} [a]	W kg^{-1} (95%) [b]	W kg^{-1} Match. Imped.	Wgt. (kg)	Vol. (L)
Maxwell[c]	2.7	2885	0.375	1.08	4.2	994	8836	0.55	0.414
Maxwell	2.7	605	0.90	0.55	2.35	1139	9597	0.20	0.211
ApowerCap[d]	2.7	55	4	0.22	5.5	5695	50,625	0.009	–
ApowerCap[d]	2.7	450	1.4	0.58	5.89	2569	24,595	0.057	0.045
Ness	2.7	1800	0.55	1.00	3.6	975	8674	0.38	0.277
Ness	2.7	3640	0.30	1.10	4.2	928	8010	0.65	0.514
Ness (cyl.)	2.7	3160	0.4	1.26	4.4	982	8728	0.522	0.38
Carbon Tech Non-ACN	2.85	1600	1.0	1.6	5.8	1026	9106	0.223	
Ashai Glass (PC)	2.7	1375	2.5	3.4	4.9	390	3471	0.210 (estimated)	0.151
Panasonic (PC)	2.5	1200	1.0	1.2	2.3	514	4596	0.34	0.245
EPCOS	2.7	3400	0.45	1.5	4.3	760	6750	0.60	0.48
LS Cable	2.8	3200	0.25	0.80	3.7	1400	12,400	0.63	0.47
BatScap	2.7	2680	0.20	0.54	4.2	2050	18,225	0.50	0.572
Power Sys. (AC and PC)[d]	2.7	1350	1.5	2.0	4.9	650	5785	0.21	0.151
Power Sys. (graphitic carbon and PC)[d]	3.3	1800	3.0	5.4	8.0	486	4320	0.21	0.15
	3.3	1500	1.7	2.5	6.0	776	6903	0.23	0.15
Fuji Heavy Industry-Hybrid (AC/graphitic carbon)[d]	3.8	1800	1.5	2.6	9.2	1025	10,375	0.232	0.143
JSR Micro (AC/ graphitic carbon)[d]	3.8	1000	4	4	11.2	900	7987	0.113	0.073
		2000	1.9	3.8	12.1	1038	9223	0.206	0.132

Source: Reprinted from *Journal of Power Sources*, 196, Burk, A., and Miller, M., The power capability of ultracapacitors and lithium batteries for electric and hybrid vehicle applications, 514–522, Copyright 2011, with permission from Elsevier.

[a] Energy density at 400 W kg^{-1} constant power, Vrated −1/2 V_{rated}.

[b] Power based on $P = 9/16 \times (1 - EF) \times V^2/R$, EF = efficiency of discharge.

[c] Except where noted, all the devices use ACN as the electrolyte.

[d] All device except those with footnote d are packaged in metal containers, where devices are in laminated pouches.

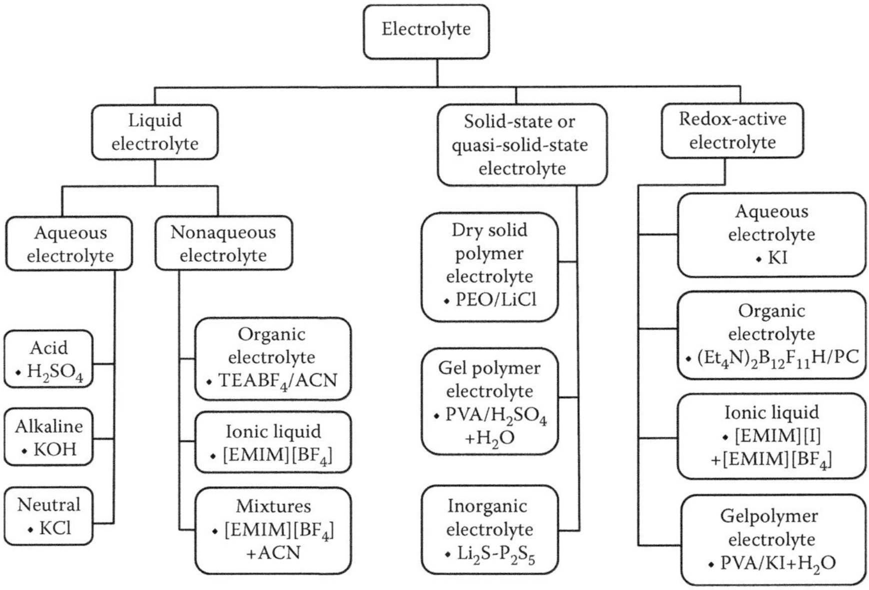

FIGURE 2.1 Classification of electrolytes for ESs. (Zhong, C. et al. 2015. A review of electrolyte materials and compositions for electrochemical supercapacitors. *Chemical Society Reviews* 44:7484–7539. Reproduced by permission of The Royal Society of Chemistry.)

resistance (ESR) of an ES is also strongly dependent on the ionic conductivity of the electrolytes, which also greatly affects its power density. In addition, the operating temperature of an ES strongly depends on the electrolyte used. Furthermore, some recently developed ESs, such as all solid-state ESs, flexible ESs, and integrated on-chip ESs, required the development of unique electrolytes. The general requirements for ES electrolytes include: (a) a wide electrochemically stable window; (b) high ionic conductivity; (c) a wide operational temperature range; (d) a low viscosity; (e) a low volatility; (f) environmental friendliness; (g) low cost; and (h) availability at high purity.

To date, various types of electrolytes have been developed and they can be mainly classified as liquid electrolytes and solid or semi-solid state electrolytes, as shown in Figure 2.1 (each group shows the most used electrolyte). Liquid electrolytes can be further divided into aqueous electrolytes, organic electrolytes, and ILs while solid or semi-solid state electrolytes can be briefly grouped into organic electrolytes and inorganic electrolytes. However, so far there is no perfect electrolyte meeting all the requirements and each electrolyte has its own advantages and disadvantages. For example, aqueous electrolytes-based ESs have high conductivity and capacitance, but a working voltage limited by the low decomposition voltage of aqueous electrolytes. Organic electrolytes and ILs can support higher cell voltages of ESs, but they suffer from much lower ionic conductivity. Therefore, extensive efforts have been devoted to the improvement of the overall performance of electrolytes and the corresponding ES performance. The research directions and the developments concerning the electrolytes so far can be briefly summarized as follows: (a) improving the performance of the electrolyte in terms of operating potential window, ionic conductivity, viscosity, working temperature range, etc., through developing new electrolytes (e.g., IL) or optimizing conventional aqueous and organic electrolytes; (b) exploring the effect of the electrolyte on ES properties such as the capacitance, energy and power densities, thermal stability, and self-discharging process; (c) understanding the fundamental role of the electrolyte on ES performance through advanced characterization, modeling, and simulation methods. This book chapter will provide a comprehensive overview of all kinds of electrolytes for ESs. The interactions between electrolytes and electrode materials are also considered. In order to understand the methods used to design and optimize electrolytes, the effects of electrolytes on the performance of ESs have been highlighted. Some comparisons among different electrolytes in terms of their ES performance are also discussed.

2.2 AQUEOUS ELECTROLYTES

Although aqueous electrolytes have a limited cell voltage window and most commercial ESs use organic electrolytes instead of aqueous electrolytes, aqueous electrolytes have been extensively used in the reported literature [2]. This is mainly due to the fact aqueous electrolytes are cheap and can be easily handled in the lab without needing special conditions, whereas organic electrolytes and ILs generally require complicated purification procedures under a strictly controlled atmosphere to avoid moisture. This characteristic of aqueous electrolytes greatly simplifies the fabrication and assembling of ESs.

As shown in Table 2.2, aqueous electrolytes exhibit high conductivity (e.g., 0.8 S cm^{-2} for 1 M H_2SO_4 at 25°C), which is generally at least one order of magnitude higher than that of the organic and IL electrolytes. This favors the decrease of the ESR, resulting in a higher power density of ESs.

The specific capacitance of carbon electrodes in aqueous electrolytes is generally higher than that obtained in nonaqueous electrolytes, owing to the smaller solvated ions and higher dielectric constant of aqueous electrolytes [7]. The selection criteria for aqueous electrolytes generally takes into account the sizes of hydrated and bare ions (including cations and anions), the mobility of ions, as well as the ionic conductivity, as shown in Tables 2.1 through 2.3. Furthermore, the electrochemical stability and the corrosive degree should also be taken into account. As shown in Figure 2.1, aqueous electrolytes can be grouped into acid, alkaline, and neutral solutions in which H_2SO_4, KOH, and Na_2SO_4 are the most frequently used electrolytes. As previously mentioned, the main drawback of aqueous electrolytes is their relatively narrow voltage window, which is restricted by the decomposition of water, resulting in hydrogen evolution at low cell potential (~0 V vs. standard hydrogen electrode) and oxygen evolution at a high potential range (~1.23 V) [6,8]. The gas evolution might cause the rupture of the ES cells, which threatens personnel safety and decreases performance. Therefore, the cell voltage of ESs with aqueous electrolytes

TABLE 2.2
Electrolyte Conductivity and Voltages of Various Electrolyte Solutions at Room Temperature

Electrolyte Solution	Density (g cm^{-3})	Conductivity (mS cm^{-1})	ΔV (V)
Aqueous, KOH	1.29	540	1
Aqueous, KCl	1.09	210	1
Aqueous, H_2SO_4	1.2	750	1
Aqueous, Na_2SO_4	1.13	91.1	1
Aqueous, K_2SO_4	1.08	88.6	1
Organic, $TEABF_4$/PC	1.2	14.5	2.5–3
Organic, $TEABF_4$/ACN	0.78	59.9	2.5–3
IL, $[Et_2MeIm]^+[BF_4]^-$	1.3–1.5	8 (25°C)	4
IL, $[Et_2MeIm]^+[BF_4]^-$		14 (100°C)	3.25

Source: Adapted from Lide, D. 2009. *CRC Handbook of Chemistry and Physics*, 89th edn. Boca Raton, FL: CRC Press; Davies, A., and A. Yu. 2011. *Canadian Journal of Chemical Engineering* 89:1342–1357; Burke, A. 2007. *Electrochimica Acta* 53:1083–1091; Yu, A., V. Chabot, and J. Zhang. 2013. *Electrochemical Supercapacitors for Energy Storage and Delivery: Fundamentals and Applications*. CRC Press.

TABLE 2.3
Sizes of Bare and Hydrated Ions, and Ionic Conductivity Values

Ion	Bare Ion Size (Å)	Hydrated Ion Size (Å)	Ionic Conductivity (S cm^2 mol^{-1})
H^+	1.15[b]	2.80[b]	350.1[a]
Li^+	0.60[c]	3.82[b,c]	38.69[a]
Na^+	0.95[c]	3.58[b,c]	50.11[a]
K^+	1.33[c]	3.31[b,c]	73.5[a]
NH_4^+	1.48[b,c]	3.31[b,c]	73.7[a]
Mg^{2+}	0.72[a,b]	4.28[b,c]	106.12[a]
Ca^{2+}	1.00[b]	4.12[b,c]	119[a]
Ba^{2+}	1.35[c]	4.04[c]	127.8[a]
Cl^-	1.81[a,c]	3.32[b,c]	76.31[a]
NO_3^-	2.64[c]	3.35[c]	71.42[a]
SO_4^{2-}	2.90[c]	3.79[c]	160.0[a]
OH^-	1.76[c]	3.00[b,c]	198[a]
ClO_4^-	2.92[c]	3.38[c]	67.3[a]
PO_4^{3-}	2.23[d]	3.39[d]	207[a]
CO_3^{2-}	2.66[c]	3.94[c]	138.6[a]

Source: Adapted from [a]Speight, J. G. 2005. *Lange's Handbook of Chemistry*, 16th edn. McGraw-Hill; [b]Volkov, A. G., S. Paula, and D. W. Deamer. 1997. *Bioelectrochemistry and Bioenergetics* 42:153–160; [c]Nightingale, E. R. 1959. *The Journal of Physical Chemistry* 63:1381–1387; [d]Kiriukhin, M. Y., and K. D. Collins. 2002. *Biophysical Chemistry* 99:155–168.

is commonly restricted to about 1 V. This limits the maximum operating voltage limit of the ESs and thereby restricts the energy and power densities. Furthermore, the operating temperature range of ESs with aqueous electrolytes is restricted due to the freezing of electrolytes at low temperature and elevated vapor pressure with increasing temperature. These shortcomings have led to extensive work so as to fully take advantage of aqueous electrolytes.

2.2.1 Strong Acid Electrolytes

H_2SO_4 is the most widely used acid electrolyte for aqueous-based ESs due to its very high ionic conductivity (0.8 S cm^{-1} for 1 M H_2SO_4 at 25°C). It is well known there is an optimum concentration to achieve the maximum ionic conductivity of the electrolyte at a given temperature. The ionic conductivity of the electrolyte decreases if the concentration is too low or too high. The majority of studies use 1.0 M H_2SO_4 electrolyte for ESs due to the maximum ionic conductivity achieved in this concentration at 25°C. H_2SO_4 electrolytes have been mostly used in carbon-based ESs, with some ESs based on pseudocapacitive materials (e.g., RuO_2) also utilizing H_2SO_4 as the electrolyte.

2.2.1.1 Acid Electrolytes for EDLCs

Table 2.4 lists some selected results from the published literature in the last 2 years (2013–2014) concerning carbon-based EDLCs with various electrolytes including acid aqueous, alkaline aqueous, neutral aqueous, and organic electrolytes. For EDLCs utilizing the strong acid electrolyte (H_2SO_4), the reported specific capacitance from Table 2.4 is mainly between 200 and 300 F g^{-1}, which is much higher than that obtained in the organic electrolyte. The specific capacitance is strongly dependent on the structure of carbon-based materials, which has been discussed in many excellent reviews [13–17]. As can be seen in Table 2.4, although the H_2SO_4 electrolyte-based EDLC has a higher specific capacitance compared to the organic electrolyte-based EDLC, the former delivers a much lower maximum energy density (generally below 10 Wh kg^{-1}) compared to the organic-based ESs (e.g., 20–30 Wh kg^{-1}). This is due to the limited cell voltage window of the H_2SO_4 electrolyte compared to the organic electrolytes.

The parametric effects (e.g., electrolyte temperature and concentration) of electrolytes on the ES's performance have been explored [66–68]. Kang et al. [66] investigated the temperature-dependent performance of ESs. The performance such as the capacitance, resistance, and energy/power densities of ES with 3 M H_2SO_4 was much less dependent on the working temperature from –30°C to 60°C compared to the organic 1 M $TEABF_4$/PC electrolyte (Figure 2.2).

Regarding the temperature effect, Niu et al. [67] found that for carbon-cloth ES, both the power and energy densities increased as the H_2SO_4 electrolyte concentration increased from 0.01 to 5 M (Figure 2.3). For the electrolyte with very low concentration (e.g., 0.01 M), the decrease in the power density at high current rates did not result in the increase in the energy density, which was quite different with that obtained for electrolytes with higher concentration.

TABLE 2.4
Some Results from the Selected Literature Published in the Last 2 Years (2013–2014) Concerning the Carbon-Based EDLCs with Various Electrolytes

Electrolyte	Material	Surface Area ($m^2 g^{-1}$)	Capacitance ($F g^{-1}$)	Voltage (V)	Energy Density ($Wh kg^{-1}$)	Power Density ($W kg^{-1}$)	ESR (Ω)	References
		Strong Acid Electrolyte						
2 M H_2SO_4	Macro/mesoporous partially graphitized carbon	340	~105	1.0	~4	~20	–	[18]
1 M H_2SO_4	Mesoporous graphene nanoball	508	206	0.8	–	–	–	[19]
0.5 M H_2SO_4	Carbon films	2610	240	0.7	–	–	–	[20]
2 M H_2SO_4	Carbon monoliths	957	292	–	38	~6	–	[21]
1 M H_2SO_4	CNT balls	200–220	80	1.0	–	–	–	[22]
1 M H_2SO_4	ACFs	3223	280	0.9	–	–	–	[23]
1 M H_2SO_4	Vertically aligned CNTs	55	124	1.0	–	–	–	[24]
1 M H_2SO_4	Graphene quantum dots–3D graphene composites	292	268	0.8	–	–	–	[25]
0.5 M H_2SO_4	Highly graphitized nanoporous carbons	943	238	1.4	19.6	700	–	[26]
1 M H_2SO_4	Monolithic hierarchical porous carbons	2096	221	1.0	6	14	–	[27]
1 M H_2SO_4	Microporous carbons	1354	~110	1.0	~4	~150	0.439	[28]
		Strong Alkaline Electrolyte						
6 M KOH	3D graphene–CNT	903	413	–	–	–	0.15	[29]
6 M KOH	3D flower-like and hierarchical porous carbons	761.5	294	1.8	15.9	317.5	0.1	[30]
6 M KOH	Highly porous graphene planes	1754	303	1.0	~6	~50	0.004	[31]

(Continued)

TABLE 2.4 (*Continued*)
Some Results from the Selected Literature Published in the Last 2 Years (2013–2014) Concerning the Carbon-Based EDLCs with Various Electrolytes

Electrolyte	Material	Surface Area ($m^2 g^{-1}$)	Capacitance ($F g^{-1}$)	Voltage (V)	Energy Density ($Wh kg^{-1}$)	Power Density ($W kg^{-1}$)	ESR (Ω)	References
6 M KOH	Mesoporous carbon	1148	102.3	–	8.34	–	0.202	[32]
6 M KOH	Porous CNT networks decorated crumpled graphene balls	162.3	~180	0.9	4.9	150	–	[33]
6 M KOH	Mesoporous carbon microspheres	659–872	289	1.0	–	–	–	[34]
6 M KOH	Sugar-derived carbon–graphene composite	414.8	273	1.0	–	–	–	[35]
2 M KOH	Interconnected CNFs	271	170.2	–	–	–	–	[36]
6 M KOH	Micro/mesoporous carbon spheres	2502	230	–	–	–	0.58	[37]
6 M KOH	3D carbon nanomembranes	1521	202	1.0	–	–	–	[38]
		Neutral Electrolyte						
1 M $NaNO_3$	AC	2066	116	1.6	–	–	0.427	[39]
0.5 M Na_2SO_4	Graphene	1700	110	–	–	–	0.2	[40]
0.5 M Na_2SO_4	Microporous carbon	1354	~60	1.8	~5	~40	1.084	[28]
4 M $NaNO_3$–EG	PICACIFF-AC	2300	22.3	2.0	16	500	1.02	[41]
0.5 M KCl	Graphene nanoribbons	–	62	–	2.7	129.6	–	[42]
0.5 M Na_2SO_4	AC coated MWCNT doped with N	1889.12	103.1	–	–	–	–	[43]
1 M Na_2SO_4	Hollow carbon fibers	778.6	181.3	–	56.7	–	–	[44]

(*Continued*)

TABLE 2.4 (*Continued*)
Some Results from the Selected Literature Published in the Last 2 Years (2013–2014) Concerning the Carbon-Based EDLCs with Various Electrolytes

Electrolyte	Material	Surface Area ($m^2\ g^{-1}$)	Capacitance ($F\ g^{-1}$)	Voltage (V)	Energy Density ($Wh\ kg^{-1}$)	Power Density ($W\ kg^{-1}$)	ESR (Ω)	References
		Organic Electrolyte						
1.5 M NEt_4BF_4/ACN	Micro/mesoporous carbon	1619	80	–	–	–	–	[45]
1 M $(C_2H_5)_3CH_3NBF_4$/ACN	Microporous carbon	1539	126	3.0	~30	~120	–	[46]
1 M $TEABF_4$/ACN	Highly porous interconnected carbon nanosheets	2200	150	2.7	~30	~13,000	–	[47]
1 M $LiPF_6$/(EC–DEC 1:1)	Porous carbon flakes	1306	126	3.0	45.33	~70	0.9	[48]
1 M $(C_2H_5)_4NBF_4$/PC	Porous network structured carbon	2653	210	2.7	41.8	67.5	11.6	[49]
1 M $LiPF_6$/(EC–DMC 1:1)	High surface area AC	3950	198	3.0	~61	~400	29	[50]
1 M $TEABF_4$/PC	Graphene–CNT composites	–	110	3.0	34.3	400	–	[51]
1 M $TEABF_4$/ACN	AC	1500–2000	124	2.5	28	–	0.81–0.95	[52]
1 M $TEABF_4$/ACN	Hollow mesoporous carbon capsules	1667	162	2.5	–	–	–	[53]
1 M $LiPF_6$/(EC–DEC 1:1)	3D rGO	1040	210	2.0	–	–	–	[54]
		Ionic Liquids						
[EMIM][TFSI]	Holey graphene	–	~45	2.5	–	–	–	[55]
[EMIM][BF_4]	CNTs	600	63	4.0	~35	~40	–	[56]
[EMIM][BF_4]	rGO–mesoporous carbon	1370	144.4	3.5	60.7	174	–	[57]

(Continued)

TABLE 2.4 (*Continued*)
Some Results from the Selected Literature Published in the Last 2 Years (2013–2014) Concerning the Carbon-Based EDLCs with Various Electrolytes

Electrolyte	Material	Surface Area ($m^2\ g^{-1}$)	Capacitance ($F\ g^{-1}$)	Voltage (V)	Energy Density ($Wh\ kg^{-1}$)	Power Density ($W\ kg^{-1}$)	ESR (Ω)	References
[EMIM][BF_4]	3D hierarchical porous graphene–carbon composite	2034	151.7	3.0	50.1	–	–	[58]
[EMIM][BF_4]	3D nanoporous carbon	2933	343	3.5	94.1	488	1.38–4	[59]
[PIP_{13}][FSI] –[PYR_{14}][FSI] (1:1 by molar ratio)	Activated graphene	2000	180	3.5	–	–	3.4	[60]
[EMIM][TFSI]	3D OM carbons	1600	146–178	3.5	–	–		[61]
[BMPY][TFSI]	Interconnected carbon nanosheets	2287	144	3.0	19–40	20,000	6	[62]
[EMIM][BF_4]	Edge-enriched porous graphene nanoribbons	1249	130	3.5	55.2	56.8	1.08	[63]
[BMIM][PF_6]	Hierarchical porous carbon	3000	213	4.0	118	100	54.3	[64]
[EMIM][BF_4]/PC	Aligned CNTs	–	260	4.0	150	~20,000	–	[65]

Abbreviations: CNT, carbon nanotube; MWCNT, multiwalled carbon nanotube; AC, activated carbon; GO, graphene oxide; rGO, reduced graphene oxide; EC, ethylene carbonate; DEC, diethyl carbonate; and DMC, dimethyl carbonate.

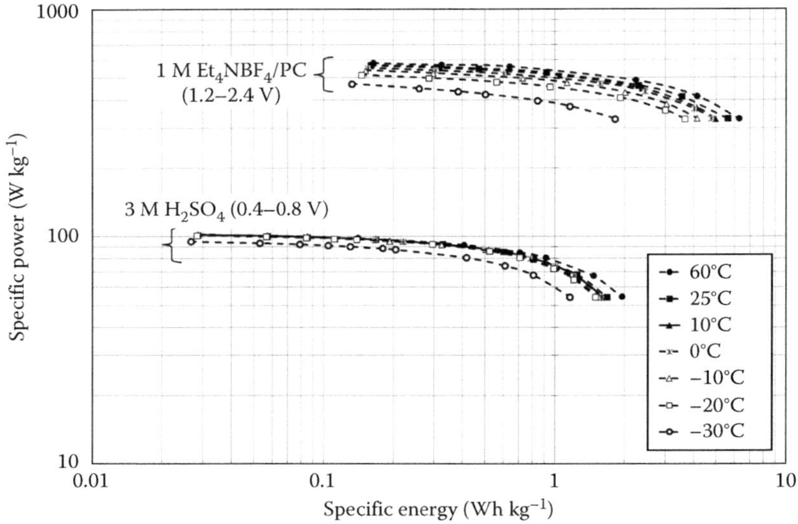

FIGURE 2.2 Ragone chart with specific power (W kg^{-1})/energy density (Wh kg^{-1}) for 3 M H_2SO_4 and 1 M $TEABF_4$/PC electrolytes operating at different temperatures. (Reprinted from *Electrochimica Acta*, 144, Kang, J. H. et al., Characterization of thermal behaviors of electrochemical double layer capacitors (EDLCs) with aqueous and organic electrolytes, 1–11, Copyright 2014, with permission from Elsevier.)

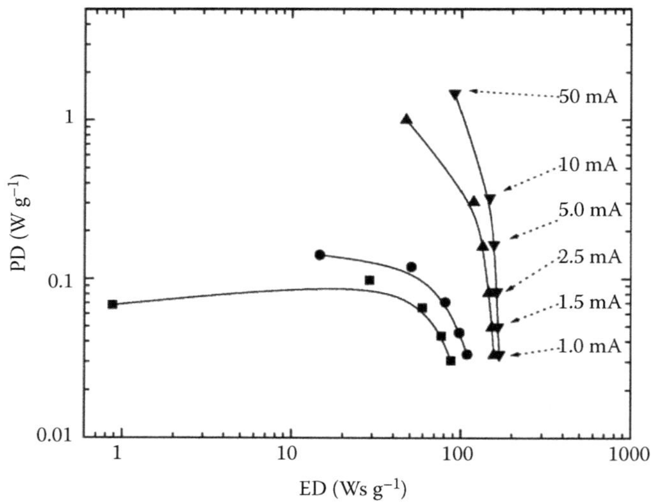

FIGURE 2.3 Ragone plots for the fully charged carbon-cloth electrode for various constant currents as marked on the figure for four electrolyte concentrations: 0.01 M (■), 0.05 M (•), 0.5 M (▲), and 5 M (▼). (Reprinted from *Journal of Power Sources*, 156, Niu, J., W. G. Pell, and B. E. Conway, Requirements for performance characterization of C double-layer supercapacitors: Applications to a high specific-area C-cloth material, 725–740, Copyright 2006, with permission from Elsevier.)

2.2.1.2 Acid Electrolytes for Pseudocapacitors

Up to date, extensive efforts have been made to increase the energy density of aqueous-based ESs and several promising approaches have been proposed. One way is to further increase the capacitance of electrode materials by introducing pseudocapacitive contributions. For carbon-based materials, it has been found that the specific capacitance in aqueous H_2SO_4 electrolyte also includes pseudocapacitance contributions apart from the EDL capacitance [69]. This is due to the fast redox reactions that occur on particular surface functionalities on carbon-based materials, such as the oxygenated species [69]. This pseudocapacitance can be further enhanced by introducing heteroatoms, such as oxygen [70,71], nitrogen [71–73], boron, and phosphorus [72,74] or certain surface functional group to the surface of carbon materials and thereby the specific capacitance can be markedly improved [74–76]. Figure 2.4 shows an example of anthraquinone (AQ)-modified activated carbon (AC), which showed a much higher specific capacitance compared to the unmodified carbon owing to the pseudocapacitance contribution from the AQ group [76].

It is worth noticing that the electrolyte nature plays a significant role in the pseudocapacitive properties of carbon-based materials because the surface functionalities can exhibit different behavior depending on the electrolytes. For example, quinone-type functionalities generate pseudocapacitive effects in the presence of acidic aqueous electrolytes such as H_2SO_4 since they require protons to proceed as shown in Reaction 2.1, while this effect is hardly observed when alkaline electrolytes are used [76,77].

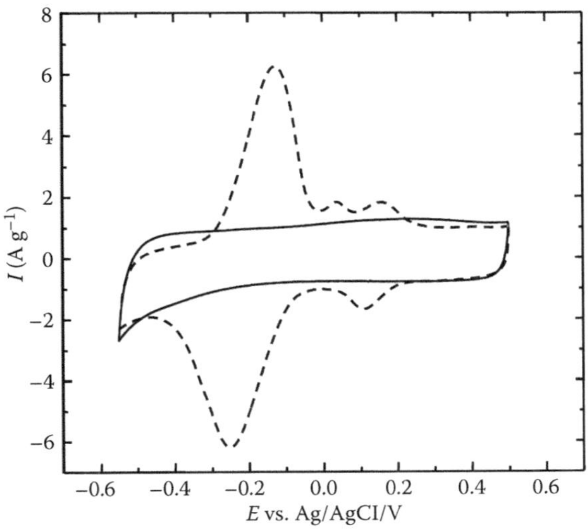

FIGURE 2.4 Cyclic voltammograms of unmodified AC (—) and AC modified with 11 wt.% (- - -) AQ, recorded in 0.1 M H_2SO_4 at a scan rate of 10 mV s^{-1}. (Reprinted from *Journal of Power Sources*, 196, Pognon, G. et al., Performance and stability of electrochemical capacitor based on anthraquinone modified activated carbon, 4117–4122, Copyright 2011, with permission from Elsevier.)

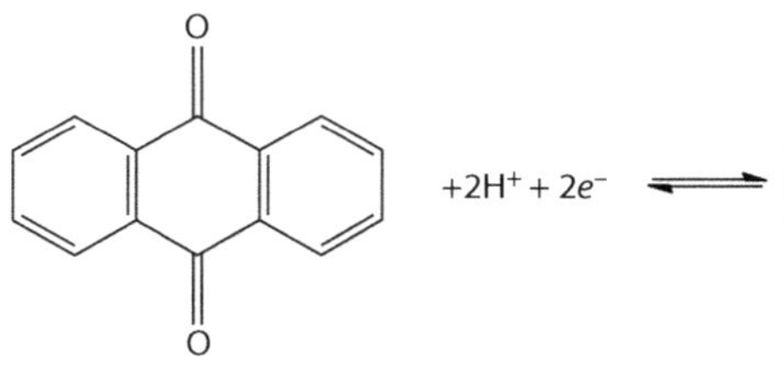

(2.1)

Calvo et al. [78] investigated the effect of electrolytes on the pseudocapacitive behavior of different oxygen surface groups on AC xerogels. For carbon xerogel with quinone-type oxygen groups, pseudocapacitive effects were found when the pH of the electrolyte was lower than the zero charge point of the electrodes, that is, when the electrode surface is positively charged. However, when alkaline electrolyte was employed, the carbon electrodes showed similar capacitance values, indicating the negative charges contributed less to the pseudocapacitance. Ruiz et al. [79] studied the electrochemical behavior of ACs with different oxygen content and confirmed the pseudocapacitance contribution in 1 M H_2SO_4 and 6 M KOH for ACs with high oxygen content. However, the pseudocapacitance contribution in 1 M H_2SO_4 is significantly higher than that in 6 M KOH. In the two-electrode system, the pseudocapacitance contribution was estimated to be about 22% in acid electrolyte and 11% in alkaline electrolyte. Yan et al. [72] studied the effects of electrolytes on the capacitive behavior of nitrogen-doped nonporous carbon nanofibers (N-NPCNF) and nitrogen/phosphorus co-doped nonporous carbon nanofibers (N/P-NPCNF). It was reported that the capacitance of the N-NPCNF electrode in neutral electrolytes was negligible while the capacitance of the N/P-NPCNF electrode increased remarkably in both acid electrolyte (H_2SO_4) and neutral electrolytes (Li_2SO_4, Na_2SO_4, and K_2SO_4). The capacitance of the phosphorus groups doped carbons in different electrolytes decreases in the order of: 1 M H_2SO_4 (102 F g^{-1}) > 0.5 M K_2SO_4 (80.5 F g^{-1}) > 0.5 M Li_2SO_4 (75.6 F g^{-1}) > 1 M Na_2SO_4 (51.5 F g^{-1}). The authors attributed the better capacitive behavior of the N/P-NPCNFs in acid electrolyte to the hydrogen bonding between oxygen-rich phosphorus groups and H^+ ions. The effect of electrolyte oxygen content and visible-light exposure on the capacitance of sulfur-doped carbon was studied by Singh et al. [80]. In this study, a significant increase in the capacitance was observed in oxygenated electrolytes after potential cycling, which was accompanied by the oxidation of the carbon surface. The oxygen in the 1 M H_2SO_4 electrolyte and visible-light irradiation was found to considerably increase the capacitance of sulfur-doped carbon [80]. However, it should be noted that the presence of oxygen in the electrolyte may increase the self-discharge rate of the ES [81]. Oda et al. [70] found that in the case of 1 M H_2SO_4 electrolyte, the specific capacitance depended more on oxygen functional groups than on the surface area. A maximum increase of 30% in the specific capacitance was obtained for AC fibers (ACFs) with different oxygen-containing functional groups (OCFGs). On the other hand, in the case of an organic electrolyte (0.5 M $LiClO_4$/PC), OCFGs on ACF with primary micropores caused a capacitance degradation. Jana et al. [82] prepared the water dispersible graphene by 6-amino-4-hydroxy-2-naphthalenesulfonic acid (ANS). The specific capacitance of ANS-modified reduced graphene oxide (ANS-rGO) was found to be 375 F g^{-1} at a current density of 1.3 A g^{-1} in 1 M H_2SO_4. This electrode was highly stable and exhibited about 97.5% retention in the specific capacitance after 1000 charging/discharging cycles. The energy density of the ANS-rGO electrode reached 213 Wh kg^{-1} at a power density of 1328 W kg^{-1}. The authors ascribed the high values of specific capacitance, energy density, and power density to the contribution by pseudocapacitance through the faradaic reactions of –OH and $-SO_3H$ functionalities with H^+ ions in H_2SO_4 electrolyte. As can be seen from the above discussion, to achieve an optimum performance, it is important to use suitable electrolytes to match the corresponding surface functionalities on the carbon-based electrode materials. Furthermore, the cycle life of the electrode materials is also an important issue. The presence of such pseudocapacitance has often been reported to decrease the cycle life. This is contrary to what occurs in organic electrolytes, where the capacitance largely comes from the electrostatic charge separation and good cycling stability is usually achieved [83]. Ruiz et al. [84] reported the cycling stability of the AC thermally treated at 1000°C was better than that of the original AC. This result was attributed to the fact that the AC was rich in surface oxygen functionalities while most of the functionalities were removed for thermally treated AC. It is worth noticing that the type of aqueous electrolytes also plays a significant role in the cycle life of carbon-based materials, especially when pseudocapacitance exists. For instance, Ruiz et al. [83] reported that in spite of the high contribution of pseudocapacitance in the 1 M H_2SO_4 electrolyte, the performance of AC electrodes during long-term cycling was good, with only 8% loss of the specific capacitance after 7000 galvanostatic cycles at 0.6 V and 16% at 1 V. On the contrary, AC presented a poor long-term cycling stability in 6 M KOH and resulted in a 72% loss in the capacitance value even at a low working voltage window of 0.6 V. The authors attributed this to the strong oxidation of the positive electrode. Interestingly, the introduction of certain surface functional groups into the carbon surface has been found to increase the stability of the electrode materials [74,85]. Hulicova-Jurcakova et al. [74] found the enrichment of carbons with phosphorus groups not only contributed to the pseudocapacitance but also improved the stability of the electrode up to 1.3–1.5 V in 1 M H_2SO_4 electrolyte (Figure 2.5).

Apart from the pseudocapacitive contribution from specific surface functionalities, the possible intercalation of electrolyte ions into certain electrode materials may also result in the increased capacitance [86–88]. For example, Soneda et al. [87] reported a large capacitance of about 555 F g^{-1} for exfoliated carbon fibers in 18 M H_2SO_4 electrolyte, which was due to the H_2SO_4 molecules intercalation into the carbon layers of exfoliated carbon fibers.

The pseudocapacitance can also be provided by other pseudocapacitive materials such as some metal oxides and electrically conductive polymers (ECPs) that have much higher theoretical capacitance than carbon-based materials. These materials have been reviewed in detail elsewhere [89,90]. Although many materials have been reported to exhibit pseudocapacitive behavior, they are very sensitive to the type and pH of the electrolytes and few of them are suitable for application in strong acid electrolytes. As previously mentioned in Section 1.3.2, RuO_2 is one of the most extensively studied pseudocapacitive materials in H_2SO_4 electrolytes.

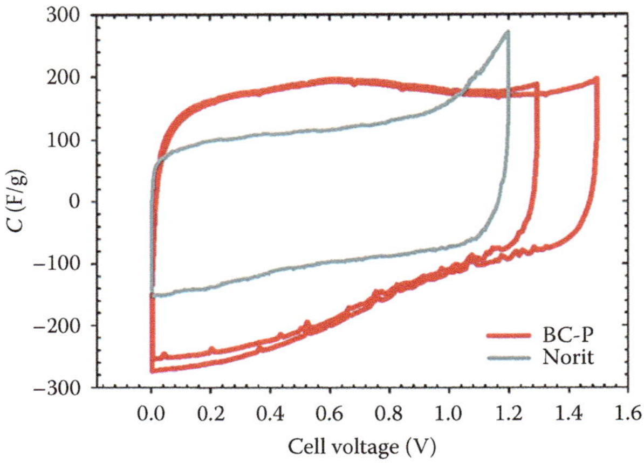

FIGURE 2.5 Cyclic voltammograms of BC-P recorded at the 5 mV s^{-1} and the cell voltage of 1.3 and 1.5 V compared with Norit at 1.2 V. Commercial microporous carbon Norit (Norit RB3) is used as comparison. (Reprinted with permission from Hulicova-Jurcakova, D. et al. 2009. Highly stable performance of supercapacitors from phosphorus-enriched carbons. *Journal of the American Chemical Society* 131, 5026–5027. Copyright 2009 American Chemical Society.)

The capacitance of amorphous RuO_2 can reach a very high value (~1000 F g^{-1}) compared to the crystalline RuO_2 due to the relatively easy insertion of protons into the amorphous structure [91]. However, since Ru is a rare metal, the high cost and limited resource restrict the wide applications of RuO_2-based ESs. Therefore, alternative materials with pseudocapacitive behavior that can be used in strong acid electrolytes have been studied. For example, Mendoza-Sánchez et al. [92] investigated the pseudocapacitive behavior of α-MoO_3 nanobelts and thin films in different aqueous electrolytes. 1 M H_2SO_4 electrolyte gave the highest charge storage compared to other electrolytes including Na_2SO_4, K_2SO_4, and Li_2SO_4. A combination of x-ray photoelectron spectroscopy (XPS) and electrochemical characterization methods revealed a partial reduction of MoO_3 to a mix of lower valence oxides with concentrations varying as Mo (5+) > Mo (4+) > Mo (6+) in 0–1 V (vs. Ag/AgCl). MoO_2 was found to be the main species below 0.185 V (vs. Ag/AgCl). A relative small capacitance of 8.8 F g^{-1} (64 μF cm^{-2}) was maintained for 720 cycles in a relatively small electrochemical window of 0.26–0.43 V (vs. Ag/AgCl). Increasing the potential window drastically lowered the cycling ability of the electrode. Purushothaman et al. [93] studied the pseudocapacitance behavior of α-$MnMoO_4$ nanorods in three different electrolytes: para toluene sulfonic acid (p-TSA), H_2SO_4, and HCl. As shown in Figure 2.6, at a low scan rate of 5 mV s^{-1}, H_2SO_4 electrolyte gave a maximum specific capacitance of 998 F g^{-1} whereas p-TSA and HCl electrolytes led to the specific capacitance of 784 and 530 F g^{-1}, respectively. However, α-$MnMoO_4$ nanorods showed better pseudocapacitive behavior in p-TSA electrolyte compared to H_2SO_4 electrolyte at higher scan rates, probably due to the different diffusion rate of anions and cations.

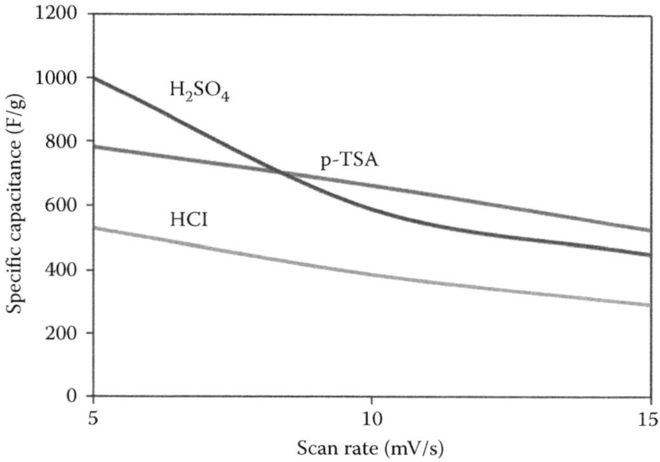

FIGURE 2.6 Specific capacitance as a function of scan rate. (Reprinted from *Materials Research Bulletin*, 47, Purushothaman, K. K., M. Cuba, and G. Muralidharan, Supercapacitor behavior of alpha-$MnMoO_4$ nanorods on different electrolytes, 3348–3351, Copyright 2012, with permission from Elsevier.)

2.2.1.3 Acid Electrolytes for Hybrid ESs

Another effective way to improve the energy and power densities of aqueous electrolyte-based ESs is to increase the operating cell voltage. ESs have been traditionally designed in a symmetric configuration with the same type and amount of the electrode materials (e.g., carbon-based materials or pseudocapacitive materials). When a symmetric ES based on ACs is used in aqueous electrolytes such as H_2SO_4 or KOH, the maximum operative voltage is restricted by the gas evolution reactions and oxidation of carbon electrodes [94]. Since the theoretical stability window of commonly used acid (1 M H_2SO_4) or basic (6 M KOH) aqueous electrolytes is about 1.2 V, the practical operating voltage hardly exceeds 1 V. On the contrary, ESs with asymmetric configurations have been proved to be an effective way to increase the working potential window of aqueous electrolytes so as to overcome the theoretical electrochemically stable window of water [95]. The key for designing asymmetric ESs with higher operating voltage in aqueous electrolytes is the combination of two different electrodes that can work complementarily in separate potential windows. For example, the operative voltage window of aqueous electrolyte-based ESs can be greatly extended by combining the pseudocapacitive electrode such as RuO_2 or battery-like positive electrode such as PbO_2 that have a high overpotential for O_2 evolution and carbon-based negative electrode that have a high overpotential for H_2 evolution [95]. To date, several types of asymmetric ESs, such as carbon//PbO_2 [96], carbon//RuO_2 [97], carbon//ECPs [98], and carbon with different mass or properties in each electrode [84,99], have been proposed for the application in strong acid electrolytes. In the case of carbon//PbO_2 asymmetric ESs, the high oxygen evolution reaction overpotential at the PbO_2 positive electrode in the H_2SO_4 electrolyte can significantly increase the effective cell voltage to above 1 V, as shown in Figure 2.7 [95].

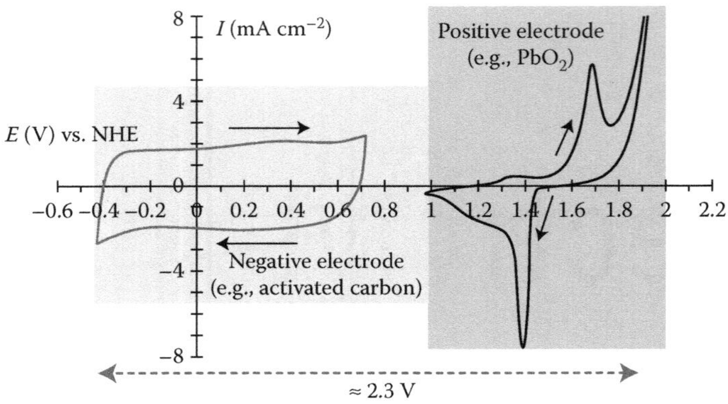

FIGURE 2.7 Schematic representation of cyclic voltammograms for asymmetric AC//PbO_2 ES in 1 M H_2SO_4. Right and left cyclic voltammograms represent the potential window of positive and negative electrode, respectively. NHE, normal hydrogen electrode; *I*, measured current; and *E*, electrode potential. (Reprinted from Long, J. W. et al. 2011. *MRS Bulletin* 36:513–522, with permission from Cambridge University Press.)

The reaction which occurs in the positive PbO_2 electrode in H_2SO_4 electrolytes is the same as the conventional Pb//PbO_2, which is described by [94,100]:

On the positive electrode:

$$PbO_2 + 3H^+ + HSO_4^- + 2e^- \rightarrow PbSO_4 + 2H_2O(1.685\,\text{V vs. ENH}) \quad (2.2)$$

On the negative electrode:

$$Pb + HSO_4^- \rightarrow PbSO_4 + H^+ + 2e^-(-0.356\,\text{V vs. ENH}) \quad (2.3)$$

while the charge and discharge mechanism on the EDL carbon-based electrode can be described by [94]

$$nC_6^{x-}(H^+)_x \rightarrow nC_6^{(x-2)-}(H^+)_{x-2} + 2H^+ + 2e^-(\text{discharged}) \quad (2.4)$$

As mentioned previously, the total capacitance of the ES is dominated by the electrode with smaller capacitance. Since the PbO_2 faradaic electrode has a very high capacity compared to the carbon electrode, the total capacitance is almost equal to that of the negative carbon electrode. Therefore, the full EDL capacitance of the carbon electrode can be achieved compared to the symmetric EDLCs [95]. The typical energy densities for carbon//PbO_2 hybrid ESs using H_2SO_4 electrolyte reported in the literature range from 25 to 30 Wh kg^{-1}, which is much higher than for symmetric carbon-based EDLCs with H_2SO_4 electrolyte (3–6 Wh kg^{-1}) [101]. In addition, the use of battery-type materials (e.g., PbO_2) with a fixed thermodynamic potential may inhibit the self-discharge process of ESs [95]. In order to maximize the effective surface area in contact with the electrolyte and increase the reaction kinetics, nanostructured

PbO_2 such as thin films and nanowires have been developed as electrode materials in H_2SO_4 electrolytes. Yu and Gao [102] prepared PbO_2 thin films on Ti/SnO_2 substrates by pulse current deposition in the $Pb(NO_3)_2$ solution. The asymmetric ES consisted of the prepared PbO_2 thin films as negative electrode and AC as positive electrode in 5.3 M H_2SO_4 solution. In the potential range of 0.8–1.8 V, the PbO_2/AC asymmetric ES delivered a high specific energy density of 29.0 Wh kg^{-1} at a power density of 1.01 kW kg^{-1} based on the total weight of both active electrode materials. This ES lost 10% of its initial capacity after 4500 deep cycles. One-dimensional (1D) nanowires generally have a larger electrochemically active surface area, which is expected to further improve the specific energy density. However, it should be noted that when PbO_2 nanowires are used in H_2SO_4 electrolytes, their stability is a major concern. Perret et al. [103] prepared PbO_2 nanowires by electrochemical deposition in an anodized aluminum oxide template and investigated the structural change of PbO_2 nanowires during electrochemical cycling. It was found that the structure of the PbO_2 nanowires changed obviously during the electrochemical potential cycling in 1 M H_2SO_4. As shown in Figure 2.8b and c, the overall morphology of nanowires was maintained during the first five cycles. However, after 10 cycles, the 1D nanowire structure disappeared (Figure 2.8d) and the resultant structure was very similar to that of PbO_2 thin films (Figure 2.8f). It was believed that the sulfation of the PbO_2 electrode in H_2SO_4 solution during repeated potential cycling resulted in its poor cycling stability [96]. Therefore, for PbO_2 nanowires, the use of H_2SO_4 as electrolyte is not suitable and other alternative electrolytes should be developed to preserve the benefit of the 1D nanostructure.

FIGURE 2.8 Scanning electron microscopy (SEM) micrographs of PbO_2 (a–d) nanowires and (e and f) thin film after cycling in 1 M H_2SO_4: (a and e) as-deposited and (b) 3, (c) 5, and (d and f) 10 cycles. The deposition charge was 40°C for the PbO_2 nanowires and thin film. All samples were examined in the charged state. (Reproduced by permission of The Electrochemical Society from Perret, P. et al. 2009. *Journal of the Electrochemical Society* 156:A645–A651.)

An alternative electrolyte composed of methanesulfonic acid (CH_3SO_3H) and lead methanesulfonate has been proposed to address this problem [96]. In this case, the redox reaction at the PbO_2 electrode becomes solid/solvated ions in methanesulfonic acid-based electrolyte instead of a solid/solid couple in H_2SO_4 electrolyte [96]:

$$PbO_2 + 4H^+ + 2e^- \leftrightarrow Pb^{2+}_{(aq)} + 2H_2O \tag{2.5}$$

PbO_2 is reduced into Pb^{2+} (aq.) during the discharging process. Pb^{2+} (aq.) is oxidized and the formed PbO_2 is electrodeposited on the surface of the electrode during the charging process. This solves the problems associated with the sulfation of PbO_2 in H_2SO_4 electrolytes and therefore an improved cycle life could be expected. Perret et al. [96] reported an asymmetric ES using an AC negative electrode, and PbO_2 nanowire array positive electrode with an electrolyte composed of 0.1 M methanesulfonic acid and 0.1 M $Pb(NO_3)_2$. The maximum energy density and specific capacity of the carbon//PbO_2 nanowire ES were 29 Wh kg^{-1} and 34 F g^{-1}, respectively. This hybrid ES did not show any sign of performance degradation up to 5000 charging/discharging cycles at 22°C rate. This was due to the much better morphology stability of the Pb nanowires in the methanesulfonic acid electrolyte compared to the H_2SO_4 electrolyte (Figure 2.9) [96]. On the contrary, the use of H_2SO_4 electrolyte generally results in the performance loss of carbon//PbO_2 asymmetric ESs [102,103]. Despite the increased energy density, the use of battery-type electrode limits the power density and cycle life of the aqueous hybrid ESs. This encourages the development of other asymmetric ESs with the combination of different carbon-based and/or pseudocapacitive electrode materials. For example, to take advantage of the high capacitance and relatively good cycling stability of RuO_2 pseudocapacitive material, asymmetric carbon//RuO_2 ES with H_2SO_4 electrolyte has been reported by Algharaibeh et al. [97]. In this study, the negative electrode was AQ-modified carbon fabric, which extended the negative potential limit relative to RuO_2. The positive potential limit is also enlarged compared to the carbon-based electrode. Besides, the redox activity of the AQ provided extra charge storage so as to match the high capacitance of RuO_2 positive electrode. The maximum energy density was 26.7 Wh kg^{-1} and an energy density of 12.7 Wh kg^{-1} was achieved at 0.8 A cm^{-1} discharge rate. Another distinct advantage of this ES is that it requires 64% less Ru compared to symmetric RuO_2 ES. Similarly, asymmetric ESs with acid electrolyte can also consist of a carbon-based electrode and an ECP electrode to widen the effective cell voltage. Suppes et al. [104] investigated the asymmetric ES using polypyrrole (PPy)/phosphomolybdic acid (PMA) as the positive electrode and poly(ethylenedioxythiophene) (PEDOT)/phosphotungstic acid (PTA) as the negative electrode in 0.5 M H_2SO_4 electrolyte. It was found that the energy density and power density of the asymmetric ES were increased by 3 and 50 times, respectively, compared to the asymmetric ES with PPy/PMA electrodes, due to the increased cell potential window of the former. However, the long-term stability should be improved. Salinas-Torres et al. [98] developed an asymmetric ES based on ACF with polyaniline (PANI) coatings inside its microporosity as the positive electrode and AC as the negative electrode in 0.5 M H_2SO_4 electrolyte. The PANI contributed the pseudocapacitance and thus the total capacitance of the ES. The presence of a thin PANI layer also protected the carbon material from

FIGURE 2.9 Scanning electron microscopy micrographs of PbO_2 nanowire after different cycling period in methanesulfonic acid: (a and b) side and top view of as-deposited nanowire, (c) in the discharged state during the first cycle, (d–f) in the charged state after 5, 30, and 100 complete cycles. (Reprinted from *Electrochimica Acta*, 56, Perret, P. et al., Carbon/PbO_2 asymmetric electrochemical capacitor based on methanesulfonic acid electrolyte, 8122–8128, Copyright 2011, with permission from Elsevier.)

oxidation and the oxygen evolution reaction occurred at more positive potentials. This ES was tested at a maximum cell voltage of 1.6 V and showed energy density of 20 Wh kg^{-1} and power density of 2.1 kW kg^{-1} with a fair cycle lifetime (90% during the first 1000 cycles) in 0.5 M H_2SO_4 electrolyte. Shen et al. [105] reported an asymmetric ES fabricated by a sulfonated graphene nanosheet/carboxylated multiwalled carbon nanotube/PANI (sGNS/cMWCNT/PANI) nanocomposites as the positive electrode and activated porous graphene (aGNS) as the negative electrode in H_2SO_4 aqueous electrolyte. Because of the complementary potential window and high capacitance of the positive and negative electrodes, the obtained ES was able to be operated at a cell voltage of 1.6 V in 1 M H_2SO_4 electrolyte and had a high energy density of 20.5 Wh kg^{-1} at a power density of 25 kW kg^{-1}. The maximum energy density was 41.5 Wh kg^{-1}, which was much higher than that of symmetric ESs based on sGNS–cMWCNT–PANI electrodes (21.1 kW kg^{-1}) and aGNS electrodes (6.5 kW kg^{-1}). The reported ES had a better cycle life with 9% performance loss after 5000 cycles compared to that reported by Salinas-Torres et al. [98].

Aqueous-based asymmetric ESs with wider potential window can also be made from different carbon-based materials, which have been discovered in recent studies [84,99]. For instance, Khomenko et al. [99] proposed an asymmetric ES with high energy density using only AC for both electrodes in 1 M H_2SO_4 or 6 M KOH electrolyte. As discussed previously, in such electrolyte, the charges were stored by both EDL and fast faradaic charge transfer processes. Through utilizing different pseudocapacitive properties of carbons in positive and negative electrodes, it was possible to shift the potentials of destructive oxidation of AC and H_2 evolution to more positive and negative values, respectively. By optimizing carbons for the positive and negative electrodes and balancing the mass of the electrodes, the ES could be operated up to 1.6 V in 1 M H_2SO_4 with excellent cycle life (15% decrease of the discharge capacitance after 10,000 cycles) (Figure 2.10a). This is a high cell voltage for aqueous-based ES considering that the cell voltage for symmetrical carbon ES in H_2SO_4 is generally about 1 V. As a result, the obtained energy density was close to 30 Wh kg^{-1} (Figure 2.10b) which was comparable to that of EDLCs with an organic electrolyte. Therefore, the application of asymmetric ESs provides a promising approach to best maintain the advantages of aqueous electrolytes while increasing the energy and power densities. However, other performance characteristics such as the rate performance and the cycle life should also be considered when developing aqueous electrolyte-based asymmetric ESs.

On the other hand, to ensure a low ESR and to maximize the power density, highly concentrated solutions are used for ESs. However, the strong acid electrolytes are extremely corrosive, especially for current collectors and packaging materials. Therefore, some corrosion-resistant current collectors (e.g., Au and Ti) have to be used in strong acid electrolytes, which will increase the cost of ESs. Besides, for some electrode materials (especially conductive polymers), the corrosive effect of the highly concentrated acid electrolytes may lead to their performance deterioration. Hou et al. [106] found that PPy suffered from the dissolution during charging/discharging cycles in a concentrated $HClO_4$ aqueous electrolyte (1 M) as compared to a diluted $HClO_4$ electrolyte (0.1 M). Therefore, the optimization of the electrolyte concentration is also required in some cases to improve the cycling stability of the electrode materials. There are other acid electrolytes that may be used such as perchloric acid, hexafluorosilicic acid, and tetrafluoroboric acid, but few of them have been studied for the application in ESs and most of them face safety issues [96].

Moreover, the self-discharge in the concentrated electrolytes especially in the presence of contamination and oxygen is also a concern. Andreas et al. [107] reported the effect of Fe concentration in H_2SO_4 electrolyte on rate of self-discharge of carbon-based ESs. No enhancement of self-discharge on the positive electrode was found for Fe concentrations up to 10^{-5} M. However, the negative carbon electrode could withstand Fe concentrations of 10^{-3} M without self-discharge increase. Kazaryan et al. [108] found that Mn and Ti ions in the H_2SO_4 electrolyte of a PbO_2//carbon ES resulted in the shuttle self-discharge. When the concentration of the Fe, Mn, and Ti was 10^{20} cm^{-3} in the electrolyte, the decrease rate of the ES's voltage during self-discharge increased 51.7, 3.06, and 6.0 times, respectively. Oickle et al. [81] investigated water electrolysis and oxygen reduction as possible self-discharge mechanisms for carbon-based ESs with an aqueous H_2SO_4 electrolyte through varying

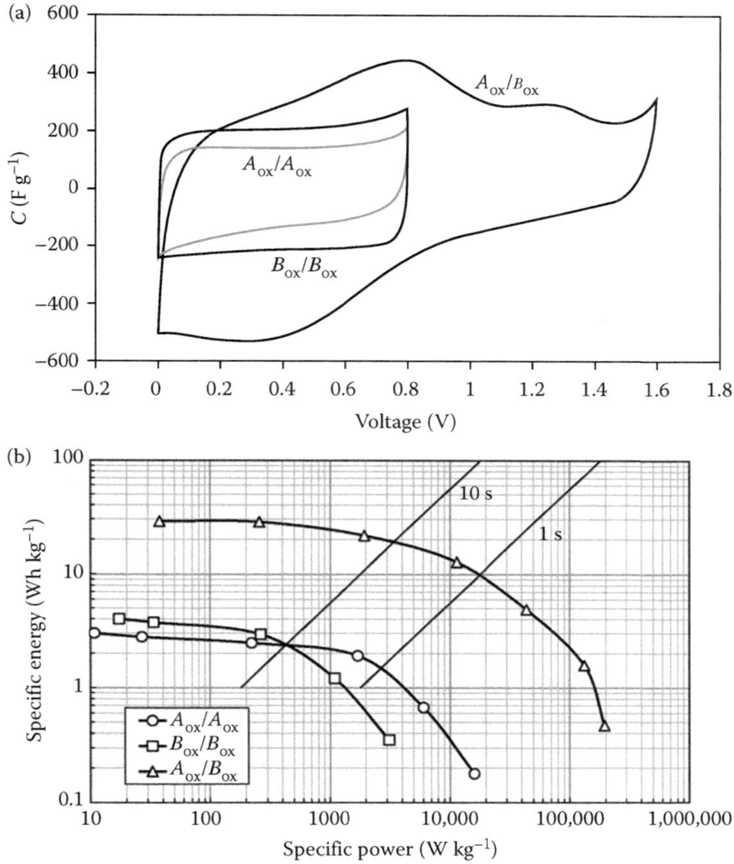

FIGURE 2.10 (a) Cyclic voltammograms of an optimized asymmetric A_{ox}/B_{ox} capacitor and of symmetric A_{ox}/A_{ox} and B_{ox}/B_{ox} capacitors in 1 M H_2SO_4. Scan rate of voltage 2 mV s^{-1}; (b) Ragone plots for the asymmetric A_{ox}/B_{ox} capacitor at a voltage of 1.6 V (-Δ-) and for the A_{ox}/A_{ox} (-O-) and B_{ox}/B_{ox} (-□-) symmetric capacitors at a voltage of 0.8 V. Electrolyte: 1 M H_2SO_4. (Reprinted from *Journal of Power Sources*, 195, Khomenko, V., E. Raymundo-Piñero, and F. Béguin, A new type of high energy asymmetric capacitor with nanoporous carbon electrodes in aqueous electrolyte, 4234–4241, Copyright 2010, with permission from Elsevier.)

the dissolved oxygen and hydrogen content in the electrolyte. It was found that water electrolysis is not a self-discharge mechanism on the positive and negative electrodes. However, oxygen reduction caused a large increase in self-discharge on the negative electrode. Therefore, it is necessary to deaerate the electrolytes for use in ESs.

2.2.2 Strong Alkaline Electrolytes

Based on statistics, it can be seen that alkaline electrolytes are the most widely used aqueous electrolytes in the reported literature [2]. Compared to the strong acid electrolyte, some cheaper metallic materials such as Ni can be used as the current collectors for ESs, which saves the cost. Among various alkaline electrolytes, KOH has been most extensively used because of its high ionic conductivity, that is, maximum value of 0.6 S cm^{-1} for 6 M at 25°C. Other types of base electrolyte, such as NaOH and LiOH, have also been investigated, which will be discussed later. These alkaline electrolytes can be used for carbon-based EDLCs, pseudocapacitors, and hybrid ESs.

2.2.2.1 Alkaline Electrolytes for EDLCs

As shown in Table 2.4, the reported specific capacitance and energy density of EDLCs with KOH electrolytes are generally comparable to that reported with H_2SO_4 electrolytes. Apart from the commonly used KOH electrolyte, other electrolytes such as LiOH have also been explored for EDLCs [109–111]. He et al. [109] found the addition of LiOH into the KOH electrolyte (KOH:LiOH weight ratio of 50:1) was beneficial for increasing the specific capacitance of AC, possibly due to the smaller hydration number of Li^+ ions. Stepniak et al. [111] reported ACF cloths-based EDLCs with LiOH electrolytes exhibited a similar capacitive behavior compared to that with KOH electrolytes, but the former had a higher power density. In addition, very recent studies showed adding small amount of ethanol to KOH electrolytes can effectively improve the wettability of carbide-derived carbons (CDCs) and consequently increase the utilization of the electrode surface area, leading to the improvement of the specific capacitance [112,113].

Similar to the research trend in the field of strong acid electrolytes, extensive effort has been devoted to enhancing the energy density of ESs with alkaline electrolytes by increasing the capacitance and/or widening the operating voltage window. These developments can be briefly summarized as [2]: (a) increasing the capacitance of carbon-based materials by the pseudocapacitance contribution; (b) exploring pseudocapacitive materials with high specific capacitance; (c) developing composite materials that combine the carbon-based materials and pseudocapacitive materials; and (d) widening the operating voltage of alkaline electrolytes by designing asymmetric ESs.

2.2.2.2 Alkaline Electrolytes for Pseudocapacitors

For carbon-based materials, the pseudocapacitance contributed from the surface functionalities on the carbons is related with the faradaic interactions between electrolyte ions and surface functional group [75]. Most studies found the KOH electrolyte is beneficial for the nitrogen-doped carbon electrode [114–116]. Therefore, the resultant pseudocapacitance is strongly dependent on the nature of electrolytes such as the type and pH. Lang et al. [117] investigated the influence of surface functional groups on the capacitive characteristics of the ordered mesoporous (OM) carbons after nitric acid modification. Nitric acid modification resulted in the increase of the pyrone-like (C–O and O–C=O) and the nitrogen-containing surface functionalities on the mesoporous carbons. In 2 M KOH electrolytes, the increased pyrone-like and the nitrogen-containing functionalities contributed to the pseudocapacitance, leading to the improved specific capacitance. On the other hand, in 2 M H_2SO_4 electrolytes, the nitric acid modification lowered the specific capacitance of OM carbons due to the decreasing amount of quinone-like functionalities. The contribution of pseudocapacitive was negligible in neutral electrolytes (2 M Na_2SO_4), even if the mesoporous carbons had plentiful surface oxygen and nitrogen-containing functionalities. Wang et al. [118] investigated the pseudocapacitive behavior nitrogen-doped porous carbon monolith in 6 M KOH electrolytes. It was found that the nitrogen-doped porous carbon electrode showed a higher capacitance (246 F g^{-1}) compared to the nitrogen-free carbon (186 F g^{-1}). Based on the calculated redox affinity of different groups and *ex situ* electrochemical-spectroscopy analysis, the authors proposed possible redox reaction pathways involving different species such as hydroxyl groups between pyridone and pyridine and nitrogen atoms in pyridine (Figure 2.11). Rennie and Hall [119] also studied the electrochemical behavior of nitrogen-enriched carbon materials in KOH electrolytes. Contrast matched-small angle neutron scattering (SANS) indicated that there was no obvious difference in the electrolyte accessibility between the nitrogen-enriched carbon materials studied. Electrochemical measurements confirmed nitrogen groups especially pyridinic nitrogen had a positive effect on the specific capacitance. It was proposed that a greater amount of charge is located at pyridinic sites during the charging/discharging process, which is possibly due to a specific adsorption mechanism coupled with enhanced electron conductivity through the carbon matrix. Wang et al. [120] found codoping the porous carbons by phosphorus and nitrogen could further improve the performance with higher specific

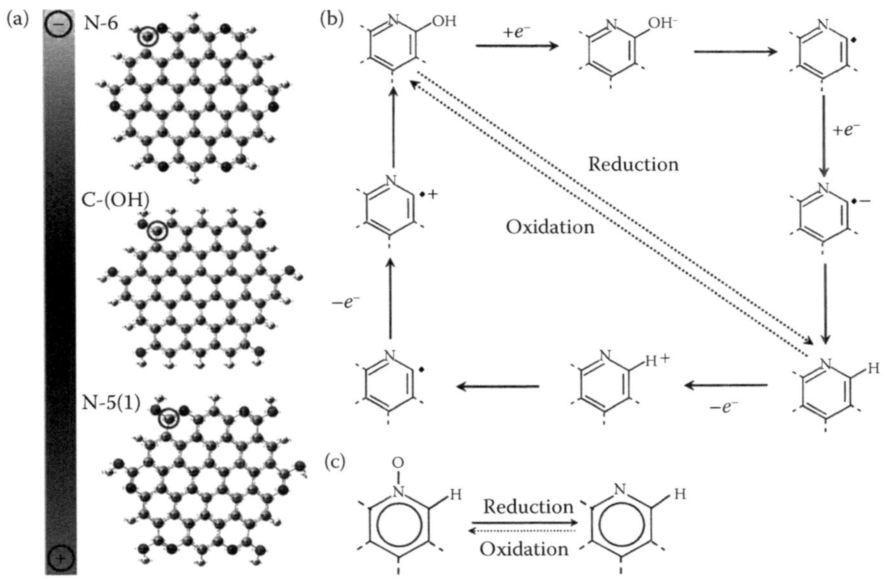

FIGURE 2.11 (a) Charge per target carbon atom (circled) in undoped and nitrogen-doped graphene planes saturated with H or OH groups. (b) Path 1 reaction between N-5 (1) and N-6, (c) Path 2 reaction between N-X and N-6 (c). (Wang, D. W. et al.: Nitrogen-doped carbon monolith for alkaline supercapacitors and understanding nitrogen-induced redox transitions. *Chemistry – A European Journal*. 2012. 18. 5345–5351. Copyright Wiley-VCH Verlag GmbH & Co. KGaA. Reproduced with permission.)

capacitance, wider potential window, and better stability compared to the phosphorus-free counterparts. It is interesting to note there is another possibility of pseudocapacitive behavior, which is observed in carbon-based materials that involved hydrogen storage through negative polarization at lower potentials [99,121,122]. This behavior also strongly depends on the type of the electrolytes and is more favorable to occur in alkaline electrolytes. Jurewicz et al. [121] proposed that when a negative polarization was applied, hydrated K^+ in the alkaline electrolyte or H_3O^+ cations in the acid electrolyte were adsorbed on the AC electrode, forming the EDL. Once the electrode potential decreased to a lower value than the equilibrium potential, hydrogen with zero oxidation state was formed by the reduction of water in alkaline electrolytes or the reduction of H_3O^+ in acidic electrolytes. Subsequently, hydrogen was physically adsorbed (H_{ad}) onto the carbon surface. The authors further pointed out the recombination of H_{ad} to the molecular hydrogen in the KOH electrolyte was less favored than in the H_2SO_4 electrolyte because of the high overvoltage value in KOH. Therefore, a noticeable hydrogen sorption was observed in the alkaline electrolyte instead of acid electrolyte. Qu et al. [123] found the majority of atomic hydrogen formed through electrolysis in the KOH electrolyte was intercalated into the interlayer of graphene sheets in carbonaceous materials. Therefore, a larger interlayer distance is beneficial for increasing the hydrogen storage capacity.

Besides, due to their high theoretical capacitance values, several transition metal oxides, hydroxides, and sulfides, such as NiO_x [124], CoO_x [125–128], MnO_2 [129,130], Mn substituted spinel cobalt oxide [131], MoO_2 [132], $Ni(OH)_2$ [133–135], $Co(OH)_2$ [126,136], and cobalt sulfide [137–139], have been investigated for use in alkaline electrolytes. The interaction between electrolyte ions and the electrode materials greatly affects the pseudocapacitive behavior of these materials. For example, Nam et al. [124] investigated the charge-storage mechanism of NiO_x in 1 M KOH using an electrochemical quartz-crystal microbalance (QCM) and probe beam deflection technique. It was shown the redox reaction of the NiO_x consisted of predominant H^+ desorption in the initial stage of oxidation and subsequently predominant OH^- adsorption in the latter stage and vice versa during the reduction process. When $Ni(OH)_2$ is used as the electrode in a KOH electrolyte, the following reversible reaction (2.6) occurs, contributing to the pseudocapacitance [135,140,141]:

$$\mathrm{Ni(OH)_2 + OH^- \rightleftarrows NiOOH + H_2O} + e^- \tag{2.6}$$

Similarly, using Co_3O_4 as pseudocapacitive materials in the KOH electrolyte, the surface faradaic reaction is involved with OH^- ions adsorption/desorption or insertion/extraction accompanied by the charge transfer process. This can be expressed as follows [127,128]:

$$\begin{aligned} \mathrm{Co_3O_4 + OH^- + H_2O} &\rightleftarrows \mathrm{3CoOOH} + e^- \\ \mathrm{CoOOH + OH^-} &\rightleftarrows \mathrm{CoO_2 + H_2O} + e^- \end{aligned} \tag{2.7}$$

It is interesting to note that Feng et al. [126] prepared sub-3 nm Co_3O_4 nanofilms (Figure 2.12), and a very high specific capacitance of 1400 F g^{-1} was reported in 2 M

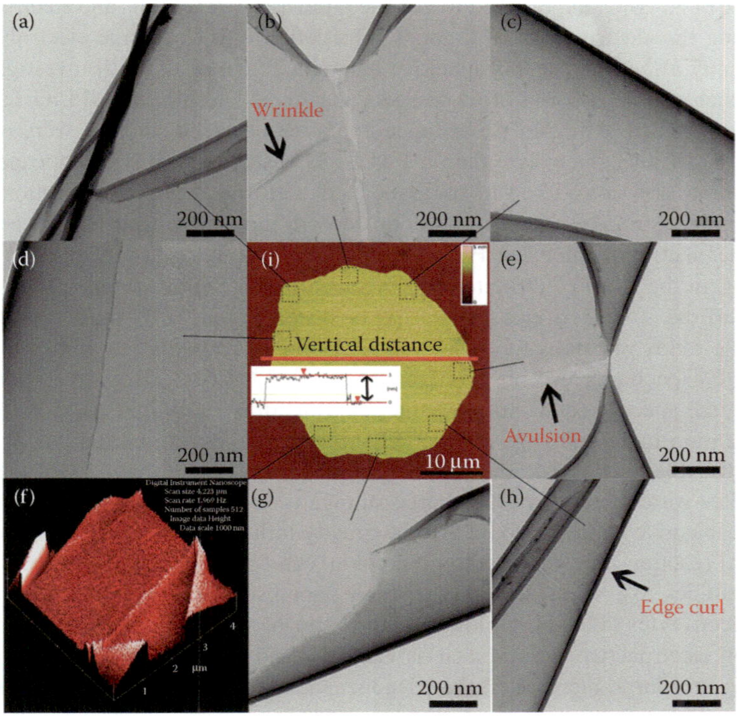

FIGURE 2.12 TEM image of (a) top left corner, (b) top, (c) top right, (d) left, (e) right, (g) bottom, and (h) right bottom of our representative Co_3O_4 nanofilms. (f) 3D surface plot of selected area of nanofilms. (i) Scanning probe microscopy (SPM) image of atomic layer Co_3O_4 nanofilms. The inset is the thickness measurement data. (Reprinted with permission from Feng, C. et al. 2015. Sub-3 nm Co_3O_4 nanofilms with enhanced supercapacitor properties. *ACS Nano* 9:1730–1739. Copyright 2015 American Chemical Society.)

KOH electrolytes. Palmas et al. [128] proposed the OH^- diffusion in the pores of the electrode materials was slower than the quasi-reversible charge transfer step, which limited the whole process. Similarly, Joseph et al. [142] attributed the redox peaks of $Ni_3(NO_3)_2(OH)_4$ electrodes in alkaline electrolytes to the intercalation/deintercalation of OH^- ions from the electrolyte into the micro and/or mesopores of the materials. Therefore, shortening the diffusion length of OH^- ions via controlling the morphology of electrode materials is beneficial for improving pseudocapacitive performance [127]. Besides, the possible volume change of the electrode materials arising from high-rate ions intercalation/deintercalation should also be considered. By using porous $NiCo_2O_4$ nanotubes, Li et al. [143] reported a high specific capacitance of 1647 F g^{-1} at 1 A g^{-1}, good rate capability (77.3% capacity retention at 25 A g^{-1}), and excellent cycling stability (6.4% loss after 3000 cycles). It was believed the hollow nature of $NiCo_2O_4$ nanotubes not only provided spaces for OH^- ions, decreasing the diffusion distance to the interior surfaces, but also accommodated the volume change caused by the repeated insertion/extraction of OH^- ions. Furthermore, the porous structure of $NiCo_2O_4$ nanotubes favored the rapid transport of ions and electrons, thus enhancing the electrochemical kinetics. The cation in the alkaline electrolytes can also greatly affect the pseudocapacitive behavior. Yuan et al. [130] investigated the electrochemical behavior of MnO_2 pseudocapacitive materials in 1 M LiOH and 1 M KOH electrolytes. A higher capacitance and a better cyclic stability were obtained using a 1 M LiOH electrolyte than a 1 M KOH electrolyte. The pseudocapacitance contribution was considered to be associated with the reversible insertion/extraction of Li^+ ions in MnO_2 electrodes coupled with the charge transfer process at the electrode/electrolyte interface. For nitrides such as VN, it has also been reported that OH^- ions in the KOH electrolyte were involved in the charge storage process at the surface of the VN electrodes [144,145]. Lucio-Porto et al. [145] found that fast and reversible redox reactions took place in KOH electrolytes for VN thin films while only EDL capacitance was observed in 1 M NEt_4BF_4/ACN electrolytes (Figure 2.13). Besides, the specific capacitance of the VN films drastically increased as the KOH concentration increased from 0.01 to 5 M. It was suggested that the use of diluted 0.01 M KOH led to a starvation OH^- ions and thus only a fraction of all the accessible redox sites were involved in the charge storage process. It should be noted that the pseudocapacitive behavior could also be greatly influenced by the potential electrochemical reactions between electrode materials and electrolytes. For example, some metal sulfides (e.g., CoS_x and NiS) exhibit poor pseudocapacitive performance in KOH electrolytes [146–148]. However, they can be electrochemically transformed into new electroactive phases such as $Co(OH)_2$ and $Ni(OH)_2$ in KOH electrolytes, resulting in a large increase of the pseudocapacitance [147,148]. It is interesting to note that all the above-mentioned pseudocapacitive materials are based on the cation

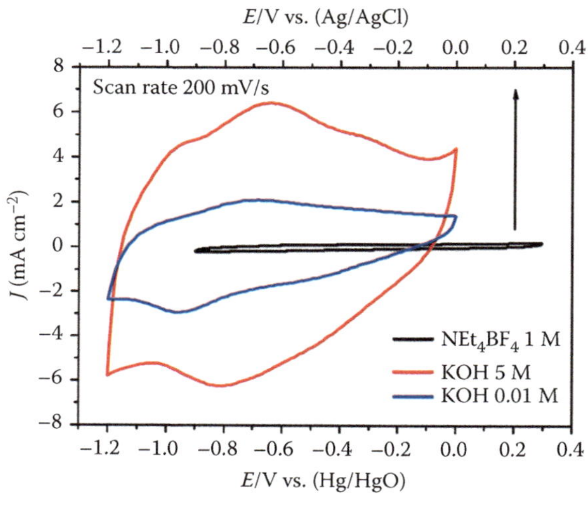

FIGURE 2.13 Cyclic voltammograms of VN film with a thickness of 480 nm at 200 mV s^{-1} with different concentrations of KOH after 100 cycles. In addition, the cyclic voltammogram of 1 M NEt_4BF_4 in ACN at 200 mV s^{-1} versus Ag/AgCl, is also shown. (Reprinted from *Electrochimica Acta*, 141, Lucio-Porto, R. et al., VN thin films as electrode materials for electrochemical capacitors, 203–211, Copyright 2014, with permission from Elsevier.)

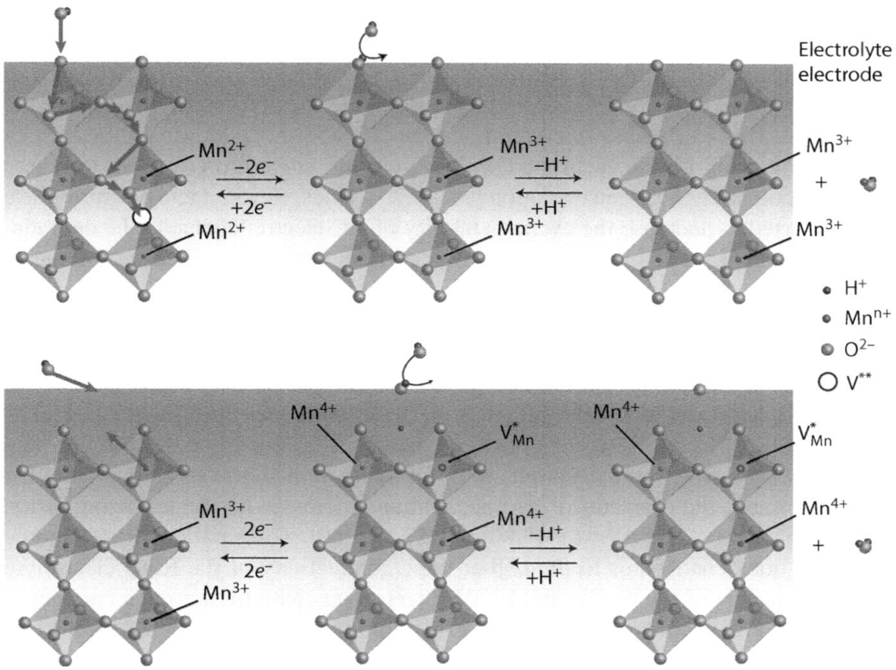

FIGURE 2.14 Mechanism of oxygen intercalation into $LaMnO_{3\pm\delta}$. Initially oxygen vacancies are filled through intercalation of an electrolyte oxygen ion and diffusion of O^{2-} along octahedral edges through the crystal concomitant with the oxidation of two Mn^{2+} centers to two Mn^{3+}. In the next step of the reaction, excess oxygen is intercalated at the surface through diffusion of manganese to the surface and oxidation of two Mn^{3+} centers to two Mn^{4+}. (Reprinted by permission from Macmillan Publishers Ltd. *Nature Materials* Mefford, J. T. et al. 2014. Anion charge storage through oxygen intercalation in $LaMnO_3$ perovskite pseudocapacitor electrodes. 13:726–732, copyright 2014.)

intercalation, however, Mefford et al. [149] illustrated $LaMnO_3$ perovskite pseudocapacitive electrodes in KOH electrolytes could exhibit the anion intercalation-based charge storage mechanism (Figure 2.14). This provides a new strategy to increase the specific capacitance of the pseudocapacitive materials.

The parametric influence of the electrolytes (e.g., concentration and temperature) and other types of alkaline electrolytes have also been studied. The concentration of the electrolyte has a great influence on the ESR, specific capacitance, and oxygen evolution reaction. Electrochemical impedance spectroscopy (EIS) measurements showed that increasing the electrolyte concentration decreases the ohmic resistance of the electrolyte, charge transfer resistance, and diffusion-type Warburg resistance, leading to a lower ESR [150,151]. Besides, a higher electrolyte concentration generally results in a higher specific capacitance of the electrode materials [146,151,152]. However, higher concentration of alkaline electrolytes also favored the oxygen evolution, which limits the operating potential window of the ESs [141,151]. The corrosion at the surface of the substrates in highly concentrated electrolytes is also a concern, which may lead to the peeling off of the electrode materials [152]. Therefore, it is essential to optimize the concentration of the electrolyte in terms of the overall performance of ESs. Similarly, the increase of the temperature of the electrolyte usually results in the decrease of ESR and increase of the specific capacitance due to the improved diffusion process of electrolyte ions [151,153–155]. However, the oxygen evolution reaction is also facilitated with increasing temperature, which decreases the onset potential of oxygen evolution [94,156]. Besides, elevated temperatures have been reported to decrease the cycling stability of AC electrodes due to the degradation of materials caused by the surface oxidation in KOH electrolytes at high temperatures (e.g., 80°C) [151].

On the other hand, the effect of the type of the alkaline electrolytes (i.e., cation in the electrolyte) on the ES performance is complicated. The different nature of the cations in various alkaline electrolytes, such as KOH, LiOH, and NaOH, are expected to have a pronounced influence on the ESR and specific capacitance. The unsolvated K^+, Na^+, and Li^+ ions have diameters of 1.33, 0.95, and 0.6 Å, respectively [157,158]. Due to the different solvation interaction ability of K^+, Na^+, and Li^+ ions with water, the diameter of hydrated cations increased in the following order: K^+ (3.31 Å) < Na^+ (3.58 Å) < Li^+ (3.82 Å) [157,159,160]. The smaller hydrated ionic size of K^+ ions contributes to the higher ionic conductivity of the KOH electrolyte compared to that of the NaOH and LiOH. Electrolytes with higher ionic conductivity contribute to a lower ESR and that is the main reason why the KOH electrolyte is the most extensively used electrolyte among various electrolytes. However, since ion intercalation/deintercalation is generally involved for pseudocapacitive materials, the bare (unsolvated) ionic size should be considered as well. For example, some studies reported that specific capacitances of MnO_2 were higher in LiOH electrolyte than in the KOH or NaOH electrolyte [129,157,161]. The authors attributed this to the smaller ionic radius of Li^+ ion compared to K^+ and Na^+ ions, favoring the ion intercalation/deintercalation process. Similarly, Inamdar et al. [162] found that the specific capacitance of NiO was nearly twice as high in the NaOH electrolyte than in the KOH electrolyte, which was attributed to the higher intercalation rate of Na^+ ions into the surface of the electrode materials. However, other comparative studies concerning different alkaline electrolytes for $Co_2P_2O_7$ [163], $MnFe_2O_4$ [160], and Bi_2WO_6 [164] showed a higher specific capacitance was achieved in the KOH electrolytes compared with NaOH or LiOH electrolytes. Since comparative studies on various alkaline electrolytes for different pseudocapacitive electrode materials have been rather limited, and it has remained unclear whether this phenomenon is associated with the nature of the electrolyte, or the type of the electrode materials, or the electrode preparation procedure and the resultant electrode material structure [2].

Furthermore, for pseudocapacitive materials, the cycle stability is a major concern. Besides the repeated ion intercalation/deintercalation-related failure mechanism, it was also found that the decrease of the capacitive performance after long-term charging/discharging cycling might also be related to the dissolution of electrode materials in alkaline electrolytes [142]. Joseph et al. [142] found the cycling stability of $Ni_3(NO_3)_2(OH)_4$ in LiOH was higher than that in KOH and NaOH electrolytes. Inductively coupled plasma atomic emission spectroscopy (ICP-AES) analysis showed the evidence of the Ni dissolution into the electrolyte and found the dissolution of

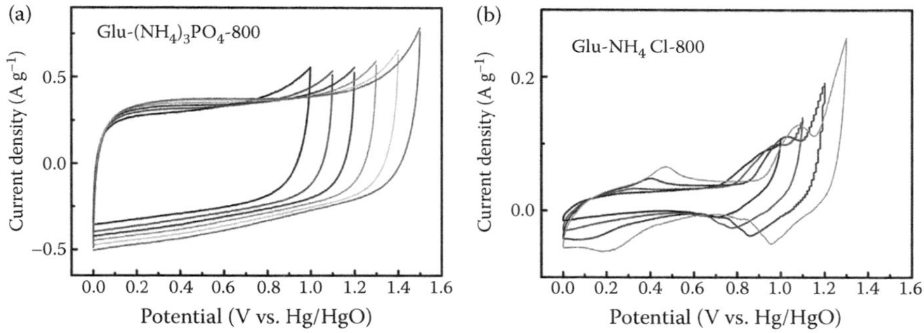

FIGURE 2.15 CVs at 5 mV s^{-1} of symmetric capacitor based on the (a) Glu-$(NH_4)_3PO_4$-800 and (b) Glu-NH_4Cl-800 in 6 M KOH. (Reprinted from *Journal of Power Sources*, 239, Wang, C. et al., Sustainable synthesis of phosphorus- and nitrogen-co-doped porous carbons with tunable surface properties for supercapacitors, 81–88, Copyright 2013, with permission from Elsevier.)

nickel in the LiOH electrolyte was lower than in the KOH and NaOH electrolyte, leading to a higher stability of electrode materials in the LiOH electrolyte.

Another strong research trend in this area is to widen the operation voltage window of ESs with alkaline electrolytes in order to significantly improve the energy density. In the case of symmetric ESs, this is generally achieved through inhibiting the side reaction in the electrolytes and/or increasing the stability of the electrodes at a large potential window [2]. For carbon-based symmetrical ESs, an operating voltage up to 1.4 V in 6 M KOH was achieved by Wang et al. [165] through codoping of phosphorus and nitrogen to the microporous carbons (Figure 2.15). It has been reported that nanostructured carbons with high oxygen content showed poor cycle performance in KOH electrolytes [166] whereas the absence of oxygen atoms bonded to carbons in phosphorus-doped carbons was the reason for the long-life cycle performances [74,165]. For pseudocapacitance-based symmetric ESs, modification of electrode materials through alloying or doping strategy has been proposed as an effective method to increase the operating working potential in alkaline electrolytes [167]. Yuan et al. [167] reported that a symmetric Ru–Cr binary oxide-based ES with the specific capacitance of about 148 F g^{-1} could work in a relatively large cell voltage of 1.4 V in 1 M KOH electrolytes, delivering an energy density of about 40 Wh kg^{-1}. This operating potential window was larger than the symmetric ESs based on pure RuO_2. The authors attributed the increased working potential to the introduction of Cr species into the RuO_2 matrix, which enhanced the electrochemical stability of Ru in the KOH electrolyte.

2.2.2.3 Alkaline Electrolytes for Hybrid ESs

As discussed in detail previously, it is also effective to widen the potential window in aqueous electrolytes through the design of asymmetric ESs by using different electrode materials with a complementary working potential window. Similarly, for alkaline electrolyte-based asymmetric ESs, a battery-type positive electrode such as $Ni(OH)_2$ [101,133,168] or a pseudocapacitive positive electrode such as RuO_2 [169]

that stores charge through faradaic reactions is paired with a carbon-based negative electrode that stores charge primarily by the EDL [95]. Besides, asymmetric ESs based on carbon electrodes with different properties on each electrode have also been reported [170]. As a result, the operative cell voltages of these asymmetric KOH electrolyte-based ESs were effectively increased, for example, 1.6–1.7 V for carbon//$Ni(OH)_2$ [168,171,172], 1.4–1.6 V for carbon//$Co(OH)_2$ [173,174], carbon//Co_3O_4 [175,176], and carbon//Co_9S_8 [138], 1.6 V for carbon//Ni_3S_2 [177], and 1.4 V for RuO_2–TiO_2//carbon [169]. Noticeably, the maximum operating voltage was found to be further increased for some asymmetric alkaline electrolyte-based ESs with composite materials such as rGO and CoAl layered double hydroxide (LDH) (1.75 V) [178] and NiO–$Ru_{0.35}V_{0.65}O_2$ (1.7 V) [179] as the positive electrode. Due to the wider operative voltage window and the use of faradaic-type electrodes with high capacity, most of these ESs exhibit higher energy densities with the range from 20 to 40 Wh kg^{-1}, and some reported value (e.g., 140 Wh kg^{-1} [172]) that is even comparable to rechargeable lithium-ion batteries (LIBs). Unfortunately, the cycling stability of these asymmetric ESs is much lower than that of the EDLCs because of the introduction of faradaic-type electrodes. For example, it has been commonly reported that the specific capacitance of some asymmetric ESs decreased more than 10% after 1000–5000 cycles [172,179,180]. In addition, compared to EDLCs with a few-second response to the charging/discharging process, the quick charging/discharging of asymmetric ESs is generally limited.

2.2.3 Neutral Electrolytes

Neutral electrolytes have also been investigated as alternatives for those extreme-pH electrolytes (highly concentrated acid or alkaline electrolytes) due to their advantages such as a larger operating potential window, less corrosion, and safety. Typical conducting salts in the neutral electrolytes include lithium salts such as LiCl, Li_2SO_4, and $LiClO_4$, sodium salts such as NaCl, Na_2SO_4, and $NaNO_3$, potassium salts such as KCl, K_2SO_4, and KNO_3, calcium salts such as $Ca(NO_3)_2$, and magnesium salts such as $MgSO_4$. Na_2SO_4 is the most commonly used neutral electrolyte, which has been shown to be a promising electrolyte for many pseudocapacitive materials especially MnO_2-based materials.

2.2.3.1 Neutral Electrolytes for EDLCs

As shown in Table 2.4, most of the reported specific capacitance values were around 100 F g^{-1} for EDLCs with neutral electrolytes, which is lower than that obtained with highly concentrated acid or alkaline electrolytes. However, it is worth noticing that carbon-based ESs with neutral electrolytes have been shown to have a large operating potential window because of the increased electrochemical stability of the electrolytes compared to acidic and alkaline aqueous electrolytes [159,181,182]. Compared to acid and alkaline electrolytes, the neutral electrolyte has lower H^+ and OH^- concentrations, resulting in a higher overpotential for hydrogen and oxygen evolution and thus increasing the electrochemical stability of the neutral electrolytes [159,182]. For example, Demarconnay et al. [182] obtained an excellent cycle life during 10,000 charging/discharging cycles up to a high cell voltage of 1.6 V for a symmetric ES with AC electrodes in 0.5 M Na_2SO_4 electrolytes. Similar results have been reported for a symmetric ES fabricated by seaweed carbons in 0.5 M Na_2SO_4 electrolytes [166] and a symmetric graphite-based ES with 1 M Na_2SO_4 electrolytes [183]. Through nitrogen and oxygen doping for carbon nanofiber electrodes, Zhao et al. [184] further increased the cell voltage to 1.8 V and achieved a high energy density of 29.1 Wh kg^{-1} at a power density of 450 W kg^{-1} for the obtained ES with 1 M Na_2SO_4 electrolytes. Using 1 M Li_2SO_4 electrolytes, Fic et al. [159] reported an even higher working voltage of 2.2 V without any significant capacitance decrease after 15,000 cycles for a carbon-based symmetrical ES (Figure 2.16). This was attributed to the strong solvation of both Li^+ cations and sulfate anions, increasing the electrolyte decomposition voltage to above 1.23 V. Noticeably, this voltage value is more than twice as high than that generally obtained for EDLCs with acid or alkaline aqueous electrolytes. Gao et al. [185] pointed out that the high cell voltage achieved in Li_2SO_4 electrolytes was attributed to the high overpotential related with the hydrogen sorption mechanism at the negative electrode. The maximum operating voltage was found to be limited by the irreversible electro-oxidation of the positive carbon electrode. If the maximum potential of the positive electrode was too high,

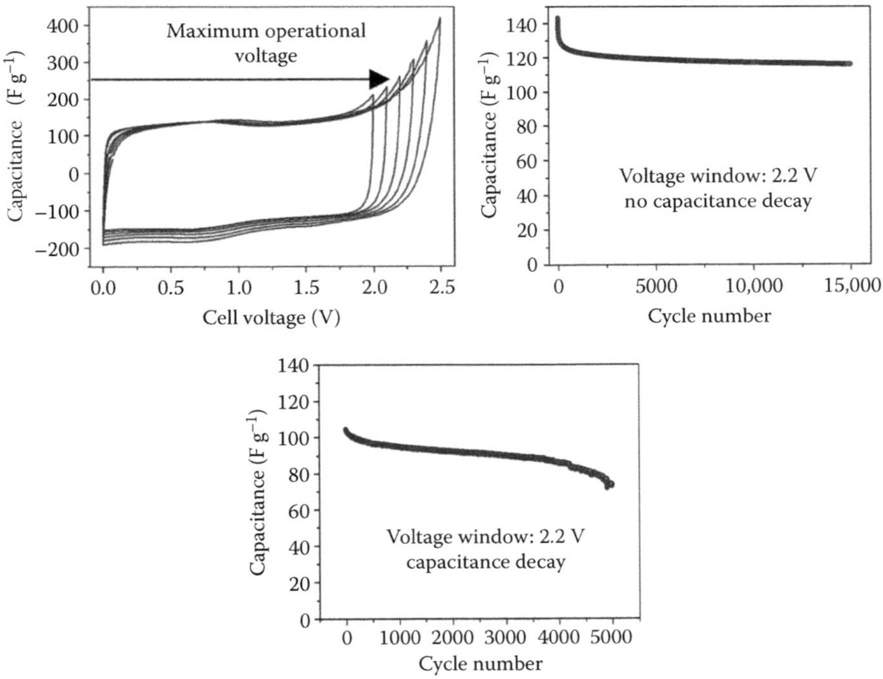

FIGURE 2.16 (a) CV at scan rate of 10 mV s^{-1} and gradual 100 mV voltage shift and cyclability (±1 A g^{-1}) at voltage range (b) 2.2 V (c) and 2.5 V. (Fic, K. et al. 2012. Novel insight into neutral medium as electrolyte for high-voltage supercapacitors. *Energy & Environmental Science* 5:5842–5850. Reproduced by permission of The Royal Society of Chemistry.)

it would result in irreversible electro-oxidation on the active sites of carbon and the evolution of CO_2. To investigate the state of health of carbon-based ESs under high voltage operation in Li_2SO_4 electrolytes and to determine the factors influencing their lifetime, Ratajczak et al. [186,187] performed accelerated aging tests. A pressure sensor was connected to the ES, allowing the measurement of the gas evolution caused by the overcharging of the cell. It was pointed out that galvanostatic cycling with limited cycles was not an accurate test to determine the maximum working voltage of ESs. The authors found gases such as CO_2 and CO actually started to evolve at a voltage higher than 1.5 V, together with oxidation of the carbon electrode material. The main oxidation occurred on the positive carbon electrode, which was covered by twice the surface oxygenated groups than that of the negative electrode after 120 h of aging at 1.7 V. It was thus concluded the carbon-based ESs with 1 M Li_2SO_4 aqueous electrolytes could operate safely up to 1.5 V, which was lower than reported by other work [159,184]. It is noticed that this study used a stainless steel current collector instead of gold current collector as reported in other studies [159,184], which may also account for the different reported maximum operating voltage. Ratajczak et al. [187] also found the corrosion of the positive collector and the ESR could be lowered by adding sodium molybdate (Na_2MoO_4) to the electrolyte. Since the operating voltage in neutral electrolytes is remarkably high compared to KOH and H_2SO_4 electrolytes and neutral electrolytes are generally less corrosive, neutral electrolytes have received considerable attention for the development of environment friendly and high energy density symmetric carbon-based ESs [2].

For neutral electrolytes, salt concentration is an important concern. Highly concentrated electrolytes are generally used for ESs to ensure good performance. This is not an issue for acid and alkaline electrolytes because they can achieve a high concentration, such as 6 M for the KOH electrolyte. However, in the case of neutral electrolytes, some salts such as K_2SO_4 cannot achieve such a high concentration. It appears that the effect of concentration of neutral electrolytes depends on the nature of the electrolyte. Fic et al. [159] investigated Li_2SO_4 electrolytes with a wide range of concentrations from 0.1 to 2.5 M. Among these concentrations, 1 M Li_2SO_4 electrolytes gave the highest capacitance values and operational voltage window for EDLCs. Diluted electrolytes (0.1 and 0.5 M) showed poor conductivity and lower capacitance. More concentrated electrolytes (1.5 and 2.5 M) resulted in unsatisfactory charge propagation of the ESs, mainly due to sterical hindrance caused by solvated molecule complexes, limiting the fast charge transfer in the blocked porous structure. For Na_2SO_4 electrolytes with a concentration in the range of 0.1–1.0 M, Tsay et al. [188] found 0.5 M Na_2SO_4 electrolytes resulted in the highest specific capacitance and energy density. The effect of halide ion (NaI) concentration of the electrolytes was also investigated [189]. It was found that the capacitance of AC-based ESs increased as the NaI concentration increased from 0.25 to 2 M, but it decreased when the NaI concentration was further increased to 5 M. Yin et al. [190] found the carbon-based ES using 5 M $NaClO_4$ electrolytes gave the highest specific capacitance, rate capability, energy and power densities compared to those carbon-based ESs using electrolytes with lower concentration. The increased performance was attributed to the decreased hydration number of ions in the electrolytes with the increasing concentration, leading to a higher specific capacitance in the Helmholtz layer. Furthermore, highly concentrated $NaClO_4$ also enhanced the low-temperature performance of the ES, due to the decrease of the freezing point in concentrated solutions [190].

To clarify the effect of different ions on the performance of carbon-based ESs, some comparative studies have focused on neutral electrolytes containing different salts. However, some contradictory results have been found in this area. For example, several studies reported the value of the specific capacitance of the ES followed the order: $Li_2SO_4 > Na_2SO_4 > K_2SO_4$ [159,191,192] whereas other research did not support this trend [193]. Fic et al. [159] related this trend with the mobility of the alkali metal ions. As mentioned previously, alkali metal ions are strongly solvated in water and the diameter of the hydrated cations increases in the order: $K^+ < Na^+ < Li^+$ [159]. Consequently, the mobility of the hydrated alkali metal ions decreases following the order: $K^+ > Na^+ > Li^+$. Due to the lower mobility and diffusion coefficient of solvated Li^+ ions, Fic et al. [159] assumed that the solvated ions remained near the electrode/electrolyte interface during charging/discharging process compared to hydrated Na^+ and K^+ cations with higher mobility and thereby contribute to a higher specific capacitance. However, Qu et al. [193] reported the specific capacitance values of AC showed no obvious differences among the three electrolytes (0.5 M Li_2SO_4, 0.5 M Na_2SO_4, and 0.5 M K_2SO_4) at the low slow scan rate. For the fast scan rate of 100 mV s^{-1}, the capacitance of AC increased in the order: $Li_2SO_4 < Na_2SO_4 < K_2SO_4$. Other factors such as the preparation method of electrode materials can also influence the conclusion, and further work is required to clearly illustrate the effect of salt on the performance of ESs. In addition, ESR was found to increase with the decreasing ionic conductivity of bulk electrolytes, following the order: $Li_2SO_4 > Na_2SO_4 > K_2SO_4$ [191,193]. As a result, the maximum power density and rate performance of carbon-based ESs increased in the order: $Li_2SO_4 < Na_2SO_4 < K_2SO_4$ [193]. With regard to the anions in the neutral electrolytes, Chae and Chen [191] reported that when the anion in the electrolytes was changed from the SO_4^{2-} ion to the Cl^- ion, the ES specific capacitance was increased due to the smaller size of the Cl^- ion than the SO_4^{2-} ion.

Other less common salts such as $Mg(NO_3)_2$ salt have also been tested in the electrolyte for EDLCs [194]. Chandrasekaran et al. [194] demonstrated the good reversibility of Mg^{2+} adsorption/desorption (electrolyte: $Mg(NO_3)_2$) and the stability of the electrolyte within the applied potential limit of 0.8 V. The maximum calculated energy densities were 9, 10.2, and 8.5 Wh kg^{-1} for the EDLCs using 0.1, 0.5, and 1.0 M $Mg(NO_3)_2$ electrolytes, respectively. Recently, Gao et al. [195] developed some new types of salts, such as lithium, sodium, and potassium salts of silicotungstic acid (Li-SiW, Na-SiW, and K-SiW) as aqueous neutral electrolytes for EDLCs. These electrolytes showed higher ionic conductivity compared to their counterparts with Cl^-, SO_4^{2-}, or NO_3^- anions (Figure 2.17), owing to more dissociated cations and greater mobility of the Keggin anion. Carbon-based EDLCs using silicotungstic acid-type electrolytes could exhibit a 1.5 V cell voltage [195].

Interestingly, mild neutral electrolytes not only allow a selection of diverse current collectors, but also favor the development of some electrode materials such as deoxyribonucleic acid (DNA)–AC composite electrodes that cannot afford strongly acidic and basic electrolytes [196].

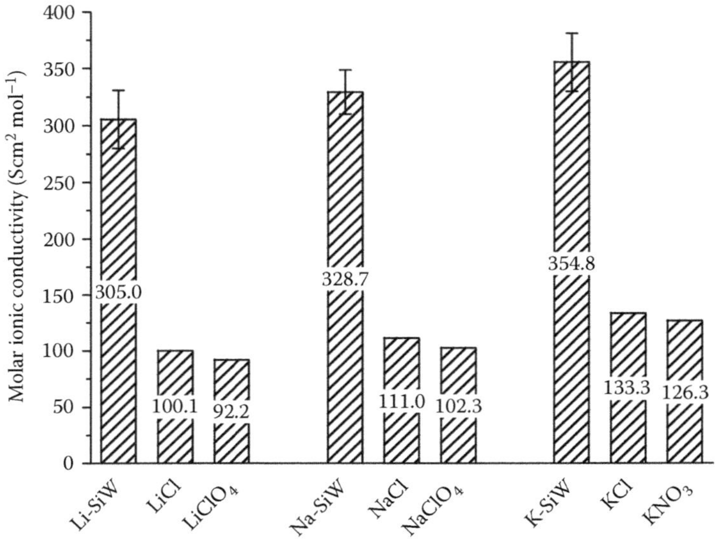

FIGURE 2.17 Comparison of molar conductivity (at 0.05 M) of SiWA salts and other common salts with the same counter cations. (Reprinted from *Electrochimica Acta*, 138, Gao, H., A. Virya, and K. Lian, Monovalent silicotungstate salts as electrolytes for electrochemical supercapacitors, 240–246, Copyright 2014, with permission from Elsevier.)

To extend the operative temperature range to the lower temperature limit of EDLCs, Yin et al. [197] proposed the addition of antifreezing additives for aqueous electrolytes. The assembled EDLCs with methanol/water-concentrated $NaClO_4$ electrolytes exhibited improved energy and power densities at temperatures as low as −40°C compared to the $NaClO_4$ electrolyte.

2.2.3.2 Neutral Electrolytes for Pseudocapacitors

Since the discovery of MnO_2 as a promising pseudocapacitive material in neutral electrolytes [198], extensive research has been focused on using neutral electrolytes for potential pseudocapacitive materials (e.g., MnO_2 and V_2O_5). MnO_2 is so far the most extensively studied pseudocapacitive materials in neutral electrolytes. During charging/discharging process, the Mn oxidation state varies between III and IV, balanced by the surface adsorption/desorption or intercalation/deintercalation of electrolyte cations M^+ such as K^+, Na^+, and Li^+ as well as proton (H^+) intercalation/deintercalation, which can be described by the following reaction [199–201]:

$$\begin{array}{lcl} \mathrm{Mn(IV)O_2} \cdot n\mathrm{H_2O} + \delta e^- & \rightleftarrows & (\mathrm{H_3O})_{\delta(1-f)}\mathrm{M}_{\delta f}[\mathrm{Mn(III)}_{\delta}\mathrm{Mn(IV)}_{1-\delta}] \\ + \delta(1-f)\mathrm{H_3O^+} + \delta f \mathrm{M^+} & & \mathrm{O_2} \cdot n\mathrm{H_2O} \end{array} \qquad (2.8)$$

Similar mechanisms have been proposed for other metal oxide-based pseudocapacitive materials, such as molybdenum oxide [202,203]. Since ions of electrolytes are directly involved in the charging/discharging process of these pseudocapacitive materials, the properties of neutral electrolytes play a significant role in the pseudocapacitive performance [2]. Various characteristics of the electrolytes, such as pH, the cation and anion species, salt concentrations, additives, and solution temperatures, are expected to have a profound influence on ES performance. Kuo et al. [201] used an *in situ* synchrotron x-ray diffraction (XRD) analysis to investigate the role of electrolytes containing various alkaline salts (i.e., LiCl, NaCl, KCl, CsCl, and $CaCl_2$) in the charge storage mechanisms of MnO_2 electrodes. For the NaCl electrolyte, the charge mechanism mainly involved the insertion/extraction of H_3O+ into/from the oxide structure whereas Na was not detected in the reduced $MnO_2 \cdot nH_2O$ electrode. The lattice expansion caused by the insertion of H_3O+ was also observed. For KCl, LiCl, CsCl, and $CaCl_2$ electrolytes, simultaneous insertion/extraction of alkaline metal ions (i.e., K^+, Li^+, Cs^+, and Ca^{2+}) accompanied by H_3O^+ was identified, but H_3O^+ played the predominant role (>60%). Therefore, there is not a simple continuous function between the ionic size of the alkali ions and their contribution to charge storage. The contribution from alkali ion first decreases and then increases with ionic size. The authors further explained this by considering two opposite factors. One would favor the insertion of smaller cations related with the lattice expansion and the other would favor the larger cations associated with the removal of water molecules from the solvated cations before the insertion process [201]. For practical use of such materials, the effect of electrolytes on the performance of the resultant ESs is an important issue. With regard to the electrolyte concentration, the specific capacitance of MnO_2 electrode materials was found to increase noticeably with the increasing of salt concentration [204–207] and a logarithmic dependence of specific capacitance on the concentration (activity) of cations in the electrolytes was reported [204–206]. This could be understood by considering the charge storage mechanism of the MnO_2 electrode materials, which involves the insertion/extraction of alkaline metal ions [199,200].

Concerning the cation species, different alkaline metal or alkaline-earth metal cations have various ionic sizes and hydrated ion sizes as well as different diffusion coefficient and ionic conductivity, greatly affecting the specific capacitance and ESR of the ESs. However, it is difficult to give a general rule for the dependence of the specific capacitance of MnO_2-based ESs on the cation species since it may also depend on the preparation methods and the resultant structure of the electrode materials. For example, Li et al. [208] found the specific capacitance and the resultant energy and power densities of mesoporous MnO_2 oxide-based electrode was in the order of $Li_2SO_4 > Na_2SO_4 > K_2SO_4$. This trend seems to follow the order of the unsolvated ion size of alkali metal ions: $Li^+ < Na^+ < K^+$ (Table 2.3), suggesting that the specific capacitance increases with the decreasing ionic size of the electrolyte. Similar to results as those for alkaline metal ions, the specific capacitance increases with the decrease of the bare alkali metal ion size have been reported by other studies including sulfate salts [209] and acetate salts (i.e., LiAc > NaAc > KAc) [210]. On the other hand, several studies showed sodium salts such as Na_2SO_4 [204,205] and NaCl [211] resulted in a higher specific capacitance for MnO_2 electrodes compared to the lithium and potassium salts. Others found K_2SO_4 salts gave a higher specific capacitance for $K_xMnO_2 \cdot nH_2O$ electrodes compared to Na_2SO_4 and Li_2SO_4 salts [212]. Besides, Wen et al. [206] found no obvious change of the specific capacitance of MnO_2

electrodes among various electrolytes including $KClO_4$, $NaClO_4$, and $LiClO_4$. It is worth noticing that most studies compared the performance of ESs with different electrolytes by the value of the mass specific capacitance. However, the mass specific capacitance value of ESs depends on several other factors, such as the preparation methods, surface morphology, structure (e.g., δ-MnO_2 or α-MnO_2), and amount of the electrode materials [2]. Therefore, it is not easy to identify the individual effect of cation species on the capacitance of an ES through the comparison of the mass specific capacitance. For example, different preparation methods generally could result in different pore structures of electrode materials, which influences the diffusion path of electrolyte ions and complicates the whole process. Therefore, further investigations on electrode materials with well-defined surfaces will be helpful for clarifying the relationship between the cation species and the intrinsic capacitance of the electrode materials. Besides, since the intercalation/deintercalation of cations is involved for pseudocapacitive electrode materials, the charging/discharging current density or scan rate of cyclic voltammetry (CV) is also expected to noticeably influence the specific capacitance value. The ion intercalation/deintercalation is generally the rate-determining step during the charging/discharging process and it is thus favored by a lower charging/discharging rate or a lower scan rate. Therefore, in this case, the bare ionic size of cations, closely related with the intercalation/deintercalation process, may play a dominant role [2]. It seems that Li^+ ion-containing electrolytes would be beneficial because the smaller ionic size of Li^+ ions favors the intercalation/deintercalation process. The intercalation/deintercalation process would be limited at a higher scan rate of CV or larger charging/discharging current density, resulting in the decrease of the capacitance. As expected, a decrease of the specific capacitance value was found for MnO_2 electrodes with the increase of the scan rate of CV or charging/discharging current [208,213]. Li et al. [208] found at slow scan rates, the obtained specific capacitance value showed a distinct difference between various electrolytes and decreased following the order: $Li_2SO_4 > Na_2SO_4 > K_2SO_4$. On the contrary, no obvious difference in the specific capacitance was found at high scan rates in various electrolytes. Qu et al. [214] also reported MnO_2 nanorods showed the largest specific capacitance at slow scan rates in the 0.5 M Li_2SO_4 electrolyte among three neutral electrolytes (Li_2SO_4, Na_2SO_4, and K_2SO_4) while they exhibited the largest specific capacitance in the 0.5 M K_2SO_4 electrolyte at the fast scan rates due to the smallest hydration size of K^+ and highest ionic conductivity. Wu et al. [213] also found the decrease of the specific capacitance of MnO_2 electrodes with the increase of the discharge current density in various electrolytes followed the order of $LiCl > NaCl > KCl$. Apart from alkaline metal ions, studies have also been carried out on the electrolytes containing alkaline-earth cations for ESs [205,209,210,215,216]. Interestingly, Xu et al. [205] reported the specific capacitance of MnO_2 was increased twice over through replacing the univalent cation such as Li^+, Na^+, and K^+ with bivalent cation such as Mg^{2+}, Ca^{2+}, and Ba^{2+} for the close cation activity and bare ion size (Figure 2.18). It was attributed to the reason that one bivalent alkaline-earth metal cation intercalated into MnO_2 could balance the valence change of two Mn ions between (IV) and (III) whereas one alkaline metal cation could only balance the valence change of one Mn ion. Similar results have also been reported by Munaiah et al. [217], which showed that the initial discharge capacities

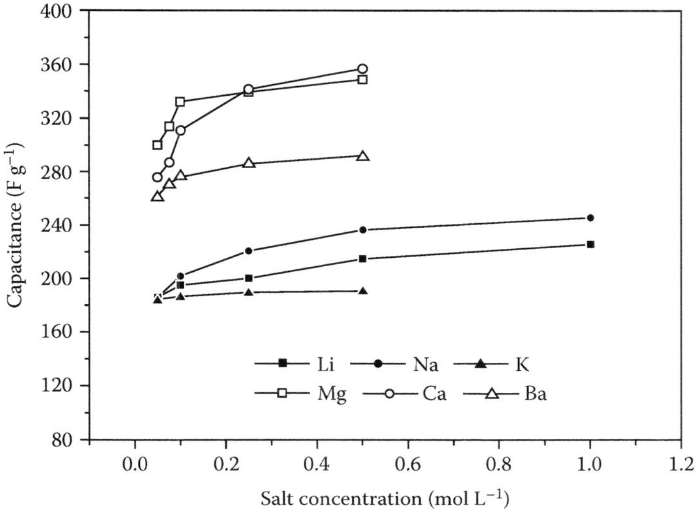

FIGURE 2.18 Dependence of capacitance on the salt concentration. (Reprinted from *Journal of Power Sources*, 196, Xu, C. J. et al., Charge storage mechanism of manganese dioxide for capacitor application: Effect of the mild electrolytes containing alkaline and alkaline-earth metal cations, 7854–7859, Copyright 2011, with permission from Elsevier.)

of hollow sphere amorphous MnO_2 in 0.1 M $Ca(NO_3)_2$ (283 F g^{-1}) was much higher than that in 0.1 M Na_2SO_4 (188 F g^{-1}), at a current density of 0.5 mA cm^{-2}.

With regard to the effect of anions, Boisset et al. [218] provided in-depth research on the effect of neutral aqueous electrolytes containing different lithium salts on the performance of birnessite-type and cryptomelane-type MnO_2 electrode materials. As shown in Figure 2.19, 10 different lithium salts based on mineral anions (e.g., NO_3^-, SO_4^{2-}, and ClO_4^-) and organic anions (e.g., $TFSI^-$, $CH_3CO_2^-$, $CH_3SO_3^-$, $Beti^-$, BOB^-, $C_7H_{12}CO_2^-$, and $Lact^-$) were investigated in this study. The basicity and the volume of the anion of each aqueous lithium salt electrolyte were found to control the electrochemical properties of MnO_2 electrodes (Figure 2.20). The potential limit (i.e., onset of the limit current of oxidation of each electrolyte solution) increased with the decrease of the pK_a. The effect of basicity of the anion could be explained based on the surface redox reaction mechanism in birnessite MnO_2 material that involves protons. On the other hand, the cathodic peak current values measured on the birnessite MnO_2 increased as the volume of the anion decreased. For larger anions such as $Beti^-$ or the octanoate, the $i(ep^C)$ decreased obviously and tended to zero due to the inhibition of the insertion. Moreover, for the anion with a diameter below 7 Å that corresponded to the space between the lamellar layer of birnessite MnO_2, the volume of the anion was no longer a determining factor of the electrochemical properties of the birnessite MnO_2. The authors finally concluded that electrolytes containing a mineral lithium salt under neutral pH condition and high concentration (up to 5 M) gave better electrochemical performances than organic ones, with larger voltage limit (up to 1.4 V) and good capacity retention. Besides, Xu et al. [205] compared

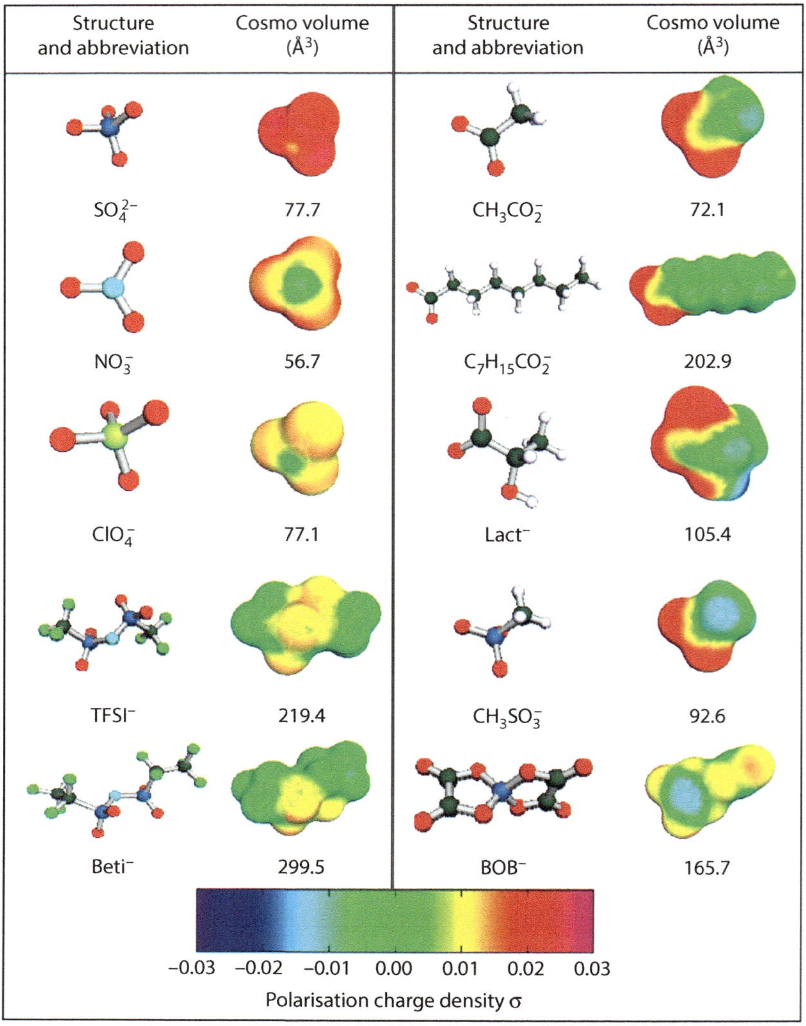

FIGURE 2.19 Structure, abbreviation, and cosmo volume evaluated by CosmothermX interface of studied anion (X^-) in lithium salt (LiX). (Reprinted with permission from Boisset, A. et al. 2013. Comparative performances of birnessite and cryptomelane MnO_2 as electrode material in neutral aqueous lithium salt for supercapacitor application. *Journal of Physical Chemistry C* 117:7408–7422. Copyright 2013 American Chemical Society.)

the specific capacitance of MnO_2 in 0.2 M ANO_3, 0.1 M A_2SO_4, and 0.2 M ACl (where A = Li^+, Na^+, or K^+) under a pH value of 6.0. It was found that the specific capacitance of MnO_2 was almost the same for the same cation species, indicating the pseudocapacitive behavior of MnO_2 was inactive to these three anions in the neutral electrolytes. The salt concentration in this study is much lower than that reported by Boisset et al. [218], which may also lead to a different result.

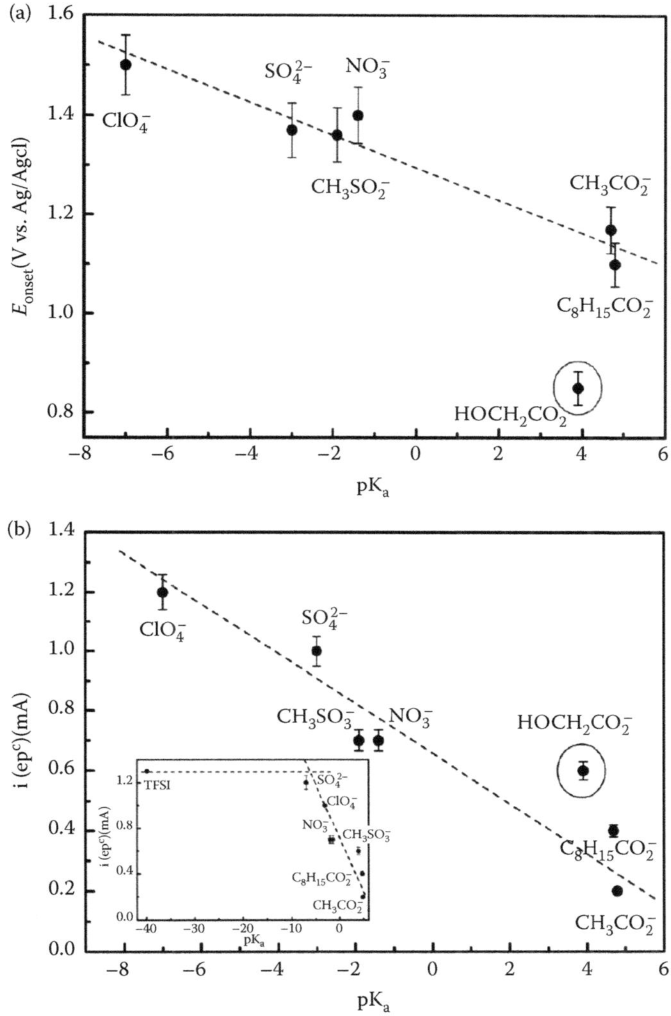

FIGURE 2.20 Variation of the potential limit (a) on oxidation, E_{onset}, or (b) in reduction, expressed by i(ep^c) as a function of the anion (denoted X^-) basicities expressed by their pK$_a$ values (HX/X^-). (Reprinted with permission from Boisset, A. et al. 2013. Comparative performances of birnessite and cryptomelane MnO_2 as electrode material in neutral aqueous lithium salt for supercapacitor application. *Journal of Physical Chemistry C* 117:7408–7422. Copyright 2013 American Chemical Society.)

Noticeably, the structure of MnO_2 electrode materials could be changed at high temperatures during the charging/discharging process [219,220]. Therefore, it is rather important to consider the effect of electrolyte temperature on the pseudocapacitive performance especially the stability of MnO_2-based ESs. Generally, it was found that a higher temperature leads to an increased specific capacitance, a better rate capability, and a decreased ESR but a reduced coulombic efficiency for MnO_2-based electrode materials, due to the improved ion diffusion, intercalation/deintercalation, and charge transfer process facilitated by the increasing temperature [220–222]. In addition, obvious morphology change and dissolution of the MnO_2 electrode materials were found at high temperatures after charging/discharging cycling [219,220]. As a result, Wang et al. [220] found the MnO_2 at CNFs composites exhibited a poorer cycling stability (82.4% capacity retention after 1000 cycles) at 75°C than at 25°C (95.3% capacity retention after 1000 cycles). It is interesting to note for an amorphous nanostructured MnO_2 anodically deposited onto a duplex stainless steel substrate, it was found that the as-prepared MnO_2 materials changed from the relatively compact films to highly porous leaf-like nanosheets after 2000 cycles at 70°C due to the repeated dissolution/redeposition of MnO_2 [219]. Since the morphology of the latter had a much larger electrochemically active surface area, this actually greatly enhanced the specific capacitance of MnO_2 materials during the cycling at high temperatures.

In order to improve the performance of ESs, the introduction of additives into neutral electrolytes has been explored [223,224]. For example, Wang et al. [224] investigated the capacitive performances of $MnFe_2O_4$-based ESs with $LiNO_3$ electrolytes containing various surfactants including anionic surfactant sodium dodecyl sulfate (SDS), nonionic surfactant p–t-octylophenol (Triton X-100), or poly(ethylene glycol)-block-poly(propylene glycol)-block-poly(ethylene glycol) (P123), respectively. The presence of surfactants was found to improve the specific capacitance owing to the decrease in the surface tension of the electrode interface for better permeation of Li^+ ions. The addition of a small amount of SDS into the $LiNO_3$ electrolyte was found to increase the specific capacitance by about 36.8% at 0.05 A g^{-1}, and the capacitive retention was about 82% after 1000 cycles. The specific capacitance improved about 22.8% and 12.8% after the addition of Triton X-100 and P123, respectively, which are lower than that obtained for SDS.

To enlarge the operating potential, Ataherian and Wu [225] explored the addition of Ti(IV) ions into KCl electrolytes for symmetric MnO_2-based ES. The capacitance loss of MnO_2 electrodes under negative polarization at low potentials was related to the irreversible formation of inactive Mn(II) surface species, and thus the working cell voltage was limited to be lower than 0.8 V [225]. The Ti(IV)/Ti(III) redox couple in the electrolyte could enhance the electrocatalytic activity of MnO_2 for the oxygen reduction reaction and thus inhibit the formation of the Mn(II) species. As a result, the lower potential limit could be extended, and the cell voltage of MnO_2 symmetric ES increased to 1.2 V with remarkable cycling stability after 10,000 cycles (Figure 2.21) [225].

Since most pseudocapacitive materials such as MnO_2 involve the intercalation/deintercalation of electrolyte ions to store charges, it is of significance to consider the structure change of the electrode material during the repeated charging/discharging process. This is strongly related to the cycle life of the electrode materials. Athouël et al. [226] found an increase of the interlayer spacing of birnessite-type MnO_2 from 0.710 to 0.720 nm upon electrochemical oxidation in the presence of Na^+ in the electrolyte (0.5 M Na_2SO_4). This was related with the deintercalation of Na^+ and the intercalation of H_2O between the interlayers. A progressive loss of the crystallinity of

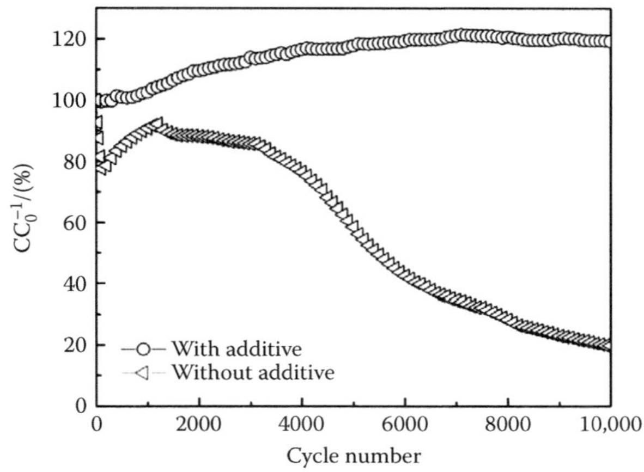

FIGURE 2.21 Capacitance retention (CC_0^{-1}), normalized based on the capacitance of the second cycle (C_0), versus cycle number for the ES cells either with (O) or without (Δ) Ti(IV) in the electrolyte (1 M KCl; scan rate = 50 mV s^{-1}). (Reprinted from *Electrochemistry Communications*, 13, Ataherian, F., and N. L. Wu, 1.2 Volt manganese oxide symmetric supercapacitor, 1264–1267, Copyright 2011, with permission from Elsevier.)

the material was observed during potential cycling. Shao et al. [212] demonstrated that the cycling ability of $K_xMnO_2 \cdot nH_2O$ electrode was greatly influenced by the electrolyte type. No obvious loss of the specific capacitance was observed after 850 cycles in the K_2SO_4 electrolyte whereas the specific capacitance decreased noticeably in the Na_2SO_4 and especially in the Li_2SO_4 electrolyte (13% decay). The best cycling stability achieved in the K_2SO_4 electrolyte was attributed to the slightest structural expansion/contraction degree of the $K_xMnO_2 \cdot nH_2O$ electrode during charging/discharging compared to that in Li_2SO_4 and Na_2SO_4 electrolytes, as revealed by XRD. Tao et al. [227] used *in situ* atomic force microscopy (AFM) in conjunction with *in situ* nanoindentation to investigate the dynamic morphology change of a nanostructured MnO_2 during the charging/discharging cycling in the 1 M KCl aqueous electrolyte. It was reported that the proton intercalation induced expansion of the prepared nanostructured MnO_2 film could be self-accommodated by the localized compression and reduction of the porosity (Figure 2.22). More interestingly, the proton intercalation induced softening was found to favor the elastic deformation of MnO_2.

Apart from MnO_2, the effect of electrolytes on other pseudocapacitive metal oxides such as Fe_3O_4 [228–230], SnO_2 [231], V_2O_5 [232], and ZnO [233], and ECPs [234,235], has also been investigated. Again, it should be noted that the effect of the electrolyte is also strongly dependent on the material system. For V_2O_5, it was found that the 0.5 M K_2SO_4 gave the largest specific capacitance value compared to that obtained in 0.5 M Li_2SO_4 and Na_2SO_4 electrolytes [232]. The authors attributed this to the most facile intercalation/deintercalation of K^+ ions into/from the interlayers of $V_2O_5 \cdot 0.6\ H_2O$. Intercalation/deintercalation of Li^+ ions was believed to be difficult since it strongly interacted with the interlayer H_2O of $V_2O_5 \cdot 0.6\ H_2O$. Similar results that 2 M KCl resulted in a higher specific capacitance compared to the 2 M LiCl were reported for

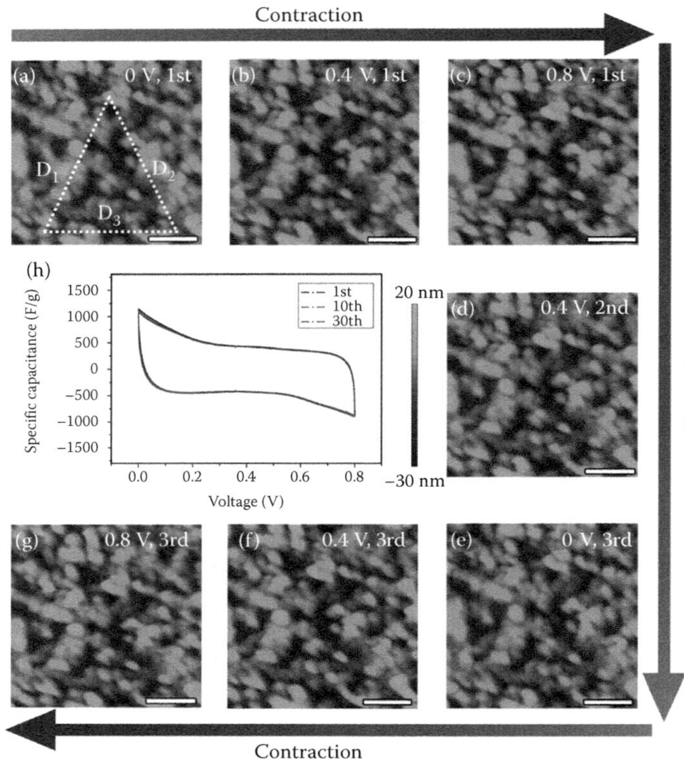

FIGURE 2.22 *In situ* AFM observation of the microstructural evolution during charging and discharging processes in aqueous electrolyte. (a–g) *In situ* AFM images recorded at different applied potentials versus Ag/AgCl. (h) Representative CV curves of the electrodeposited MnO_2 for the 1st, 10th, and 30th cycles in 0.1 M aqueous electrolyte. Scale bar: 500 nm. (Tao, X. Y. et al.: *Advanced Functional Materials*. 4745–4751. 2013. Copyright Wiley-VCH Verlag GmbH & Co. KGaA. Reprinted with permission.)

a porous V_2O_5 electrode [236]. For an ES based on ZnO–rGO nanocomposites, it was reported that the capacitive performance was better in KCl than Na_2SO_4, which was attributed to the easier access of K^+ ions toward pores or voids in the nanocomposites compared to Na^+ ions [233]. In the case of Fe_3O_4-based ESs, the Na_2SO_3 electrolyte has been reported to be beneficial for achieving a good performance [228,229] due to the reactions (2.9) and (2.10) [237], leading to a pseudocapacitive behavior.

$$FeO + SO_3^{2-} \leftrightarrow FeSO_4 + 2e^- \tag{2.9}$$

$$2FeO^{II}O + SO_3^{2-} \leftrightarrow (Fe^{III}O)^+SO_3^{2-}(Fe^{III}O)^+ + 2e^- \tag{2.10}$$

Although RuO_2-based electrodes are generally used in acid electrolytes, it was reported that RuO_2-based ESs using 1 M Na_2SO_4 neutral electrolyte could reach a high cell voltage of 1.6 V [238]. As a result, a relatively high energy density of 18.77 Wh kg^{-1} at a power density of 500 W kg^{-1} could be achieved. Besides, neutral electrolytes are usually suitable to be used for various conductive polymer-based ESs due to the mild pH condition. For example, Ammam and Fransaer [235] reported a relatively high specific capacitance of about 401 F g^{-1} for the MWCNTs/PANI in the physiological electrolyte and about 326 F g^{-1} in serum, which made it promising for biomedical applications. Sun et al. [239] investigated performance of the PPy-based ESs with the 0.5 M NaCl or 0.5 M sodium *p*-toluenesulfonate (TOSNa) electrolytes. The PPy electrode showed a better pseudocapacitive behavior and much larger specific capacitance in the NaCl electrolyte than that in the TOSNa electrolyte, due to the smaller size and higher mobility of Cl^- compared with TOS^-, allowing Cl^- to go into the deeper location of the PPy film.

2.2.3.3 Neutral Electrolytes for Hybrid ESs

Neutral electrolytes have also been widely used in asymmetric ESs, so as to enlarge the operational voltage and thus to increase the energy density. Despite a higher specific capacitance, most pseudocapacitive materials such as MnO_2 have a limited potential window. This restricts the cell voltage and thus the energy density of symmetrical ESs based on these materials. As mentioned previously, ESs with an asymmetric electrode configuration is an effective way to increase the cell voltage window. This is also the case for asymmetric ESs with neutral electrolytes. The cell voltage of MnO_2-based symmetric ES is generally limited to about 1 V. By replacing the negative electrode with a complementary potential window to that for MnO_2, such as AC, the cell voltage can be significantly extended to a more negative voltage (Figure 2.23) [95]. If the cell voltage could be increased from 1 to 2 V, then the energy density would increase by four times theoretically.

Compared to the previously discussed asymmetric ESs based on a battery-type positive electrode such as AC//PbO_2 and AC//$Ni(OH)_2$ in the strong acid or alkaline electrolyte, asymmetric AC//MnO_2 ESs have a great advantage in terms of a much longer cycling life because of the pseudocapacitive nature of MnO_2. Therefore, since the pioneering work on asymmetric MnO_2//AC ESs using neutral electrolytes reported by Hong et al. [240] and Brousse et al. [241], considerable work has been focused on the development of symmetric ESs with neutral electrolytes [242,243]. Besides carbon-based negative electrodes, other electrode materials, such as $LiFeO_2$ [244], SnO_2 [245], and TiO_2 [246], can also be used as the negative electrode materials in the neutral electrolytes. As for the positive electrode, other materials such as V_2O_5 [247,248], MnO_2-based composites [249,250], and Co-based composites [251] have been reported. Table 2.5 shows some selected data on neutral aqueous electrolytes-based asymmetric ESs from the literature published in 2014. It can be seen that most of these aqueous electrolytes use sulfate salts. Most of these asymmetric ESs can reach an operation voltage of about 2 V, which is higher than reported for strong acid or alkaline electrolyte-based asymmetric ESs. Due to the enlarged cell voltage, most studies reported high energy density values above 20 Wh kg^{-1} and some reported values were even higher than 50 Wh kg^{-1}. It is interesting to note these reported energy density values from the literature published in 2014 are generally higher than those energy density values reported in earlier studies, as reviewed in previous work [252].

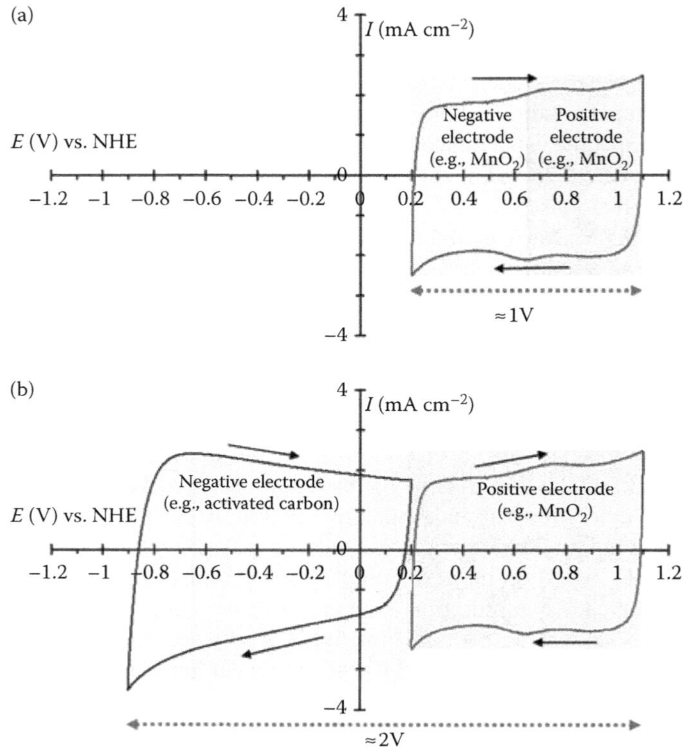

FIGURE 2.23 Schematic representation of cyclic voltammograms for different configurations of aqueous-based ESs. Right and left cyclic voltammograms represent the potential window of the positive and negative electrode, respectively for: (a) symmetric MnO_2// MnO_2 ES with α-MnO_2 electrodes in 0.5 M K_2SO_4 and (b) asymmetric AC//MnO_2 ES in 0.5 M K_2SO_4. NHE, normal hydrogen electrode; *I*, measured current; and *E*, electrode potential. (Reprinted Long, J. W. et al. 2011. *MRS Bulletin* 36:513–522, with permission from Cambridge University Press.)

More excitingly, most of these reported values of the energy density are comparable to or even higher than those of EDLCs using organic electrolytes. Therefore, if other concerns such as the cycle life can be well addressed, these neutral electrolyte-based asymmetric ESs can be promising alternatives for commercial EDLCs with organic electrolytes.

It should be noted that although a high cell voltage can be achieved for these asymmetric ESs, the possible gas evolution at high potentials may hinder the practical use. For example, Brousse et al. [241] found the hydrogen evolution on the AC electrode and the oxygen evolution on the MnO_2 electrode in the 0.65 M K_2SO_4 electrolyte at high operating voltages (e.g., 2.2 V). The effect of the electrolytes on the performance of asymmetric ESs generally shares a similar rule as discussed above for symmetrical ESs. For example, regarding the effect of cation and anion species, Mosqueda et al. [266] found that for sulfate and nitrate salts, the specific capacitance

TABLE 2.5
Summary of Selected Literature Data on Neutral Aqueous Electrolyte-Based Asymmetric ESs

Electrolyte	Negative Electrode	Positive Electrode	Cell Voltage (V)	Capacitance (F g^{-1})	ESR (Ω)	Energy Density (Wh kg^{-1})	Power Density (kW kg^{-1})	Cycle Numbers	References
1 M Na_2SO_4	AC	AC	1.8	200	~0.3	~25	~220 (maximum value)	4000	[253]
1 M Li_2SO_4	γ-FeOOH nanosheets	MnO_2 nanosheets	1.85	310.3	–	37.4	16	–	[254]
1 M Na_2SO_4	HPCs	MnO_2–HPCs nanocomposites	2.0	196	–	21.7	0.6	–	[255]
1 M Na_2SO_4	AC	GO–PPy	1.6	332.6	1.1	21.4	0.4539	5000	[256]
1 M Na_2SO_4	GO	CuO@MnO_2	1.8	276	–	22.1	85.6 (maximum value)	10,000	[257]
1 M Na_2SO_4	AC	SiC–N–MnO_2	1.9	59.9	–	30.06	0.11392	1000	[258]
0.5 M Na_2SO_4	CNT–CNF	MnO_2	2.0	93.99	–	52.22	0.1	2000	[259]
0.5 M Li_2SO_4	AC	$LiMn_2O_4$	1.8	415	–	29.8	0.09	1000	[260]
1 M Na_2SO_4	rGO	3D MnO_2–rGO	1.6	242	–	21.2	0.82	1000	[261]
1 M Na_2SO_4	AC	MnO_2	2.0	67	–	37.22	0.178	1000	[262]
0.5 M Na_2SO_4	AC	$Na_{0.35}MnO_2$ NWs	1.0	157	–	42.6	0.1298	5000	[263]
1 M Na_2SO_4	Graphene	Graphene–MnO_2	1.8	274	–	23.9	0.225	1000	[243]
1 M Na_2SO_4	Activated graphene	MnO_2–TiO_2	2.0	120	–	29.8	–	3000	[264]
0.5 M Na_2SO_4	Carbon black modified graphene	Cobalt hexacyanoferrate nanoparticles	~2.4	>250	–	34.4	2.5	5000	[265]

Abbreviations: AC, activated carbon; HPCs, hierarchically porous carbons; GO, graphene oxide; PPy, polypyrrole; CNTs, carbon nanotubes; CNFs, carbon nanofibers; rGO, reduced graphene oxide; and NWs, nanowires.

of the MnO_2 electrode increased by 20% in lithium salts compared to sodium or potassium salts, probably due to the partial lithium intercalation into the α-MnO_2 structure. It was further pointed that the specific capacitance of the AC//MnO_2 with the 5 M $LiNO_3$ electrolyte is slightly higher than that with the 0.5 M K_2SO_4 electrolyte and the cycle stability was also similar between the $LiNO_3$ and K_2SO_4 electrolytes, but the $LiNO_3$ electrolyte gave a better low-temperature performance due to its higher solubility compared to the K_2SO_4. Chen et al. [267] reported the AC//$LiCoO_2$ hybrid ES showed better capacitive performance in 1 M Li_2SO_4 electrolytes compared to 1 M Na_2SO_4, K_2SO_4, and $(NH_4)_2SO_4$. This was probably due to the facile insertion/deinsertion of Li^+ ions into the $LiCoO_2$ electrode. Besides, Xu et al. [268] investigated the effect of electrolytes with different alkaline-earth salts (i.e., $Mg(NO_3)_2$, $Ca(NO_3)_2$, or $Ba(NO_3)_2$) on the performance of a AC//MnO_2 asymmetric ES. It was pointed out that Ca^{2+} cation was the most suitable for the asymmetric ES because of the suitable bare ion size and the smallest hydrated ion size among three alkaline-earth cations (i.e., Mg^{2+}, Ca^{2+}, and Ba^{2+}). The energy density of the AC//MnO_2 asymmetric ES with the 0.1 M $Ca(NO_3)_2$ electrolyte was reported to be 21 Wh kg^{-1} at a current density of 0.3 A g^{-1}. Similar results were also reported by Tomko et al. [269], where the energy density of carbon//MnO_2 with different electrolytes followed the order of $CaCl_2 > MgCl_2 > KCl$ and was 22.5, 20, and 18 Wh kg^{-1}, respectively, at a power density of 1 kW kg^{-1}. Recently, it has been reported, by using a battery-type Li negative electrode coupled with a suitable positive electrode material such as AC and MnO_2, a very high cell voltage of about 4 V can be achieved in the neutral electrolytes such as Li_2SO_4 and LiCl [270,271]. In this case, the negative electrode potential can be greatly decreased below the limit of the hydrogen evolution and thus further increase the cell voltage [271]. Since the metallic Li electrode fails to work directly in the aqueous electrolyte, a multilayered Li negative electrode, which is water stable, has been used [270,271].

Besides the asymmetric ESs based on pseudocapacitive or battery-type materials, ESs can also consist of carbon materials with different properties, capacitance, or mass of the carbon materials on each electrode (i.e., carbon-based asymmetric ESs) [181,272,273]. In this way, the cell voltage can be further increased compared to the symmetric carbon-based ESs with the neutral electrolytes. For example, a cell voltage of 1.9 V was reported for carbon//0.3 M K_2SO_4//carbon asymmetric ESs with unequal electrode capacitances and more than 85% of the initial cell capacitance could be retained after 10,000 charging/discharging cycles [272]. Therefore, the utilization of the neutral aqueous electrolytes can not only address the corrosion concerns but also provide an environmental-friendly and cost-effective method to effectively improve the working voltage and thus the energy density of the ESs [2]. However, there is still room for the further improvement in the performance of ESs with neutral electrolytes in terms of specific capacitance, ESR, etc.

2.2.4 Comparative Studies between Different Aqueous Electrolytes

To get a clearer picture about the electrolyte effect on the performance of ESs, some typical comparative studies concerning various electrolytes including strong acid, neutral, and strong alkaline electrolytes are reviewed. For EDLCs, the majority of

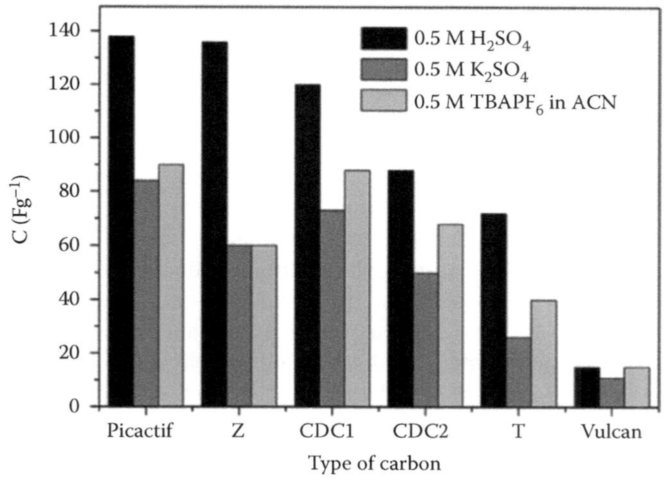

FIGURE 2.24 Specific capacitance of the ACs in different electrolytes calculated from CV at 20 mV s^{-1}. (Reprinted from *Electrochimica Acta*, 86, Vaquero, S. et al., Insights into the influence of pore size distribution and surface functionalities in the behavior of carbon supercapacitors, 241–247, Copyright 2012, with permission from Elsevier.)

comparative studies found the specific capacitance obtained in the H_2SO_4 electrolyte or KOH electrolyte is higher than that in neutral electrolytes such as K_2SO_4, Na_2SO_4, $NaNO_3$, KNO_3, and KCl [28,274–280]. An example is shown in Figure 2.24 [281].

In addition, due to the higher ionic conductivity of the H_2SO_4 or KOH compared to the neutral electrolytes, the ESR of ESs with H_2SO_4 or KOH are generally lower than those with the neutral electrolytes [28,279,280]. Actually, various studies have found there is a relationship between the specific capacitance of the ACs and the electrolyte conductivity, that is, the specific capacitance increased with the increase of the electrolyte conductivity [274]. For example, Torchala et al. [274] reported the specific capacitance of the activated mesocarbons followed the order: 6 M KOH > 1 M H_2SO_4 > 4.5 M KCl > 2 M KNO_3 > 0.5 M Na_2SO_4, which basically corresponded to the trend of the electrolyte conductivity (note that 1 M H_2SO_4 has a higher conductivity than the 6 M KOH). This can be understood by considering the ion mobility which is related to the electrolyte conductivity. However, some controversial results lie between the specific capacitance obtained in the KOH or H_2SO_4 electrolyte. For example, some studies found the 6 M KOH electrolyte gave a higher specific capacitance for AC [274] whereas others reported the specific capacitance obtained in the 1 M H_2SO_4 electrolyte is higher than that in the 6 M KOH electrolyte [28].

Besides, the possible surface functional groups on the AC might also have an influence on the obtained specific capacitance, which will be discussed below. Although the specific capacitance obtained in neutral electrolytes is lower than that in the H_2SO_4 or KOH electrolyte, EDLCs with neutral electrolytes can generally achieve a higher working voltage, as mentioned previously. Since the cell voltage has a greater influence on the energy density of ESs than the cell capacitance, a higher energy density is expected to be achieved in the neutral electrolyte-based EDLC. For example, Sun et al. [282] reported the EDLC with the Li_2SO_4 electrolyte exhibited an energy density of 16.9 Wh kg^{-1} at the power density of 200 W kg^{-1}, which was 80% higher than that with the KOH electrolyte, due to its larger potential window despite the lower specific capacitance.

Regarding the effect of electrode wettability on ES performance, Fic et al. [283] compared the effect of surfactant additions into the acidic (1 M H_2SO_4) and alkaline (6 M KOH) electrolytes on the specific capacitance of an EDLC. Three types of surfactants, including SDS, tetrapropylammonium bromide (TPAB), and nonionic ones such as polymer of polyethylene glycol and p–t-octylophenol (Triton X-100), were investigated in this study. It was found that a more significant performance improvement (e.g., good charging/discharging propagation, higher capacitance values, better cycling stability, smaller leakage current, and self-discharge values) was achieved in the 6 M KOH electrolyte particularly for the nonionic surfactant (Triton X-100). The authors attributed the improved capacitance performance to the decrease of the surface tension of the electrolyte after the surfactant addition.

As for the carbon-based pseudocapacitive materials, different surface functional groups on the carbon materials have different natures and thus are very sensitive to the type of electrolyte. Andreas and Conway [77] investigated the CV response of a carbon-cloth electrode in the electrolytes with a wide pH range from 0 to 14. As the pH increased progressively from 0 to 11, a 30% loss in the specific capacitance was observed, partly due to the disappearance of the pseudocapacitance arising from the oxidation/reduction of the surface quinone groups that required the proton participation. As the pH increased above 11, the specific capacitance increased about 20%, as other surface functionalities such as pyrone derivative species became activated in the alkaline electrolyte. Other studies also support that carbon materials with quinone-type oxygen groups showed a higher specific capacitance in the acidic electrolyte through comparison with the alkaline and neutral electrolytes [78]. However, the nitrogen-containing surface functionalities have a different behavior. For instance, Hulicova et al. [75] and Wu et al. [284] found nitrogen-enriched carbons have a higher specific capacitance in the KOH electrolyte as compared to that in the H_2SO_4 or K_2SO_4 electrolytes.

With regard to the metal oxide or hydroxides-based ESs, since protons play a significant role in the charge storage mechanism of some pseudocapacitive materials such as RuO_2, it is expected that pH of electrolytes would have a strong influence on their electrochemical behavior. Furthermore, the electrolyte effect is also strongly dependent on the electrode materials [160,164,206,237,285–288]. For example, Zhang et al. [289] found the tubular RuO_2 electrode showed a much higher maximum specific capacitance of 861 F g^{-1} in the H_2SO_4 electrolyte compared to the Na_2SO_4 electrolyte. The improved specific capacitance was attributed to the participation of protons in the charge/discharge processes. Wen et al. [206] investigated the effect of different electrolytes on the MnO_2 and RuO_2 electrode, respectively. The specific capacitance (normalized by the surface area) of the RuO_2 electrode in the H_2SO_4 electrolyte (pH = 2.04) was found to be larger than that obtained in the KCl electrolyte (pK = 2.00). The specific capacitances of the RuO_2 electrode in the H_2SO_4 and KCl electrolytes showed the linear dependences upon the pH and pK, respectively (Figure 2.25), indicating the charging processes were strongly related with the

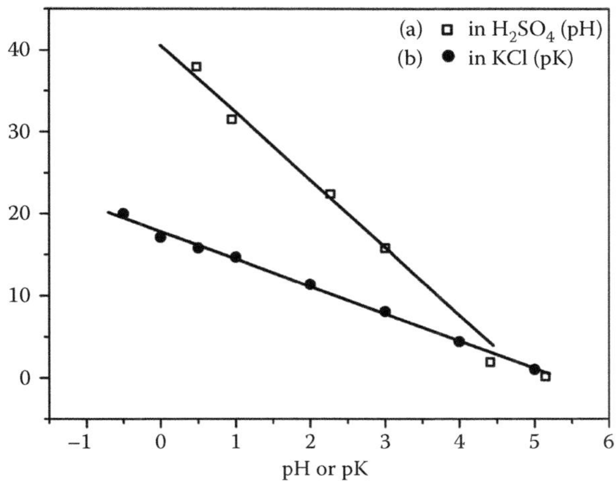

FIGURE 2.25 Linear regression fit of (a) capacitances against the pH of H_2SO_4 electrolyte and (b) capacitances against the pK of KCl electrolyte solutions for RuO_2 electrode. (Reprinted from *Electrochimica Acta*, 50, Wen, S. et al., The role of cations of the electrolyte for the pseudocapacitive behavior of metal oxide electrodes, MnO_2 and RuO_2, 849–855, Copyright 2004, with permission from Elsevier.)

proton concentration and the potassium concentration, respectively. For the MnO_2 electrode, the obtained specific capacitance in the aprotic solvents (CH_3CN) was drastically decreased to a smaller value than that in the aqueous neutral electrolytes, suggesting a large contribution of the proton to the pseudocapacitance. It has been found that the pH of the electrolytes containing certain alkaline or alkaline-earth salt significantly affects the specific capacitance value of MnO_2 electrode materials [206]. As pH value is decreased to a lower value (e.g., 3.0), the capacitance deceased sharply and the CV plot was less rectangular, which was attributed to dissolution of manganese from the electrode to the electrolyte [206,290]. As for the $Co(OH)_2$-based ESs, Fedorov et al. [285] found that although the initial specific capacitance value was higher in the 1 M NaOH electrolyte, the 1 M Na_2SO_4 gave a better cycling stability. Wang et al. [160] reported the specific capacitance of the $MnFe_2O_4$ electrode at the current density of 0.1 A g^{-1} in various electrolytes (2 M) followed the order: KOH (97.1 F g^{-1}) > NaOH (93.9 F g^{-1}) > LiOH (74.2 F g^{-1}) > $LiNO_3$ (51.1 F g^{-1}) > Na_2SO_4 (47.4 F g^{-1}). The authors explained this trend by considering the hydrated ion size and the related ion mobility. Wang et al. [291] demonstrated the specific capacitance of the Fe_3O_4 electrode were 170, 25, and 3 F g^{-1} in Na_2SO_3, Na_2SO_4, and KOH, respectively. Based on electrochemical QCM analysis, XPS, and electrochemical characterizations, the authors pointed out the capacitance of Fe_3O_4 electrode in Na_2SO_3 originated from both EDLC capacitance and the pseudocapacitance that involved successive reduction of the specifically adsorbed sulfite anions. The obtained capacitance in the Na_2SO_4 electrolyte was entirely attributed to the EDLC capacitance. In the KOH electrolyte, the formation of an insulating layer on the Fe_3O_4 electrode resulted in the very low capacitance of the electrode. In the case of the tungsten oxynitride in the various electrolytes including 1 M H_2SO_4, 1 M KOH, 3 M KCl, 3 M NaCl, 1 M LiCl, and 1 M $CaCl_2$, the highest specific capacitance value was obtained in the H_2SO_4 while the highest cyclic stability was achieved in the 1 M KOH electrolyte [292]. Sankar et al. [287] found that the specific capacitance of the Mn_3O_4 electrode showed the highest value in the 6 M KOH electrolyte among various electrolytes including 6 M KOH, 1 M KCl, 1 M Na_2SO_4, and 1 M $NaNO_3$.

As for the asymmetric active carbon-based ESs concerned, Sun et al. [293] reported the specific capacitance of the negative electrode was remarkably larger than that of the positive one in the KOH electrolyte, whereas similar specific capacitance values for each electrode were observed in the H_2SO_4 electrolyte. This was attributed to the different redox reactions arising from the surface functionalities on the carbon electrodes in the positive and negative potential regions [293,294]. For the asymmetric ESs using Mn_3O_4–graphene composites as the positive and AC as negative electrodes, Xiao et al. [295] found a maximum energy density of 34.6 Wh kg^{-1} at power density of 0.5 kW kg^{-1} was achieved in the 1 M Na_2SO_4 electrolyte, which was much higher than that with the 6 M KOH electrolyte. This was due to the larger operating voltage (2 V) in the Na_2SO_4 electrolyte compared to that in the KOH electrolyte (1.6 V).

2.3 ORGANIC ELECTROLYTES

Although aqueous electrolyte-based ESs have been widely studied in the academic research, organic electrolytes currently dominate the commercial ES market due to their high operative cell voltage (typically 2.5–2.8 V), as shown in Table 2.1. Typical commercial organic electrolytes consist of conductive salt such as tetraethylammonium tetrafluoroborate ($TEABF_4$) dissolved in the organic solvent such as ACN and PC. The achieved operating cell voltage of the organic electrolyte-based EDLCs is generally much higher than that reported for the aqueous electrolyte-based ESs including most of the asymmetric ones, as discussed in Section 2.2. The increase of the operating potential provides a significant enhancement in the energy and power densities that industry demands. Besides, cheaper materials such as Al can be used for current collectors and packages when organic electrolytes are used for ESs.

However, it is necessary to consider other issues related with the use of organic electrolytes, such as high cost, smaller specific capacitance, low conductivity limiting the power performance, safety concerns related to flammability, volatility, and toxicity compared to the aqueous electrolyte. Furthermore, the use of organic electrolytes requires complicated purification and assembling procedures under a strictly controlled atmosphere to avoid the introduction of impurities such as water that can lead to performance degradation [69].

2.3.1 General Composition and Properties of Organic Electrolytes

Organic electrolytes for the ESs are typically composed of conducting salts dissolved in organic solvents. Tables 2.6 and 2.7 show some commonly used salts and solvents, respectively. For convenience, the abbreviations of these salts and solvents are also given. The nature of the salts and the solvents, such as the ion size, interaction

TABLE 2.6

Summary of Abbreviations, Physical Properties, Limiting Electrochemical Reduction, and Oxidation Potentials of Commonly Used Organic Solvents Containing 0.65 M $TEABF_4$ at 25°C

Solvent		ϵ_r	η (cp)	bp (°C)	mp (°C)	MW	σ (mS cm^{-1})	E_{red} (V vs. SCE)	E_{ox} (V vs. SCE)
Propylene carbonate (PC)	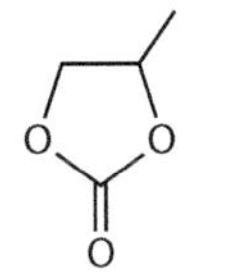	65	2.5	242	–49	102	10.6	–3.0	3.6
Butylene carbonate (BC)	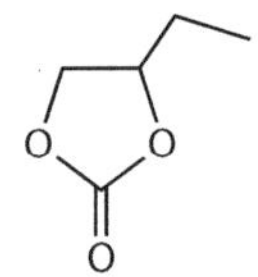	53[a]	3.2[a]	240[a]	–53[a]	116	7.5	–3.0	4.2
γ-butyrolactone (GBL)	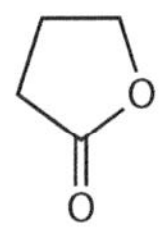	42	1.7	204	44	86	14.3	–3.0	5.2
γ-valerolactone (GVL)	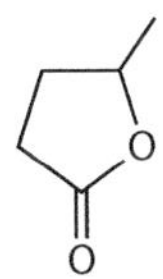	34[a]	2.0[a]	208[b]	–31[b]	100	10.3	–3.0	5.2
Acetonitrile (ACN)		36	0.3	82	–49	41	49.6	–2.8	3.3

(Continued)

TABLE 2.6 (*Continued*)
Summary of Abbreviations, Physical Properties, Limiting Electrochemical Reduction, and Oxidation Potentials of Commonly Used Organic Solvents Containing 0.65 M $TEABF_4$ at 25°C

Solvent		ϵ_r	η (cp)	bp (°C)	mp (°C)	MW	σ (mS cm^{-1})	E_{red} (V vs. SCE)	E_{ox} (V vs. SCE)
Propionitrile (PN)	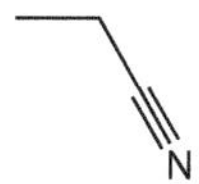	26[a]	0.5[a]	97	−93	55	Insoluble		
Glutaronitrile (GLN)	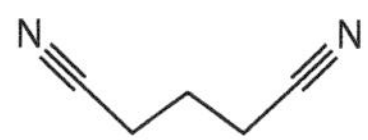	37[a]	5.3[a]	286[b]	−29[b]	94	5.7	−2.8	5.0
Adiponitrile (ADN)	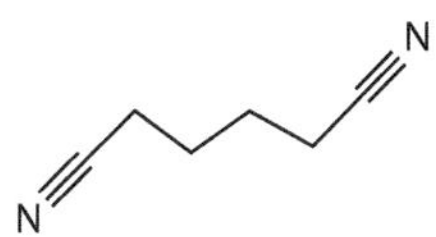	30[a]	6.0[a]	295[b]	2[b]	108	4.3	−2.9	5.2
Methoxyacetonitrile (MAN)	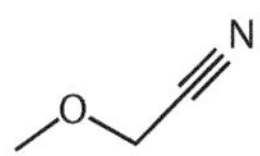	21[a]	0.7[a]	120[b]	−35[a]	71	21.3	−2.7	3.0
3-methoxypropionitrile (MPN)	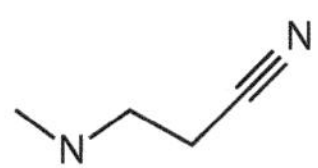	36[a]	1.1[a]	165[b]	−57[a]	85	15.8	−2.7	3.1

(*Continued*)

TABLE 2.6 (*Continued*)
Summary of Abbreviations, Physical Properties, Limiting Electrochemical Reduction, and Oxidation Potentials of Commonly Used Organic Solvents Containing 0.65 M $TEABF_4$ at 25°C

Solvent		ϵ_r	η (cp)	bp (°C)	mp (°C)	MW	σ (mS cm^{-1})	E_{red} (V vs. SCE)	E_{ox} (V vs. SCE)
N,N-dimethylformamide (DMF)	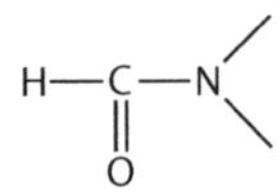	37	0.8	153	–61	73	22.8	–3.0	1.6
N,N-dimethylacetamide (DMA)	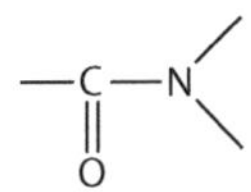	38	0.9	166	–20	87	15.7		
N-methylpyrrolidinone (NMP)	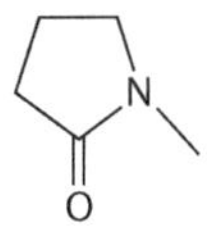	32	1.7	202	–24	99	8.9		
N-methyloxazolidinone (NMO)	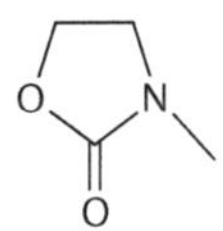	78[c]	2.5[c]	270[b]	15[b]	101	10.7	–3.0	1.7

(*Continued*)

TABLE 2.6 (*Continued*)

Summary of Abbreviations, Physical Properties, Limiting Electrochemical Reduction, and Oxidation Potentials of Commonly Used Organic Solvents Containing 0.65 M $TEABF_4$ at 25°C

Solvent		ϵ_r	η (cp)	bp (°C)	mp (°C)	MW	σ (mS cm^{-1})	E_{red} (V vs. SCE)	E_{ox} (V vs. SCE)
N,N′-dimethylimidazolidinone (DMI)	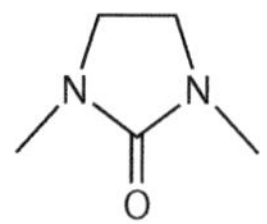	38[d]	1.9[d]	226[d]	8[d]	114	7.0	−3.0	1.2
Nitromethane (NM)	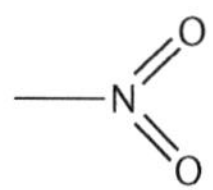	38	0.6	101	−29	61	33.8	−1.2	2.7
Nitroethane (NE)	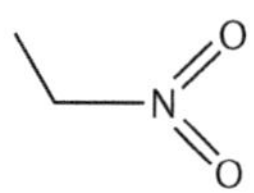	28[a]	0.7[a]	115[b]	−90[b]	75	22.1	−1.3	3.2
Sulfolane (TMS)	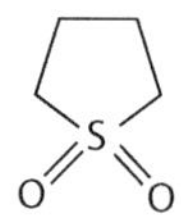	43	10.0 (30°C)	287	28	120	2.9	−3.1	3.3

(*Continued*)

TABLE 2.6 (*Continued*)
Summary of Abbreviations, Physical Properties, Limiting Electrochemical Reduction, and Oxidation Potentials of Commonly Used Organic Solvents Containing 0.65 M $TEABF_4$ at 25°C

Solvent		ϵ_r	η (cp)	bp (°C)	mp (°C)	MW	σ (mS cm^{-1})	E_{red} (V vs. SCE)	E_{ox} (V vs. SCE)
3-methylsulfolane (3MS)	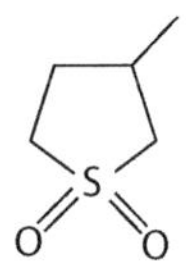	29[e]	11.7[e]	276[b]	6[b]	134	Insoluble		
Dimethylsulfoxide (DMSO)	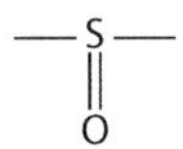	47	2.0	189	19	78	13.9	−2.9	1.5
Trimethyl phosphate (TMP)	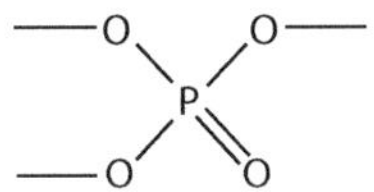	21[a]	2.2[a]	197[b]	−46[b]	140	8.1	−2.9	3.5

Source: Physical properties are cited from With kind permission from Springer Science+Business Media: *Non-Aqueous Electrolyte Solutions in Chemistry and Modern Technology*, 1983, Barthel, J. et al.; Coetzee, J. E. 1982. *Recommended Methods for Purification of Solvents and Tests for Impurities*. New York: Pergamon Press, Commission on Electroanalytical Chemistry International Union of Pure and Applied Chemistry. [a]Ue, M., K. Ida, and S. Mori. 1994. *Journal of the Electrochemical Society* 141:2989–2996; [b]1993. *Catalog Handbook of Fine Chemicals*. Aldrich Chemical Co.; [c]Kelly, M. J., and W. R. Heineman. 1988. *Journal of Electroanalytical Chemistry* 248:441–446; [d]Rosenfarb, J., and J. A. Caruso. 1976. *Journal of Solution Chemistry* 5:345–354; [e]Rosenfarb, J. et al. 1976. *Journal of Solution Chemistry* 5:311–318.

Abbreviations: ϵ_r, relative permittivity; η, viscosity; bp, boiling point; mp, melting point; MW, molecular weight; σ, ionic conductivity; E_{red}, limiting reduction potential; and E_{ox}, limiting oxidation potential (glassy carbon electrode was used as a working electrode to measure the E_{red} and E_{ox}).

TABLE 2.7
Summary of Abbreviations, Electrolyte Conductivities, and Limiting Electrochemical Reduction and Oxidation Potentials of Some Quaternary Ammonium or Phosphonium Tetrafluoroborate Conducting Salts (0.65 M) in PC at 25°C

Electrolyte	σ (mS cm^{-1})	E_{red} (V vs. SCE)	E_{ox} (V vs. SCE)	Electrolyte	σ (mS cm^{-1})	E_{red} (V vs. SCE)	E_{ox} (V vs. SCE)
Methyl ammonium tetrafluoroborate (Me_4N BF_4)	2.41[a]	−3.10[a]	3.50[a]	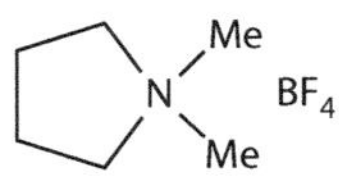	10.36	−3.00	3.65
Triethylmethylammonium tetrafluoroborate (Me_3EtN BF_4)	10.16	−3.00	3.60	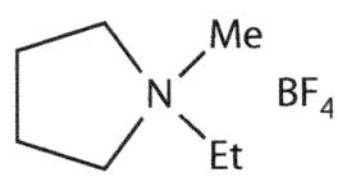	10.82	−3.00	3.70
Tetrafluoroboric acid dimethyldi ethylammonium (Me_2Et_2N BF_4)	10.34	−3.00	3.65	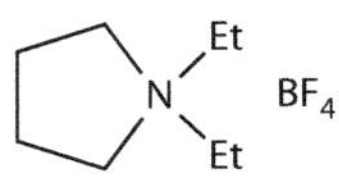	10.40	−3.00	3.60
Triethylmethylammonium tetrafluoroborate ($MeEt_3N$ BF_4)	10.68	−3.00	3.65	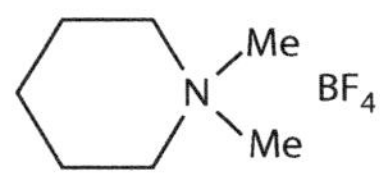	10.20	−3.05	3.65
Tetraethylammonium tetrafluoroborate (Et_4N BF_4 or TEA BF_4)	10.55	−3.00	3.65	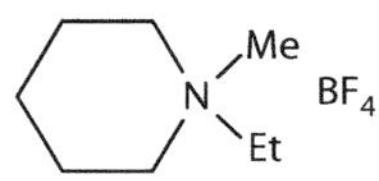	10.40	−3.05	3.70
Tetrapropyl ammonium tetrafluoroborate (Pr_4N BF_4)	8.72	−3.05	3.65	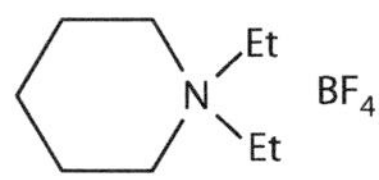	10.17	−3.05	3.60
Methyltributylammonium tetrafluoroborate ($MeBu_3N$ BF_4)	7.80			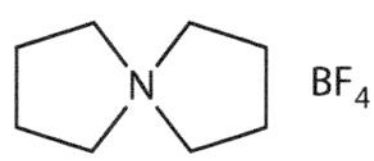	10.94	−3.00	3.60
Tetrabutylammonium tetrafluoroborate (Bu_4N BF_4)	7.23	−3.05	3.65		9.67	−3.00	3.60
Tetrahexylammonium tetrafluoroborate (Hex_4N BF_4)	5.17	−3.10	3.85	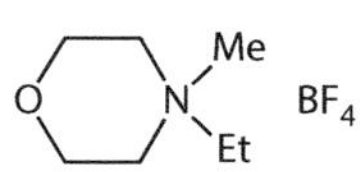	8.78	−3.00	3.60
Tetramethylammonium tetrafluoroborate (Me_4P BF_4)	9.21	−3.05	3.60				
Tetraethyl phosphonium tetrafluoroborate (Et_4P BF_4)	10.52	−3.00	3.60				

(Continued)

TABLE 2.7 (*Continued*)
Summary of Abbreviations, Electrolyte Conductivities, and Limiting Electrochemical Reduction and Oxidation Potentials of Some Quaternary Ammonium or Phosphonium Tetrafluoroborate Conducting Salts (0.65 M) in PC at 25°C

Electrolyte	σ (mS cm^{-1})	E_{red} E_{ox} (V vs. SCE)	Electrolyte	σ (mS cm^{-1})	E_{red} E_{ox} (V vs. SCE)
Tetrapropylphosphonium tetrafluoroborate (Pr_4P BF_4)	8.63	−3.05 3.60			
Tetrabutylphosphonium tetrafluoroborate (Bu_4P BF_4)	7.14	−3.05 3.80			

Source: Adapted from Ue, M., K. Ida, and S. Mori. 1994. *Journal of the Electrochemical Society* 141:2989–2996.

Abbreviations: σ, ionic conductivity; E_{red}, limiting reduction potential; and E_{ox}, limiting oxidation potential (glassy carbon electrode was used as a working electrode to measure the E_{red} and E_{ox}).

[a] 0.1 mol L^{-1}.

between ions and solvents, conductivity, viscosity, and electrochemical stability, plays a significant role in the performance of ESs using organic electrolytes. This will be discussed below for the specific electrolyte. Most studies have investigated organic electrolytes for EDLCs. With the development of electrode materials and electrolytes in recent years, increasing research has been focused on organic electrolyte-based pseudocapacitors, hybrid ESs, and lithium-ion capacitors (LICs).

2.3.1.1 Organic Electrolytes for EDLCs

For carbon-based ESs, it has been generally found that the specific capacitance achieved in organic electrolytes is lower than that in aqueous electrolytes. As demonstrated by extensive comparative studies between aqueous and organic electrolytes, this is the case for ESs based on a wide variety of carbon materials, including commonly used AC [281,303,304], porous carbons [305–308], CNFs [309,310], and recently developed onion-like carbons (OLCs) [311], hierarchical porous carbons [312–314], carbon nanocoils [315], carbon nanohorns [316], and graphene-based materials [317,318]. The organic electrolytes generally have larger solvated ion size and lower dielectric constant, leading to a lower EDL capacitance. Furthermore, as mentioned previously, due to the presence of surface functionalities on the carbon-based materials especially for those carbon-based materials modified by heteroatoms (e.g., oxygen, nitrogen, and phosphorus), the resultant pseudocapacitance in the aqueous electrolytes such as H_2SO_4 and KOH also contributes to the overall specific capacitance. However, the contribution of the pseudocapacitance is very small or negligible in the organic electrolytes (e.g., $TEABF_4$/ACN) [281,309,319–321]. This can be explained based on the nature of the surface functionalities. For example, proton participation is involved in the pseudocapacitance of the AQ-modified carbons, which is not favored in the aprotic organic electrolytes. This also results in the lower specific capacitance obtained in the organic electrolytes. Although the specific capacitance of the EDLC with organic electrolytes is generally lower than that with aqueous electrolytes, the former has higher energy and power densities due to the larger operating cell voltage. For example, Fernández et al. [321] found that the CDCs-based EDLC using 2 M H_2SO_4 electrolyte exhibited a higher gravimetric capacitance of 200 F g^{-1} whereas EDLC using 1 M $TEABF_4$/ACN showed a specific capacitance of 150 F g^{-1}. However, the organic electrolyte resulted in a larger energy and power densities compared to the aqueous electrolyte (Figure 2.26) since the former gave a higher cell voltage of 2 V compared to the aqueous electrolyte (0.8 V). As shown in Figure 2.27, Mladenov et al. [322] compared the capacitance performance of AC-based EDLC with different electrolytes: $TEABF_4$/PC, $LiBF_4$/PC, and $LiPF_6$/DMC–EC (dimethyl carbonate–ethylene carbonate). At low discharge current, EDLCs with three different electrolytes had similar capacitance values. However, as the discharge current increased, the specific capacitance of EDLCs using Li^+ ions containing electrolytes decreased much more severely compared to EDLCs using $TEABF_4$ electrolytes.

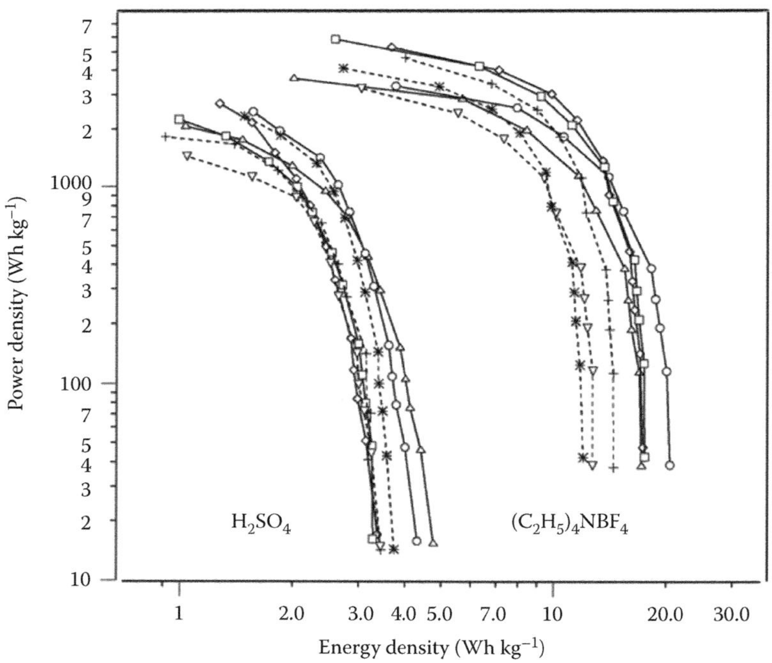

FIGURE 2.26 Ragone-type plots for CDCs: TiC (○), TiC/SiC (□), TiC/TiO_2 (△), and SiC (◇); commercial ACs for supercapacitors: Super DLC-30 (▽) and Picactif SC (+); mesoporous carbon: MP-55 (⁎). The data correspond to the carbon mass in the capacitor. (Reprinted from *Electrochimica Acta*, 53, Fernández, J. A. et al., EDLC performance of carbide-derived carbons in aprotic and acidic electrolytes, 7111–7116, Copyright 2008, with permission from Elsevier.)

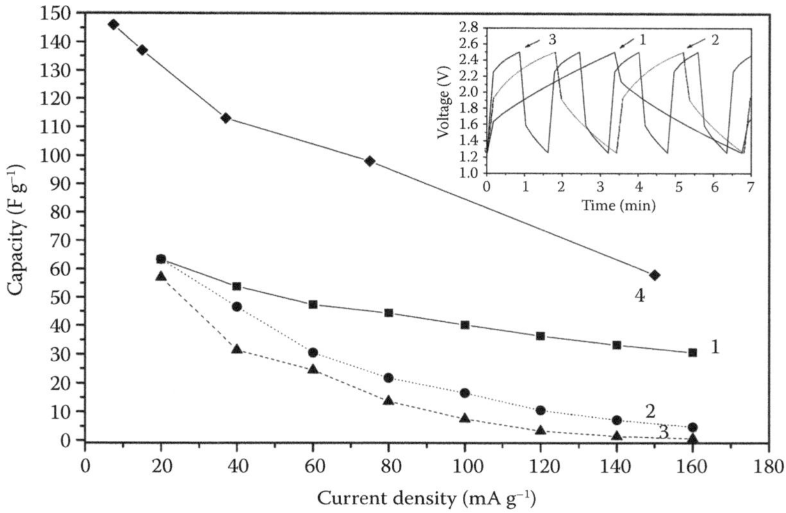

FIGURE 2.27 Dependence of capacity on discharge current for symmetric cells with carbon CF-50 and different electrolytes—Et_4NBF_4–PC (*curve 1*), $LiPF_6$–DMC/EC (*curve 2*) and $LiBF_4$–PC (*curve 3*), and for asymmetric cell CF-50H//composite CF-50–$Li_4Ti_5O_{12}$ with electrolyte $LiPF_6$–DMC/EC (*curve 4*). The *inset*—charge/discharge voltage profiles for symmetric cells at current rate 100 mA g^{-1} after 1050 cycles. (With kind permission from Springer Science+Business Media: *Journal of Solid State Electrochemistry*, Synthesis and electrochemical properties of activated carbons and $Li_4Ti_5O_{12}$ as electrode materials for supercapacitors, 17, 2013, 2101–2108, Mladenov, M. et al.)

As mentioned in Section 1.3.1, the specific capacitance of EDLCs is dependent not only on the specific surface area but also the pore size and its size distribution within the carbon materials. The pore accessibility is associated with the nature of the organic electrolytes such as the sizes of cation and anion species, and the interaction between ions and solvents. The presence of pores with very small size in the carbons may increase the specific surface area, but it can also limit the accessibility of the ions in the electrolytes especially for the larger organic ions, which has a negative effect on improving the specific capacitance. For example, Xu et al. [323] found that for poly(vinylidene fluoride)-derived porous carbons with a very narrow pore size, distribution peaked at 0.55 nm, and it had a high specific capacitance (264 F g^{-1}) in 6 M KOH electrolyte. However, the porous carbons showed a very low specific capacitance of less than 7 F g^{-1} in the 1 M $TEABF_4$/PC, indicating these pores are too small to be accessible even for the desolvated ions in the organic electrolyte. Taking into account the sizes of the typical cation and anion species in the organic electrolytes are relatively larger and vary a great deal between different salts, it is very important to match the size of the electrolyte ions with the pore size of carbon materials in order to maximize the specific capacitance. Therefore, considerable research has focused on the effect of the electrolyte ion size on the capacitance of carbons with different pore sizes, and the optimization of the matching between the electrolyte ion size and carbon pore size [311,316,324–327]. Besides, one should consider both the sizes of solvated ions and desolvated ions. As mentioned in Section 1.3, the possible desolvation or distortion of solvation shells of the ions when entering into the pores with the size close to the desolvated ion size could remarkably increase the specific capacitance [328]. Vix-Guterl et al. [329] also proposed that for $TEABF_4$ in the ACN, the solvated ions become desolvated to be adapted to the pore size when reaching the micropores.

It is worth noticing the interaction between the ions and the solvents in the organic electrolytes also has a profound effect on the performance of ESs based on porous carbon electrodes [330,331]. Decaux et al. [330] found for the $TEABF_4$ or LiTFSI in the ACN, the specific capacitance increased continuously as the average pore size of carbons decreased from 1.0 to 0.75 nm. On the contrary, in the case of LiTFSI in the EC–PC–DMC (1:1:3 by volume), the specific capacitance was low for carbons with pore sizes in the range 0.7–0.8 nm but was found to be increased with the increase of the average pore size. It was proposed that ions in the ACN solvents are relatively easy to be desolvated and can penetrate the micropores matching the bare ion size. However, ions are not completely desolvated in the alkylcarbonate solvents. The authors explained this by considering the alkylcarbonates solvents such as EC are more polarizable than ACN and thus tend to bind more strongly to Li^+ ions. This makes it difficult for the complete removal of the solvation shell. Lin et al. [331] reported the size of the TEA^+ cation and BF_4^- anion adsorbed in the carbon pores was also dependent on the solvent (i.e., ACN or PC), and the size of the adsorbed ion in the organic electrolytes was found to decrease in the following order: TEA^+ in PC > TEA^+ in ACN > BF_4^- in PC > BF_4^- in ACN. Furthermore, since the dimensions of the cations and anions are generally different, the matching between the pore size of carbons and electrolyte ion size should be considered for the positive and negative electrodes, respectively [325]. Using EIS, Segalini et al. [332] found the effective size of the TEA^+ was larger than that of the BF_4^- in the ACN or PC, and the effective ion size was larger in the PC than in the ACN. It was thus found for the 1.5 M $TEABF_4$ in the ACN-based electrolyte, the best performances would be achieved for a negative electrode with 0.76 nm pore size and a positive electrode with 0.68 nm pore size. Apart from the specific capacitance, other properties such as the ESR, charging/discharging rate, and power density should also be considered when optimizing the pore structures and designing the electrolytes. For example, Lin et al. [331] showed the pore sizes of carbon electrodes should be optimized according to whether (a) the maximum energy storage (capacitance) or (b) charging/discharging rate or (c) high cycling stability, were required.

Recently, modeling and simulations of the EDL in organic electrolytes have received increasing attention, thus providing useful guidance for the design of EDLCs from a microscopic view [333]. Feng et al. [334] investigated the solvation of ions, and the EDL structure and capacitance in the 1.2 M $TEABF_4$/ACN electrolyte using molecular dynamics (MD) simulations complemented with density functional theory (DFT) calculations. It was found that the solvation of TEA^+ and BF_4^- ions in the ACN was much weaker than that of typical small inorganic ions in aqueous electrolytes owing to the large size, charge delocalization, and irregular shape (in the case of TEA^+ cation) of the ions. As shown in Figure 2.28 [334], distinct counter-ion concentration peaks were observed near the charged electrodes. These peaks

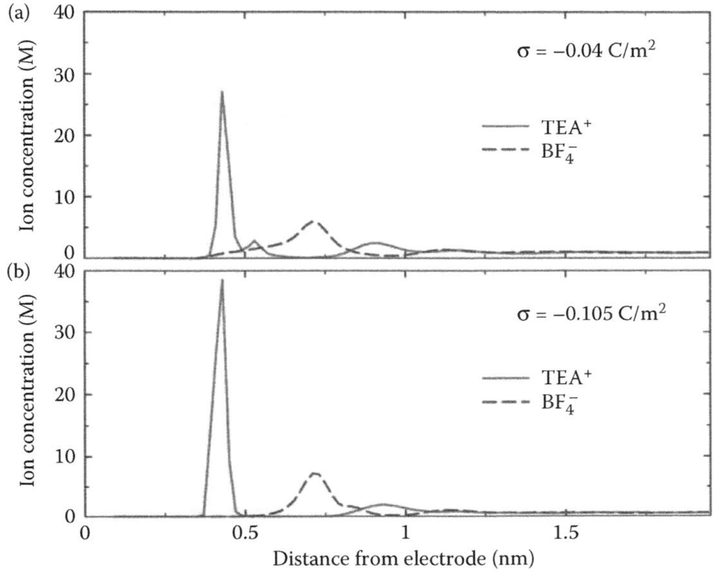

FIGURE 2.28 Concentration profiles of TEA^+ cations and BF_4^- anions near electrodes with charge densities of (a) −0.04 C m^{-2} and (b) −0.105 C m^{-2}. (Feng, G. et al. 2010. Structure and dynamics of electrical double layers in organic electrolytes. *Physical Chemistry Chemical Physics* 12:5468–5479. Reproduced by permission of The Royal Society of Chemistry.)

could not be explained by the Helmholtz model or the combination of the Helmholtz and Poisson–Boltzmann models because the number of counter-ions adsorbed on the electrode exceeded the electron charge on the electrode. The EDL capacitance increased from 0.065 to 0.092 F m^{-2} with the increase of the electrode charge density from −0.105 to 0.105 C m^{-2}, which agreed with the experimental measurements. Takana et al. [335] investigated the effect of the $TEABF_4$ addition on the structure of the PC in slit-shape carbon nanopores of ACF using synchrotron XRD and Monte Carlo (MC) simulations. As shown in Figure 2.29, in the absence of $TEABF_4$, PC molecules were closely packed in the pore and the distribution along the pore width was quite uniform [335]. On the other hand, the addition of the salts ($TEABF_4$) promoted a distinct ordering of the PC molecules along the pore walls. Similarly, Fukano et al. [336] found the vertical alignment of PC molecules interacting with TEA^+ and BF_4^- which were confined in the slit pores with an extremely narrow width of about 0.7 nm. Based on the XRD and electron radial distribution function, the nearest PC–PC distance was found to be 0.05–0.06 nm, which is shorter than that in the bulk solution. Ohba and Kaneko [337] investigated the structures and stabilities of TEA^+ and BF_4^- ions in the PC solvent in the noncharged carbon nanopores by MC simulations. Figure 2.30 shows the simulated adsorbed molecular distributions of TEA^+, BF_4^-, and PC against the pore walls [337]. TEA^+ and PC molecules basically formed mono, double, triple, and quadruple layers in the 0.7, 1.0, 1.5, and 2.0 nm nanopores, respectively, while BF_4^- molecules had more layers due to its smaller

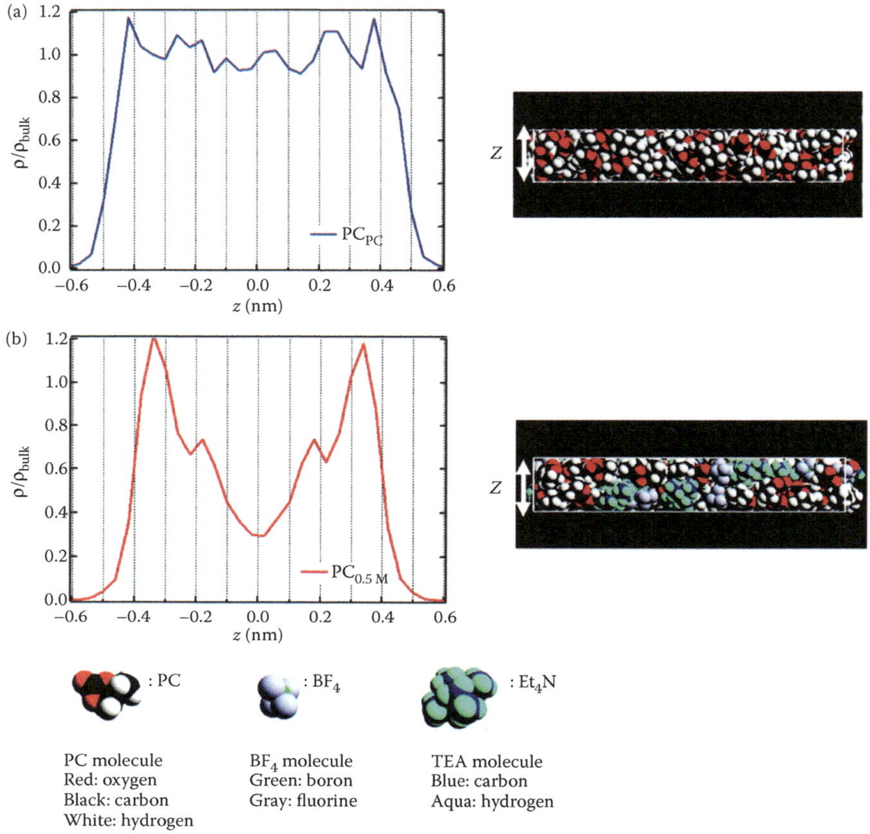

FIGURE 2.29 Snapshots and distribution of PC molecules and electrolytes along the pore width. (a) Pure PC and (b) Et_4NBF_4–PC solution corresponding to 0.5 M bulk solution. The coordinate z along the pore width is shown by an arrow. (Reprinted with permission from Takana, A. et al. 2010. Effect of a quaternary ammonium salt on propylene carbonate structure in slit-shape carbon nanopores. *Journal of the American Chemical Society* 132:2112–2113. Copyright 2010 American Chemical Society.)

molecular size. It was also found the solvation numbers of TEA^+ and BF_4^- were 14.5 and 9.0 in the bulk electrolyte, respectively, and were slightly decreased in the 1.5 and 2.0 nm nanopores. The solvation numbers of TEA^+ and BF_4^- decreased to 12 and 7, respectively, in the 1.0 nm nanopores and significantly decreased to 7 and 5, respectively, in the 0.6–0.8 nm nanopores. Desolvation reduced the stabilized energy of the ions in the nanopores and TEA^+ became unstable in the nanopores narrower than 0.7 nm owing to its larger molecular size while BF_4^- and PC were still stabilized. The nanopore potential fields provided a compensation for the stabilization loss. It was finally concluded the most suitable nanopore widths were 0.7 – 1.2 nm for TEA^+ and 0.6 – 0.9 nm for BF_4^-. Jiang et al. [338] used DFT to investigate the dependence of the surface area-normalized capacitance on the pore size from 0.5 to 3.0 nm for

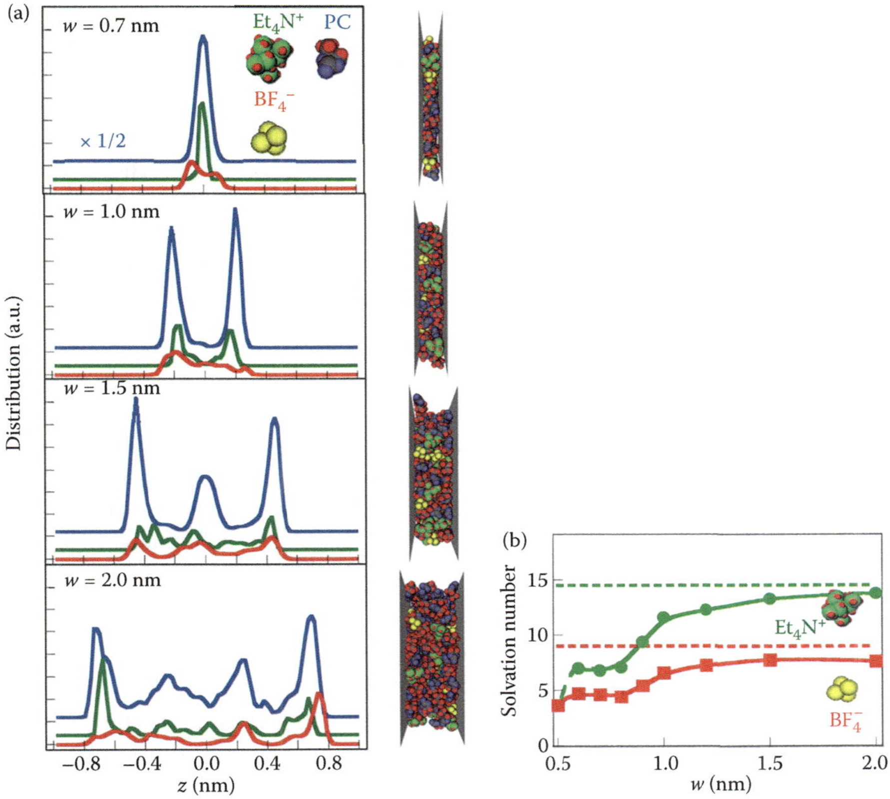

FIGURE 2.30 (a) Molecular distribution of Et_4N^+, BF_4^-, and PC in the perpendicular direction against a pore wall for w = 0.7–2.0 nm (left), and snapshots (right). (b) Solvation numbers of Et_4N^+ and BF_4^- as a function of nanopore width. Dashed lines represent solvation numbers in the bulk. (Reprinted with permission from Ohba, T., and K. Kaneko. 2013. Competition of desolvation and stabilization of organic electrolytes in extremely narrow nanopores. *Journal of Physical Chemistry C* 117:17092–17098. Copyright 2013 American Chemical Society.)

a 1.5 M $TEABF_4$/ACN electrolyte. Figure 2.31a shows an important example concerning the relationship between the capacitance and the pore width at a surface potential of 1.5 V. A sharp increase in the capacitance was observed at the 0.525 nm pore, which was accompanied by a significant decrease in the density of the solvent. However, the peak capacitance was only about 17% higher than that for the larger pores, and the degree of increase was much lower than that suggested by the previous experiment [328]. Besides, it is interesting to note the capacitance became relatively constant for the pores larger than 1.0 nm. The authors attributed this to the EDL formation that consists of both counterions and highly organized solvent molecules, as shown in Figure 2.31b.

Besides the modeling and simulations for EDLs in the nanopores, other morphologies of carbon electrodes have also been studied [339,340]. Wang et al. [339]

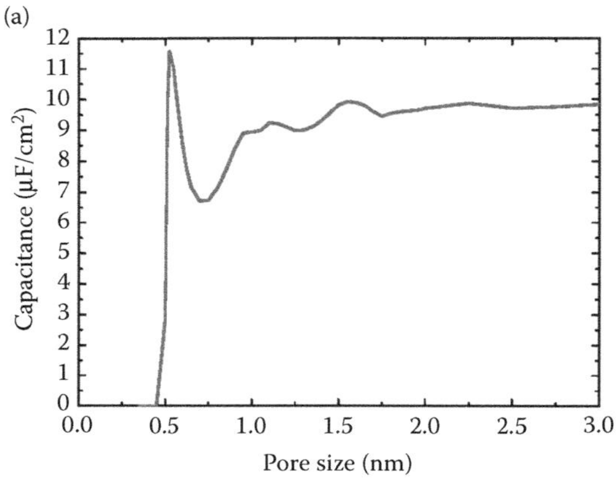

FIGURE 2.31 (a) Capacitance of a nanoporous electrode as a function of the pore size for a model system at a surface potential of 1.5 V, (b) approximate structure of the ions and the solvent molecules inside the 1.5 nm slit pore with a surface potential of 1.5 V according to the density profiles in (a) but not in proportion. (Reprinted with permission from Jiang, D. E. et al. 2012. Solvent effect on the pore-size dependence of an organic electrolyte supercapacitor. *Journal of Physical Chemistry Letters* 3:1727–1731. Copyright 2012 American Chemical Society.)

numerically simulated the EDLCs with mesoporous electrodes consisting of closely packed monodispersed mesoporous carbon spheres in the $TEABF_4$/PC electrolyte. The authors took into account the electrode packing morphology (i.e., simple cubic, body-, and face-centered cubic, as shown in Figure 2.32) and the effect of the local electric field on electrolyte dielectric permittivity. It was demonstrated that the predicted Stern and diffuse layer capacitances of EDLCs were strongly dependent on the field-dependent permittivity. For a given packing of the carbon spheres, the capacitance of the diffuse layer increased with the increase of the sphere diameter due to the increased electric field at the electrode surface. It was also found for sphere diameters larger than 40 nm, the packing morphology had no significant influence on the capacitance of the diffuse layer. Whereas, for sphere diameters smaller than 40 nm, simple cubic packing showed a larger diffuse layer capacitance compared to the body- and face-centered cubic packings, due to its larger electrolyte volume fraction, leading to a larger electric field at the electrode surface [339]. Yang et al. [340] simulated the EDLCs based on carbon nanotube (CNT) forests in the $TEABF_4$/PC electrolyte. The capacitance of the single electrode was determined to be 80 F g^{-1} by the computations, which agreed with the experimental data.

A number of experimental works concerning the electrolyte/electrode interactions were also carried out in parallel, providing a deeper understanding of the behavior of the electrolyte ions in the electrodes especially under working conditions [335,341–348]. Levi et al. [344] demonstrated the application of the QCM as a gravimetric probe to measure ionic fluxes during the charging/discharging process in the highly porous carbon electrodes. This study revealed the ionic fluxes strongly depended on the electrode's point of zero change (PZC), pore width, ion size, and cycling conditions such as the polarization amplitude and charging/discharging depth

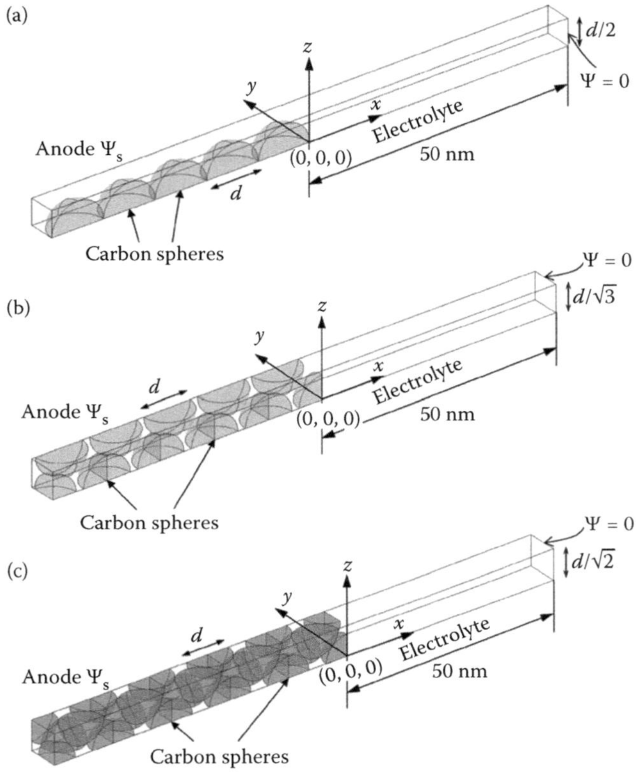

FIGURE 2.32 Schematic, dimensions, and coordinate system of the computational domain simulated for carbon spheres with (a) simple cubic, (b) body-centered cubic, and (c) face-centered cubic packings. Shaded areas represent carbon spheres of diameter *d*. Five unit cells are shown here for illustration purposes. (Reprinted from *Electrochimica Acta*, 56, Wang, H. N., J. Varghese, and L. Pilon, Simulation of electric double layer capacitors with mesoporous electrodes: Effects of morphology and electrolyte permittivity, 6189–6197, Copyright 2011, with permission from Elsevier.)

(Figure 2.33). Bonhomme et al. [349] applied confocal Raman microspectrometry to *in situ* characterize the charging and discharging processes occurring in an EDLC with a 1 M $TEABF_4$/PC electrolyte. The intensity variations of the G and D bands in the Raman spectra were related with electronic conductivity changes of the carbon electrodes. The salt concentration was simultaneously measured at a distance of about 10 μm from the electrode and significant ionic depletions were observed during the charging/discharging process (Figure 2.34). It was also found that the EDL interface at the two electrodes did not work in a symmetrical way [349]. Nuclear magnetic resonance (NMR) spectroscopy can provide molecular-level, quantitative, and element selective information concerning the local structure and dynamics of the ions in the electrolyte and at the electrolyte/electrode interface [343]. Furthermore, the element selective feature of NMR allows the separate observation of different

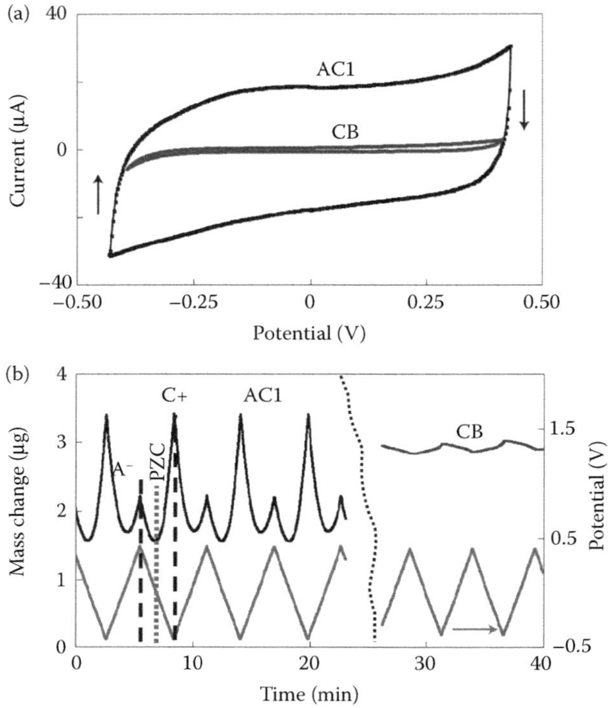

FIGURE 2.33 (a) CV curves measured at a scan rate of 5 mV s^{-1}. Ag/AgCl in aqueous 0.5 M CsCl was used as the reference electrode (the potentials on the CV curves in this figure were rescaled with respect to the immersion potential of the fresh electrode). (b) Simultaneously measured QCM responses. The lower straight lines relate to the potential-time dependencies; the upper left and the upper right curves represent the related mass changes of the cations and anions during potentiodynamic cycling of the electrodes. The dashed straight lines mark the mass change peaks of the cation (C^+) and the anion (A^-) at the negative and positive vertex potentials, respectively. The position of the PZC is shown by the dotted red line. CB: carbon black. (Reprinted by permission from Macmillan Publishers Ltd. *Nature Materials* Levi, M. D. et al. 2009. Application of a quartz-crystal microbalance to measure ionic fluxes in microporous carbons for energy storage. 8:872–875, copyright 2009.)

chemical species (e.g., cations and anions) in the ESs [342,346]. For example, by using ^{1}H and ^{11}B imaging, the anion and cation of commercial TEABF$_4$/ACN electrolyte can be probed individually [350]. The *in situ* NMR study can also provide information at the electrode/electrolyte interface in an ES under the working condition while the *ex situ* NMR approach has the advantage of high resolution provided by the magic angle spinning (MAS) [351]. Lee et al. [341] used the solid-state ^{11}B NMR spectroscopy to study the behavior of the BF_4^- anion in 1 M TEABF$_4$/PC electrolyte for an EDLC with AC electrodes. ^{11}B MAS-NMR distinguished two or three kinds of BF_4^- species depending on the type of ACs. Two of these species were found to be located on the outer surface and in the large pores of carbons while the third species was observed only in very small pores. Wang et al. [343] applied *in situ*

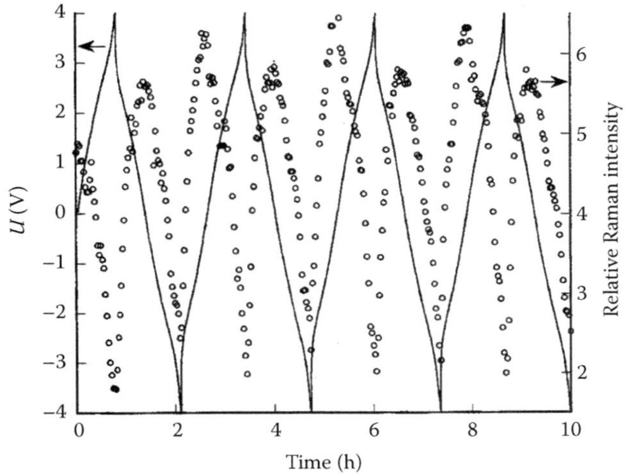

FIGURE 2.34 Galvanostatic cycling between +4 and −4 V at $I = \pm 1$ mA cm^{-2}. The solid curves represent the potential variation and the circles the total Raman intensity of the carbon spectrum integrated between 1800 and 1000 cm^{-1}. (Reproduced by permission of The Electrochemical Society from Bonhomme, F., J. C. Lassègues, and L. Servant. 2001. *Journal of the Electrochemical Society* 148:E450–E458.)

^{11}B NMR spectroscopy to investigate the behavior of BF_4^- anions (1.5 M $TEABF_4$/ACN) in a highly porous and high surface area carbon during the charging/discharging process. Most of the BF_4^- anions in the electrolyte were found to be located nearby the carbon, due to (a) the high internal surface area of the investigated carbons (YP-17); (b) the relatively nonpolar solvent; and (c) the limited volume of electrolyte. When the carbon electrode was charged from −2 to 2 V, the loss of strongly bound anions in the negatively charged electrode and the movement of anions to the positively charged electrode were observed. Deschamps et al. [345] used *ex situ* ^{13}C and ^{11}B MAS-NMR to investigate the electrolyte ($TEABF_4$/ACN) organization in the ES electrode. For AC powders soaked with the electrolyte, two distinct adsorption sites (type I and II sites) were detected by NMR. The type I sites were associated with the species adsorbed between graphene layers corresponding to a −6 ppm diamagnetic shift [352,353] and the type II sites were associated with adsorbed ACN and TEA^+ with a −3 ppm shift [354]. Two-dimensional ^{13}C and ^{11}B NMR exchange spectra that showed electrolyte species were exchanged between each site, that is, type I ⇆ free electrolyte molecules ⇆ type II exchange. This indicates the type II sites are not located between the graphene interlayer space and larger pores, but are related to sites of different natures, such as rings made with five or seven carbons or adsorption sites perpendicular to graphene planes. It was also found a large excess of TEA^+ ions were adsorbed between graphene layers compared to the BF_4^- ions. When the ES was charged, anions were substituted by the cations in the negative electrode while cations were substituted by anions in the positive electrode, and ACN molecules were expelled from the adsorption sites at the negative electrode. Based on solid-state ^1H, ^{11}B, and ^{13}C MAS-NMR spectroscopy, Borchardt et al. [347] investigated different porous carbon materials of well-defined, variable pore size in 1 M $TEABF_4$/ACN or pure ACN electrolyte. Only one single NMR signal was detected for all nuclei located at molecules adsorbed inside the pores, indicating the adsorbed molecules move and exchange rapidly inside the pores and between different pore systems. Interestingly, the characteristic diamagnetic shift was found to be affected by the hybridization state of the carbon atoms. For carbon materials with sufficiently large pores, TEA^+ and BF_4^- molecules were closer to the pore walls compared to the ACN molecules. This study also supports the idea that the ions in the electrolyte do not exhibit an intact solvent shell in pores with a diameter smaller than 1 nm. Forse et al. [346] investigated the effect of pore size on the electrolyte ion adsorption for porous carbon with different pore size by multinuclear NMR. Two main environments inside the carbons could be identified. One was related with the ions adsorbed inside the pores and another corresponded to the ions outside the pores (i.e., in the spaces between carbon particles). In addition, it was found that careful choice of the sample orientation could improve the NMR spectral resolution. Wang et al. [342] applied *in situ* NMR spectroscopy to investigate the changes at the electrode/electrolyte interface in the working ESs, which could provide a fundamental insight into the charge storage mechanisms. For YP50F microporous carbon electrodes in 1.5 M $TEABF_4$/ACN, ^{19}F NMR spectra revealed two distinct charge storage regimes. When the cell voltage was below 0.75 V, the number of electrolyte anions changed little in the positive electrode while the anions population decreased in the negative electrode, indicating that the charging was achieved mainly through short-scale rearrangements and the ejection of co-ions from the micropores instead of the adsorption of counterions. On the contrary, at higher cell voltages above 0.75 V, the amount of adsorbed anions in the positive electrode increased dramatically whereas anions continued to be desorbed from the negative electrode, suggesting the involvement of both counterion adsorption and co-ion ejection in this regime. Boukhalfa et al. [348] utilized *in situ* SANS to investigate the electroadsorption of organic electrolyte ions in carbon pores of different sizes. The investigated electrolyte was 1 M $TEABF_4$/deuterated ACN. The ion adsorption was observed to be strongly enhanced in subnanometer pores under the applied potential, despite incomplete wetting by the electrolyte.

The electrochemical stability window of the organic electrolyte is a very important property since it determines the operative cell voltage, which in turn plays a significant role in the energy and power densities of the ESs. The electrochemical stability window of the organic electrolytes is dependent on several factors, such as the conducting salts including the cation and anion nature, solvents, interaction between electrolyte ions and solvents, and impurities, especially the trace amount of water. This is discussed in more detail for specific conducting salts and solvents in the following sections. The electrochemical stability window of various organic electrolytes consisting of different quaternary onium salts (e.g., triethylmethylammonium, ethylmethylpyrrolidinium, and tetramethylenepyrrolidinium tetrafluoroborate salts) and different solvents (e.g., PC, ACN, and γ-butyrolactone [GBL]) was investigated by Ue et al. [298] on a glassy carbon electrode. It was pointed out the PC-based electrolyte containing tetraethylammonium cation and tetrafluoroborate anion showed a wide potential window and good durability against hydrolysis. In addition to the experimental methods to determine the electrochemical stability window of the organic

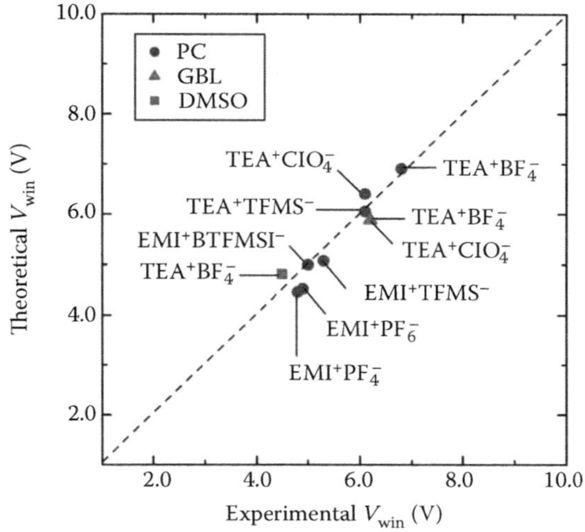

FIGURE 2.35 Comparison of the electrochemical potential windows of 10 organic electrolytes determined by CV measurements (Experimental V_{win}) and the sum of calculated redox potentials of the most reactive species in each electrolyte (Theoretical V_{win}). The different solvent species are indicated by circles (PC), triangles (GBL), and squares (DMSO). (Reproduced by permission of The Electrochemical Society from Maeshima, H. et al. 2014. *Journal of the Electrochemical Society* 161:G7–G14.)

electrolytes, computational methods for quantitatively predicting the potential window based on *ab initio* calculations have been proposed [355,356]. By using *ab initio* molecular orbital calculations and taking into account the intermolecular interactions (cation–anion interactions and solvent effects), Maeshima et al. [356] predicted the potential windows of 10 organic electrolytes. The calculated values of the electrochemical potential windows were found to agree well with those obtained by the experimental methods (CV measurements on glassy carbon electrodes), as shown in Figure 2.35. It is worth noticing the experimental measurements for determining the electrochemical window were generally carried out on smooth platinum (Pt) or glassy carbon electrode, which is quite different from actual electrodes such as porous carbons of EDLCs. The obtained potential window for a Pt or glassy carbon electrode is usually much higher than that obtained for EDLCs with the same electrolyte.

In addition, impurities in the organic electrolytes can have a strong influence on the electrochemical potential window of the electrolytes [94,357,358]. For example, a very small amount of the water in the 1 M $TEABF_4$/ACN could significantly decrease the electrochemical stability of the organic electrolyte [359].

For an ideal EDLC, the charge storage is realized entirely through an electrostatic process occurring at the electrode/electrolyte surface. However, it has been reported the intercalation of electrolyte ions into the carbon electrodes can also occur in the ESs especially at high cell voltages, which is suggested to result in the degradation of carbon electrodes and thus reduce the lifetime of the ESs [360–367]. Furthermore, electrolyte ion intercalation into the AC electrodes has also been reported as a possible method to further enhance the operational voltage and charge storage capability of ESs [361]. Therefore, investigations on the electrolyte ion intercalation process have received attention not only to find its effect on the degradation of the electrode materials but also to develop this as a possible way to further increase the charge storage performance of the ESs. The intercalation of cations and anions from the organic electrolytes into the graphite or AC electrodes have been investigated by *in situ* AFM [364], *in situ* Raman microscopy [360], *in situ* XRD [361], and XPS [368]. Campana et al. [364] used an *in situ* AFM to characterize the surface morphology of a highly oriented pyrolytic graphite (HOPG) negative electrode in the 1 M $TEABF_4$/PC electrolyte. Cation intercalation was found to begin at 1.2 V versus Li/Li^+ and led to a significant increase of the step height of the HOPG. A surface film on the electrode was also detected, which started to grow preferentially at the HOPG steps at about 0.9 V versus Li/Li^+. Through an *in situ* Raman study, the intercalation/deintercalation of the TEA^+ and BF_4^- into and out of the graphite electrode was reported by Hardwick et al. [360]. The TEA^+ intercalation into the graphite negative electrode was found to occur at 1.0 V versus Li/Li^+ and BF_4^- started to insert into the graphite positive electrode above 4.3 V versus Li/Li^+ during CV measurements. The intercalation of TEA^+ and BF_4^- in the ACN and PC solvents into the graphite electrodes during CV tests was also identified by *in situ* XRD reported by Ruch et al. [361]. The TEA^+ intercalation resulted in the disappearance of the graphite (0 0 2) diffraction peak, indicating the accumulation of cations between the graphene layers. Foelske-Schmitz et al. [368] used XPS to analyze the HOPG electrodes which were electrochemically treated in the 1 M $TEABF_4$/PC. The binding energies for the matrix and the intercalating species increased for the negatively polarized electrodes whereas they were decreased for the positively polarized electrodes. The authors attributed this to the shift of the Fermi level of the intercalation compound. Besides the graphite electrodes, intercalation induced dimensional change of the AC electrodes during the charging/discharging process was also observed in the 1 M $TEABF_4$/ACN electrolyte below a cell voltage of 2 V, as reported by Hahn et al. [362].

The ionic conductivity of the organic electrolyte is also an important concern, which has a strong influence on the ESR and the power performance of ESs. The ionic conductivity is generally not an issue for aqueous electrolytes especially for the H_2SO_4 or KOH electrolyte. However, it is a problem for organic electrolytes in most cases. Organic electrolytes generally have a significantly lower ionic conductivity than aqueous electrolytes. For instance, the commonly used 1 M $TEABF_4$/ACN electrolyte has the ionic conductivity of 0.06 S cm^{-1}, which is more than one magnitude lower than the ionic conductivity of the 30 wt.% H_2SO_4 electrolyte (0.8 S cm^{-1} at 25°C) [369]. The low conductivity of the organic electrolyte can lead to a higher ESR of ESs compared to ESs using aqueous electrolytes. For instance, Vaquero et al. [281] found the ESR of the carbon-based ESs with 0.5 M $TBAPF_6$/ACN electrolyte ranged from 7 to 12 Ω cm^{-2} depending on the type of carbons, whereas the ESR was only 0.75 Ω cm^{-2} in the 0.5 M H_2SO_4 electrolyte. The higher ESR value also limits the maximum power density of the ESs.

Since the ESs generally operate under a wide temperature range, the influence of the electrolyte on the ES performance at different temperatures is an important

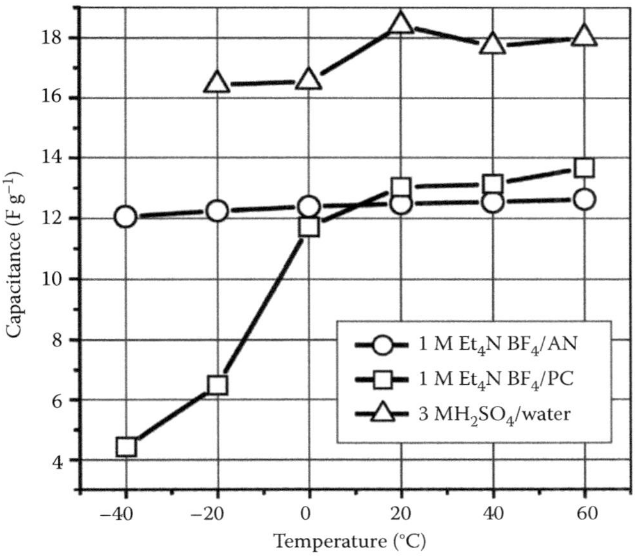

FIGURE 2.36 Capacitance as a function of temperature measured at 0.1 Hz and $U = 0.0$ V. Measurements performed in coin-type laboratory cell. *Electrolytes:* 3 M H_2SO_4 (triangles); 1 M Et_4NBF_4 in AN (circles); and 1 M Et_4NBF_4 in PC (squares). (Reprinted from *Journal of Power Sources*, 154, Kötz, R., M. Hahn, and R. Gallay, Temperature behavior and impedance fundamentals of supercapacitors, 550–555, Copyright 2006, with permission from Elsevier.)

issue. Kötz et al. [366] reported the specific capacitance of an EDLC with 1 M $TEABF_4$/ PC decreased significantly as the temperature decreased from 60°C to −40°C whereas the specific capacitance did not change obviously in the $TEABF_4$/ ACN electrolyte (Figure 2.36). For both electrolytes, the ESR of the EDLC increased with the decreasing temperature, but the ESR increased more slowly for the ACN-based electrolyte compared to the PC-based electrolyte. Similar results that the ESR of the ESs increased with the decrease of the temperature were also reported for commercial 2700 and 3700 F ESs with organic electrolytes [370]. The lower temperature limit of typical organic electrolyte-based ESs is −40°C [69,94]. Some studies have focused on the development of organic electrolytes with the temperature limit lower than −40°C so as to further widen the usages for some even lower temperature operations such as space-rated electronics [94,371]. This can be achieved through the rational selection and design of conducting salts, solvents, and/or additives, which is further discussed in the following sections.

As for the self-discharge process of the ESs concerned, the used organic electrolyte also has an influence in this process [358,372]. The presence of an impurity such as water in the organic electrolyte has been identified as playing an important role in the self-discharge process [358]. Zhang et al. [372] reported that self-discharge mechanisms depended on the type of the organic electrolyte. The authors found that for the ESs with AC fabric electrodes, the self-discharge behavior was controlled by the diffusion process in the 1 M $LiPF_6$/EC–DEC (1:1 by volume) electrolyte while the self-discharge was governed by the dual mechanism (potential driving model and diffusion model) in the 1 M $TEABF_4$/PC electrolyte. The achieved energy retention was more than 70% in the $TEABF_4$/PC electrolyte after 36 h, which was much higher than that obtained in the $LiPF_6$/EC–DEC electrolyte.

2.3.1.2 Organic Electrolytes for Pseudocapacitors

Apart from the EDLCs, organic electrolytes have been used for pseudocapacitors based on pseudocapacitive materials such as metal oxides, ECPs, and composite materials. In order to promote the ion intercalation/deintercalation, most of the organic electrolytes used for the pseudocapacitive materials contain lithium ions in the electrolyte due to the smaller bare ion size. Typically used salts in these organic electrolytes include $LiClO_4$ [373,374], $LiPF_6$ [375–382], LiTFSI, etc. The commonly used organic solvents are PC [373,374], ACN, or a mixture of different solvents such as EC–DEC [377], EC–DMC [375,383–385], EC–ethyl methyl carbonate (EMC) [378], EC–DMC–EMC [380], and EC–DMC–diethyl carbonate (DEC) [376,379,381,382,386]. It can be seen that most of these organic electrolytes have also been extensively used in LIBs. A number of transition metal oxides, such as MnO_2 [387], V_2O_5 [388,389], SnO_2 [380], MoO_3 [373], and TiO_2 [379,390], have been studied as electrode materials owing to their Li intercalation capacity. Several studies found the obtained specific capacitance of the V_2O_5-based electrodes in the Li-containing organic electrolyte (e.g., $LiClO_4$/PC and LiTFSI/ACN) was much higher than that obtained in aqueous electrolytes (e.g., KCl) [388,389]. Similar results have been reported for MnO_2 electrodes, which showed higher specific capacitance in the 1 M $LiPF_6$/PC–DMC (455 F g^{-1}) electrolyte compared to that obtained in the 1 M NaOH electrolyte (342 F g^{-1}) [387]. This is possibly due to the Li intercalation-enhanced pseudocapacitance. Due to the larger potential window, the organic electrolyte-based pseudocapacitors generally exhibit a higher energy density compared to the aqueous electrolyte-based pseudocapacitors. For example, Nam et al. [391] reported the specific energy density of a pseudocapacitor based on the MnO_2–CNT electrodes in the 1 M $LiClO_4$/PC electrolyte was about six times higher than that in the 1 M KCl electrolyte, due to the larger potential window achieved in the organic electrolyte. However, the $LiClO_4$/PC electrolyte resulted in a lower cycling stability compared to the KCl electrolyte. In addition, ECPs, such as the poly (3,4-ethylenedioxythiophene) (PEDOT) [392,393], PANI [394,395], poly[3-(3,4-difluorophenyl) thiophene] (MPFPT) [396], and poly(3,4-propylenedioxypyrrole) (PProDOP) [397], have also been investigated as electrode materials. Although ECPs have been considered as promising pseudocapacitive electrode materials due to their light weight, low cost, mechanical flexibility, ease of processing, and relatively fast ion insertion/extraction (doping/dedoping) process, they generally suffer from a poor cycling stability [2]. It has been found water plays an important role in the low cycling stability of the ECPs in aqueous electrolytes [394,398,399]. For example, Planes et al. [399] reported the nucleophilic attack of water to the imine nitrogen of PANI could break the polymer chain and produce ammonium ion and benzoquinone as soluble products. This resulted in the degradation of the PANI, leading to a poor cycling stability in the aqueous electrolytes. Therefore, the application of appropriate water-free organic electrolytes may minimize the degradation of the ECPs and thus improve the cycling stability of the ECP-based ESs. For example, Zhang et al. [394] reported the specific capacitance of the PANI-based electrodes maintained stable for 30,000 cycles in the 0.5 M $LiClO_4$/PC electrolyte up to upper potential limit of 1 V versus the saturated calomel reference electrode (SCE). This, together with the wider potential window, makes organic electrolytes very promising for ECP-based ESs.

2.3.1.3 Organic Electrolytes for Hybrid ESs

Asymmetric ESs with organic electrolytes are currently attracting increasing interest with the objective of significantly increasing their energy density. Since the pioneering studies on the organic electrolyte-based asymmetric ES reported by Amatucci et al. [400], a wide variety of organic electrolyte-based asymmetric ESs, such as asymmetric carbon//carbon (e.g., graphite//AC) [401–403], carbon//MnO_2 [391,404], carbon//TiO_2 [381,405], carbon//V_2O_5 [389], carbon//$Li_4Ti_5O_{12}$ [322,378,406–408], carbon//ECP [409,410], and asymmetric ECP//ECP [392,411–413], have been investigated. Table 2.8 shows some examples of the asymmetric ESs based on organic electrolytes, selected from the literature published in the last 2 years (2013–2014). Although these asymmetric ESs with organic electrolytes have lower specific capacitance compared to aqueous-based asymmetric ESs, they exhibit much higher energy density because of the much larger operating cell voltage (generally 3–4 V).

Among various asymmetric ESs, LICs have drawn particular interest [322,378, 405–408,414–416]. As mentioned in Section 1.3.3, typical LICs consist of an EDLC-type positive electrode such as AC and a LIB-type negative electrode for Li ion insertion/extraction in an organic electrolyte containing Li salt. Typical negative electrodes include carbon-based (mostly graphite) and lithium titanate ($Li_4Ti_5O_{12}$) electrodes. It can also be seen that most organic electrolytes for the LICs are composed of lithium salts (e.g., $LiPF_6$ and $LiClO_4$), and a mixtures of two or more solvents (e.g., EC–DMC). An electrolyte based on a single solvent for LICs has been very rare. In some cases, additives have also been used in the electrolyte [417]. Actually, to date, the compositions of the majority of reported LIC electrolytes are almost the same as LIB electrolytes, which have already been extensively reviewed elsewhere (e.g., Reference 418). Fuji Heavy Industries have announced a LIC consisting of pre-lithiated AC positive electrode and AC negative electrode [94,419]. Although extensive efforts have been paid to the development of electrode materials, the influence of the electrolytes on the performance of LICs has received limited attention [101,322,415,419]. Zheng [101] reported on asymmetric ESs based on AC and a lithium intercalation compound (i.e., $Li_xTi_5O_{12}$ or WO_2) using $LiPF_6$/EC–DMC as the electrolyte. The maximum energy density was largely determined by the ion concentration of the organic electrolyte and the low ion concentration limited the maximum energy density. It was thus pointed out that an effective way to increase the energy density of organic electrolyte-based ESs was to use an electrolyte with a high ion concentration and a large operating potential window. The electrolyte-dependent performance of the LICs has been found to be strongly related to ambient temperature [415]. The performance such as the energy and power densities of LICs using carbonate-based electrolytes has been found to deteriorate severely at low temperatures below 0°C [415]. This is different with ACN-based EDLCs, whose performance is relatively stable with the temperature range between –30°C and 40°C.

TABLE 2.8
Summary of Selected Literature Data on Organic Electrolyte-Based Asymmetric ESs Including LICs

Electrolyte	Negative Electrode	Positive Electrode	Capacitance ($F\ g^{-1}$)	Cell Voltage (V)	Energy Density ($Wh\ kg^{-1}$)	Power Density ($W\ kg^{-1}$)	Cycle Numbers	References
$LiPF_6$/EC–DMC	Activated graphitized carbon	AC–$Li_4Ti_5O_{12}$ oxide composite	150	2.6	–	–	600	[322]
1.5 M $TEMABF_4$/PC	Nonporous activated MCMBs	Graphitized carbon	–	4	~60	~30	700	[401]
1 M $LiPF_6$/EC–DMC (1:1)	A mesoporous Nb_2O_5–carbon nanocomposite	Commercial AC (MSP-20)	–	3.5	74	18,510	1000	[375]
1 M $TEABF_4$/PC	AC	PMT–MWCNTs nanocomposites	38.5	2.5	33.4	–	1000	[409]
1 M $LiPF_6$/EC–EMC (1:1)	$Li_4Ti_5O_{12}$	Partially graphitic AC	77	2.8	–	–	1000	[378]
1 M $LiPF_6$/EC–DEC–DMC (1:1:1)	TiO_2 mesoporous microspheres	AC	72.8	2.8	79.3	178.1	1000	[381]
1 M $LiPF_6$/EC–DMC (1:1)	Soft carbon	AC	–	3.8	80.6	1700	15,000	[414]
1 M $LiPF_6$/EC–DMC	$LiMnBO_3$ nanobead	PANI nanofibers	125	3	42	1500	30,000	[420]
1 M $LiPF_6$/EC–DMC	$TiNb_2O_7$	AC	–	3	43	3000	3000	[384]
1.2 M $LiPF_6$/EC–DEC (1:1)	Pre-lithiated mesocarbon microbeads	AC	102.87	4	92.3	5500	1000	[421]

(Continued)

TABLE 2.8 (*Continued*)
Summary of Selected Literature Data on Organic Electrolyte-Based Asymmetric ESs Including LICs

Electrolyte	Negative Electrode	Positive Electrode	Capacitance ($F\ g^{-1}$)	Cell Voltage (V)	Energy Density ($Wh\ kg^{-1}$)	Power Density ($W\ kg^{-1}$)	Cycle Numbers	References
1 M $LiPF_6$/EC–DMC (1:1)	AC	$LiNi_{0.4}Co_{0.6}O_2$ nanofibers	72.9	3	180.2	248	2000	[422]
1 M $LiPF_6$/EC–DMC	$Li_4Ti_5O_{12}$	High surface area porous carbon	~120	3	~55	500	2000	[423]
1 M $LiPF_6$/EC–DMC (1:1)	AC	$Li_{1.2}(Mn_{0.32}Ni_{0.32}Fe_{0.16})O_2$ nanoparticles	86	3	36	1670	1000	[424]
1 M $LiPF_6$/EC–DEC–DMC (1:1:1)	Fe_3O_4–graphene nanocomposite	3D graphene	–	4	147	150	1000	[386]
1 M $LiPF_6$/EC–DMC (1:1)	$Li_4Ti_5O_{12}$	Porous graphitic carbon	–	3	55	6474.7	–	[425]
1 M LiTFSI/ACN	V_2O_5 NWs–rGO	MnO_2 nanorods–rGO	36.9	2	15.4	436.5	50	[426]
1 M $LiPF_6$/EC–DMC (1:1)	AC	MgO–MWCNTs composite	66	3	30	220	35,000	[427]

Abbreviations: AC, activated carbon; NWs, nanowires; rGO, reduced graphene oxide; PMT, poly(3-methyl thiophene); PANI, polyaniline; and MWCNTs, multiwalled carbon nanotubes.

Smith et al. [415] also found the self-discharge rate of the LIC over a 72 h period started to increase significantly at higher temperatures above 40°C. LICs also have a disadvantage of relatively low rate capability due to the use of the battery-type negative electrode. Development of both new electrode materials and electrolytes will be required to solve this problem. Considering the safety problem of the LIBs is a major concern, the safety issue of the LICs must also be carefully addressed [415]. Furthermore, electrolyte decomposition and formation of a solid electrolyte interphase (SEI) on the negative electrode surface is also an important issue, which may affect long-term LIC stability and requires further investigations [406,414].

2.3.2 Investigation and Development of Organic Solvents

An ideal organic solvent should meet the following criteria: good solvation ability for a certain conducting salt; high electrochemical stability; low viscosity within the operating temperature range; and safety measured by nonflammability and nontoxicity [2]. To date, ACN and PC are the most frequently used organic solvents for EDLCs. As shown in Table 2.6, ACN has lower viscosity, lower melting point, and a higher dielectric constant compared to PC. This can contribute to a higher conductivity of the ACN-based electrolytes and thus a lower ESR and a higher power performance of the corresponding EDLCs—even at low temperatures—compared to EDLCs with PC-based electrolytes. Nevertheless, the toxicity and low flash point of the ACN are two critical drawbacks. Because of its high toxicity, the application of ACN in ESs has been prohibited in Japan [428–430]. Due to the higher flash point and lower toxicity of PC compared to ACN, there has been an increasing interest in the development of the PC-based electrolyte as an alternative. Besides, the PC-based electrolyte generally has a wider potential window compared to ACN-based electrolytes [8,199,298,431,432]. However, the conductivity of the PC-based electrolyte is generally lower than the ACN-based electrolyte, which has a negative effect on the power density and energy efficiency of PC-based EDLCs. This problem is more serious at low temperature due to the high viscosity of the PC. For example, Liu et al. [433] found the performance (e.g., specific energy and capacitance) of the EDLC with 1 M TEA/ACN electrolyte was nearly insensitive to the temperature between –30°C and 45°C, whereas the performance of the EDLC with 1 M TEA/PC electrolyte deteriorated significantly at low temperatures—especially at –30°C (Figure 2.37). Furthermore, as previously mentioned, the operating cell voltage of commercial ACN and PC-based EDLCs is limited to 2.5–2.8 V. Therefore, it is highly desirable to further improve the operative voltage and in turn increase the energy density of ESs. Based on the above considerations, extensive studies have focused on the development of other organic electrolytes with high performance for ESs.

2.3.2.1 Single Organic Solvents

For single organic solvents, several studies investigated GBL and compared the performance of GBL-based EDLCs with ACN- and PC-based EDLCs [434–437]. Ue et al. [298] found GBL-based electrolyte had an advantage of higher oxidative stability compared to the PC-based electrolyte (both with 0.65 M $TEABF_4$). However, the lower viscosity and ionic conductivity of the GBL-based electrolyte resulted in a lower power performance. Arulepp et al. [435] showed the power performance of

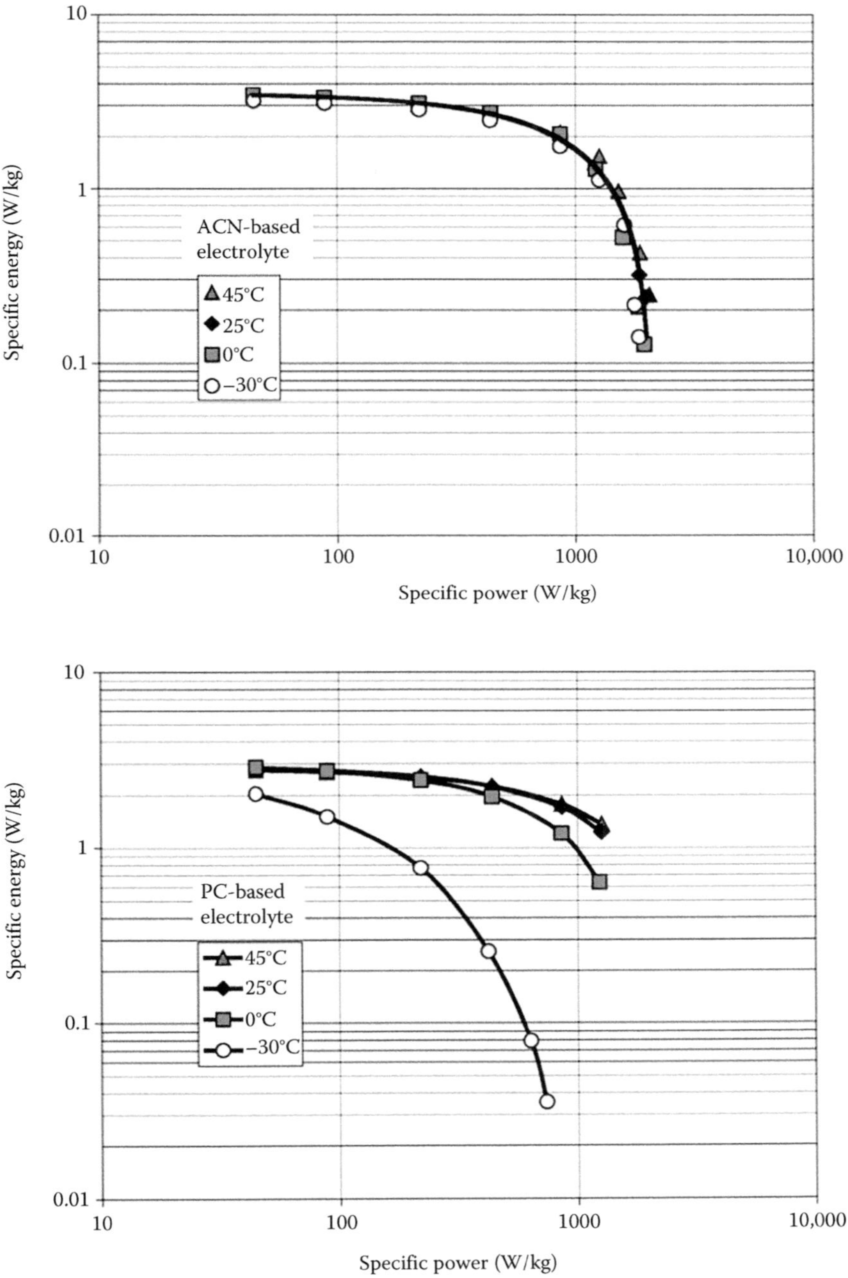

FIGURE 2.37 Ragone plots for ACN (top) and PC (bottom)-based capacitors at different temperatures. (Reprinted from *Journal of Power Sources*, 156, Liu, P., M. Verbrugge, and S. Soukiazian, Influence of temperature and electrolyte on the performance of activated-carbon supercapacitors, 712–718, Copyright 2006, with permission from Elsevier.)

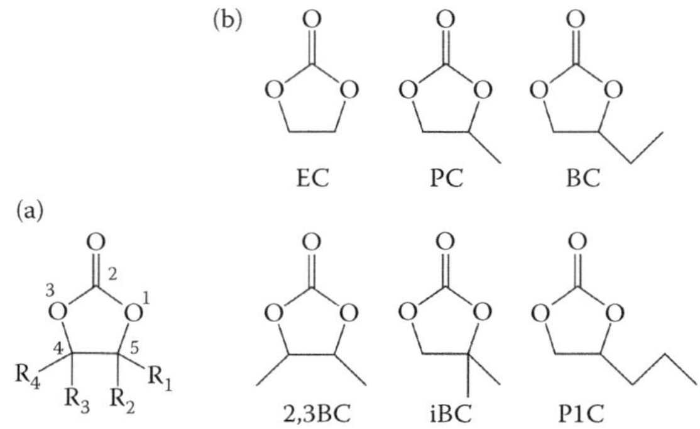

FIGURE 2.38 Structure of (a) fluorinated carbonates and (b) cyclic carbonates studied in Reference 438. (Reproduced by permission of The Electrochemical Society from Chiba, K. et al. 2011. *Journal of the Electrochemical Society* 158:A1320–A1327.)

EDLCs with various solvents (1 M $TEMABF_4$) increased in the order: PC < DMK (acetone) < GBL < ACN, which was in agreement with the order of the series resistance for various solvents: ACN < GBL < DMK < PC. Lust et al. [434] also found that for an EDLC with various solvents containing 1 M $TEMABF_4$ salt, relaxation time constants obtained by EIS increased in the order: ACN < GBL < PC, due to the decreasing of the solvent viscosity and conductivity from ACN to PC.

Inspired by the finding indicated by Ue et al. [298] that butylene carbonate (BC) possessed a higher oxidative stability than PC, Chiba et al. [438] postulated the substituent at the fourth and/or fifth position in the five-membered rings of carbonates (Figure 2.38a) would increase the resistance to oxidation. Thus, it may be used as a promising way to further increase the operating voltage of the organic electrolyte-based EDLCs. In this study, six different types of alkylated cyclic carbonates, including ethylene carbonate, PC, BC, 2,3BC, isobutylene carbonate (iBC), and pentylene carbonate (P1C) were investigated (Figure 2.38b). It was found that the alkyl substituent at the fourth position and/or the fifth position of the five-membered rings of the carbonates resulted in a higher cell voltage compared to the PC-based electrolytes, due to the protection from the OH^- or RO^- attack in the electrolytes. Particularly, for the 2,3BC whose fourth and fifth positions were substituted by methyl, it had a much higher withstand voltage up to 3.5 V (containing spiro-(1,1′)-bipyrrolidinium tetrafluoroborate [$SBPBF_4$] salt) compared to PC solvent-based EDLCs (2.5–2.7 V). Besides, the boiling point of the 2,3BC is high (243°C), which is comparable to the PC (242°C).

In addition to carbonate solvents, partially fluorinated organic solvents have also been developed in parallel. It has been reported the chemical and electrochemical stability of solvents could be remarkably improved by the substitution of C–H bonds with C–F bonds [439,440]. Nanbu et al. [441] found that although the ionic conductivity of fluoroethylene carbonate (FEC) electrolyte containing 1 M $TEABF_4$ was slightly lower than the PC-based electrolyte, the specific capacitance and rate performance achieved in the FEC-based electrolyte were improved compared to the PC-based electrolyte. It was also found the solubility of $TEMABF_4$ in the FEC could reach a high value to 3.0 M at 25°C, which was higher than in the PC solvent, due to the higher relative permittivity of FEC (about 107 at 25°C) [442] compared to the PC (64.92 at 25°C) [443]. Francke et al. [440] explored a highly fluorinated solvent (1,1,1,3,3,3-hexafluoropropan-2-ol [HFIP]), owing to its nonflammable nature. It was found that a higher specific capacitance of the EDLC could be achieved with the HFIP-based electrolyte compared to the PC-based electrolyte (both with 1 M $TEABF_4$ salt), possibly due to the low surface tension of HFIP that increased the wettability of AC electrodes. Because of its better electrochemical stability, the HFIP-based electrolyte for the AC electrodes was stable up to almost 2.0 V versus AC whereas ACN-based electrolyte degraded at a lower voltage of 1.4 V versus AC. However, the HFIP has a small liquid range (–3°C to 59°C) compared to the ACN and PC, which limits its working temperature range. Suzuki et al. [444] investigated the fluoroacetonitorile (FAN) solvent for EDLCs. Although the 1 M $TEABF_4$/FAN electrolyte exhibited a slightly lower ionic conductivity than the 1 M $TEABF_4$/ACN electrolyte, it showed much higher conductivity than the 1 M $TEABF_4$/PC electrolyte. Notably, the FAN-based EDLC showed a higher gravimetric capacitance compared to the ACN and PC-based EDLCs.

To further increase the operative voltage of EDLCs using organic electrolytes, other novel solvent systems such as sulfonated solvents have been developed. For example, it has been reported higher operation voltage can be achieved for sulfolane (SL)- and dimethyl sulfide-based electrolytes compared to carbonates-based electrolytes [430]. However, the melting point of the SL (28.5°C) is much higher than the PC (–49°C) and ACN (–45°C), which limits its practical use [445]. With the aim to decrease the melting point and viscosity of SL-based electrolytes, Chiba et al. [445] investigated the use of linear sulfones and in particular, eight different types of linear sulfones with relatively low molecular weights, including dimethyl sulfone (DMS), ethyl methyl sulfone (EMS), methyl isopropyl sulfone (MiPS), ethyl isopropyl sulfone (EiPS), ethyl isobutyl sulfone (EiBS), isopropyls-butyl sulfone (iPsBS), isopropyl isobutyl sulfone (iPiBS), and butyl isobutyl sulfone (BiBS) (Figure 2.39). It was found that the melting point and viscosity could be lowered dramatically when the molecular structure of SL was changed from the cyclic structure to the

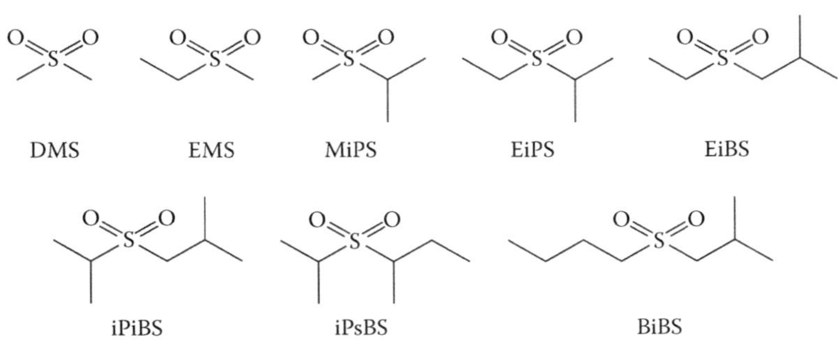

FIGURE 2.39 Structures of sulfones investigated in Reference 445. (Reproduced by permission of The Electrochemical Society from Chiba, K. et al. 2011. *Journal of the Electrochemical Society* 158:A872–A882.)

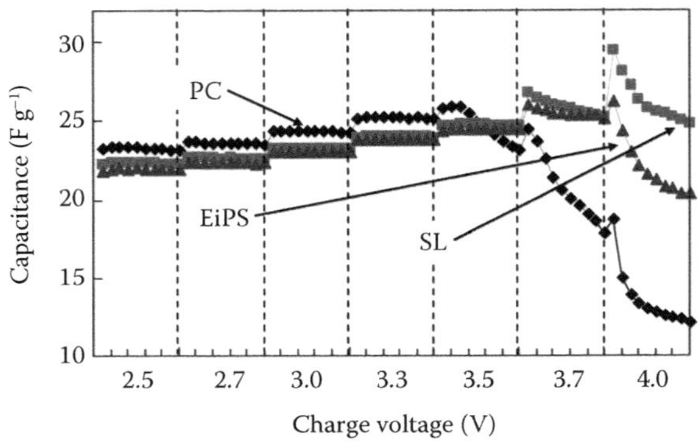

FIGURE 2.40 Withstand voltage of each solvent system. (Reproduced by permission of The Electrochemical Society from Chiba, K. et al. 2011. *Journal of the Electrochemical Society* 158:A872–A882.)

linear structure. For example, a relatively low melting point of the EiPS (−8°C) was achieved. It was interesting to note the melting point of the $SBPBF_4$/EiPS electrolyte was below −40°C whereas cyclic sulfones are easily solidified at such low temperature. Furthermore, a higher boiling point of 265°C was obtained for the EiPS compared to the PC (242°C). The iPiBS, iPsBS, and BiBS were found not to be practical alternatives since they exhibited low solubility (<1 M) for typical conducting salts including $TEABF_4$, $TEMABF_4$, and $SBPBF_4$. On the contrary, the EiPS was able to dissolve the conducting salt ($TEMABF_4$ and $SBPBF_4$) at 1.5 M or higher (at 25°C). Noticeably, the high electrochemical stability of the SL-based electrolyte was maintained. As shown in Figure 2.40 [445], EDLCs with $SBPBF_4$/EiPS showed a higher operating voltage (3.3–3.7 V) compared to the PC-based EDLCs (2.5–2.7 V), which could lead to an approximate doubling of the energy density compared to that with PC-based electrolytes. The high electrochemical stability of the EiPS was attributed to its low reactivity with H_2O. It was thus concluded the EiPS was the most suitable alternative for the solvent of EDLCs among the investigated linear sulfones. However, this electrolyte system still needs to be improved because EiPS has a much higher viscosity (5.6 mPa s at 25°C) compared to the PC (2.5 mPa s at 25°C) and ACN (0.3 mPa s at 25°C).

In an effort to overcome the disadvantages of ACN-based electrolytes such as relatively low electrochemical stability and low flash point, Brandt et al. [446] proposed the use of adiponitrile (ADN) as a possible alternative solvent. The flash point of the ADN is much higher than that of the ACN [446,447]. Brandt et al. [446] used 0.7 M $TEABF_4$/ADN as the electrolyte for the EDLCs. Due to the high electrochemical stability of the ADN, ADN-based EDLCs could achieve an operative cell voltage as high as 3.75 V. EDLCs with the 0.7 M $TEABF_4$/ADN electrolyte showed high capacitance retention over 35,000 cycles carried out at a cell voltage of 3.5 V. To increase the affinity of AC electrodes for ADN-based electrolytes, the same group treated the commercial ACs with CO_2 [447]. Based on the treated AC electrodes and 0.7 M $TEABF_4$/ADN electrolyte, the EDLC achieved a high cell voltage of 3.5–3.6 V. This

EDLC delivered an average energy and power of 45 and 1.5 kW kg^{-1}, respectively, at the operation voltage of 3.6 V and the current density of 1 A g^{-1}. In addition, a capacitance retention of more than 80% was achieved after 50,000 cycles. However, the solubility of the common conducting salts such as the $TEABF_4$ in the ADN solvent is lower than in the ACN. The ionic conductivity of the ADN-based electrolyte is also much lower than that of the ACN-based electrolytes, which limits the power performance of the ADN-based EDLCs.

It can be seen from the above discussion so far no ideal single solvent has been found that can replace commercially used solvents such as ACN in terms of overall performance such as viscosity, conductivity, and thermal stability. Although some promising solvents with higher electrochemical stability have been explored, they generally suffered from poorer conductivity, lower salt solubility, and higher viscosity compared to ACN-based electrolytes.

2.3.2.2 Mixtures of Solvents

In an attempt to solve the problems associated with single solvent-based electrolytes, solvents based on a mixture of different organic solvents or additives have been explored. As previously mentioned, PC has been considered a promising alternative for the ACN solvent, but PC-based electrolytes generally have lower viscosity, conductivity, and thermal stability compared to ACN-based electrolytes. Recently, a number of studies have focused on the development of PC-based solvent mixtures to improve performance [448]. Nambu et al. [449] investigated the effect of adding trimethylene carbonate (TMC) and EC to the PC solvents. It was found that the relative permittivity of the solvents decreased in the following order: PC–TMC (1:1 by molar ratio) > PC–EC (1:1 by molar ratio) > PC > BC. The viscosities of the electrolytes decreased in the order: PC–TMC > BC > PC–EC > PC. PC–TMC containing 1 M $TEABF_4$ showed a lower ionic conductivity compared to the PC and PC–EC electrolytes containing the same salt. However, the specific capacitance of an EDLC with 1 M $TEABF_4$/PC–TMC electrolyte was improved. The authors explained this by assuming the TMC might weaken solvation of electrolyte ions, resulting in a decreased distance between the electrode surface and the electrolyte ions in the EDLs. Due to the high dielectric constant of the EC and similar molecular structures between EC and PC, Tian et al. [450] proposed the use of EC as a co-solvent for the PC solvents to increase the ionic conductivity of the electrolytes based on tetramethyl ammonium difluoro(oxalato)borate (TMADFOB) salts. The ionic conductivity of the 1 M TMADFOB/PC–EC electrolyte increased gradually with the increasing amount of the EC addition from 0 to 80 by vol.%. As a result, this enhanced the power density of the formed EDLC and the optimum EC content was found to be about 30 by vol.% in the mixed PC–EC solvents. For the asymmetric AC//graphite EC, the EC addition to the PC solvents inhibited the $DFOB^-$ intercalation into the graphite electrodes and consequently decreased the reversible capacity of the ES, as indicated by *in situ* XRD patterns. Jänes et al. [451] found the addition of FEC into the PC solvents changed the viscosity and specific conductivity of the electrolyte and thus affected the ES performance with TiC–CDC electrodes. Kim et al. [452] found a small amount of FEC addition (2 by vol.%) to the 0.2 M $TEABF_4$/PC–DMC electrolyte resulted in an increase of the specific capacitance of the EDLC.

In addition to the carbonates, sulfites such as the diethyl sulfite (DES) and 1,3-propylene sulfite (PS) were studied by Jänes et al. [453] as the additives or co-solvents for binary PC-based solvents. The performance of EDLCs with 1 M $TEMABF_4$ in various solvent mixtures including pure PC, PC–DES (95:5 by volume), PC–DES (90:10), PC–PS (95:5), PC–PS (90:10), and PC–PS (80:20) were investigated. For all the mixed electrolytes and the investigated temperatures (–45°C to 60°C), the EDLCs showed ideal polarizability Delta E ≥ 2.7 V. The addition of DES or PS into PC solvent noticeably changed the viscosity and the conductivity of electrolytes, and thereby had an influence on the limiting capacitance, the characteristic time constant and complex power and energy values of the obtained ESs. Ragone plots showed the EDLC with 1 M TEMA/PC–PS (95:5) delivered a high energy density of about 25–27 Wh kg^{-1} at moderate power densities of about 1 kW kg^{-1} at T ≥ –20°C.

Besides the PC, there has been a growing interest in solvent mixtures based on other carbonates such as EC, DEC, and DMC. Similarly, these solvent mixtures have already been widely studied as battery solvents for LIBs and currently have attracted increasing attention for ESs. Laheäär et al. [454] pointed out that an EC–DMC (1:1 by volume) mixture containing 1 M $LiPF_6$ could be used for high energy and power density ESs. In this study, EDLCs consisting of the microporous carbon and the 1 M $LiPF_6$/EC–DMC electrolyte showed a high limiting capacitance (129 F g^{-1}), moderate characteristic time constant (≥2.7 s), and an energy density of 36.7 Wh kg^{-1} at a power density of 83.6 kW kg^{-1}. The effect of organic ester addition into EC has also been reported by Jänes and Lust [448], who investigated the performance of the EDLCs with 1 M $(C_2H_5)_3CH_3NBF_4$ in various solvent mixtures including EC–methyl acetate (MA) (1:1 by volume), EC–ethyl acetate (EA) (1:1), and EC–methyl formate (MF) (1:1). The nanoporous carbon electrodes showed ideal polarizability region of Delta E ≥ 3.0 V for all the mixed solvents. The series capacitance increased with the increase of specific conductivity of the electrolyte, following the order: EC–DEC (1:1) < EC–DMC (1:1) < EC–EMC (1:1) < EC–EA (1:1) < EC–MA (1:1) < EC–MF (1:1). Binary, ternary, and quaternary solvent mixtures based on the organic carbonates (e.g., PC, EC, DMC, EMC, and DEC) and organic esters (e.g., MF, MA, and EA) have also been investigated by the same group [455]. Depending on the composition, these solvent mixtures exhibited different properties in terms of the relative macroscopic dielectrical permittivity, viscosity, melting temperatures, and dipole moment, resulting in different solubility of salts and ionic conductivity. The addition of organic esters was believed to increase the solubility of $TEMABF_4$ salts and decrease the viscosity of the electrolyte, increasing the conductivity of the electrolyte in the carbon nanopores compared to the PC-based electrolyte, although the specific conductivity was lower than the ACN-based electrolyte. It was also found the conductivity of the electrolyte containing 1 M $TEMABF_4$ increased in the following order: EC–DEC–EA < EC–EMC–EA < EC–DMC–EA (all with 1:1:1 molar ratio), which was in agreement with the decrease of the molar mass of asymmetric carbonate. For all the investigated solvent mixtures containing 1 M $TEMABF_4$, EDLCs based on AC showed nearly ideal capacitor behavior at potential scan rates <5 mV s^{-1} and cell voltage <3.0 V. It should be noted that due to their low flash point and high volatility, the use of some organic esters such as EA and MA may introduce safety problems. To take this into account, Perricone et al. [456] added organic ester co-solvents with methoxy or fluorinated groups so as to address the safety issue. Methylmethoxyacetate (MMOA) and ethyl difluoroacetate (EDFA) were found to be good co-solvents for EC which resulted in similar performances as compared with EA and exhibited higher level performances than those obtained with PC-based electrolytes. Taking into account the lower viscosity and intermediate dielectric constant (36) of methoxypropionitrile (MPN), Perricone et al. [457] investigated the effect of an MP addition into EC on the performance of the EDLCs with the objective to improve the viscosity and conductivity of the EC-based electrolytes. It was found that the ionic conductivity increased by about 1.2–1.5 times by adding MP to EC, due to the decrease of the electrolyte viscosity. The highest conductivity values obtained for the MP–EC-based electrolytes (with both 1 M $TEABF_4$ and $SBPBF_4$ salts) were 20 mS cm^{-1} at 30°C, which were much higher than obtained in the PC-based electrolytes (14 mS cm^{-1}). The addition of MP to EC increased the liquefied range over a wide temperature range and thus extended the lower operating temperature limit of the EDLCs.

In spite of the toxicity of the ACN, the ACN-based mixture solvents have also been investigated in order to increase the operating cell voltage and/or to extend the lower temperature limit of ESs with ACN-based electrolytes. Ding et al. [436] found the liquid range of ACN (conducting salt: 1 M $TEMABF_4$) could be significantly expanded after the GBL addition. The oxidation resistance of the solvents also increased as the GBL content increased. However, the addition of the GBL decreased the ionic conductivity of the electrolytes and thus increased the ESR of the EDLCs with GBL–ACN mixture electrolytes although the operating cell voltage was increased. With the aim to expand the low-temperature operation limit of the ACN-based EDLCs to below −40°C, Brandon et al. [371] proposed the integration of low melting co-solvents such as formates, esters, and cyclic ethers into the ACN solvents. It was shown that the concentration of the salts or properties of the co-solvent greatly influenced the ESR of the EDLCs. The EDLCs using electrolytes based on these solvents containing 0.75 M $TEABF_4$ could be operated as low as −75°C in some cases but with decreased specific capacitance. However, the much increased ESR at the low temperatures (e.g., at −55°C) limited the practical operation of these EDLCs. Further research is required to improve the ESR, increase the power density at lower temperatures, and to evaluate the lifetime of the EDLCs at lower operating temperatures. Iwama et al. [458] found the addition of MA into an ACN-based electrolyte could extend the low-temperature operation range of commercial 600F ES from −40°C to −55°C (Figure 2.41). West et al. [459] added the lower melting point 1,3-dioxolane (DIOX) to the common ACN solvent, and the low-temperature operational range of commercial mesoporous AC-based EDLCs was greatly extended to −70°C. By controlling the concentration of the tetraethylammonium tetrafluoroborate (TEATFB) salt, the ESR could be minimized to allow the optimal power delivery at a given temperature. In a later study, the same group synthesized microporous carbon electrodes and further improved the low-temperature performance of EDLCs [460]. With the use of microporous carbons and the carefully designed electrolyte ($SBPBF_4$ in 1:1 ACN–MF), the assembled EDLC showed rapid charging/discharging ability and high energy density at −70°C (up to 86% of the energy density available at room temperature) (Figure 2.42) [460].

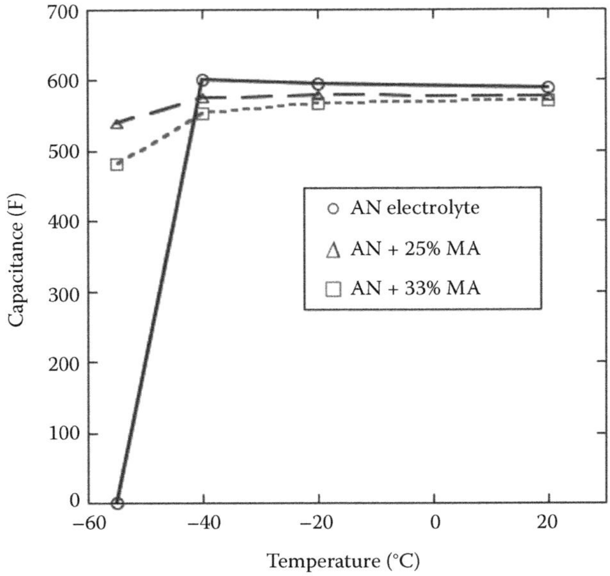

FIGURE 2.41 Evaluated capacitance from discharge curves for 1 M $TEABF_4$ in (○) AN, (△) AN + MA (75:25), and (□) AN + MA (67:33) at +20°C, −20°C, −40°C, and −55°C. (Reprinted from *Journal of Power Sources*, 219, Iwama, E. et al., Characterization of commercial supercapacitors for low temperature applications, 235–239, Copyright 2012, with permission from Elsevier.)

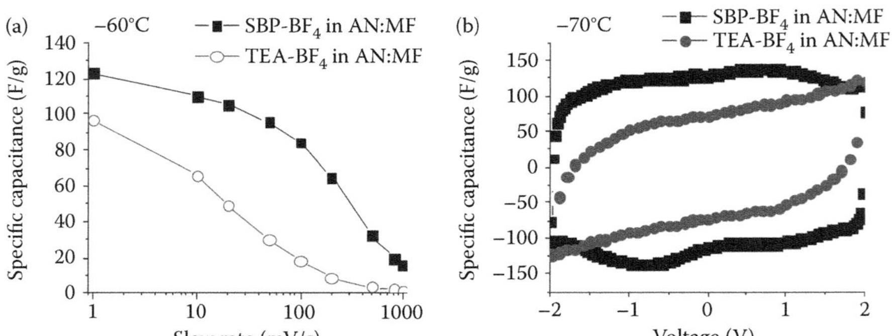

FIGURE 2.42 Effect of the cation on the low-temperature performance of EDLCs utilizing AN:MF = 1:1 solvent:specific capacitance calculated from the CV curves as a function of sweep rate at (a) −60°C and (b) cyclic voltammograms at a slew rate of 1 mV s^{-1} recorded at −70°C. Zeolite NaY-templated carbons synthesized at 700°C were used as electrodes and $TEABF_4$ and $SBPBF_4$ salts were used in the electrolyte. (Korenblit, Y. et al.: *Advanced Functional Materials* 22. 1655–1662. 2012. Copyright Wiley-VCH Verlag GmbH & Co. KGaA. Reprinted with permission.)

There have been also some studies focused on the optimization of electrolytes for LICs [417]. As mentioned previously, LICs have large operating voltage and high energy density, but they generally suffer from a poor rate performance. Based on the previous findings that the electrolyte additive (i.e., 1,3,5-trifluorobenzene [TFB]) could enhance the Li^+ ions transfer reaction at the $LiCoO_2$ electrode interface of the LIB [461], Lee et al. [417] reported the use of TFB additive as a potential way to enhance the rate performance of LICs. The addition of the TFB (3.0 wt.%) into the 1.0 M $LiBF_4$/EC–DEC (3:7 by volume) was found to increase the high rate performance of the microporous AC//Li ESs.

2.3.3 Investigation and Development of Conducting Salts

2.3.3.1 Effect of the Conducting Salt on the Performance of ESs

The dissolved conducting salt in the organic solvent provides charge carriers (i.e., cations and anions) in the electrolyte. Accordingly, the concentration and mobility of the ions play a significant role in the ionic conductivity of the electrolyte. Furthermore, as will be discussed below, the conducting salts also remarkably affect the electrochemical stability and the thermal stability of the organic electrolytes, and the capacitance of the ESs. Several factors such as the solubility, conductivity, stability, safety, and cost should generally be considered for choosing or designing appropriate salts for certain solvents. Because of good overall properties, $TEABF_4$ has been widely used for commercial EDLCs and is currently the most frequently reported salt in the academic research. Many other salts have also been explored in order to further improve the properties of the electrolytes such as solubility, conductivity, stability, and temperature performance [2].

The salt concentration, that is, ion concentration, influences not only the ionic conductivity of the electrolyte but also the maximum energy density of the ESs [462,463]. Figure 2.43 shows an example that there is an optimum concentration value for the maximum ionic conductivity of certain electrolyte [429]. As can be seen in this figure, $TEABF_4$ has a limited solubility up to 1 M in PC while other salts show a better solubility (above 2 M). A sufficiently high ion concentration is essential for achieving a high maximum energy density of the EDLC [462,463]. Zheng and Jow [463] pointed out that the performance of the organic electrolyte-based EDLC strongly depended on the salt concentration in the electrolyte. For low concentration electrolytes, the maximum energy density decreased with the decrease of the salt concentration, due to the depletion of free ions in the organic electrolyte (Figure 2.44).

Since ESs usually operate under a wide temperature range for many applications, the solubility of the salts at lower temperature should be considered. A high salt concentration may lead to the precipitation in the organic electrolytes [94]. Concerning the conductivity, it is dependent on several factors such as the salt concentration, the dissociation degree of the dissolved salt to provide free charge carriers, the diffusion coefficient of dissociated ions, the solvents (as mentioned previously), and the temperature [2]. Common organic electrolytes have a much lower dissociation degree than aqueous electrolytes, leading to the lower ionic conductivity of the electrolytes and larger ESR of the organic electrolyte-based ESs compared to the aqueous electrolyte-based ESs. The dissociation degree and ion mobility are dependent on the structure of the organic salts such as ionic size of the cation and anion, and symmetric or asymmetric structure,

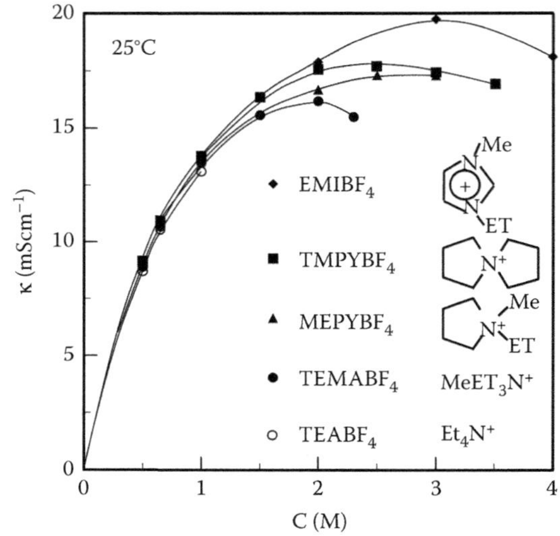

FIGURE 2.43 Concentration dependence of electrolytic conductivity of quaternary ammonium tetrafluoroborate salts in PC. (Reproduced by permission of The Electrochemical Society of Japan from Ue, M. 2007. *Electrochemistry* 75:565–572.)

which affect the conductivity and viscosity of the electrolytes. Table 2.9 summarizes the conductivities of some common salts in different solvents including PC, GBL, *N,N*-dimethylformamide (DMF), and ACN [298]. With reference to the anions, the conductivity decreases in the order: $BF_4^- > PF_4^- > ClO_4^- > CF_3SO_3$ while for the cations, the conductivity decreases in the following order: $TEA^+ > Pr_4N^+ > Bu_4N^+ > Li^+ > Me_4N^+$. Owing to the smaller anion size and moderate dissociation degree, $TEABF_4$ shows the highest conductivity among these salts [298].

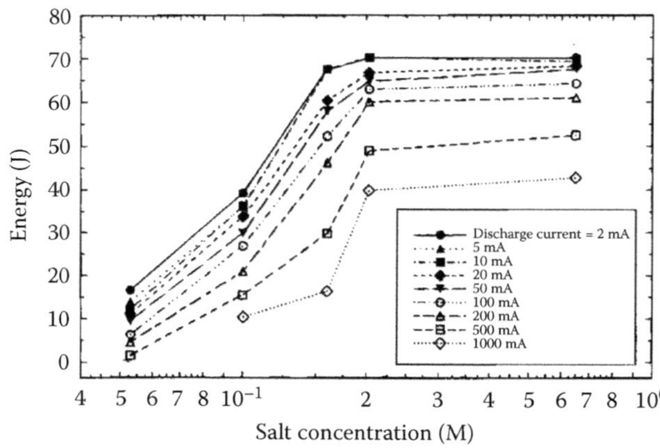

FIGURE 2.44 Maximum energy as a function of the salt concentration in electrolyte at different discharge currents. (Reproduced by permission of The Electrochemical Society from Zheng, J. P., and T. R. Jow. 1997. *Journal of the Electrochemical Society* 144:2417–2420.)

TABLE 2.9
Conductivities (mS cm⁻¹) of Organic Liquid Electrolytes at 1 M, 25°C

Electrolyte	PC	GBL	DMF	ACN
$LiBF_4$	3.4	7.5	22	18
Me_4NBF_4	2.7[a]	2.9[a]	7.0[a]	10[a]
Et_4NBF_4	13	18	26[b]	56[b]
Pr_4NBF_4	9.8	12	20[b]	43[b]
Bu_4NBF_4	7.4	9.4	14[b]	32[b]
$LiPF_6$	5.8	11	21	50
Me_4NPF_6	2.2[a]	3.7[a]	11[a]	12[a]
Et_4NPF_6	12	16	25	55
Pr_4NPF_6	6.4[a]	11	19	42
Bu_4NPF_6	6.1	8.6	13	31
$LiClO_4$	5.6	11	20	32
Me_4NClO_4	2.9[a]	3.9[a]	7.8[a]	7.7[a]
Et_4NClO_4	11	16	24[b]	50[b]
Pr_4NClO_4	6.3[a]	11[a]	17[a,b]	35[a,b]
Bu_4NClO_4	6.0	8.1	12[b]	27[b]
$LiCF_3SO_3$	1.7	4.3	16	9.7
$Me_4NCF_3SO_3$	9.0[a]	14	24	46
$Et_4NCF_3SO_3$	11	15	21[c]	42[c]
$Pr_4NCF_3SO_3$	7.8	11	15[c]	31[c]
$Bu_4NCF_3SO_3$	5.7	7.4	11[c]	23[c]

Source: Adapted from Ue, M., K. Ida, and S. Mori. 1994. *Journal of the Electrochemical Society* 141:2989–2996; House, H. O., E. Feng, and N. P. Peet. 1971. *The Journal of Organic Chemistry* 36:2371–2375.

[a] Saturated solution (<1 M).
[b] Sources from Reference 464.
[c] Sources from Reference 465.

In addition to the solvent, the conducting salt also strongly affected the electrochemical stability of the organic electrolytes. Ue [429] investigated the electrochemical stability of a series of salts in PC at glassy carbon (GC) working electrodes and found the electrochemical window for the cations decreased in the following order: $Hex_4N^+ > R_4N^+$, R_4P^+ (P = Me, Et, Pr, Bu) (no remarkable difference) and for the anions with ET_4N^+ cation, it decreased in the order: $SbF_6^- > AsF_6^- > PF_6^- > BF_4^- > C_4F_9SO_3^-, (CF_3SO_2)_3C^-, (CF_3SO_2)_2N^- > ClO_4^-, CF_3SO_3^- > B(C_2H_5)_4^-$ [437] This trend was later found to be in agreement with a DFT study concerning the anodic stability of the anions [466]. Xu et al. [467] investigated the electrochemical stability of many salts at the AC electrodes and glassy carbon electrodes in the EC–DMC solvent (1:1 by volume). The electrochemical stable window obtained at the AC electrodes was much smaller than that at the GC electrodes (Figure 2.45). The electrochemical stability order of the cations at the AC electrodes was $Pr_3MeN^+ > Et_4N^+ > Bu_3MeN^+ > Et_3MeN^+ > iPr^2MeEtN^+ > Me_3EtN^+ > Bu_3MeP^+ > Et_3MeP^+$ while that of the anions at the glassy carbon electrode followed: $AsF_6^- \sim BF_4^- > PF_6^- > Tf^- \sim Im^-$ (solvent: EC–DMC).

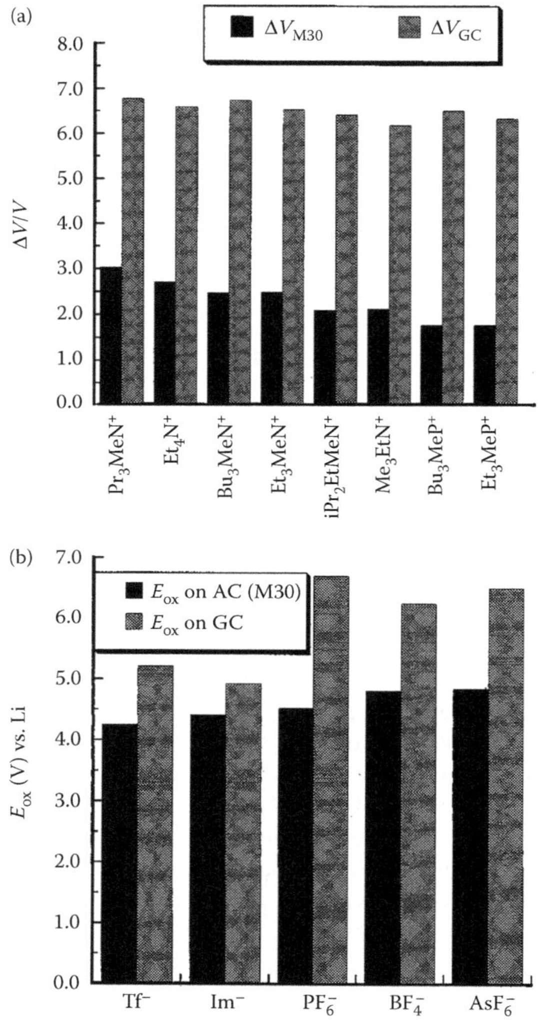

FIGURE 2.45 Comparison of electrochemical stability limits measured on a nonporous electrode. (a) Stability window of PF_6^- salts on AC composite and on GC; (b) anion dependence of anodic limits, electrolytes: 1.0 M Et_3MeN^+ salts of corresponding anions in EC–DMC. (Reproduced by permission of The Electrochemical Society from Xu, K. et al. 2001. *Journal of the Electrochemical Society* 148:A267–A274.)

Since various salts have diverse cations and anions with different bare and solvated ion size, it is rather important to match the size of the electrolyte ions with the pore size of the carbon-based electrodes, and also to consider the distortion or partial removal of the solvation shell of ions when the size of the bare electrolyte ion is close to the pore size of carbon electrodes (as mentioned previously). This has a great influence on the capacitance and power performance of the EDLCs. Based on the consideration that the smaller ions can decrease the distance (thickness) of the EDL and may thus contribute to a higher capacitance, Koh et al. [468] investigated the effect of ionic size of different salts on the specific capacitance of the EDLCs. The investigated salts included $TEABF_4$, $TEMABF_4$, trimethylpropylammonium BF_4 ($TMPABF_4$), and diethyldimethylammonium BF_4 ($DEDMABF_4$). Note that the cations of these salts have different lengths of hydrocarbon chains and thereby have different cation sizes. It was demonstrated that the specific capacitance of the EDLCs increased as the cation size of quaternary ammonium salts decreased. $TEMABF_4$ and $TMPABF_4$ salts resulted in an approximately 10% higher specific capacitance compared to the $TEABF_4$ salt for the AC electrodes, at all the investigated currents. Furthermore, the charged state (valence) of ions has also been reported to influence the capacitance and energy density of EDLCs [469,470]. Jänes et al. [469] found that by using doubly charged cation-based electrolytes, that is, *N*,*N*-dimethyl-1,4-diazabicyclo[2,2,2]octanediium tetrafluoroborate ($DMDABCO(BF_4)_2$) in GBL, the energy density of the EDLC was increased by nearly 18% compared to the EDLC with the $TEMABF_4$/GBL electrolyte (both with 0.4 M salt concentration). This was attributed to the higher electrode surface charge density caused by the adsorption of doubly charged $DMDABCO^{2+}$ cations. Zheng et al. [470] also found that the divalent salts could enhance the charge storage ability of EDLCs compared with the monovalent salts (Figure 2.46). In this study, the energy density of EDLCs with 1,1,4,4-tetramethyl piperazine ($TMP(BF_4)_2$) dissolved in PC was shown to be higher than that with the 1 M $TEABF_4$/PC electrolyte. However, the highly condensed charge density in the divalent cation resulted in a strong interaction with the organic solvent (PC), which may lead to the co-adsorption of extra solvent molecules and decrease the mobility of the solvated cations. This would negatively affect the performance (e.g., rate capability) of the EDLCs.

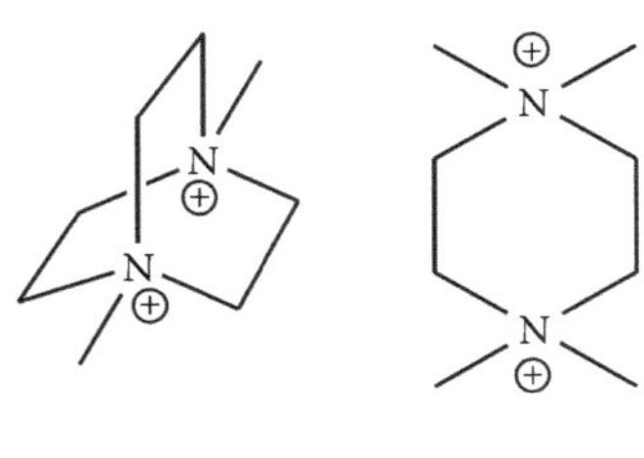

FIGURE 2.46 Chemical structures of divalent quaternary alkyl ammonium cations (the left $DMDABCO^{2+}$ and the right TMP^{2+} in this study). (Reprinted from *Journal of Power Sources*, 220, Zheng, C. et al., A divalent quaternary alkyl ammonium salt as the electrolyte for high-energy electric double-layer capacitors, 169–172, Copyright 2012, with permission from Elsevier.)

The effect of the salt (cation and anion) on the performance of the asymmetric ESs has also been investigated [471,472]. Yokoyama et al. [471] found for the asymmetric carbon-based ESs, the type of electrolyte anion strongly affected the capacity and the capacity decreased in the order: $PF_6^- > BF_4^- > ClO_4^-$ (cation: TEA, solvent: PC–EMC). The largest capacity obtained for electrolytes containing PF_6^- anion was due to the reason that PF_6^- was the most stable anion among all the investigated anions. XRD measurements suggested the intercalation of PF_6^- into the graphite electrode. By changing the cation from the TEA^+ to $TEMA^+$ (i.e., a smaller one), the capacity of the cell was slightly increased. A hybrid ES with 0.71 M $TEMAPF_6$/PC–EMC (1:2 by volume) electrolyte showed a capacity of 95 mAh g^{-1}. It was also found that the overall capacity of the hybrid ESs was limited by the anodic capacity and thus an increase in anodic capacity was effective for increasing the total capacity. Wang and Yoshio [472] reported the intercalation of the biggest cation (TBA^+) led to the largest degree of the graphite crystal lattice expansion, but it also showed the best cycling performance of the asymmetric graphite//AC ESs among the four quaternary alkyl ammonium cations (i.e., TEA^+, $TEMA^+$, TBA^+, and dodecylethyldimethylammonium [$DEDMA^+$]). *In situ* XRD results indicated the intercalation of these cations into the graphite crystal lattice during the charging process. It is worth noticing that the use of electrolytes based on quaternary alkyl ammonium salts for asymmetric ESs are very interesting since they may overcome some disadvantages of electrolytes containing Li salts (e.g., Li deposition on the negative electrode).

2.3.3.2 Development of New Conducting Salts

As mentioned previously, $TEABF_4$ has a limited solubility especially in PC. To increase the solubility of conducting salt in the organic solvents, especially in PC, some asymmetric tetraalkylammonium salts and cyclic quaternary ammonium salts, such as $TEMABF_4$, 1-ethyl-1-methylpyrrolidinium ($MEPYBF_4$), and tetramethylene-pyrrolidinium ($TMPYBF_4$), have been studied. As shown in Figure 2.43, the solubility of these salts was found to be much higher in PC in comparison with $TEABF_4$ [429]. Nambu et al. [473] reported the specific capacitance of the EDLCs slightly increased with the increasing concentration of $TEMABF_4$ in FEC from 2.0 to 3.0 M. However, since the salts constitute an important part of the total cost of the electrolytes, one should consider the balance between the salt concentration and the ES performance.

Recently, spiro-type quaternary ammonium salt, spiro-(1,1′)-bipyrrolidinium tetrafluoroborate ($SBPBF_4$), has received much attention owing to its smaller cation size and higher solubility than the conventional $TEABF_4$ [474]. The solubility of $SBPBF_4$ in PC was found to be three times higher than that of $TEABF_4$ while its electrochemical stability was similar to that of the $TEABF_4$ and TEMABF4 [474]. The high solubility of $SBPBF_4$ has also been found in other solvents besides carbonate solvents. For example, Chiba et al. [445] found the solubility of $SBPBF_4$ in various linear sulfones was much higher than that of $TEABF_4$ and $TEMABF_4$. Furthermore, Chiba et al. [474] found the electrolyte composed of $SBPBF_4$ and PC showed better performance in terms of solubility, conductivity, and viscosity of the electrolytes—especially at lower temperatures—as compared with the $TEABF_4$/PC and $TEMABF_4$/PC electrolytes. Consequently, the EDLCs with the $SBPBF_4$/PC electrolyte demonstrated a lower ESR and higher specific capacitance at low temperatures or at high discharging rates compared to that with the

$TEABF_4$/PC and $TEMABF_4$/PC electrolyte. To further improve the performance of the electrolytes based on $SBPBF_4$, Chiba et al. [475] investigated the effect of DMC, DEC, and EMC additions into the $SBPBF_4$/PC electrolyte. It was found that the addition of DMC into PC (~50 vol.%) increased the conductivity of the electrolyte to 21 mS cm^{-1}, due to the decrease of the viscosity contributed by the DMC. However, the addition of DEC and EMC had a negative effect on the conductivity owing to their poor solubility of the $SBPBF_4$ salt. As a result, the $SBPBF_4$/PC–EMC led to a lower ESR and higher specific capacitance of the EDLCs compared to the $SBPBF_4$/PC electrolyte. Yu et al. [476] found the distortion degree of the CVs obtained for the EDLC with $SBPBF_4$/PC electrolyte was lower than that of the $TEMABF_4$/PC at a high scan rate of 200 mV s^{-1}, due to the higher mobility of the SBP cations. This suggests the $SBPBF_4$/PC electrolyte is very attractive for the development of high power EDLCs. By using 1.5 M $TEMABF_4$/PC electrolyte, EDLCs with AC electrodes showed a specific capacitance of 120 F g^{-1}, an energy density and power density of 31 Wh kg^{-1} and 6938 W kg^{-1}, respectively, at an operating voltage of 2.7 V. The electrical conductivity of $SBPBF_4$/PC (17.0 mS cm^{-1}) was higher than that of $TEMABF_4$/PC (14.6 mS cm^{-1}). The application of the $SBPBF_4$ salt has also been reported for the asymmetric ESs composed of mesophase carbon microbeads (MCMBs) negative electrode and AC positive electrode, as reported by Zheng et al. [477]. In this study, a series of quaternary alkyl ammonium-based electrolytes, including $TEABF_4$, $TEMABF_4$, diethyldimethyl ammonium BF_4 ($DEDMABF_4$), tetrapropyl ammonium ($TPABF_4$), tetrabutyl ammonium ($TBABF_4$), and $SBPBF_4$, have been studied. It was demonstrated the intercalation of the quaternary alkyl ammonium cations in the electrolytes resulted in an irreversible expansion of interlayer spaces between adjacent carbon layers, favoring subsequent adsorption of ions. Among the investigated cations, SBP^+ cation was found to increase the storage ability of the negative electrode most significantly. In spite of so many advantages, the wide application of $SBPBF_4$ salt is limited due to its high cost compared to $TEABF_4$.

In addition, Kurig et al. [478] developed several substituted phosphonium cations for EDLC electrolytes. The investigated electrolytes included tetrakis(diethylamino)phosphonium tetrafluoroborate ($TDENPBF_4$), tetrakis(diethylamino)-phosphonium hexafluorophosphate ($TDENPPF_6$), and tetrakis(dimethylamino)phosphonium tetrafluoroborate ($TDMNPBF_4$) in ACN. The C(TiC) electrodes could be ideally polarizable in the 1 M $TDENPPF_6$/ACN up to 3.2 V whereas deviations from ideal capacitive behavior occurred in other electrolyte systems above 2.7 V. The EDLC using 1 M $TDENPPF_6$/ACN electrolyte exhibited the gravimetric capacitance of 85 F g^{-1}, characteristic relaxation time of 0.9 s, and gravimetric energy density of 27 Wh kg^{-1} at power density of 269 kW kg^{-1} at the cell voltage of 3.2 V. No visible loss in the discharging capacitance was observed after 10,000 charging/discharging cycles. However, it should be noted that these substituted phosphonium cations containing electrolytes had much lower conductivity compared to $TEABF_4$ and $TEMABF_4$ containing electrolytes with the same solvent. This may negatively affect the ESR, rate performance, and thermal stability of the EDLCs and requires further studies.

Regarding the anion, other types of anions free of halogen have been investigated. As can be seen from the above discussion, the frequently used anions such as BF_4 contain halogen. Several research groups pointed out that toxic hydrogen fluoride (HF) might be formed due to the possible hydrolysis of these fluoro-complex

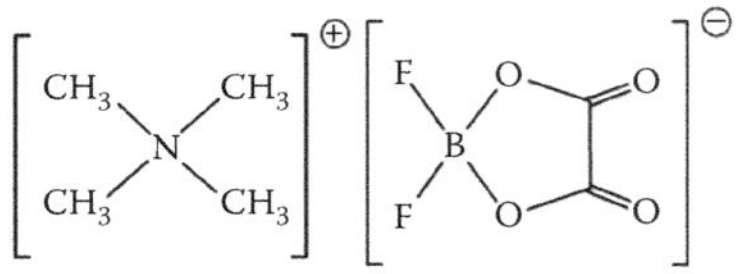

Tetramethylammonium difluoro(oxalato)borate
(TMADFOB)

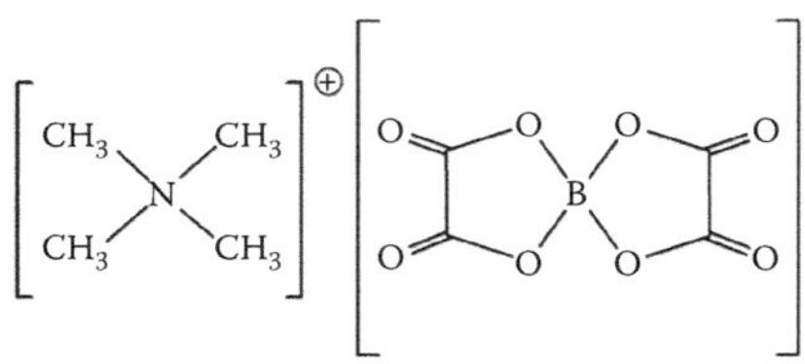

Tetramethylammonium bis(oxalato)borate
(TMABOB)

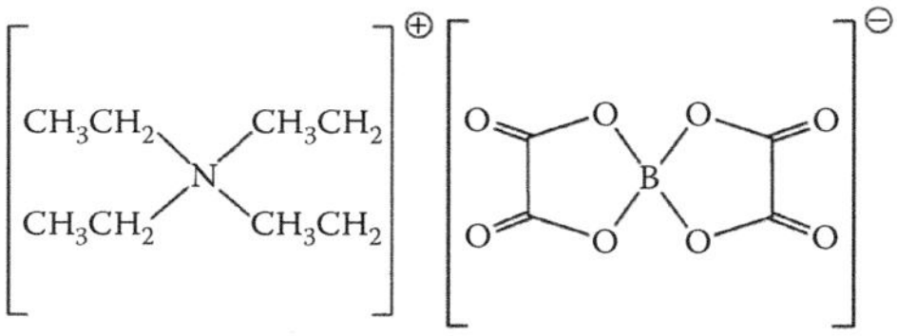

Tetraethylammonium bis(oxalato)borate
(TEABOB)

FIGURE 2.47 Structure of TMADFOB, TMABOB, and TEABOB, which was studied in Reference 481.

salts [479–481]. Based on this consideration, they developed bis(oxalato)borates-based salts. Nanbu et al. [481] investigated the properties of a series of quaternary ammonium bis(oxalato)borates (QABOBs) including TMABOB, ETMABOB, TEMABOB, and TEABOB (Figure 2.47). The viscosities and surface tensions of QABOBs/PC electrolytes were larger than that of $TEABF_4$, and the conductivity became lower. However, the electrochemical potential windows obtained on the AC electrode and the gravimetric capacitances of the EDLCs were comparable to those obtained with the $TEABF_4$/PC electrolyte. Sasaki et al. [482,483] reported that the solubility of the salt could be increased when the bis(oxalato)borate anion was changed to the asymmetric difluoro(oxalato)borate anion (TMADFOB). The authors found the conductivity of 1 M TMADFOB in PC was higher than those obtained for TMABOB and TEABOB, and it was only slightly lower than that of 1.6 M $TEABF_4$ in PC. The gravimetric capacitance of the EDLCs with TMADFOB/PC was found to be the highest among all these investigated salts due to the small ionic size. Lai et al. [484] also reported the TEAODFB had good solubility in PC compared to $TEABF_4$. The TEAODFB/PC electrolyte exhibited good conductivity of 14.46 mS cm^{-1} when the concentration of TEAODFB was 1.6 M at 25°C, which was higher than that of the $TEABF_4$ with 1 M concentration (12.94 mS cm^{-1}). Francke et al. [485] found the stepwise introduction of fluorosubstituents extended the electrochemically stable window for bis(2,2′-biphenoxy)borates more than 600 mV. AC-based EDLCs using these fluorinated bis(2,2′-biphenoxy)borates could reach a cell voltage of above 3.0 V.

Apart from the organic cations, some attention has been focused on the inorganic cations such as Li^+, Na^+, Mg^+, and Cs^+. As mentioned previously, due to the small ionic size of Li^+, lithium salt-based organic electrolytes have been extensively used in hybrid ESs such as LICs [486]. Considerable studies have also reported the application of Li salt-based electrolytes for pseudocapacitors, and there are also some studies on EDLCs [330,487–491]. Considering the popular use of $LiPF_6$ and $LiClO_4$ for LIBs, these Li salts have also been widely studied for the ESs. Considering the wide use of the $LiPF_6$/EC–DMC electrolyte in LIBs, Laheäär et al. [454] reported the application of this electrolyte for EDLCs. The EDLCs based on microporous carbons showed a wide region of ideal polarizability up to a cell voltage of 3.2 V in 1 M $LiPF_6$/EC–DMC (1:1 by volume). Besides, relatively high limiting capacitance (129 F g^{-1}), moderate characteristic time constant (≥2.7 s), high energy density (36.7 Wh kg^{-1}), and power density (83.6 kW kg^{-1}) of the EDLC were obtained in this electrolyte. Noticeably, the SEI was found to be formed at the negative electrode, which was similar to that observed in LIBs. This may have some negative effects on the performance (e.g., blocking the active surface area and decreasing the capacitance) of the EDLCs especially for long-term operation. Li et al. [488] reported the use of $LiPF_6$/acetamide as the electrolyte for EDLCs. It was found that the $LiPF_6$ and acetamide mixtures with the molar ratios between 1:4 and 1:6 existed as a liquid below 25°C. The operating voltage of the EDLC with $LiPF_6$/acetamide electrolyte could achieve 2.25 V. The ESR of the EDLC decreased from 38.9 Ω at 30°C to 10.3 Ω at 70°C. Since the price of $LiPF_6$ is high, Laheäär et al. [487] investigated the performance of EDLCs based on $LiClO_4$ salt. The operating voltage window obtained in the $LiClO_4$/EC–DMC (1:1 by volume) was similar to that in the $LiPF_6$/EC–DMC. However, the capacitance of the EDLC with 1 M $LiClO_4$/EC–DMC electrolyte remarkably decreased after only 800 charging/discharging cycles due to the continuous blocking of the micropores on the carbon surface. This indicated this electrolyte was a comparatively weak electrolyte candidate for long lifetime ESs. Kim et al. [486] compared the performance of $LiBF_4$/PC-based electrolyte with the $TEABF_4$/PC electrolyte (both with 1.0 M concentration) for the asymmetric ESs consisting of the alkali-treated soft carbon negative electrode and ACF positive electrode. It was found that $TEABF_4$-based electrolytes exhibited higher specific capacitance and good cycling stability whereas $LiBF_4$-based electrolytes showed poor capacitance behavior probably owing to the electrolyte decomposition at the positive electrode. Laheäär et al. [490] investigated the 1 M lithium bis(oxalato)borate (LiBOB) in EC–DMC (1:1 volume ratio) solvent mixture as an electrolyte for ESs. The ideal polarizability region for the EDLC with porous carbon electrodes was about 3.0 V. However, it appeared that the performance of the EDLCs based on this electrolyte was not very promising. For example, it showed a lower gravimetric capacitance compared to the EDLCs based on other Li-salt electrolytes ($LiPF_6$ and $LiClO_4$) and had a low cycling stability (33% loss in the capacitance during 5000 cycles). Regarding pseudocapacitors, Chen et al. [492] compared the performance of ESs based on MnO_2–CNT electrodes with three different electrolytes (i.e., 1 M Na_2SO_4/H_2O, $TEABF_4$/PC, and $LiClO_4$/PC). The energy density of the ESs with 1 M $TEABF_4$/PC was about three times higher than that with the aqueous electrolyte due to the much larger operating cell voltage (2.5 V vs. 0.85 V). Interestingly, the energy density was further improved twice over in the 1 M $LiClO_4$/PC electrolyte compared to the $TEABF_4$/PC. It was explained that the much smaller ionic sizes of Li^+ (0.09 nm) and ClO_4^- (0.215 nm) compared to the TEA^+ (0.337 nm) and BF_4^- (0.232 nm) contributed to a higher pseudocapacitance achieved in the $LiClO_4$/PC. However, the cycling stability obtained in the $LiClO_4$/PC was lower than that in the

$TEABF_4$ (61.6% capacitance retention vs. 75.5% capacitance retention after 10,000 cycles), which was related to the morphology change of the electrode materials after long cycling tests.

Owing to their much greater abundance, Na-salts (e.g., $NaPF_6$ and $NaClO_4$) based electrolytes have also been proposed to be used for the ESs [493]. Laheäär et al. [493] found the EDLCs with 1 M $NaClO_4$/EC–DMC and $NaPF_6$/EC–DMC electrolytes showed similar limiting capacitance values (~130 F g^{-1}) and charging/discharging time constant (from ~2 s). The maximum operative cell voltage for $NaClO_4$ and $NaPF_6$/EC–DMC (1:1 by volume) based EDLCs was 3.2 and 3.4 V, respectively. However, although $NaClO_4$-based EDLCs showed a higher cycling stability than the $NaPF_6$-based EDLCs (15% loss in capacitance vs. 22% during 5000 galvanic cycles), both of them had limited cycling stability (Figure 2.48) [493]. Therefore, the cycling stability was required to be further improved for the practical use of these Na-based salts. To improve ionic conductivity and viscosity of the Na-salt-based electrolytes at low temperatures, Vali et al. [494] used a quaternary solvent mixture of EC, DMC, PC, and EA PC and EA (1:1:1:0.5 by volume ratio) as the solvent for the 1 M $NaPF_6$, $NaClO_4$, and $NaN(SO_2F)_2$ (NaFSI). The maximum cell voltage of the EDLCs with various Na-salt-based electrolytes decreased in the following order: $NaPF_6$ (3.4 V) > $NaClO_4$ (3.2 V) > NaFSI (2.5 V). The low cell voltage obtained in the NaFSI-based electrolyte was attributed to the corrosion of Al current collectors in this electrolyte. The limiting capacitance could be retained down to –30°C for the $NaPF_6$-based EDLCs whereas it decreased noticeably below 0°C for the $NaClO_4$-based EDLCs. Floating potential measurements at 3.0 V showed the $NaPF_6$-based

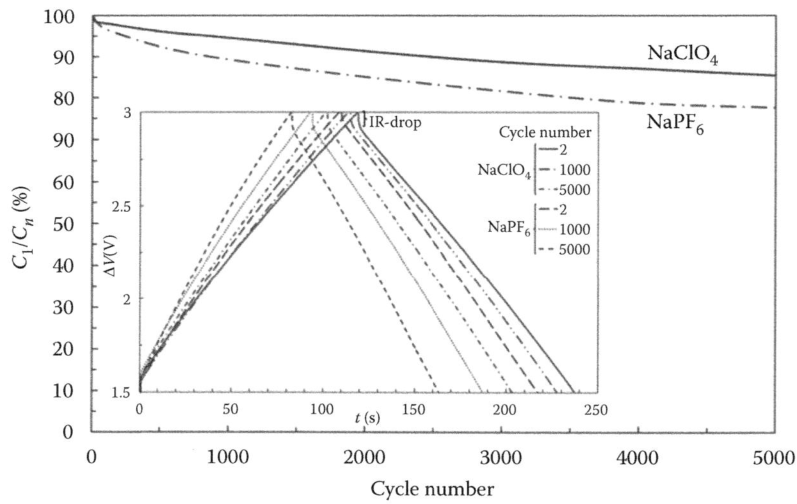

FIGURE 2.48 Ratio of discharge capacitances of first and *n*th cycle versus cycle number and galvanic cycles (figure inset) for $NaClO_4$-TC and $NaPF_6$-TC at charge/discharge current density of 4 mA cm^{-2}. (Reprinted from *Electrochimica Acta*, 82, Laheäär, A., A. Jänes, and E. Lust, $NaClO_4$ and $NaPF_6$ as potential non-aqueous electrolyte salts for electrical double layer capacitor application, 309–313, Copyright 2012, with permission from Elsevier.)

electrolyte had much better electrochemical stability compared to the $NaClO_4$. Unfortunately, the electrochemical decomposition of electrolytes was found to occur at 60°C, resulting in the subsequent performance degradation of the EDLCs.

Chandrasekaran et al. [495] reported the $Mg(ClO_4)_2$ salt-based EDLCs showed similar or better performances than the $LiClO_4$ salt-based electrolytes. In particular, the rate capability of Mg-based ACN and PC electrolytes was much better than the Li-based electrolytes, suggesting the high accessibility and utility of AC pores by solvated Mg ions. The electrochemical behavior of the cesium carborane ($CsCB_{11}H_{12}$) salt-based ESs were also investigated by Laheäär et al. [496]. Specific adsorption of Cs^+ ions on micro–mesoporous carbon surface and carbide-derived carbon powder was observed and it was dependent on the electrolyte concentration and on the solvent. Stronger specific adsorption with partial charge transfer was observed in the ACN-based electrolytes compared to the EC–DMC (1:1 by volume ratio).

2.4 IL-BASED ELECTROLYTES

2.4.1 General Composition and Properties of IL Electrolytes

ILs, also known as low-temperature or room temperature molten salts, are generally defined as salts composed solely of ions with melting points below 100°C [497]. ILs consist entirely of ions and are usually composed of a large asymmetric organic cation and an inorganic or organic anion. This special combination of a certain cation and anion results in a low melting point [498–500]. ILs have recently received considerable attention as electrolytes for ESs due to their distinct potential advantages such as high chemical, electrochemical and thermal stability, negligible volatility, and nonflammability (depending on the type of cations and anions) [498,501]. Furthermore, a large variety (virtually unlimited) of combinations of cations and anions makes the physical and chemical properties of ILs highly tunable [502,503]. Thus, ILs are also regarded as "designer solvents" [504]. This makes it very attractive for ILs to be used as electrolytes for ESs since they can be optimized or customized to meet certain requirements of ESs in terms of operative cell voltage, working temperature range, ESR, etc. Table 2.10 summarizes the conductivity, viscosity, melting point, cathodic limit, and anodic limit (obtained at smooth Pt or GC electrodes), and electrochemical stability of a number of common ILs reported for ESs [369,505].

According to their composition, ILs can be briefly grouped into aprotic, protic, and zwitterionic types, as shown in Figure 2.49 [498]. To date, the ILs used in ESs represent only a very small portion of the large family of ILs. The investigated ILs for ESs are commonly based on imidazolium, pyrrolidinium and ammonium, sulfonium, phosphonium cations, etc (Figure 2.50). Typical anions of ILs include tetrafluoroborate (BF_4^-), hexafluorophosphate (PF_6^-), bis(trifluoromethanesulfonyl) imide ($TFSI^-$), bis(fluorosulfonyl)imide (FSI^-), and dicyanamide (DCA^-).

As shown in Table 2.10, imidazolium-based ILs generally provide higher ionic conductivity whereas pyrrolidinium-based ILs exhibit larger electrochemical stability. It appears that a trade-off between the ionic conductivity and electrochemical stability of ILs exists. As mentioned previously, the operative cell voltages of commercial organic electrolyte-based EDLCs are generally restricted to 2.5–2.8 V,

TABLE 2.10
Electrochemical Stability and Conductivity Measured at Room Temperature for a Range of ILs [369,505]

	Electrochemical Stability					Conductivity
ILs	Cathodic Limit (V)	Anodic Limit (V)	ΔU (V)	Working Electrode	Reference	σ at 25°C ($mS\ cm^{-1}$)
		Imidazolium				
$[EtMeIm]^+[BF_4]^-$	−2.1	2.2	4.3	Pt	Ag/Ag^+, DMSO	14
$[EtMeIm]^+[CF_3SO_3]^-$	−1.8	2.3	4.1	Pt	I^-/I_3^-	8.6
$[EtMeIm]^+[N(CF_3SO_2)_2]^-$	−2	2.1	4.1	Pt. GC	Ag	8.8
$[EtMeIm]^+[(CN)_2N]^-$	−1.6	1.4	3.0	Pt	Ag	–
$[BuMeIm]^+[BF_4]^-$	−1.6	3	4.6	Pt	Pt	3.5
$[BuMeIm]^+[PF_6]^-$	−1.9	2.5	4.4	Pt	Ag/Ag^+, DMSO	1.8
$[BuMeIm]^+[N(CF_3SO_2)_2]^-$	−2	2.6	4.6	Pt	Ag/Ag^+, DMSO	3.9
$[PrMeMeIm]^+[N(CF_3SO_2)_2]^-$	−1.9	2.3	4.2	GC	Ag	3.0
$[PrMeMeIm]^+[C(CF_3SO_2)_3]^-$	–	5.4	5.4	GC	Li/Li^+	–
		Pyrrolidinium				
$[n\text{PrMePyrrol}]^+[N(CF_3SO_2)_2]^-$	−2.5	2.8	5.3	Pt	Ag	1.4
$[n\text{BuMePyrrol}]^+[N(CF_3SO_2)_2]^-$	−3.0	2.5	5.5	GC	Ag/Ag^+	2.2
$[n\text{BuMePyrrol}]^+[N(CF_3SO_2)_2]^-$	−3.0	3.0	6.0	Graphite	Ag/Ag^+	–
		Tetraalkylammonium				
$[nMe_3BuN]^+[N(CF_3SO_2)_2]^-$	−2.0	2.0	4.0	Carbon	–	1.4
$[nPrMe_3N]^+[N(CF_3SO_2)_2]^-$	−3.2	2.5	5.7	GC	Fc/Fc^+	3.3
$[nOctEt_3N]^+[N(CF_3SO_2)_2]^-$	–	–	5.0	GC	–	0.33
$[nOctBu_3N]^+[N(CF_3SO_2)_2]^-$	–	–	5.0	GC	–	0.13
		Pyridinium				
$[BuPyr]^+[BF_4]^-$	−1	2.4	3.4	Pt	Ag/AgCl	1.9
		Piperidinium				
$[MePrPip]^+[N(CF_3SO_2)_2]^-$	−3.3	2.3	5.6	GC	Fe/Fe^+	1.5
		Sulfonium				
$[Et_3S]^+[N(CF_3SO_2)_2]^-$	–	–	4.7	GC	–	7.1
$[nBu_3S]^+[N(CF_3SO_2)_2]^-$	–	–	4.8	GC	–	1.4

Source: Reprinted from *Journal of Power Sources*, 173, Lewandowski, A., and M. Galinski, Practical and theoretical limits for electrochemical double-layer capacitors, 822–828, Copyright 2007, with permission from Elsevier.

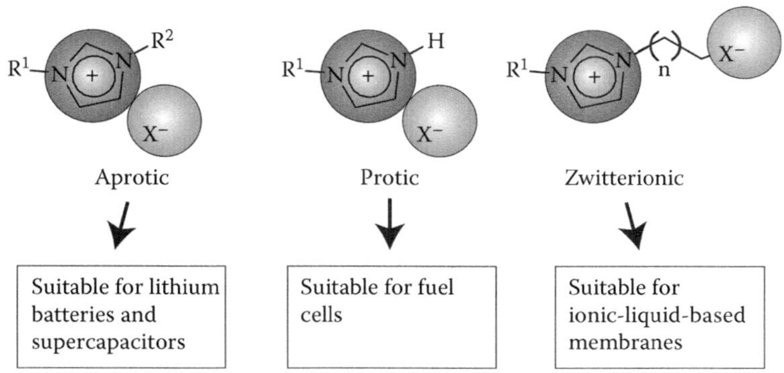

FIGURE 2.49 Basic types of ILs: aprotic, protic, and zwitterionic types. (Reprinted by permission from Macmillan Publishers Ltd. *Nature Materials* Armand, M. et al. 2009. Ionic-liquid materials for the electrochemical challenges of the future. 8:621–629, copyright 2009.)

and serious electrochemical decomposition of organic solvents would occur if the cell voltage increases beyond this limit. On the contrary, many investigations have shown that EDLCs with IL-based electrolytes can reach high operative cell voltages above 3 V [501,505,506]. The high cell voltage of EDLCs with several typical IL electrolytes can also be clearly seen in a recent review by Brandt et al. (see Table 1 in Reference 501). In addition, commercial organic solvents such as ACN also suffer from safety issues because of their volatile and flammable nature especially at high temperatures. Therefore, the application of solvent-free ILs may also solve the safety problems related with these organic solvents. The usage of ILs as electrolytes could significantly increase the operative temperature well beyond the limit of organic electrolytes (e.g., ~60°C for ACN), which makes IL-based ESs applicable for high-temperature applications. For example, Borges et al. [507] recently reported an ES which could operate even at 200°C, with [BMMI][TFSI] IL electrolyte, naturally clay separator, and graphitic carbon electrodes. Balducci et al. [508] found the cell resistance (9 Ω cm^2) of EDLCs with [PYR_{14}][TFSI] IL electrolyte did not change after 40,000 cycles at a cell voltage up to 3.5 V at 60°C, indicating a high cycling stability. Fletcher et al. [509] found increasing temperature decreased the IL electrolyte viscosity and increased the self-diffusion coefficients of both the anion and the cation, resulting in the decrease of both electrolyte resistance and ESR of EDLCs with 1,2-dimethyl-3-propylimidazolium bis(trifluoromethylsulfonyl)imide ([DMPIM][TFSI]) electrolyte.

Unfortunately, most ILs have several main disadvantages, such as low ionic conductivity, high viscosity, and high cost. This greatly limits the practical use of IL-based electrolytes for ESs. As shown in Table 2.10, among frequently used ILs, [EMIM][BF_4] has a relatively high ionic conductivity of 14 mS cm^{-1} at 25°C, but this value is still much lower than the conductivity of commonly used organic electrolytes such as $TEABF_4$/ACN (59.9 mS cm^{-1} at 25°C). Besides, ILs generally suffer from much higher viscosities compared to organic electrolytes. For example, [EMIM][BF_4] IL had the viscosity of 41 cp [510,511], which is more than one magnitude higher than the ACN organic electrolyte with the viscosity of only 0.3 cp [296]. As a result, this greatly increases the ESR of the ESs using IL electrolytes

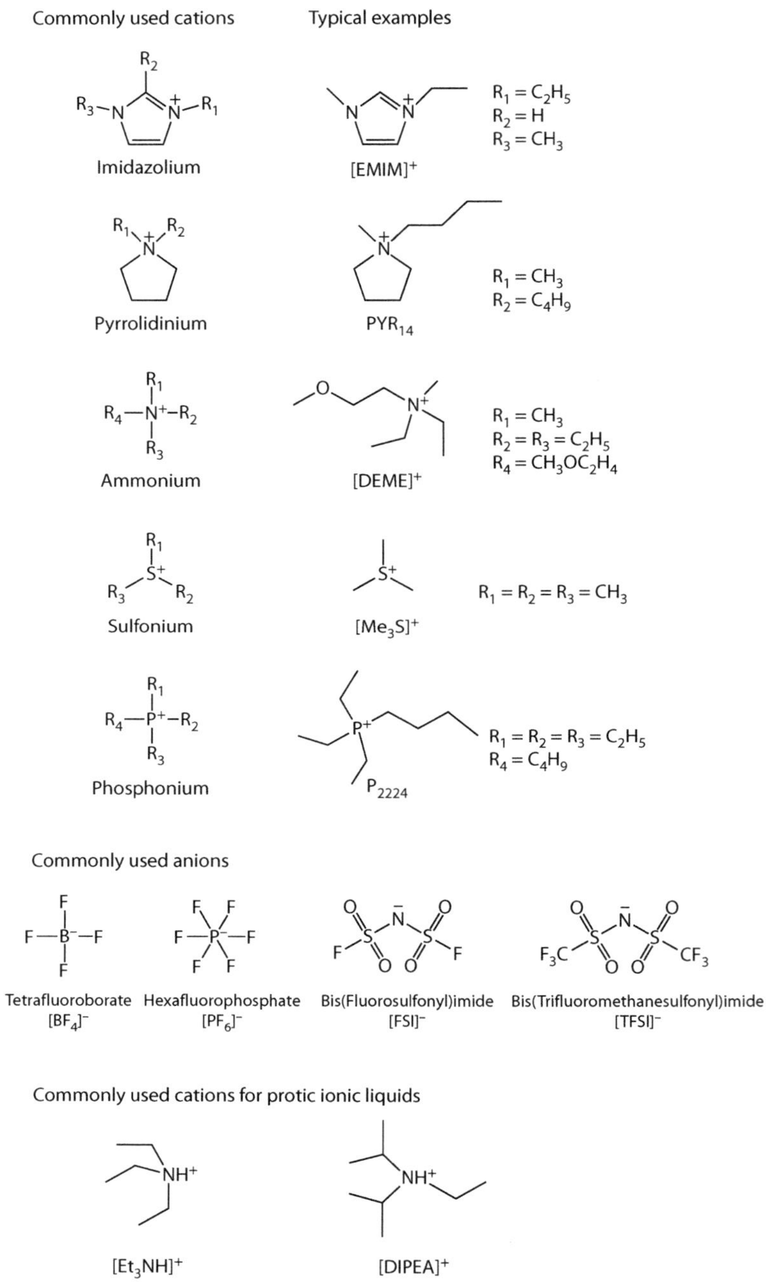

FIGURE 2.50 Commonly used cations, anions of ILs for ESs, and some typical examples of ILs. (Zhong, C. et al. 2015. A review of electrolyte materials and compositions for electrochemical supercapacitors. *Chemical Society Reviews* 44:7484–7539. Reproduced by permission of The Royal Society of Chemistry.)

and thus limits the power performance. Some studies including comparative studies between organic and IL electrolytes showed that this problem is usually more serious at room temperature or even lower temperature [501,505,506,512,513]. Besides, due to the high viscosity of ILs, EDLCs generally exhibit smaller specific capacitance values compared to aqueous and organic electrolyte-based EDLCs especially at high charging/discharging rate [512,514,515].

To date, considerable efforts have been devoted to the understanding of the IL/electrode interface behavior and the improvement of the properties of IL-based electrolytes for ESs. As will be discussed below, these achievements can be briefly summarized as follows: (1) fundamental understanding of the structure and capacitance of EDL at the IL/electrode interface; (2) optimizing the properties of ILs such as ionic conductivity, viscosity, and electrochemical stability by rationally choosing or designing the cations or anions or both; (3) exploring the mixture of ILs or the introduction of organic solvents to improve the overall performance of IL-based electrolytes for ESs; and (4) developing ILs for pseudocapacitive electrode materials to improve the energy density of ESs.

2.4.1.1 EDL at the IL/Electrode Interface

The optimum selection and design of IL-based electrolytes for EDLCs will benefit from the understanding of the EDLs in ILs [516–518]. Because of the particularity of ILs such as solvent-free and strong ion–ion interaction compared to aqueous and organic electrolytes, extensive experimental and theoretical work has been carried out to provide fundamental insights into the electrochemical behavior, EDL structure, and corresponding capacitive behavior at the IL/electrode interface. Significant achievements have recently been made, and several excellent reviews have been published in this area [333,518–522]. Briefly, owing to the relatively strong cation–anion correlations, extensive experimental [523,524] and theoretical (e.g., DFT, MD, and MC simulation) [523,525–531] studies found that the EDL of ILs shows alternating layers of cations and anions (i.e., ion densities form an oscillatory pattern) near a charged planar surface. It has also been demonstrated that overscreening is a universal characteristic of EDLs originating from the high ionic density and strong ion correlation in ILs [523,525,527,532–534]. Using the surface force apparatus, Smith et al. [535,536] and Perkin et al. [537], Perkin [538] investigated the structural and dynamic properties of imidazolium and pyrrolidinium-based ILs in confined geometries (between two identical atomically smooth and charged surfaces with varied thickness). It was found that the interaction force across the investigated IL films was an oscillatory function of the film thickness, revealing the ion packing and dimensions of layered structures of ILs. ILs with shorter alkyl chain substituents such as [C_4C_1Pyrr][NTf_2] form alternating cation–anion monolayers, whereas ILs with longer alkyl chain such as [$C_{10}C_1$Pyrr][NTf_2]) results in the bilayer formation, as shown in Figure 2.51 [536]. Using AFM, Black et al. [539] observed the reconstruction of the Stern layer that consisted of a reorientation of molecules under the applied bias for the [EMIM][Tf_2N] IL electrolyte on highly ordered pyrolytic graphite (HOPG), which was in agreement with the MD simulation.

In addition to the investigations on the EDL structure at the IL/electrode interface, extensive experimental and theoretical work has focused on the relationship between the

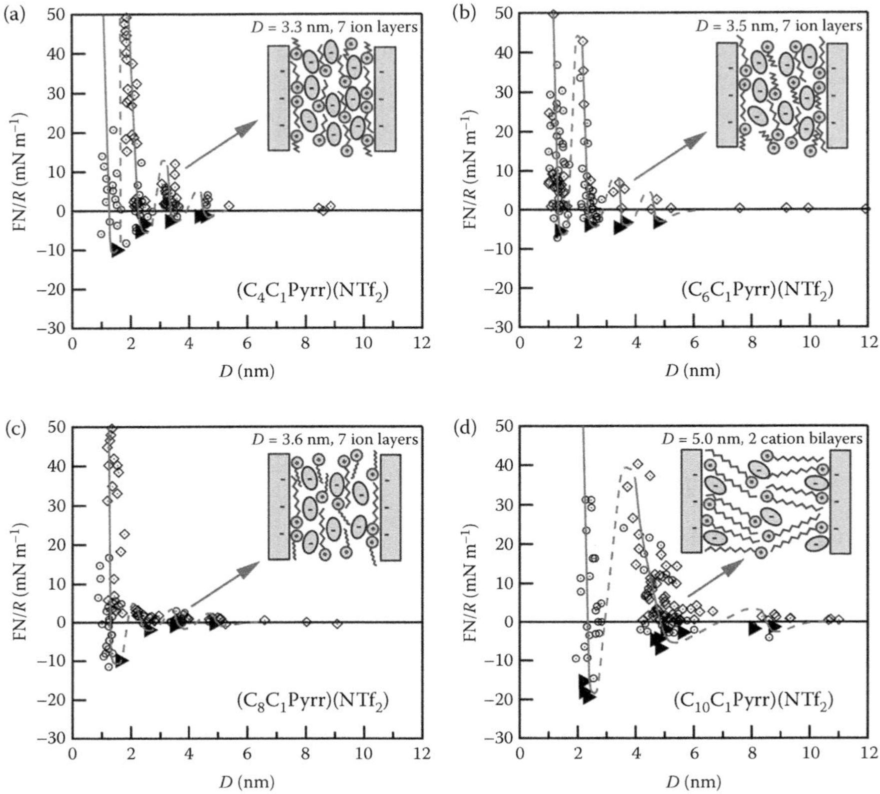

FIGURE 2.51 Measured normal force, FN, normalized by radius of curvature, *R*, between mica surfaces across (a) [C_4C_1Pyrr][NTf_2], (b) [C_6C_1Pyrr][NTf_2], (c) [C_8C_1Pyrr][NTf_2], and (d) [$C_{10}C_1$Pyrr][NTf_2] as a function of surface separation, D. Schematics indicate the likely ion layering structures, showing example film configurations consisting of alternating cation/anion monolayers for [C_nC_1Pyrr][NTf_2], where $n = 4$, 6, 8, and interdigitated cation bilayers for [$C_{10}C_1$Pyrr][NTf_2]. (Smith, A. M., K. R. J. Lovelock, and S. Perkin. 2013. Monolayer and bilayer structures in ionic liquids and their mixtures confined to nano-films. *Faraday Discussions* 167:279–292. Reproduced by permission of The Royal Society of Chemistry.)

differential capacitance of the EDL and the electrode potential (*C*–*V*). Experimentally, various shapes of *C*–*V* curves, such as the U, bell, and camel shapes, were observed at the IL/electrode interface [522,540,541]. Considerable theoretical modeling and simulation work has been carried out to understand the physical meaning and to predict the *C*–*V* in IL-based EDLCs [542–544]. For example, Kornyshev [532] proposed a mean-field theory to explain various *C*–*V* curves by taking into account the packing structure of the IL. Based on this theory, a bell-shaped *C*–*V* curve was observed for a densely packed IL while a camel-shaped *C*–*V* curve was found for a sparsely packed IL. Using MC simulations, Fedorov et al. [545] reported for ILs with charged heads and neutral counterparts, the neutral counterparts resulted in the camel shape of the *C*–*V* curve (Figure 2.52). Jiang et al. [528] demonstrated the DFT was effective for

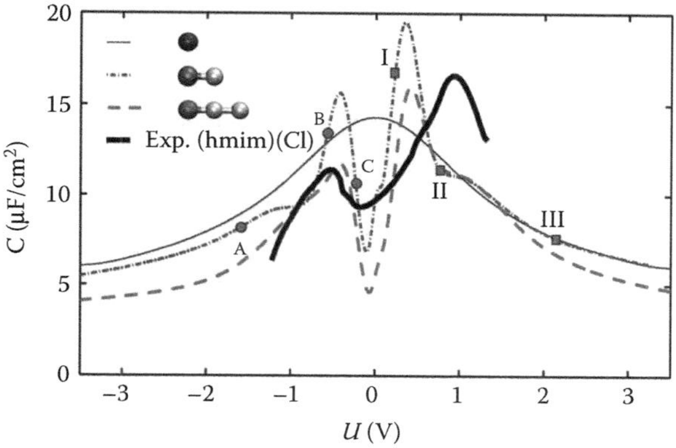

FIGURE 2.52 Differential capacitance obtained from simulations. The maxima are located at 0 V for 1BM; 0.35 V and −0.4 V for 2BM; and 0.4 V and −0.5 V for 3BM. The black solid curve is the experimental data for [hmim][Cl] from Reference 541 given with permission. (Reprinted from *Electrochemistry Communications*, 12, Fedorov, M. V., N. Georgi, and A. A. Kornyshev, Double layer in ionic liquids: The nature of the camel shape of capacitance, 296–299, Copyright 2010, with permission from Elsevier.)

illustrating the *C*–*V* curve in ILs and its correlation with the ion size and concentration. DFT studies [528,546] were able to reproduce the transition of the *C*–*V* curve from the bell shape and camel shape as indicated by other studies [532]. In an effort to bridge the gap between oversimplified theoretical assumptions and complex real IL systems, Breitsprecher et al. [547,548] proposed a hierarchy of generic coarse-grained models of ILs and took into account the ion size, shape and charge asymmetry, and so on. The general shape of *C*–*V* curve was found to be mainly controlled by the charge per occupied volume in the very first layer [548]. Merlet et al. [549] reported that the coarse-grained model was able to calculate the differential capacitances of the positive and negative graphite electrodes in [BMIM][PF_6], and also accurately reproduced the structure of ILs at the graphite electrode surface. The effect of electrode surface topography on the EDL structure and the corresponding *C*–*V* curve has also been considered [550,551]. For instance, using MD simulations, Bedrov et al. [550] very recently demonstrated the significant effect of the atomic scale topography of the electrode surface on the *C*–*V* curve. It was found that for atomically flat surfaces, a number of different ILs exhibited similar values for capacitance with a weak dependence on the electrode potential. On the contrary, in the same ILs on atomically corrugated surfaces with surface roughness comparable with ion dimensions, the magnitude of the capacitance and its relationship with the electrode potential could be varied and complex depending on the roughness of the electrode surface. The authors' results provide a possible explanation why different results exist from many experimental measurements obtained for a similar IL/planar electrode [550].

A series of factors, such as the chain length of cations [540,552–557], anion size and nature [525,552,558,559], ion shape [534], size ratio of the cation to anion [560], specific adsorption [560,561], ion valence [552,562], temperature [509,540, 553,554,563–565], electrode surface morphology, and electrode materials [558], have been investigated and found to have a profound influence on the value and shape of the *C–V* curve. Ion size affected the EDL thickness and alkyl chain length influenced the relative permittivity [566], which played an important role in the obtained differential capacitance value [540,552,555,558,566]. Using mean-field theory, the unequal size of the cation and anion was found to lead to asymmetric *C–V* curves [560]. The EDL structure and *C–V* curves of imidazolium-based dicationic ILs with different chain lengths ([C–n(MIM)$_2$][(BF$_4$)$_2$] (n = 3, 6, 9) and [C-6(MIM)$_2$][(Tf$_2$N)$_2$]) at the OLC electrode surface were investigated and compared with their monocationic counterparts by Li et al. [552] using MD simulation. Higher concentrations of dicationic IL anions were found near the electrode surface since each dication needs charge balance by two anions. The differential capacitance obtained in dicationic [C$_6$(MIM)$_2$] [(Tf$_2$N)$_2$] was observed to be higher compared with that obtained in monocationic [C$_6$MIM][Tf$_2$N], which was also supported by the CV measurements at low scan rates.

Besides the planar electrode surface, the effects of electrode curvature [567,568] and cylindrical electrode [525] on the EDL structure and capacitance of IL electrolyte were investigated by MD simulations. A nearly flat shape of the *C–V* curve was observed for [EMIM][Tf$_2$N] IL at the spherical OLCs by MD simulations, which was different with that reported for planar electrodes (e.g., bell and camel shape),

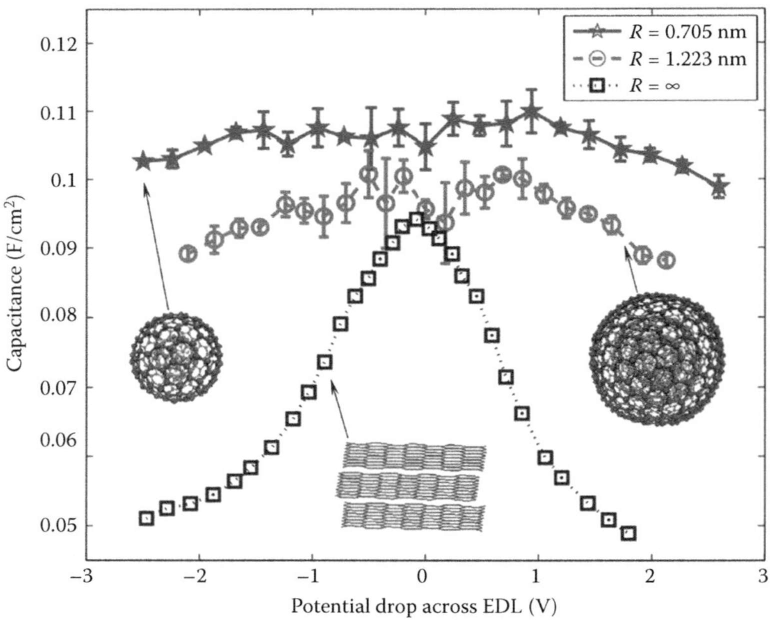

FIGURE 2.53 Influence of electrode curvature on differential capacitance. R is the radius of an OLC and is infinity for the planar electrodes. (Reprinted with permission from Feng, G., D. E. Jiang, and P. T. Cummings. 2012. Curvature effect on the capacitance of electric double layers at ionic liquid/onion-like carbon interfaces. *Journal of Chemical Theory and Computation* 8:1058–1063. Copyright 2012 American Chemical Society.)

as shown in Figure 2.53 [567]. The authors attributed this to the dominance of an almost constant charge overscreening over the simulated potential range. The MD simulations also showed the EDL capacitance on OLCs increased with the increasing curvature or size of OLCs, which was explained by the enhanced ion density per unit area of electrode surface [567]. Similar results that the EDL capacitance was higher at the cylindrical electrodes compared with the planar electrode for the same IL electrolyte were observed by MD simulations [525]. Interestingly, using MD simulations, Xing et al. [569] demonstrated that the EDL capacitance could be significantly improved by nanopatterning the electrode surface. Electrode surface patterns or rough edges with groove widths matching the ion size were found to exhibit a larger EDL capacitance for the investigated IL electrolyte ([PYR_{13}][FSI]) due to the promotion of ion separation in the EDL.

The effect of temperature on the differential capacitance obtained at IL/electrode interface is relatively complicated. Some experimental works showed the differential capacitance increased with the increasing temperature [509,540,553,558,563] while other studies found the differential capacitance decreased as the temperature increased [570]. Similar controversy also exists in the simulation work concerning the temperature effect on the EDL and capacitance properties in ILs [565,571–573]. A comprehensive discussion on the temperature effect on EDL properties and differential capacitance in ILs has been given by Fedorov and Kornyshev in a recent review [518].

Since electrode materials for EDLCs are generally based on nanoporous carbon-based materials, it is essential to investigate the technologically important nanoporous carbon electrodes. ILs generally have relatively large anions and cations. The matching between the electrolyte ion size and the pore size distribution of the carbonaceous materials is a very important issue determining the performance such as the capacitance and power density of EDLCs. The behavior and capacitance of the EDL in ILs for carbons with different pore size [521,574–583], and different structure such as CNT [568,584–586], carbon xerogel [580], and OLC [552], have been studied both experimentally [574–580] and by theoretical simulation [521,552,568,581–586] methods so as to understand the way to optimize matching between the IL electrolytes and carbon electrodes. Solvent-free ILs can be used as probes for investigating the relationship between the ion size and carbon pore size for EDLCs [577,587]. As mentioned previously, experimental studies found an anomalous increase of the specific capacitance in porous carbon with pore size approaching the electrolyte ion size [328]. This was later experimentally supported by a solvent-free IL electrolyte (i.e., [EMIM][TFSI]), as shown in Figure 2.54 [587]. The maximum capacitance was achieved when the carbon pore size was similar to that of the electrolyte ions.

Since then, extensive modeling and simulation work has been focused on providing a microscopic insight into the effect of the pore size on the EDL structure and capacitance [516,581,585,586,588,589]. Merlet et al. [516] used MD simulation to explain the molecular mechanism of capacitance increase at the subnanometer pores. The electrode was found to be wetted by the IL electrolyte even at null potential (Figure 2.55) and the charging mechanism involved the exchange of ions between the electrodes and the bulk electrolyte instead of the filling of the pores by the liquid IL (Figure 2.56). The volume of liquid IL inside the electrode remained almost constant.

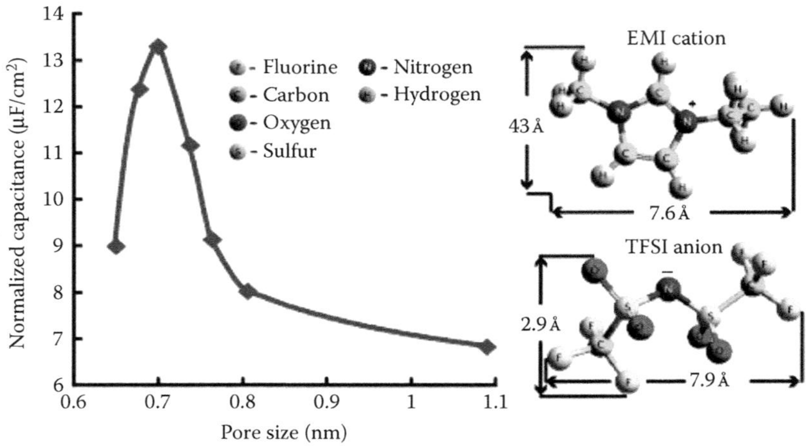

FIGURE 2.54 Normalized capacitance change versus the pore size of the CDC samples prepared at different temperatures; normalized capacitance is obtained by dividing the specific capacitance by the SSA. HyperChem models of the structure of EMI and TFSI ions show a size correlation. (Reprinted with permission from Largeot, C. et al., 2008. Relation between the ion size and pore size for an electric double-layer capacitor. *Journal of the American Chemical Society* 130:2730–2731. Copyright 2008 American Chemical Society.)

Using DFT, Jiang et al. [581] predicted the capacitance of an IL electrolyte inside a nanopore oscillated (having multiple peaks) with a decaying envelope with the increasing pore size (Figure 2.57). This capacitance oscillation behavior was different from that reported in a simulation study for organic electrolytes [338]. The authors attributed this to the interference of the overlapping EDLs. Further experiments will be required to prove this DFT prediction. Feng and Cummings [588] also reported the oscillatory behavior of the capacitance as a function of the nanopore size for the [EMIM][Tf_2N] IL electrolyte in the slit-shaped micropores. Based on atomistic simulations, Wu et al. [589] reported the capacitance of slit-shaped nanopores in IL as a function of pore size exhibiting a U-shaped behavior. Using MD simulations for [EMIM][TFSI] IL in nanoporous electrodes, Xing et al. [582] showed the capacitance obtained in subnanometer slit-geometry nanopores could be increased twice over compared with the flat electrode. Besides, this capacitance enhancement was observed at the negatively charged nanopore whereas almost no enhancement was found at the positively charged electrode. Regarding the CNT electrode, Shim and Kim [586] found the internal solvation of small (6,6) and (7,7) CNTs with an electrified interior wall was effected almost exclusively via counterions in the [EMIM][BF_4] IL electrolyte, while counterions and co-ions formed multilayer solvation structures for larger (10,10) and (15,15) CNTs by using MD simulations. As the CNT diameter decreased from (15,15) to (7,7), the specific capacitance (normalized by the pore surface area) increased. However, further decreasing the CNT diameter from (6,6) to (5,5) reduced the specific capacitance. Jover et al. [585] recently presented a screening methodology for the efficient pairing of different ILs and carbonaceous

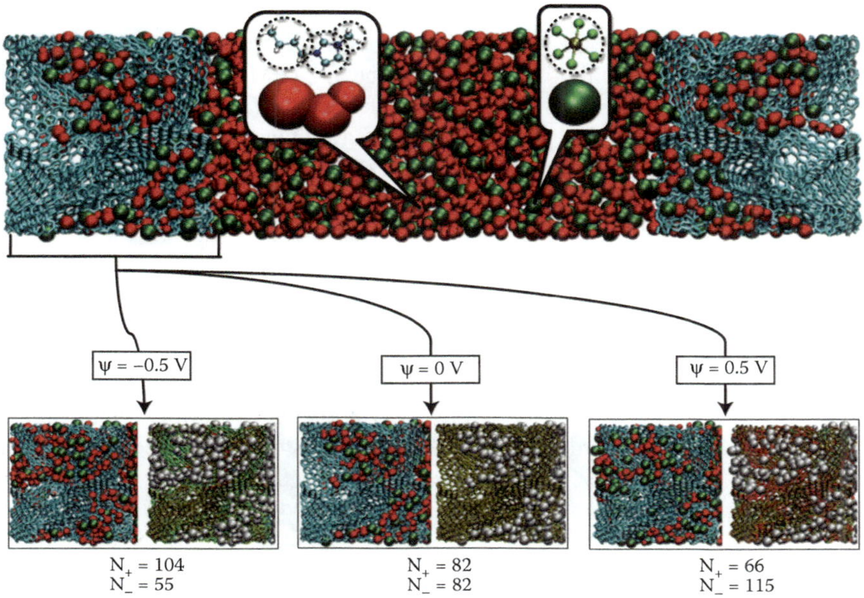

FIGURE 2.55 EDLC simulation cell. Upper panel: the simulation cell consists of a [BMIM][PF_6] IL electrolyte surrounded by two porous electrodes (CDC-1200) held at constant electrical potentials (blue: C atoms, red: the three sites of BMI^+, and green: PF_6^- ions; a coarse-grained model is used to describe these ions). Lower panel: structure of the electrode for various voltages. For each value, the same snapshot is shown twice: the ionic distribution is shown on the left. The degree of charging of the electrode atoms is shown on the right, where the carbon atoms are colored according to the charge q they carry (green: $q < 0$, red: $q > 0$, and yellow: $q \approx 0$). The charging mechanism involves the exchange of ions between the bulk and the electrode. (Reprinted by permission from Macmillan Publishers Ltd. *Nature Materials* Merlet, C. et al. 2012. On the molecular origin of supercapacitance in nanoporous carbon electrodes. 11:306–310, copyright 2012.)

electrodes (single-wall CNTs). This study related the capacitance with the desolvation, dissociation, and adsorption energy of ILs into the CNT electrode. The most appropriate pore size to achieve the maximum capacitance was given for each IL. Using MD simulations, Vatamanu et al. [590] systematically compared the specific capacitance of carbon nanoparticles of various shapes and dimensions. As shown in Figure 2.58, a noticeable capacitance enhancement was observed when the electrode curvature and the length scale of the surface roughness were comparable to the electrolyte ion dimensions.

The ion size to carbon pore width ratio (IS/PW ratio) was also reported to influence the electrochemical stability of AC electrodes as reported by Shilina et al. [591]. In this study, decreasing the IS/PW ratio was found to significantly increase the electrochemically stable window of AC electrodes in organic electrolytes and IL electrolytes ([P_{2228}][TFSI] and [PIP_{13}][FSI) although it led to the loss of the reversible capacitance.

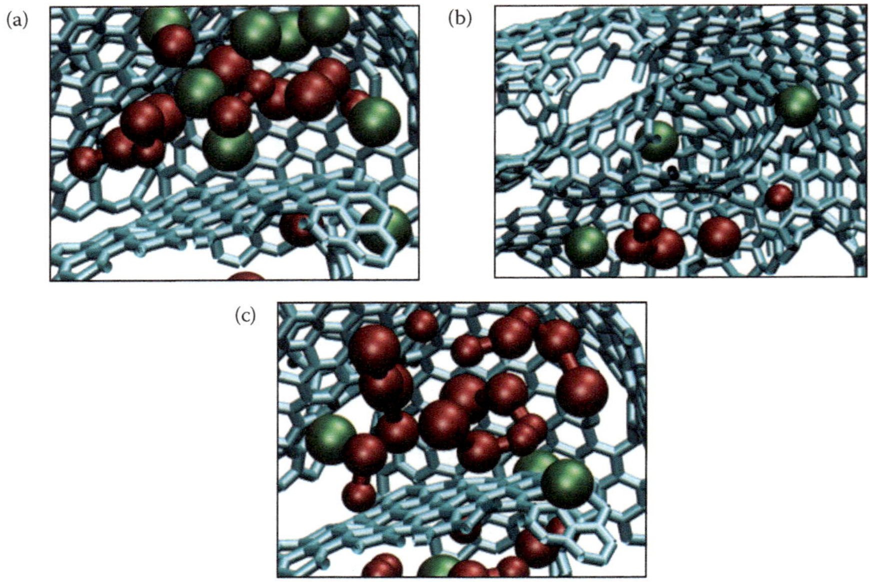

FIGURE 2.56 Typical structure of the IL inside electrified pores of the CDC-1200 material. Blue: C–C bonds, red: BMI^+, and green: PF_6^-. (a) Local structure near a positive surface (+0.5 V), the anionic density is enhanced. (b) A single anion in a nanotube-like pore positively polarized (+0.5 V). (c) Same as (a) but near a negative surface (−0.5 V). (Reprinted by permission from Macmillan Publishers Ltd. *Nature Materials* Merlet, C. et al. 2012. On the molecular origin of supercapacitance in nanoporous carbon electrodes. 11:306–310, copyright 2012.)

Apart from the EDL structure and capacitance, the dynamical properties of the IL electrolytes confined in nanopores and its relationship with the effect of pore size and structure have also been investigated [592,593]. MD simulation demonstrated the confined cations moved faster than the anions of the [BMIM][PF_6] IL electrolyte inside MWCNTs, and the confinement remarkably decreased the ion dynamics [593].

In addition to the EDL capacitance, the contribution of the quantum capacitance to the interfacial capacitance at the IL/electrode interface has been investigated by DFT and MD simulations [568,594–597]. Different electrode materials including the CNT [568,597] and graphene [594–596], and various influencing factors such as the electrode curvature [568], doping of heteroatoms [595,596], and graphene edges [594], have been investigated.

Besides the capacitance, high power density is also one of the key characteristics of ESs. Therefore, the related charging/discharging rate and electrolyte ion diffusion in the porous carbons are important issues. Although the specific capacitance value was reported to achieve a maximum value for the subnanometer pore size approaching the electrolyte ion size, some studies pointed out that this tends to decrease the rate performance and power density, especially considering the high viscosity and relatively large ions size of ILs [61,580,598]. For example, Sillars et al. [580] reported the pore size of AC xerogel for optimum capacitance and optimum rate performance

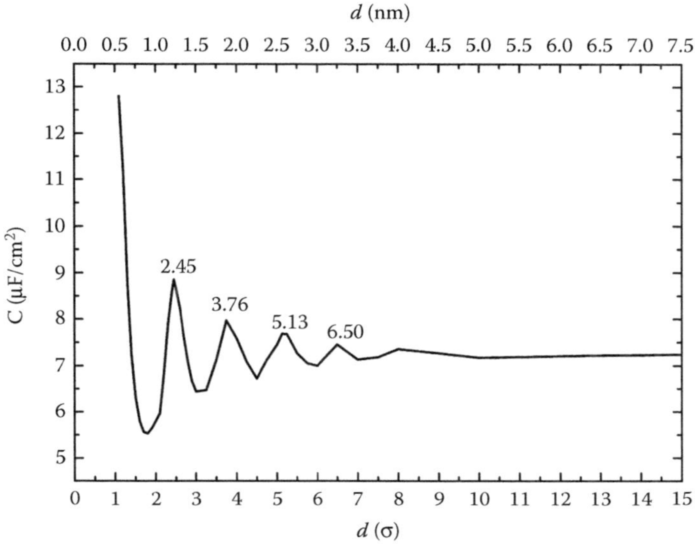

FIGURE 2.57 Capacitance (C) of a nanoporous electrode as a function of the pore width (*d*) up to 15σ. Peak positions are labeled in terms of the ion diameter (σ = 0.5 nm). (Reprinted with permission from Jiang, D. E., Z. Jin, and J. Wu. 2011. Oscillation of capacitance inside nanopores. *Nano Letters* 11:5373–5377. Copyright 2011 American Chemical Society.)

in IL electrolytes was different. In an effort to increase the capacitance (energy density) of the IL electrolyte-based EDLCs without sacrificing the power density, carbon electrodes with various structures such as hierarchical porous carbons [64,599] and three-dimensional (3D) OM carbons [61], have been developed. For hierarchical porous carbons, the meso/macroporous network was proposed to promote the IL transfer while a large number of small nanopores were beneficial for the ion storage [64,599]. This contributes to a high energy density and power density. Vu et al. [61] reported the synthesis of 3D OM carbons with well-defined geometry, 3D interconnected pore structure and tunable pore size for EDLCs using the [EMIM][TFSI] IL electrolyte. This study identified the optimal pore sizes that could provide a large interfacial area for high capacitance while maintaining efficient ion transport for high power performance. 3D OM carbons with 21–29 nm pore diameters showed similar high specific capacitance (146–178 F g^{-1} at 0.5 A g^{-1}) as typical AC-based EDLCs but exhibited much better high-rate performance (80–123 F g^{-1} at 25 A g^{-1}). The surface chemical properties of the carbon electrodes affected the wettability of the carbon surface, which was also found to influence the ion diffusion in mesopore channels in ILs [600]. Using *in situ* infrared spectroelectrochemistry, Richey et al. [601,602] investigated ion dynamics of ILs in the operating EDLCs with different carbonous electrode materials (Figures 2.59 and 2.60). For porous CDCs electrodes, both cations and anions of [EMIM][TFSI] IL electrolyte were directly observed entering and exiting CDC nanopores during the charging/discharging process [601] (Figure 2.59). On the contrary, in the case of OLC electrodes, electrolyte ions were

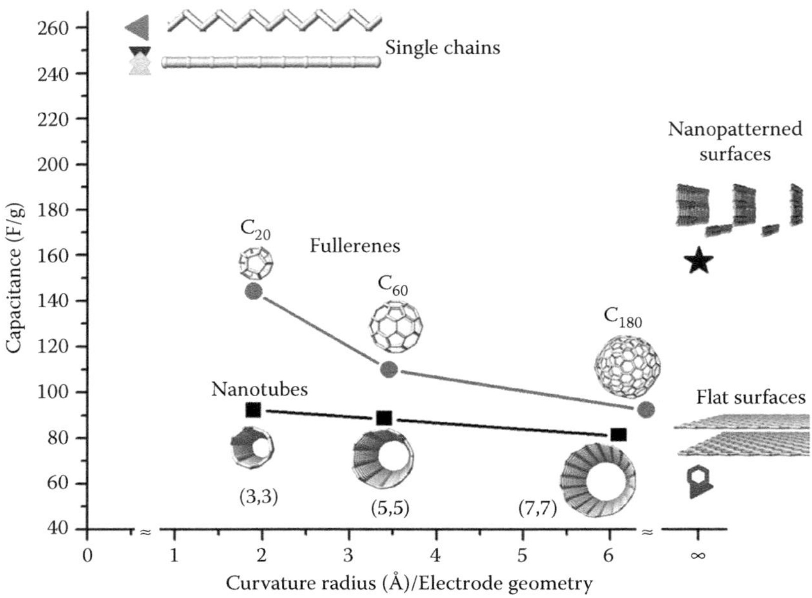

FIGURE 2.58 Capacitance as a function of electrode surface curvature radius and geometry for various open structure electrode geometries as obtained from MD simulations at the 4 V potential difference between electrodes. (Reprinted with permission from Vatamanu, J. et al., 2013. Increasing energy storage in electrochemical capacitors with ionic liquid electrolytes and nanostructured carbon electrodes. *Journal of Physical Chemistry Letters* 4:2829–2837. Copyright 2013 American Chemical Society.)

observed within close proximity to the OLC surface and transported only small distances as they adsorb on the OLC surface [601] (Figure 2.60). This explained why the OLC electrode is beneficial for EDLCs with fast charging/discharging rate. The effect of KOH activation on the ion dynamics for EDLCs with nanoporous CNFs electrodes in the [EMIM][TFSI] IL electrolyte during the charge/discharge process were investigated by the same group using *in situ* infrared spectroelectrochemistry [602]. KOH activation was found to cause cations to desorb from the nanopores while anions were adsorbed into the nanopores, providing direct experimental evidence. The surface chemistry and ionophilicity of the carbon material played an important role in the charge storage behavior of IL electrolytes. Theoretical and simulation work concerning the charging dynamics [603,604] and ion diffusion [605] for ILs in nanopores has also been carried out in parallel [604]. Using MD simulations and a phenomenological model, Kondrat et al. [598] showed the charging of ionophilic pores, with widths comparable to the ion diameter, followed a diffusion law. Interestingly, it was shown that the ion diffusion during charging could be an order of magnitude faster than in the bulk, which was in contrast to conventional expectations. Making pores ionophobic was found to further accelerate the charging by an order of magnitude by avoiding the overfilling/de-filling. Van Aken et al. [605] investigated the effect of different cations ([EMIM][Tf_2N], 1-hexyl-3-methylimidazolium

(a)

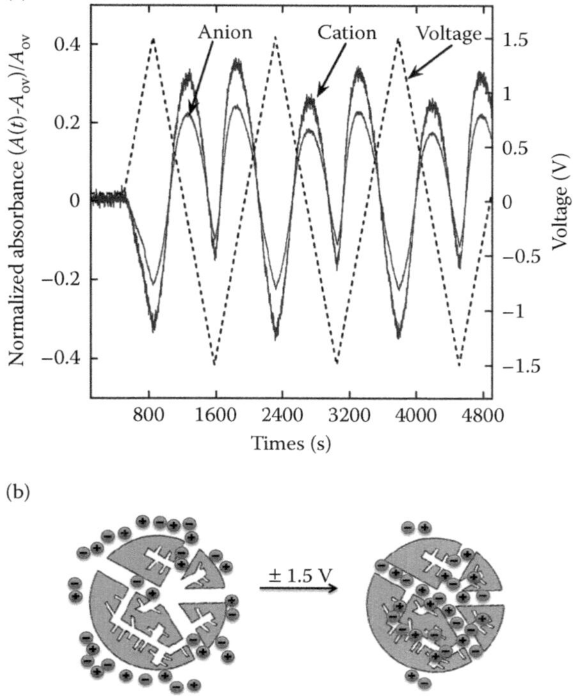

(b)

FIGURE 2.59 Results for CDC electrodes: (a) normalized absorbance of the infrared bands of the cation and anion of [EMIM][TFSI] as a function of time during three CV cycles from −1.5 to +1.5 V at a scan rate of 5 mV s^{-1} (open circuit voltage 0.068 V); (b) schematic illustrating the ion dynamics within pores of CDC particles during CV experiments. (Reprinted with permission from Richey, F. W. et al. 2013. Ion dynamics in porous carbon electrodes in supercapacitors using *in situ* infrared spectroelectrochemistry. *Journal of the American Chemical Society* 135:12818–12826. Copyright 2013 American Chemical Society.)

bis(trifluoromethylsulfonyl) imide ([HMIM][Tf_2N]), and 1,6-bis(3-methylimidazolium-1-yl) bis(trifluoromethylsulfonyl)imide ([BMIH][Tf_2N])) on the electrochemical performance of OLC-based EDLCs using both experimental and MD simulation methods. Increasing the cation size was found to increase the ionic resistance due to the decreased diffusion coefficient, which also resulted in a decrease of the capacitance value at higher charging/discharging rates.

The surface chemistry property of the carbonaceous electrode materials has also been found to have a great influence on the specific capacitance and the rate performance of the IL electrolyte-based EDLCs [606]. Pinkert et al. [606] reported that introducing oxygen functionalities to the OM CDCs decreased the specific capacitance at higher power densities while the surface chlorination significantly enhanced the rate capability and increased the specific capacitance by at least 27% at scan rates ≥10 mV s^{-1}. In an effort to enhance the accessibility of highly viscous ILs into the porous electrodes, IL-functionalized (-modified) carbonaceous materials have

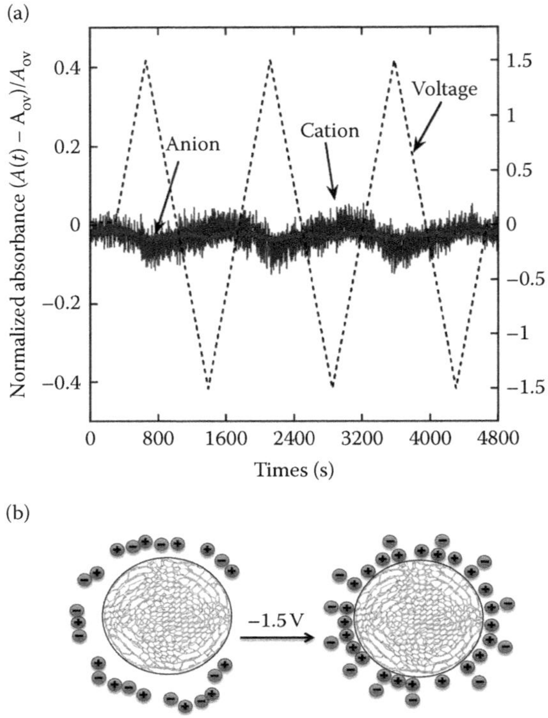

FIGURE 2.60 Results for OLC electrodes: (a) normalized absorbance of the infrared bands of the cation and anion of [EMIM][TFSI] as a function of time during three CV cycles from −1.5 to +1.5 V at 5 mV s^{-1} and (b) schematic illustrating the ion dynamics around OLC particles during CV experiments. (Reprinted with permission from Richey, F. W. et al. 2013. Ion dynamics in porous carbon electrodes in supercapacitors using *in situ* infrared spectroelectrochemistry. *Journal of the American Chemical Society* 135:12818–12826. Copyright 2013 American Chemical Society.)

been developed [607–611]. For example, Tamilarasan and Ramaprabhu [607] found CNTs functionalized with an IL ([BMIM][TFSI]) showed a specific capacitance value of 242 F g^{-1} at a current density of 2 A g^{-1}, which was 85% higher than that of the EDLCs with unfunctionalized CNTs electrodes in [BMIM][TFSI] IL electrolyte. EIS measurements showed that the IL functionalization remarkably reduced the charge transfer resistance. Kim et al. [611] reported the electrodes based on rGO modified with poly(1-vinyl-3-ethylimidazolium) bis(trifluoromethylsulfonyl) amide provided better compatibility with the IL electrolyte ([EMIM][NTf_2]) due to the improved wettability, which contributed to a relatively high specific capacitance.

In addition, some cations of investigated ILs possess a structure similar to surfactants [612,613]. It is important to take into account the possible aggregation behavior of cations like surfactants, which would influence the ion mobility and ionic conductivity of the electrolyte [612–614]. Using NMR and Raman methods, Kunze et al. [613] investigated the aggregation behavior of ILs consisting of the bis(trifluoromethanesulfonyl) imide [TFSI] anion and different *N*-alkyl-*N*-methylpyrrolidinium cations, with the alkyl chain varied from C_3H_7 to C_8H_{17}, also in the presence of Li^+ cations. It was found that the pyrrolidinium cations tended to form aggregates surrounded by anions starting from $[PYR_{14}]$[TFSI]. The aggregation became more pronounced as the length of the alkyl side chain increased or in the presence of Li^+ cations. $[PYR_{13}]^+$ cations did not show any aggregation in pure IL, however, they formed aggregates in the presence of Li^+ cations. In a later study, Kunze et al. [612] found the introduction of an oxygen atom into the side chain of the pyrrolidinium cation reduced the self-aggregation but only in the presence of Li^+ cations.

The compatibility of ILs with other ES components such as the current collector is also a concern, which is discussed in detail in Chapter 3.

2.4.2 Solvent-Free ILs

2.4.2.1 Solvent-Free ILs for EDLCs

A strong research trend in this area is to choose or design ILs with high ionic conductivity, larger electrochemical stability, and wider temperature range for EDLCs. To date, [EMIM]$[BF_4]$ IL has been the most extensively studied electrolyte for EDLCs due to its relatively high conductivity, but it has lower electrochemical stability and has a relatively high melting point of 15°C that limits its use in applications requiring high cell voltages or lower temperatures. This inspired extensive work to explore alternative ILs for EDLCs. Properties of ILs can be tuned by changing the structure of either the cation, or the anion, or both [503].

2.4.2.1.1 Aprotic ILs

For 1-ethyl-3-methylimidazolium ([EMIM])-based ILs, Ue et al. [615] synthesized 1-ethyl-3-methylimidazolium hexafluorotantalate ([EMIM]$[TaF_6]$) and hexafluoroniobate ([EMIM]$[NbF_6]$) ILs, and found that they exhibited lower melting points and higher decomposition temperatures than the [EMIM]$[BF_4]$ IL because of their larger anion size. However, the usage of these ILs with low melting point did not improve the low-temperature performance such as the capacitance of the EDLCs. In a later study, the same group [616] found that 1-ethyl-3-methylimidazolium fluoride ((EMIF) · 2.3HF) showed much higher ionic conductivity and significantly lower viscosity compared with the [EMIM]$[BF_4]$ IL electrolyte, which greatly improved the low-temperature capacitance value of the EDLCs. Nevertheless, EDLCs with (EMIF) · 2.3HF IL electrolyte exhibited a lower operative cell voltage (~2 V). Sun et al. [617] explored the 1-ethyl-3-methylimidazolium thiocyanate ([EMIM][SCN]) IL electrolyte for AC-based EDLCs. Compared to the [EMIM]$[BF_4]$ IL, [EMIM][SCN] IL exhibited higher ionic conductivity and lower viscosity. Handa et al. [618] found the [EMIM][FSI] IL showed similar ionic conductivity compared to the $TEMABF_4$/PC organic electrolyte. The rate capability of the EDLC using [EMIM][FSI] IL was comparable to that using $TEMABF_4$/PC organic electrolyte. The [EMIM][FSI] IL electrolyte also led to a lower self-discharge rate of EDLCs compared to $TEMABF_4$/PC organic electrolyte. Matsumoto and Hagiwara [619] reported the use of 1-ethyl-3-methylimidazolium difluorophosphate ([EMIM]$[PO_2F_2]$) IL as the electrolyte for EDLCs with AC electrodes. It was found that the cell voltage has a strong influence on the specific capacitance obtained for the [EMIM][PO_2F_2] IL electrolyte compared with that obtained for the [EMIM][BF_4] IL electrolyte. The [EMIM][PO_2F_2] electrolyte provided a higher specific capacitance compared with the [EMIM][BF_4] electrolyte at a cell voltage of 2.5 V. However, compared to the [EMIM][BF_4] IL electrolyte, the use of the [EMIM][PO_2F_2] IL electrolyte resulted in a lower operative cell voltage of the EDLC (below 3 V). This was explained by the lower anodic stability of $[PO_2F_2]^-$ anion compared with that of $[BF_4]^-$ anion. Pandey and Hashmi [620] proposed the application of 1-ethyl-3-methylimidazolium tetracyanoborate ([EMIM][TCB]) as the electrolyte for AC-based EDLCs since it had a lower viscosity of about 22 cP and higher ionic conductivity of about 1.3×10^{-2} S cm^{-1} at 20°C compared to the [EMIM][BF_4]. Kurig et al. [621] investigated a series of ILs composed of [EMIM] cation and different anion as electrolytes for carbon cloth electrode-based EDLCs. The investigated anions include tetrafluoroborate, tetracyanoborate, tris(pentafluoroethyl) trisfluorophosphate, bis(trifluoromethylsulfonyl)imide, bis(fluorosulfonyl)imide, and thiocyanate anions. The electrochemically stable window of the electrolyte (Figure 2.61) and the operative cell voltage of EDLCs (Figure 2.62) were greatly influenced by the chemical composition and structure of the anion of ILs. EDLCs using various ILs exhibited quite a different relaxation time (from 8 to 27 s at 2.7 V) and power density (from 15 to 35 kW kg^{-1} at 2.7 V). The relaxation time and power density were strongly influenced by the conductivity, viscosity, electrical charge delocalization, and weighted average positive values of ILs, depending on the chemical composition and polarity of anions. Among all the investigated IL electrolytes, the [EMIM][BF_4] and [EMIM][$B(CN)_4$] ILs demonstrated the best overall

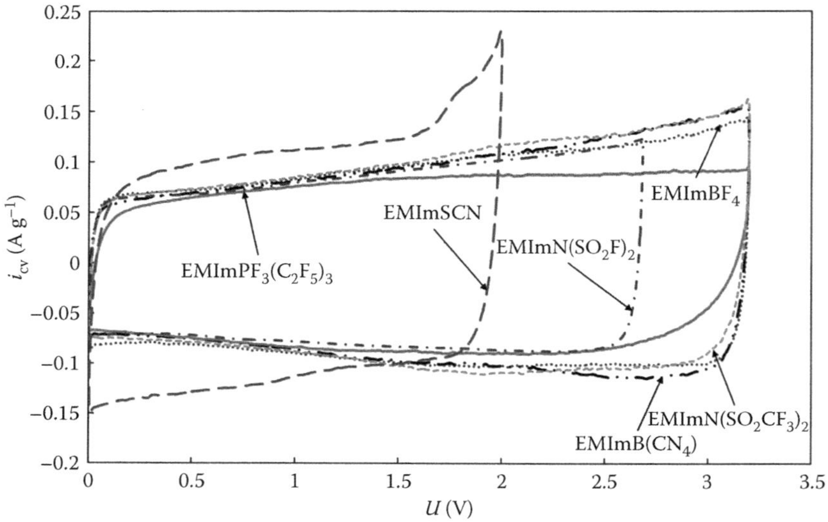

FIGURE 2.61 Cyclic voltammograms measured at cell voltage scan rate 1 mV s^{-1} for ES test cells with various ILs. (Reproduced by permission of The Electrochemical Society from Kurig, H. et al. 2012. *Journal of the Electrochemical Society* 159:A944–A951.)

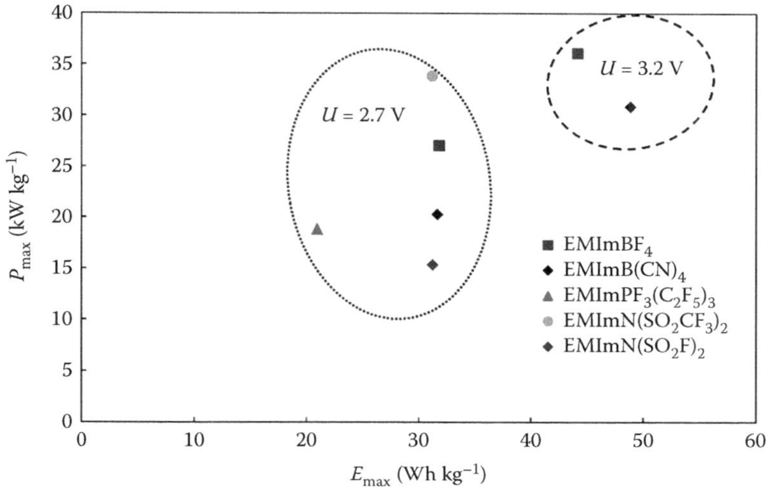

FIGURE 2.62 Power density versus energy density plots at 2.7 V (in left oval) and at 3.2 V (in right oval) for ES test cells filled with ILs studied. (Reproduced by permission of The Electrochemical Society from Kurig, H. et al. 2012. *Journal of the Electrochemical Society* 159:A944–A951.)

performance. Shi et al. [622] systematically investigated different ILs consisting of $[EMIM]^+$ cation and various anions including $[BF_4]^-$, $[NTF_2]^-$, $[DCA]^-$, $[EtSO_4]^-$, and $[OAc]^-$. Different combination of $[EMIM]^+$ with the anion changed the viscosity and the ionic conductivity of ILs, determining the performance of EDLCs. The viscosity of ILs was found to be associated with the hydrogen-bond-accepting property. Graphene-based EDLCs using the [EMIM][DCA] IL electrolyte exhibited highest specific capacitance, highest rate capability, and smallest ESR (Figure 2.63), due to the lowest viscosity, ion size, and molecular weight among all the investigated ILs. However, the operative cell voltage of the EDLC using [EMIM][DCA] IL (2.3 V) was much smaller than that using [EMIM][BF_4] IL (Figure 2.64). Consequently, the [EMIM][BF_4] IL-based EDLC exhibited a highest specific energy density of 67 Wh kg^{-1} at a current density of 1 A g^{-1}.

Other approaches have also been proposed to improve the properties of ILs such as the ionic conductivity and electrochemically stable window [56,623–625]. Nagao et al. [623] found the introduction of $BaTiO_3$ powder into the AC electrodes could enhance the dissociation of [EMIM][BF_4] IL electrolyte in EDLCs, which improved the rate performance of the EDLCs. On the contrary, no positive effect of the $BaTiO_3$ addition was observed for EDLCs with TEMABF$_4$/PC organic electrolyte. Kong et al. [56] added small amounts (e.g., 0.1 and 0.5 wt.%) of single-walled CNTs into the [EMIM][BF_4] IL, and found it increased the ionic conductivity of the electrolyte. Consequently, there was an increase in the values of the specific capacitance, energy density, and cycling stability of EDLCs using CNT electrodes in this electrolyte (Figure 2.65). Matsumoto et al. [625] found that the addition of Li salts shifted the cathodic potential limit of [EMIM][TFSI] toward the negative direction [625], that

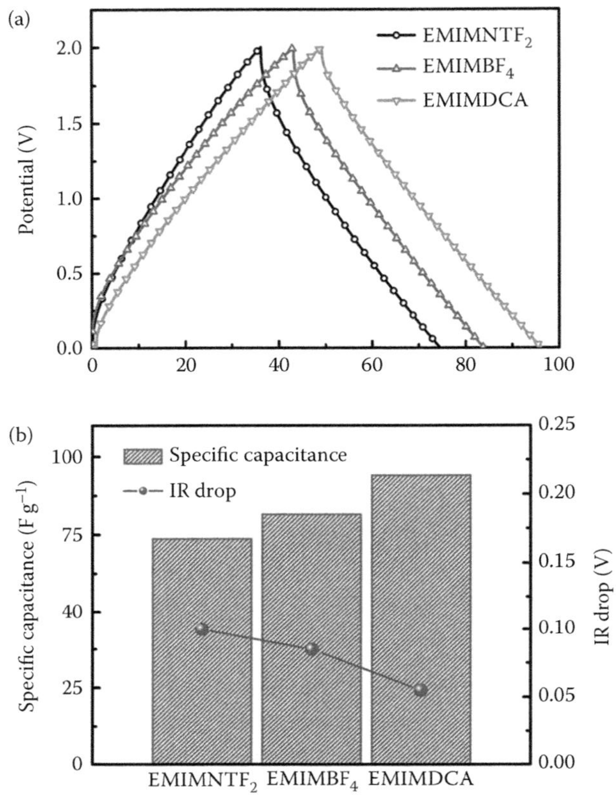

FIGURE 2.63 (a) Galvanostatic charge/discharge (GCD) curves of the GS electrodes in [EMIM][NTF_2], [EMIM][BF_4], and [EMIM][DCA] at a current density of 1 A g^{-1}. (b) The calculated specific capacitance and IR drop at 1 A g^{-1}. (Shi, M., S. Kou, and X. Yan: Engineering the electrochemical capacitive properties of graphene sheets in ionic-liquid electrolytes by correct selection of anions. ChemSusChem. 7. 3053–3062. Copyright Wiley-VCH Verlag GmbH & Co. KGaA. Reproduced with permission.)

is, the cathodic stability was increased. Using GC electrodes and electrochemical methods, Fukuda et al. [626] attributed the improvement of the cathodic stability of the [EMIM][BF_4] IL containing Li salt to the "inhibiting effect" caused by the interaction between Li^+, $[EMIM]^+$, and BF_4^-. Murayama et al. [624] found the addition of $LiBF_4$ into the [EMIM][BF_4] IL electrolyte for EDLCs increased the working cell voltage to 3.6 V and thus improved the energy density.

Apart from $[EMIM]^+$ cation, researchers have explored other imidazolium-based cations such as 1-butyl-2,3-dimethylimidazolium bis(trifluoromethylsulfonyl) imide ([BDMIM]), 1-butyl-3-methylimidazolium bis(trifluoromethylsulfonyl)imide ([BMIM]), and 1-dodecyl-3-methylimidazolium bis(trifluoromethylsulfonyl)imide ([C_{12}MIM]) [627,628]. For example, Bettini et al. [628] investigated four ILs including [BMIM][NTf_2], [C_{12}MIM][NTf_2], [EMIM][NTf_2], and 1-butyl-1-methylpyrrolidinium

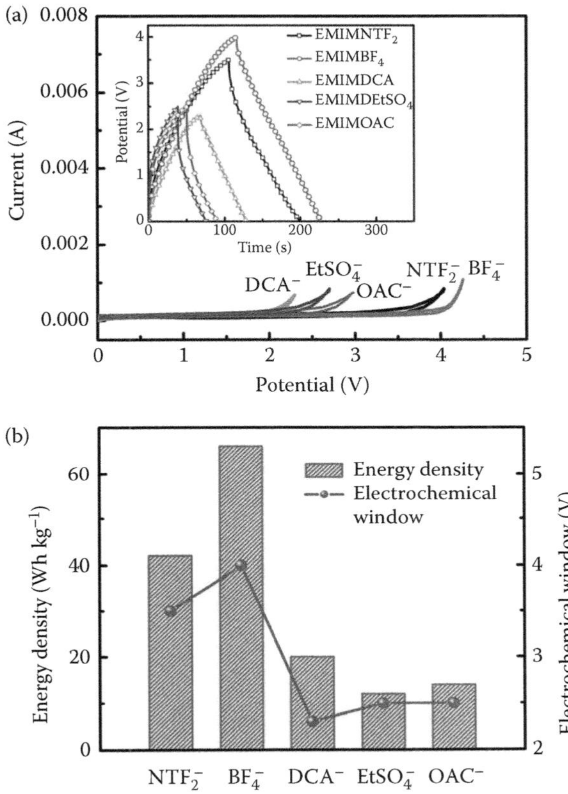

FIGURE 2.64 (a) Linear-sweep voltammograms of the blank, foam, nickel electrodes in five IL electrolytes recorded at a scan rate of 100 mV s^{-1}, and the GCD curves of GS electrodes in the corresponding IL electrolytes at a current density of 1 A g^{-1} (inset). (b) The relationship between the electrochemical window and the energy density of the GS electrodes measured in these IL electrolytes. (Shi, M., S. Kou, and X. Yan: Engineering the electrochemical capacitive properties of graphene sheets in ionic-liquid electrolytes by correct selection of anions. *ChemSusChem*. 7. 3053–3062. Copyright Wiley-VCH Verlag GmbH & Co. KGaA. Reproduced with permission.)

bis(trifluoromethylsulfonyl)imide ([BPYR][NTf_2]). It was found that the specific capacitance of ES using the [BMIM][NTf_2] electrolyte was highest (75 F g^{-1}), due to the shortest alkyl chain of [BMIM].

There has also been extensive research on ILs with other cations such as pyrrolidinium-based [629–631], ammonium-based [632–634], sulfonium-based [635], and fluorohydrogenate ILs (FHILs) [634]. Regarding pyrrolidinium cation-based ILs, Lazzari et al. [629] investigated *N*-butyl-*N*-methylpyrrolidinium bis(trifluoromethanesulfonyl)imide ([PYR_{14}][TFSI]) IL electrolyte for EDLCs based on mesoporous carbon xerogel electrodes. Although [PYR_{14}][TFSI] had a lower ionic conductivity and gave a lower capacitance, it exhibited a wider electrochemically stable window compared with [EMIM][TFSI]. In order to fully take advantage

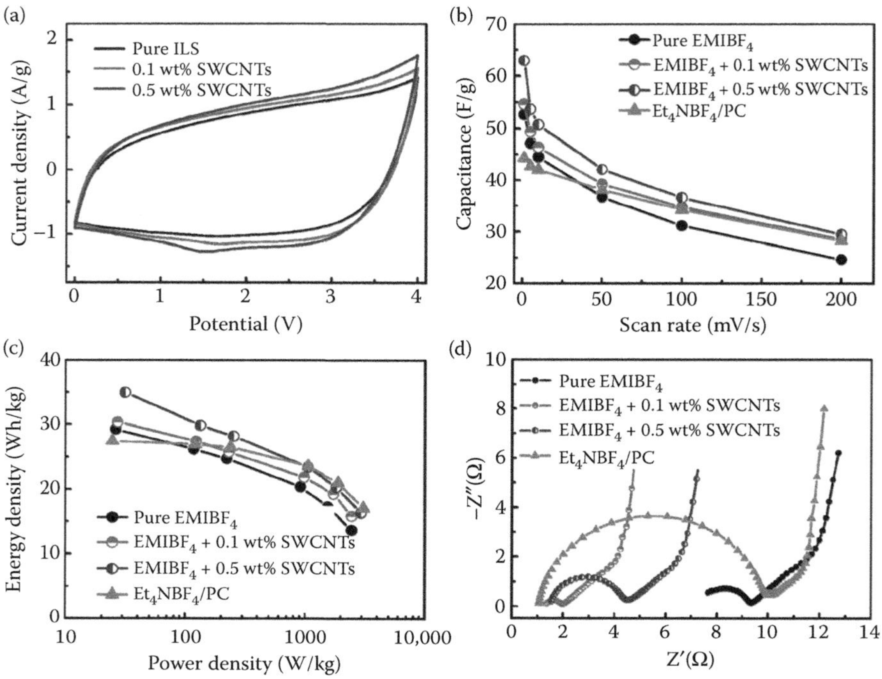

FIGURE 2.65 Capacitance performance of the electrolyte at different electrolytes. (a) CV curves of the CNT electrode in different electrolytes at a scan rate of 200 mV s^{-1}. (b) Specific capacitance of the CNT electrode in different electrolytes at different scan rates. (c) Energy density of the CNT electrode in different electrolytes. (d) EIS. (Kong, C. Y., W. Z. Qian, C. Zheng, Y. T. Yu, C. J. Cui, and F. Wei. 2013. Raising the performance of a 4 V supercapacitor based on an $EMIBF_4$-single walled carbon nanotube nanofluid electrolyte. *Chemical Communications* 49:10727–10729. Reproduced by permission of The Royal Society of Chemistry.)

of the high electrochemically stable window of $[PYR_{14}]$[TFSI] IL, the EDLCs was made by the asymmetric configuration, that is, the positive electrode had higher carbon loading than the negative electrode. In this way, it was possible to charge each electrode up to the limit potentials of the IL. The asymmetric EDLC using the $[PYR_{14}]$[TFSI] electrolyte exhibited a maximum cell voltage of 3.7 V, which was higher than using the [EMIM][TFSI] electrolyte (3.4 V). *N*-methoxyethyl-*N*-methylpyrrolidinium bistrifluoromethanesulfonyl imide $[PYR_{12}O_1]$[TFSI] IL has a big advantage that it freezes below −90°C due to the introduction of a methoxyethyl group on the nitrogen atom of the pyrrolidinium ring [631,636,637]. Lazzari et al. [630] studied the $[PYR_{12}O_1]$[TFSI] IL electrolyte for asymmetric EDLCs with different amount of carbon on the positive and negative electrodes. The asymmetric EDLCs with the $[PYR_{12}O_1]$[TFSI] electrolyte had a wide operating temperature range (−30°C to 60°C) (Figure 2.66), a high operative cell voltage of 3.7 V, and a high cycling stability with a capacitance loss of 2% after 27,000 cycles at 60°C [631].

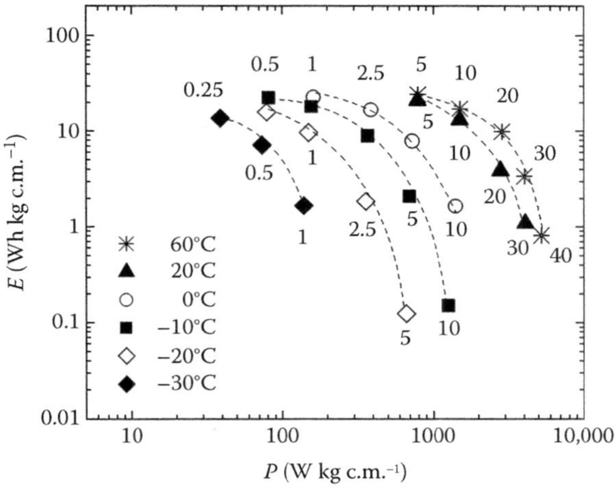

FIGURE 2.66 Traditional Ragone plot of the $PYR_{1(201)}$TFSI-based asymmetric EDLC (AEDLC) supercapacitor at different temperatures. Specific energy (E) and power (P) values have been evaluated referring to the total carbon-binder composite loading from galvanostatic discharges at different currents with $V_{max} = 3.7$ V and $V_{min} = 2.035$ V; the labels indicate the current densities in mA cm^{-2}. (Reproduced by permission of The Electrochemical Society from Lazzari, M., F. Soavi, and M. Mastragostino. 2009. *Journal of the Electrochemical Society* 156:A661–A666.)

Taking into account that azepanium-based ILs are cheaper than pyrrolidinium-based ILs, Pohlmann et al. [638] studied the use of two azepanium-based ILs, that is, *N*-methyl, *N*-butyl-azepanium bis(trifluoromethanesulfonyl)imide ($[AZP_{14}]$[TFSI]) and *N*-methyl, *N*-hexyl-azepanium bis(trifluoromethanesulfonyl)imide ($[AZP_{16}]$ [TFSI]), as electrolytes for EDLCs. Both $[AZP_{14}]$[TFSI] and $[AZP_{16}]$[TFSI] provided an operative voltage of 3.5 V, which was comparable to $[PYR_{14}]$[TFSI]. However, the energy density of EDLCs using the $[AZP_{14}]$[TFSI] or $[AZP_{16}]$[TFSI] IL electrolyte was lower than that using the $[PYR_{14}]$[TFSI] electrolyte, especially at higher current density (Figure 2.67). The authors attributed this to the larger size of cation and thereby less accessibility to the total surface area of the electrodes.

Concerning the ILs with ammonium-based cations, Kim et al. [633] found that *N,N*-diethyl-*N*-methyl-*N*-(2-methoxyethyl)ammonium tetrafluoroborate ([DEME]$[BF_4]$) and [DEME][TFSI] ILs had larger electrochemically stable window compared to the $TEABF_4$/PC organic electrolyte and conventional [EMIM]$[BF_4]$ IL. The specific capacitance of AC electrodes in both ILs increased as the applied voltage increased from 2.5 to 3.5 V, which was different with that observed for the $TEABF_4$/PC organic electrolyte. Sato et al. [632] investigated [DEME]$[BF_4]$ and [DEME][TFSI] as electrolytes for EDLCs. [DEME]$[BF_4]$ IL showed the electrochemically stable window of 6 V, which was obtained on the Pt electrode. This value was larger than the electrochemically stable window of the [EMIM]$[BF_4]$ (4.5 V) under the same testing condition. The high-temperature performance of the [DEME]$[BF_4]$ IL-based EDLC was much better compared to the [EMIM]$[BF_4]$ IL-based or 1 M $TEABF_4$/PC organic electrolyte-based EDLCs.

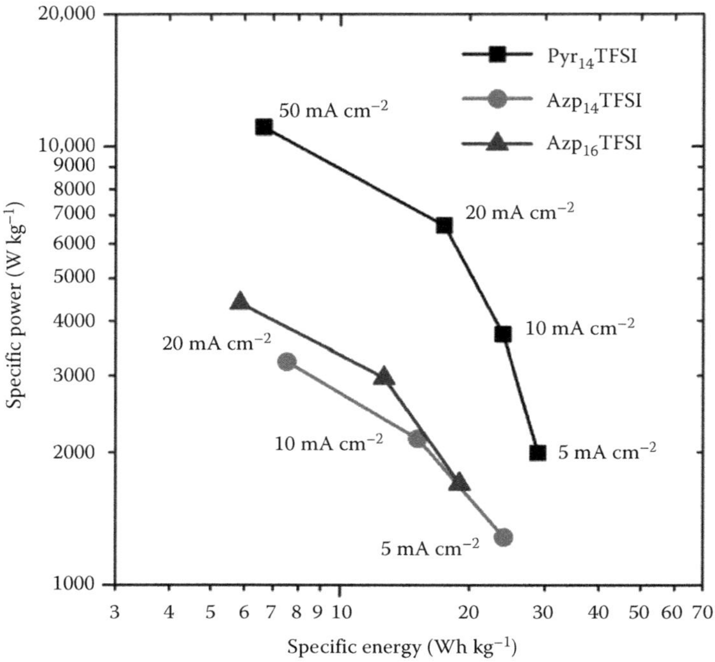

FIGURE 2.67 Ragone-like plot of EDLCs containing Azp_{14}TFSI, Azp_{16}TFSI, and Pyr_{14}TFSI as electrolytes. The values have been derived from charge–discharge experiments at varying current densities. (Reprinted from *Journal of Power Sources*, 273, Pohlmann, S. et al., Azepanium-based ionic liquids as green electrolytes for high voltage supercapacitors, 931–936, Copyright 2014, with permission from Elsevier.)

Concerning the sulfonium-based cations, Anouti et al. [635] explored the trimethylsulfonium bis(trifluorosulfonimide) $[Me_3S]$[TFSI] IL electrolyte for EDLCs with AC electrodes. $[Me_3S]$[TFSI] IL had a melting point of 45.5°C, and was thermally stable up to 280°C. The ionic conductivity and viscosity of $[Me_3S]$[TFSI] at 80°C were 20.42 mS cm^{-1} and 3 cP, respectively. EDLCs using AC electrodes in $[Me_3S]$[TFSI] IL exhibited a cell voltage up to 3 V and could operate at high temperatures, which provided a maximum energy density of 44.1 Wh kg^{-1} at 80°C.

Senda et al. [634] investigated FHILs based on five different cations, that is, 1,3-dimethylimidazolium ($[DMIM]^+$), 1-ethyl-3-methylimidazolium ($[EMIM]^+$), 1-butyl-3-methylimidazolium ($[BMIM]^+$), 1-ethyl-1-methylpyrrolidinium ($[EMPYR]^+$), and 1-methoxymethyl-1-methylpyrrolidinium ($[MOMMPYR]^+$, as electrolytes for AC-based EDLCs. The use of these FHILs resulted in higher specific capacitance compared to the [EMIM]$[BF_4]$ IL and 1 M $TEABF_4$/PC organic electrolyte at the investigated cell voltage (1–3.2 V). For some imidazolium-based FHILs, increasing the size of the cation decreased the maximum specific capacitance of the resultant ES. However, in spite of the larger size of $[MOMMPYR]^+$ than $[EMPYR]^+$, the maximum capacitance obtained in [MOMMPYR]$[(FH)_{2.3}F]$ (152 F g^{-1}) was larger than that achieved in [EMPYR]$[(FH)_{2.3}F]$ (134 F g^{-1}), which was due to the introduction of the methoxy group.

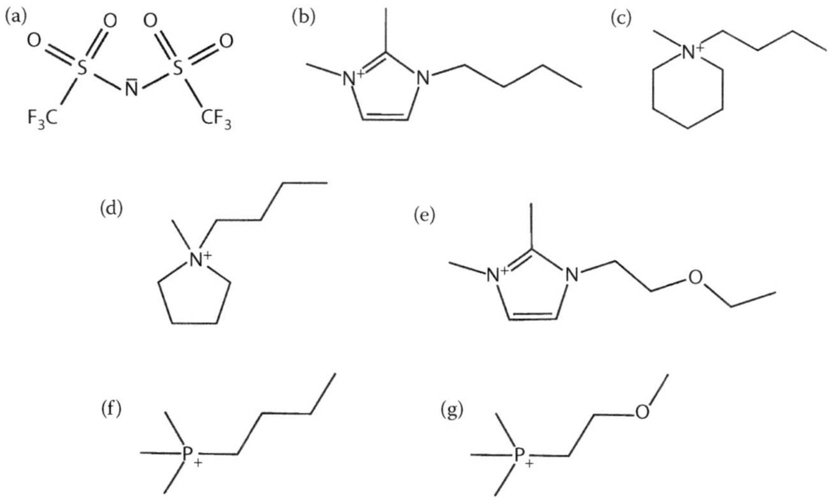

FIGURE 2.68 Schematic structure of the constituent ions in the ILs under study in Reference 640, (a) Tf_2N, (b) 1-n-butyl-2,3-dimethylimidazolium (BMIM), (c) *N*-n-butyl-*N*-methylpiperidinium (BMPi), (d) *N*-n-butyl-*N*-methylpyrrolidinium (BMPy), (e) 1-(2-ethoxyethyl)-2,3-dimethyl-1H-imidazol-3-ium ($EtO(CH_2)_2MMIm$), (f) butyltrimethylphosphonium (P_{2225}), and (g) (2-methoxyethyl)trimethylphosphonium ($P_{222(201)}$). (Reprinted with permission from Rennie, A. J. R. et al. 2013. *Journal of Physical Chemistry Letters* 4:2970–2974, open access.)

Some researchers performed comparative studies on different ILs [639,640]. As shown in Figure 2.68, Rennie et al. [640] investigated ILs with different kinds of cations including imidazolium-based cation (1-n-butyl-2,3-dimethylimidazolium ($[BMMIM]^+$)), pyrrolidinium-based cation (*N*-n-butyl-*N*-methylpyrrolidinium ($[BMPY]^+$)) piperidinium-based cation (*N*-n-butyl-*N*-methylpiperidinium ($[BMPI]^+$)), phosphonium-based cation (butyltrimethylphosphonium ($[P_{2225}]^+$)), as well as ether-bond-containing cations (i.e., 1-(2-ethoxyethyl)-2,3-dimethyl-1H-imidazol-3-ium ($[EtO(CH_2)_2MMIM]^+$) and (2-methoxyethyl)trimethylphosphonium ($[P_{222(201)}]^+$). It was found that the introduction of the ether bond into the alkyl side chain of the cation could effectively increase the specific capacitance and reduce the ESR of EDLCs (Figure 2.69). The authors attributed this to the introduction of a small electronegative region to the cation structure by the ether bond, which facilitated a denser packing of ions at the interface between the electrolyte and electrode and thus increased the charge amount. Sillars et al. [639] investigated various ILs for AC-based EDLCs, and compared them with 1 M $TEABF_4$/PC organic electrolytes. The investigated ILs were: $[EMIM][BF_4]$, 1-ethyl-3-methylimidazolium dicyanamide ($[EMIM][N(CN)_2]$), 1,2-dimethyl-3-propylimidazolium bis(trifluoromethylsulfonyl) imide ([DMPIM][TFSI]), and 1-butyl-3-methylpyrrolidinium tris(pentafluoroethyl) trifluorophosphate ([BMPY][FAP]). As shown in Figure 2.70, the highest working voltage of 3.5 V was obtained for EDLCs using [BMPY][FAP] IL. The specific capacitance of EDLCs with these ILs decreased in the following order: $[EMIM][BF_4]$ > [DMPIM]

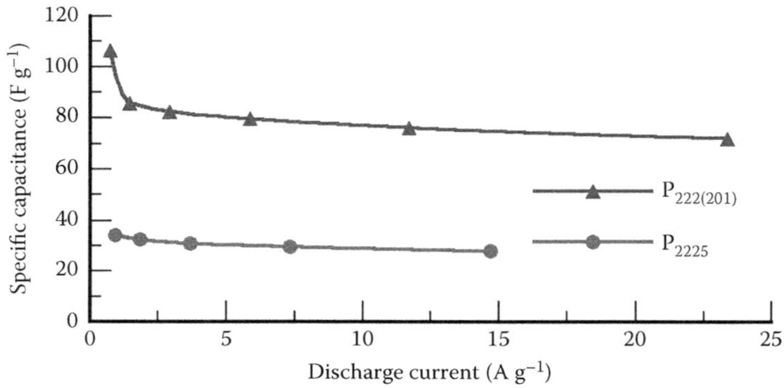

FIGURE 2.69 Specific capacitance determined at different rates of constant current discharge between 0 and 2.5 V for cells using phosphonium-based ILs. (Reprinted with permission from Rennie, A. J. R. et al. 2013. *Journal of Physical Chemistry Letters* 4:2970–2974, open access.)

[TFSI] > [BMPY][FAP] > [EMIM][N(CN)$_2$], which was found to be dependent on the ion size and viscosity of the IL electrolyte. As the viscosity of the IL was lowered, the ESR decreased in the following order: [BMPY][FAP] > [DMPIM][TFSI] > [EMIM][BF$_4$] > [EMIM][N(CN)$_2$]. [EMIM][BF$_4$] IL-based EDLCs provided much higher energy and power densities compared to 1 M TEABF$_4$/PC organic electrolyte-based EDLCs. Cho et al. [641] tested a series of ILs with single or double cation for AC-based EDLCs. It was found that the double-cationic ILs

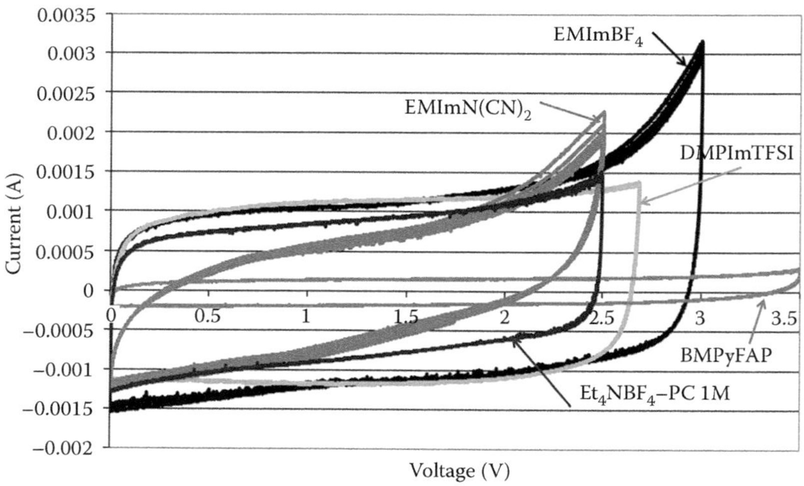

FIGURE 2.70 Cyclic voltammograms for [EMIM][BF$_4$] at 5 mV s^{-1}. (Sillars, F. B. et al. 2012. Variation of electrochemical capacitor performance with room temperature ionic liquid electrolyte viscosity and ion size. *Physical Chemistry Chemical Physics* 14:6094–6100. Reproduced by permission of The Royal Society of Chemistry.)

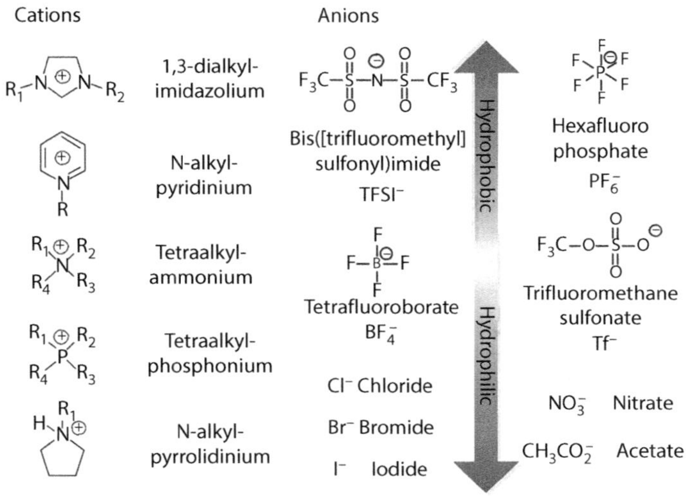

FIGURE 2.71 Examples of cations and anions used in the formation of ILs, together with changes in hydrophilic–hydrophobic properties associated with anion type. (Ye, Y. S., J. Rick, and B. J. Hwang. 2013. Ionic liquid polymer electrolytes. *Journal of Materials Chemistry A* 1:2719–2743. Reproduced by permission of The Royal Society of Chemistry.)

coupled with BF_4^- and $TFSI^-$ exhibited the most promising capacitive properties, which was attributed to the more actively charged sites and faster mobility of counter anion for double-cationic ILs compared to single-cationic ILs.

Anions play an important role in the hydrophilic/hydrophobic properties of ILs. As shown in Figure 2.71, the hydrophobicity increases following the order: $CH_3CO_2^-$ (acetate), NO_3^- (nitrate) < Tf^-, BF_4^- =< PF_6^-, $TFSI^-$ (hydrophobic) [642,643]. Trace amount of water in ILs has a significant negative effect on the electrochemical stability of ILs (i.e., decreasing the electrochemically stable window) [644]. Therefore, hydrophobic anions such as $TFSI^-$ and FSI^- have attracted increasing attention for ILs used in ESs [60]. Other less common anion such as 1,1-dicyano-1-acetylmethanide anion ($[C(CN)_2(COCH_3)]^-$ ($[DCNAcC]^-$) has also been selected to prepare ILs (Figure 2.72), as reported by Winoto et al. [645]. It was found that the specific capacitance of EDLCs using the [EMIM][DCNAcC] or [BMIM][DCNAcC] IL electrolyte was higher than that using [EMIM][Tf_2N], which might due to the smaller size of $[DCNAcC]^-$ anion.

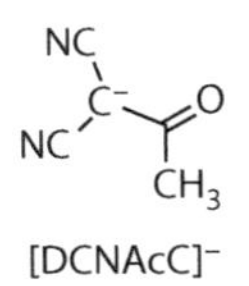

FIGURE 2.72 Structure of $[DCNAcC]^-$ anion used in Reference 645.

2.4.2.1.2 Protic ILs

Besides the extensively studied aprotic ILs, protic IL electrolytes have also received attention for EDLCs [646–649]. Demarconnay et al. [646] reported that for AC-based ESs using protic pyrrolidinium nitrate ($PyNO_3$) IL as electrolyte, the specific capacitance increased from 121 to 208 F g^{-1} as the electrolyte pH increased from 7 to 11. However, the presence of small amount of water in this protic IL limited the maximum voltage of symmetric ESs to 1.2 V. The use of the triethylammonium bis(trifluoromethylsufonyl)imide ([NEtaH][TFSI]) protic IL as the electrolyte for ESs was able to achieve maximum cell voltages up to 2.5 V. Timperman et al. [647] found that the ionic conductivity of [Et_3NH][TFSA] increased from 0.7 to 20 mS cm^{-1} as the temperature increased from –10°C to 80°C. The passivation of the Al current collector in [Et_3NH][TFSA] IL after the second CV was observed (Figure 2.73). The electrochemically stable window of the [Et_3NH][TFSA] IL was estimated to be 4 V on Pt electrodes and 2.5 V on AC using a three-electrode cell. In a later study, the same group [648] did a comparative study of two protic ILs based on diisopropylethyl-ammonium cation ([DIPEA][TFSI]) or pyrrolidinium cation ([Pyrr][TFSI]) as electrolyte for AC-based EDLCs. The ionic conductivity increased to 8.4 and 17.3 mS cm^{-1} as the temperature increased to 80°C while the viscosity decreased to 20 and 12 mPa s at 70°C for [DIPEA][TFSI] and [Pyrr][TFSI], respectively. It was found that the nature of the cation in protic ILs affected the pH, the electrochemically stable window, and the wettability on the electrode, particularly on the binder. EDLCs with [Pyrr][TFSI] had a better cycling capacitance (120 F g^{-1}) while EDLCs using [DIPEA][TFSI] exhibited a larger electrochemically stable window of 2 V and higher specific energy density. The long-term cycling stability of protic IL-based EDLCs was investigated by Brandt et al. [649]. EDLCs using [Et_3NH][TFSI] IL showed good cycling stability (30,000 cycles) at the cell voltage of 2.4 V in a relatively wide

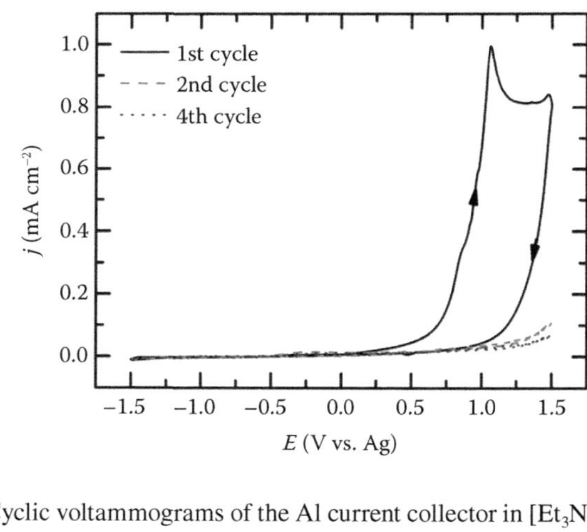

FIGURE 2.73 Cyclic voltammograms of the Al current collector in [Et_3NH][TFSA] at 20°C. Scan rate: 5 mV s^{-1}. (Timperman, L. et al. 2012. Triethylammonium bis(tetrafluoromethylsulfonyl) amide protic ionic liquid as an electrolyte for electrical double-layer capacitors. *Physical Chemistry Chemical Physics* 14:8199–8207. Reproduced by permission of The Royal Society of Chemistry.)

temperature range from 10°C to 60°C. However, the low operating cell voltage of $[Et_3NH][TFSI]$ IL-based EDLCs strongly limited the energy density.

2.4.2.1.3 Mixture of ILs

Since most commonly studied ILs have a melting point above 0°C, their use as electrolytes for low-temperature applications is greatly limited. Therefore, eutectic mixtures of ILs have been investigated so as to expand the lower limit of the working temperature for IL-based EDLCs [60,650–653]. Lin et al. [651] developed eutectic IL mixtures containing propylpiperidinium bis[fluorosulfonyl]imide ($[PIP_{13}]$ [FSI]) and *N*-butyl-*N*-methylpyrrolidinium bis[fluorosulfonyl]imide ($[PYR_{14}]$[FSI]) (1:1 by weight or molar ratio), which resulted in an IL with two different cations and one anion. The liquid range of this IL mixture was lowered to −80°C. Based on exohedral nanostructured carbon (i.e., nanotubes and onions) electrodes, the operating temperature range of EDLCs with $[PIP_{13}]$[FSI] − $[PYR_{14}]$[FSI] IL mixture electrolyte was greatly enlarged from −50°C to 100°C (Figure 2.74). By optimizing the carbon/IL electrolyte system, Tsai et al. [60] found that EDLCs with porous KOH-activated microwave exfoliated graphite oxide electrodes in an eutectic IL mixture of $[PIP_{13}]$[FSI] and $[PYR_{14}]$[FSI] (1:1 by molar ratio) could achieve a high specific capacitance (above 100 F g^{-1}) below room temperature. This specific capacitance value was much higher compared to the exohedral nanostructured carbon (nanotubes and onions) electrodes in the same eutectic mixture of ILs. Palm et al. [654] found the melting point of $[EMIM][BF_4]$ IL decreased from 13.5°C to

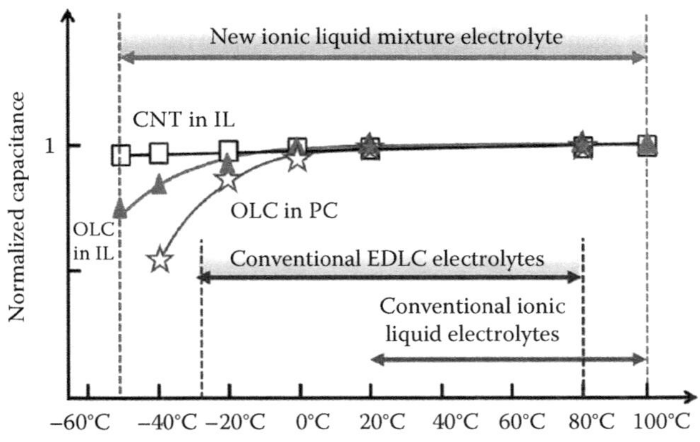

FIGURE 2.74 Normalized capacitance ($C/C_{20°C}$) for the OLC and VA-CNT electrodes in $([PIP_{13}][FSI])_{0.5}([PYR_{14}][FSI])_{0.5}$ IL mixture and PC + 1 M $TEA\text{-}BF_4$ electrolytes. Capacitances were calculated at 100 mV s^{-1}, except for the −50°C (1 mV s^{-1}) and −40°C (5 mV s^{-1}) experiments. This plot shows that the use of the IL mixture extends the temperature range for supercapacitors into the −50°C to 100°C range while conventional electrolytes using PC as solvent are limited to the −30°C to 80°C range. $C_{20°C}$ was 80 and 4 mF, respectively, for OLC and VA-CNT cells. All cells were cycled from 0 up to 2.8 V. (Reprinted with permission from Lin, R. Y. et al. Capacitive energy storage from −50 to 100°C using an ionic liquid electrolyte. *Journal of Physical Chemistry Letters* 2:2396–2401. Copyright 2011 American Chemical Society.)

–6°C after adding 1 M 1-butyl-3-methylimidazolium tetrafluoroborate ([BMIM][BF_4]). However, EDLCs using this mixed ILs showed a noticeable decrease in the specific capacitance, energy and power densities and an increase in the characteristic relaxation time compared with EDLCs with pure [EMIM][BF_4]. Using MD simulation, Li et al. [655] investigated an IL mixture of 1-methyl-1-propylpyrrolidinium bis(trifluoromethylsulfonyl)imide ([C_3MPY][Tf_2N]) and 1-methyl-1-butylpiperidinium bis(trifluoromethylsulfonyl)imide ([C_4MPIP][Tf_2N]) (1:1 by molar ratio) on an OLC electrode. The IL mixture showed lower glass transition temperature than either of the pure ILs. The IL mixture was also found to have a higher ionic conductivity compared to the pure ILs at lower temperatures. The differential capacitance of the OLC electrode showed a positive temperature dependence, which was dominated by the EDL thickness.

2.4.2.2 Solvent-Free ILs for Pseudocapacitors

In order to take advantage of the pseudocapacitive contribution to the total charge storage, ILs have also attracted great interest as electrolytes for pseudocapacitive materials for pseudocapacitors. Mysyk et al. [656] identified the pseudocapacitance of AC in pyrrolidinium-based protic ILs (pyrrolidinium nitrate ([Pyrr][NO_3]) and pyrrolidinium formate ([Pyrr][HCOO])), and that increasing the surface functionalities on AC could enhance the pseudocapacitive contribution. As mentioned previously, the anions of ILs play an important role in the hydrophobicity of ILs. Zhou et al. [600] found the hydrophobicity of ILs had an influence on the pseudocapacitive performance of mesoporous carbons with a certain amount of surface oxygenated groups. The carbons showed visible pseudocapacitance due to the rapid redox reactions of the oxygenated groups in hydrophilic [EMIM][BF_4] while no obvious pseudocapacitance was found in hydrophobic [EMIM][TSFI]. The ESs with protic ILs had a larger potential window than those with a H_2SO_4 aqueous electrolyte and were able to operate at a lower temperature of –10°C. Demarconnay et al. [646] pointed out most of the pseudocapacitive reactions between the electrolyte and AC with certain oxygenated surface groups observed in protic ILs ($PyNO_3$ in this study) were due to the free water. Rochefort and Pont [657] provided the experimental evidence that the pseudocapacitance could arise on RuO_2 electrodes in a protic IL consisting of 2-methylpyridine (α-picoline) and trifluoroacetic acid (TFA). The CVs obtained on the RuO_2 electrode in the protic IL exhibited distinct redox peaks assigned to the faradaic reactions while there was no obvious pseudocapacitance (only EDL capacitance) in [EMIM][BF_4] IL. RuO_2 electrode in this protic IL showed similar shape of CVs, and also similar specific capacitance value compared to those obtained in 0.1 M H_2SO_4 aqueous electrolyte (Figure 2.75) [657]. However, the charging rate of the RuO_2 electrode was limited, due to the high viscosity and slow proton transfer in the IL electrolyte. Mayrand-Provencher and Rochefort [658] investigated the effect of the ionic conductivity and viscosity of protic ILs on the pseudocapacitive performance of thermally prepared RuO_2 electrodes. The studied protic ILs composed of various heterocyclic amines and TFA with different base:acid ratios (1:1 and 1:2), and exhibited different conductivity and viscosity. It was found that the specific capacitance values obtained in all the investigated protic ILs were much higher than that obtained in the aprotic [BMIM][PF_6] IL (only 4.1 F g^{-1}). The specific

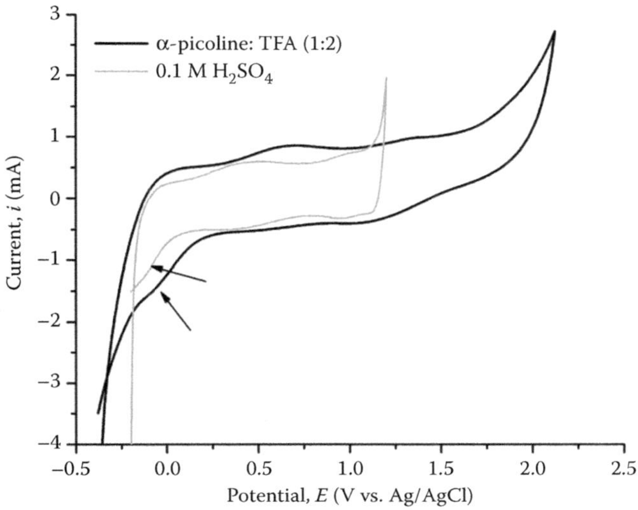

FIGURE 2.75 Cyclic voltammograms (v = 10 mV s^{-1}) obtained with the thermally prepared RuO_2 working electrode in 0.1 H_2SO_4 (gray) and in the 2-methylpyridine:trifluoroacetic acid (1:2) IL (black). (Reprinted from *Electrochemistry Communications*, 8, Rochefort, D., and A. L. Pont, Pseudocapacitive behavior of RuO_2 in a proton exchange ionic liquid, 1539–1543, Copyright 2006, with permission from Elsevier.)

capacitance of the RuO_2 electrode was strongly dependent on the conductivity of the protic IL, except for the IL based on 2-methylpyridine. However, the high viscosity of some protic ILs led to a remarkable loss of the specific capacitance with long-term cycling, which was not observed in the H_2SO_4 aqueous electrolyte. In an effort to understand the relationships between the IL structure and pseudocapacitive properties of RuO_2 electrodes, Mayrand-Provencher et al. [659] also studied a series of pyridinium-based protic ILs and investigated the effect of the alkyl chain length of the cation's substituent and the substituent position. The conductivity, viscosity, and electrochemically stable window of the electrolyte, the specific capacitance and the cycling stability of RuO_2 electrodes in these protic ILs were found to be influenced by the alkyl chain length of the cation's substituent and the position of the substituent. Richey and Elabd [660] used *in situ* infrared spectroscopy technique to characterize the behaviors of individual cation and anion of 1-ethyl-3-methylirnidazolium triflate ([EMIM][Tf]) IL in a working pseudocapacitor with RuO_2 electrolytes. It was found that cations and anions transported together as aggregates and cations were the main charge carriers. As the working voltage and RuO_2 content in the electrode increased, the ratio of cations to anions at the cathode increased during the charging process. With the increasing porosity of the RuO_2 electrode, the concentration of both cations and anions, and the rate of the change of cations and anions increased (Figure 2.76), corresponding to the increase in the volumetric capacitance. The cation and anion transport was found to follow a diffusion-limited behavior.

Other less expensive metal oxides such as Mn oxide have also been investigated as electrode materials in ILs. Chang et al. [661] found that it was possible to achieve the

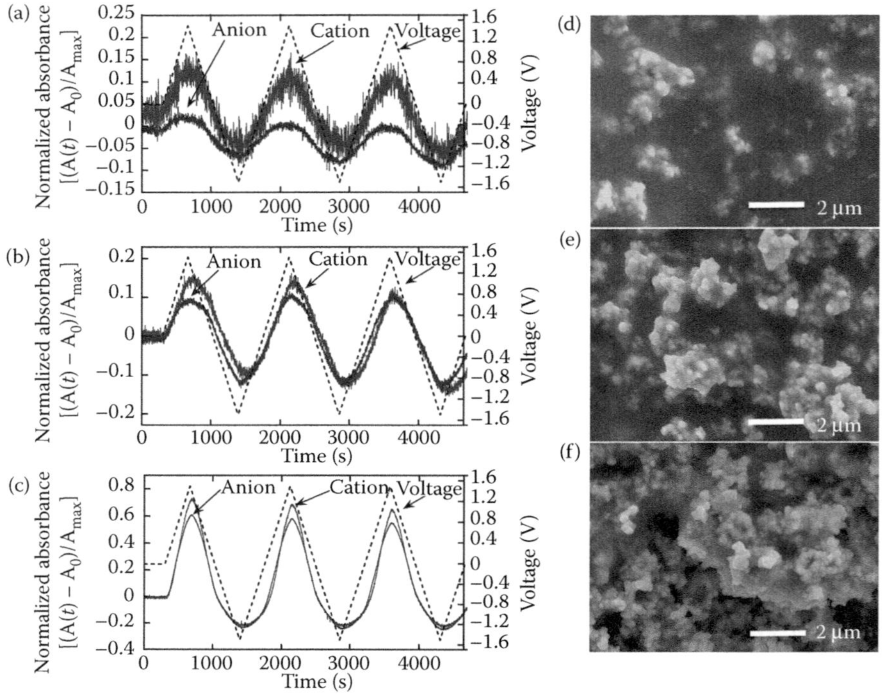

FIGURE 2.76 Normalized infrared absorbance of the cation and anion of the IL in electrodes with RuO_2 contents of (a) 22, (b) 35, and (c) 48 wt.% during CV scans from −1.5 to +1.5 V at a scan rate of 5 mV s^{-1}. Cross-sectional SEM images showing the structure of porous electrodes with RuO_2 contents of (d) 22, (e) 35, and (f) 48 wt.%. (Reproduced by permission of The Electrochemical Society from Richey, F. W., and Y. A. Elabd. 2013. *Journal of the Electrochemical Society* 160:A862–A868)

pseudocapacitance of Mn oxide in the aprotic 1-ethyl-3-methylimidazolium dicyanamide ([EMIM][DCA]) aprotic IL, which did not involve protons and alkali cations in the electrolyte. Based on XPS and *in situ* X-ray absorption spectroscopy (XAS), Chang et al. [662] found that $[DCA]^-$ anions were reversibly inserted/deserted into the tunnels between the MnO_6 octahedral units, compensating the valent state variation between the Mn^{3+} and Mn^{4+} during the charging/discharging process in the [BMP][DCA] IL electrolyte:

$$MnO_{2-x}(DCA)_{2x} + 2xe^- \leftrightarrow MnO_{2-x} + 2xDCA^-(x \leq 0.5) \tag{2.11}$$

The reaction thickness was found to be about 50 nm by XPS depth profiling analyses. A potential window of 3 V could be achieved for Mn oxide electrode in the [BMP][DCA] IL. Similar results were found for the [EMIM][DCA] IL electrolyte, and the penetration depth of $[DCA]^-$ into Mn oxide was found to be about three times deeper than that of $[BF_4]^-$. This contributed to a much higher specific capacitance of the Mn oxide electrode in [EMIM][DCA] (72 F g^{-1}) compared with that obtained in the [EMIM][BF_4] IL (28 F g^{-1}). 1-Ethyl-3-methylimidazolium thiocyanate ([EMIM][SCN]) IL was also found to be a potential electrolyte for Mn oxide pseudocapacitive materials [663,664]. Adding a small amount of NaDCA (~0.01 M) into [BMP][DCA] IL electrolyte was reported to further improve the specific capacitance, specific energy and power densities of the MnO_2 electrode, due to the participation of the Na^+ along with the $[DCA]^-$ in the pseudocapacitive process [665]. A specific capacitance of 100 F g^{-1} at 25°C was achieved for the MnO_2 electrode in this IL. Owing to a larger potential window of 3 V, the energy and power densities obtained in the IL electrolyte were significantly higher than those obtained in the Na_2SO_4 aqueous electrolyte [665]. Similarly, the introduction of small amount of Li^+ in the IL electrolyte (e.g., [BMMI][Tf_2N] and [BMP][DCA]) was also found to have a beneficial effect on increasing the specific capacitance of MnO_2 electrodes [666,667]. Based on [BMP][DCA] IL electrolyte containing 0.01 M LiTFSI, Lee et al. [667] reported a symmetric MnO_2 pseudocapacitor with a high cell voltage of 3 V could deliver a maximum energy density of 37 Wh kg^{-1} at a power density of 200 W kg^{-1}.

In addition to the increase of the working voltage, the use of IL as the electrolyte have also been found to improve the performance of the MnO_2-based pseudocapacitors at high temperatures (e.g., 80°C) whereas MnO_2-based pseudocapacitors using aqueous electrolyte showed a deteriorated performance [668].

Besides the aprotic ILs, Ruiz et al. [669] found protic ILs composed of 2-methoxypyridinium and trifluoroacetate could provide pseudocapacitance of MnO_2 electrodes, as investigated by using *in situ* UV–visible spectroscopy. Besides MnO_2, other combinations of pseudocapacitive electrode materials and IL electrolytes have been investigated, such as the N-doped carbon in [PYR_{14}][TFSI] and [PYR_{14}][FSI] [670], porous Ni in [EMIM][BF_4] [671], TiO_2 in [EMIM][TFSI] [672], Ru-doped Cu oxide in 3-carboxymethyl-1-methylimidazolium bisulfate [CMIM][HSO_4] [673], Fe_2O_3 in [EMIM][BF_4] [674], poly(3-methylthiophene) in [EMIM][TFSI] [675], and PAN in [EMIM][BF_4] [676] and [EMIM][TFSI] [677]. For example, Chen et al. [676] reported a high energy density of 84 Wh $kg_{(cell)}^{-1}$ and a power density of 182 kW $kg_{(cell)}^{-1}$ for PANI nanotubes-based pseudocapacitors using [EMIM][BF_4] IL as electrolytes. Owing to the much wider operative cell voltage, the specific energy and power densities obtained in the IL electrolyte were significantly improved by about five times compared with those achieved in the 1 M H_2SO_4 aqueous electrolyte.

2.4.2.3 Solvent-Free ILs for Hybrid ESs

IL electrolytes have also been investigated for hybrid ESs, such as AC//MnO_2 [678], AC//graphene-supported Fe_2O_3 [674], and AC//poly(3-methylthiophene) (PMeT) [679–682]. Balducci et al. [680,681] found the use of ILs ([BMIM][PF_6] and [PYR_{14}][TFSI]) as electrolytes for AC//PMeT hybrid ESs was favorable for increased the cell voltage and long cycling life at high temperatures compared with the ESs using $TEABF_4$/PC organic electrolytes. The AC//pMeT hybrid ES using [PYR_{14}][TFSI] could provide maximum energy and power densities of 24 Wh kg^{-1} and 14 kW kg^{-1}, respectively. Sun et al. [674] explored the [EMIM][BF_4] IL electrolyte for asymmetric hybrid ESs. The positive and negative electrodes were Fe_2O_3 nanoparticles supported on graphene sheets and activated PANI-derived carbon nanorods, respectively. A wide working voltage window of 0–4 V could be achieved for this hybrid

ES, and a very high maximum energy density of 177 Wh kg^{-1} was obtained and the energy density still exhibited a relatively value of 62.4 Wh kg^{-1} at a high power density of 8 kW kg^{-1}. The cycling stability of the hybrid ESs is also an important issue. For example, Arbizzani et al. [631] reported the AC//pMeT hybrid ES with [PYR_{14}][TFSI] IL electrolyte had a much lower cycling stability compared to the carbon electrode-based asymmetric EDLCs. There was a 50% loss in the specific capacitance of this hybrid ES after 5000 cycles because of the performance degradation of the pMeT electrode.

2.4.3 Mixtures of ILs and Organic Solvents

Mixtures composed of ILs and organic solvents have been studied as electrolytes for ESs in order to improve the ionic conductivity and decrease the viscosity of ILs. Similarly, due to their relatively high conductivity among common ILs, imidazolium-based ILs have also been the most frequently studied ILs in organic solvents. McEwen et al. [683] found higher concentrations (>3.5 M) could be achieved with [EMIM][PF_6] IL in PC compared with $TEABF_4$ salt (1 M). The ionic conductivities of carbonate electrolytes containing IL with [EMIM] cation ([EMIM][PF_6] and [EMIM][BF_4]) were about 25% higher than those with $TEABF_4$. The highest conductivity of 27 mS cm^{-1} was obtained for 2 M [EMIM][PF_6] in mixed carbonate solvent (EC–DMC). In addition, the ES with 2 M [EMIM][PF_6]/PC electrolyte exhibited a higher specific capacitance and better thermal stability compared to that with 1 M $TEABF_4$/PC electrolyte. Orita et al. [684] investigated various allyl-functionalized ILs in organic solvents. It was found that EDLCs using 1.4 M ILs composed of imidazolium cations with allyl groups (particularly diallylimidazolium ([DAIM]) cation) and $[TFSA]^-$ anion in PC as electrolytes showed higher specific capacitance and lower ESR under a wide temperature range compared to EDLCs using ILs based on imidazolium with saturated alkyl groups ([EMIM]). But the EDLC with 1.4 M [DAIM][BF_4]/PC had lower stability than the EDLC with [EMIM][BF_4]/PC. The addition of DMC to PC solvent could improve the stability.

Using the Cavity-MicroElectrode technique on TiC-derived carbon (TiC–CDC) electrodes, Lin et al. [685] found that for [EMIM][TFSI] in ACN organic solvent, $[EMIM]^+$ and $[TFSI]^-$ exhibited different affinities for ACN and thus led to different solvation degree. As a result, despite similar bare sizes of the $[EMIM]^+$ and $[TFSI]^-$, they exhibited different sizes of solvated ions in ACN and followed the order: $[TFSI]^-$ in ACN > $[EMIM]^+$ in ACN > $[EMIM]^+ \approx [TFSI]^-$. Therefore, it is necessary to take into account the effect of the organic solvent on the size of solvated ions when optimizing the pore size structure of porous carbon electrodes for ILs mixed with organic solvents. Interestingly, Mysyk et al. [686] found that the effect of organic solvent in the ILs on the capacitive behavior also strongly depended on the porous structure of the carbon electrode materials.

Although increasing the electrolyte temperature was found to have a positive effect on some properties of the porous carbon-based EDLCs using [EMIM][BF_4]/ACN electrolyte [687,688], the volatilization of ACN at elevated temperatures is a concern. For example, Chen et al. [689] noticed that the volatilization of ACN at high temperatures (especially at 80°C) decreased the cycling stability of the EDLC using [EMIM][BF_4]/ACN electrolyte. As discussed in Section 2.3.2, ADN organic solvent has higher flashing point than the ACN solvent and also could provide higher operative cell voltage of EDLCs [446,447]. However, the solubility of common conducting salt (e.g., $TEABF_4$) in ADN solvent is much lower compared to ACN solvent. This results in a much smaller ionic conductivity of the ADN-based organic electrolytes than that of the ACN-based electrolytes (e.g., about one order of magnitude smaller as reported by Brandt et al. [446]). This limits the practical use of ADN as a solvent. The use of ILs as salts in ADN could provide an approach to address this issue. Ghamouss et al. [690] found that the 2 M 1-ethyl-3-methylimidazolium bis[(trifluoromethyl)sulfonyl]imide ([C_2MIM][TFSI]) dissolved in ADN gave a higher ionic conductivity of 10.95 mS cm^{-1} (at 25°C) than the 0.7 M $TEABF_4$/ADN electrolyte (4.58 mS cm^{-1}). As a result, the AC-based EDLC using 2 M [C_2MIM][TFSI]/ADN showed a comparable ESR (13.24 Ω) compared with that using 1 M $TEABF_4$/ACN, which was smaller than that using 0.7 M $TEABF_4$/ADN (28.00 Ω). The 2 M [C_2MIM][TFSI]/ADN-based EDLC also exhibited a higher specific capacitance value compared with the 0.7 M $TEABF_4$/ADN-based EDLC.

Noticeably, Palm et al. [691] found that the introduction of organic solvent (ACN, PC, and GBL) to the pure [EMIM][BF_4] as electrolyte had negative effects on the performance of porous carbon cloth-based EDLCs although it could decrease the viscosity and the melting point and increase the conductivity compared with pure [EMIM][BF_4]. For example, the addition of 1 M ACN to the pure [EMIM][BF_4] IL decreased the electrochemical stability, maximal specific energy (from 49 to 39 Wh kg^{-1}), and the power densities (from 28 to 25 kW kg^{-1}), and increased the characteristic relaxation time (from 17 to 21 s). Therefore, a major challenge in this area is that the addition of organic solvents to the ILs as electrolytes for ESs should not sacrifice the important advantages of pure ILs.

Besides the porous carbons, ILs dissolved in organic solvents have also been studied as electrolytes for graphene-based electrodes [692,693]. For example, Liu et al. [693] found that specific capacitance of graphene nanosheets (GNSs) in the 0.5 M [EMIM][BF_4]/*N,N*-dimethylformamide (DMF) electrolyte was found to show a higher value than that obtained in the same IL dissolved in CH_3CN or CH_3COCH_3 organic solvents. The performance (e.g., specific capacitance and internal resistance) of the GNS electrode in the 1-methyl-3-methylimidazolium tetrafluoroborate ([MMIM][BF_4])/DMF electrolyte were better than that in the [EMIM][BF_4]/DMF and 1-butyl-3-methylimidazolium tetrafluoroborate ([BMIM][BF_4])/DMF electrolytes, due to the shorter length of the alkyl chain in [MMIM][BF_4].

Furthermore, some pseudocapacitors and hybrid ESs also used mixture electrolytes containing imidazolium-based ILs and organic solvents [678,694,695]. Zhang et al. [678] found the specific capacitance of asymmetric AC//MnO_2 ES was increased and the internal resistance was decreased after the addition of DMF into the [BMIM][PF_6] IL electrolyte, due to the improved ion mobility and penetration of the electrolyte compared with the pure IL. The optimal volume ratio of [BMIM][PF_6] and DMF was 1:1, and further increasing the DMF concentration would decrease the performance. For intercalation-type carbon materials (e.g., partially reduced graphite oxide), the dilution of [EMIM][BF_4] with ACN was found to increase the specific capacitance from below 100 F g^{-1} up to approximately 180 F g^{-1}.

Due to the high electrochemical stability, pyrrolidinium-based ILs have also been investigated as mixtures with organic solvents for ES electrolytes [696–699]. Similarly, the organic solvent could provide salvation ability and thus decrease the viscosity and increase the ionic conductivity of the pyrrolidinium IL-based electrolyte [576,696]. Ruiz et al. [696] reported that EDLCs using mixtures of $[PYR_{14}]$ [TFSI] and butyronitrile (BuCN) provided a much higher specific capacitance and could safely operate within a wider temperature range (–20°C to 80°C) compared with those using pure $[PYR_{14}]$[TFSI] as the electrolyte. It was also found that the nitrile-derived organic solvents (BuCN and ACN) reduced the electrochemically stable window of the pure IL to a greater extent than the carbonated solvents due to their lower electrochemical stability. Compared to EDLCs with the common $TEABF_4$/PC organic electrolyte, EDLCs with mixture electrolyte containing pyrrolidinium-based ILs and PC exhibited a higher operative cell voltage of 3.2–3.5 V [697–699]. For example, Pohlmann and Balducci [698] found that the 2.3 M $[PYR_{14}][BF_4]$ in PC showed ionic conductivity and viscosity values comparable with those of the 1 M $TEABF_4$/PC electrolyte. Owing to the increased electrochemical stability of the electrolyte, EDLCs using 2.3 M $[PYR_{14}][BF_4]$/PC showed a better cycling stability, larger operative voltage of 3.2 V and thereby higher energy and power densities compared with that using 1 M $TEABF_4$/PC. The cycling stability of the $[PYR_{14}][BF_4]$/PC-based EDLC was also better than that of the $TEABF_4$/PC-based EDLC. EDLCs using $[PYR_{14}]$[TFSI]/PC mixture as electrolyte also exhibited a higher operative cell voltage (3.5 V), energy and power densities compared with those using $TEABF_4$/PC electrolyte, although the ESR value of the former was slightly higher than that of the latter [699].

Phosphonium-based ILs have also been studied to be mixed with organic solvents as electrolytes for EDLCs [700,701]. As shown in Figure 2.77, Frackowiak et al. [701] reported the specific energy and power densities of EDLCs was strongly dependent on the ACN amount in trihexyl(tetradecyl)phosphonium bis(trifluoromethylsulfonyl) imide ($[(C_6H_{13})_3(P(C_{14}H_{29}))[Tf_2N]]$ or trihexyl(tetradecyl) phosphonium dicyanamide ($[(C_6H_{13})_3(P(C_{14}H_{29}))[(CN)_2N]]$. Among all the investigated concentration of ACN, 25 wt.% ACN mixed with $[(C_6H_{13})_3(P(C_{14}H_{29}))[Tf_2N]$ IL provided a highest operative cell voltage of 3.4 V highest energy density of about 40 Wh kg^{-1}, and best cycling stability for the EDLC using AC electrodes.

For ammonium-based ILs, Abdallah et al. [702] added [TMPA][TFSI] IL into ACN, and found that it could reduce the flammability and the volatility of the mixture electrolytes. The [TMPA][TFSI]/ACN mixture showed a similar electrochemical stability and specific capacitance compared with the conventional $TEABF_4$/ACN electrolyte.

Concerning the sulfonium-based ILs, Brandt et al. [703] reported the use of 3.8 M trimethylsulfonium bis[(trifluoromethyl)sulfonyl]imide ($[Me_3S]$[TFSI]) mixed with PC (PC–$[Me_3S]$[TFSI](86)) resulted in the operative cell voltage of EDLC up to 2.9 V. As shown in Figure 2.78, the relatively high concentration of $[Me_3S]$[TFSI] (i.e., 3.8 M $[Me_3S]$[TFSI]/PC) in the electrolyte was necessary to provide a stable performance of EDLC at 20°C and 60°C. However, when the concentration of $[Me_3S]$[TFSI] in PC was lower at 1.9 M (PC–$[Me_3S]$[TFSI](50)), the resultant ES did not exhibit a stable performance, due to the lower ability to inhibit the anodic oxidation of the Al current collector by the electrolyte.

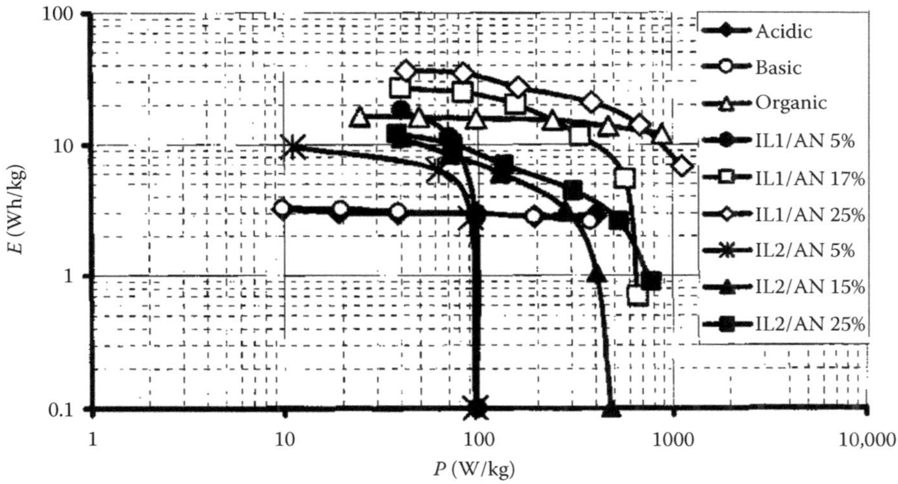

FIGURE 2.77 Ragone plot for supercapacitors built from AC in different electrolytes. (Reprinted with permission from Frackowiak, E., G. Lota, and J. Pernak. Room-temperature phosphonium ionic liquids for supercapacitor application. *Applied Physics Letters* 86:1–3. Copyright 2005, American Institute of Physics Publishing LLC.)

There have been some comparative studies between different types of ILs mixed with organic solvents [697,704–706]. Yuyama et al. [705] investigated the effect of a series of ILs in PC on the performance of AC-based EDLCs. In this study, different kinds of cation species were investigated: *N,N*-diethyl-*N*-(2-methoxyethyl)-*N*-methylammonium (DEME), *N,N*-diethyl-*N*-methoxymethyl-*N*-methylammonium (DEMM), and *N*-methoxymethyl-*N,N,N*-trimethylammonium (MMTM), *N*-(2-methoxyethyl)-*N*-methylpyrrolidinium (MEMP), and *N*-methoxymethyl-*N*-methylpyrrolidinium (MMMP). Three types of anion species including BF_4^-, PF_6^-, and $TFSI^-$ were selected for the counter anion. EDLCs using electrolytes containing the BF_4^- anion exhibited a higher capacitance and lower ESR than those containing PF_6^- and $TFSI^-$ anions. The authors concluded that [MEMP][BF_4] was the preferred IL in PC for EDLC electrolyte with respect to the capacitance, power density, low-temperature performance, and thermal durability. As shown in Figure 2.79, Orita et al. [706] investigated different sulfonium- and thiophenium-based ILs in PC as electrolytes for EDLCs. Among all the investigated ILs with [TFSA] anion, the ionic conductivity of the [DEMS][TFSA] IL mixed with PC was the highest. The [DEMS][BF_4]/PC electrolyte had higher ionic conductivity compared to the [DEMS][PF_6]/PC and [DEMS][TFSA]/PC electrolytes. The specific capacitance of EDLC with the [DEMS][BF_4]/PC electrolyte was higher than that with the conventional [EMIM][BF_4]/PC or [TEMA][BF_4]/PC electrolyte. As shown in Figure 2.80, Brandt and Balducci [707] found the use of the 1.9 M [PYR_{14}][BF_4]/PC electrolyte resulted in a higher energy and power densities of EDLCs compared to other electrolyte systems including 1.5 M [PYR_{14}][TFSI]/PC, 3.8 M [Me_3S][TFSI]/PC, and conventional 1 M $TEABF_4$/PC, due to the higher operative cell voltage. With respect

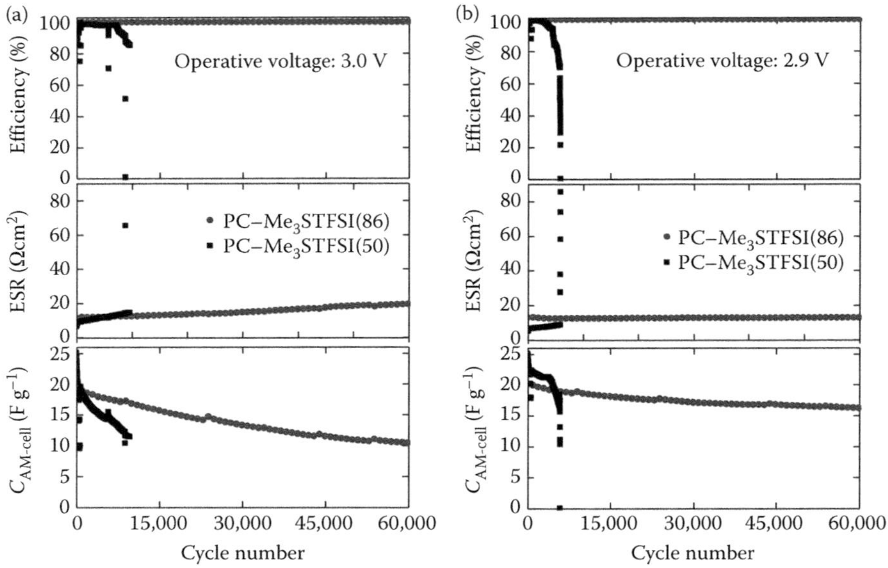

FIGURE 2.78 Comparison of the long-term stability of EDLCs containing the electrolytes PC-Me_3STFSI(86) and PC–Me_3STFSI(50) operating at voltages of (a) 3.0 V and (b) 2.9 V, respectively. The charge–discharge process was carried out at 1.5 A g^{-1} (per total active material in the cell). For each EDLC, the evolution of coulombic efficiency, ESR, and specific capacitance (C_{am}) is presented. (Brandt, A. et al. 2013. An investigation about the use of mixtures of sulfonium-based ionic liquids and propylene carbonate as electrolytes for supercapacitors. *Journal of Materials Chemistry A* 1:12669–12678. Reproduced by permission of The Royal Society of Chemistry.)

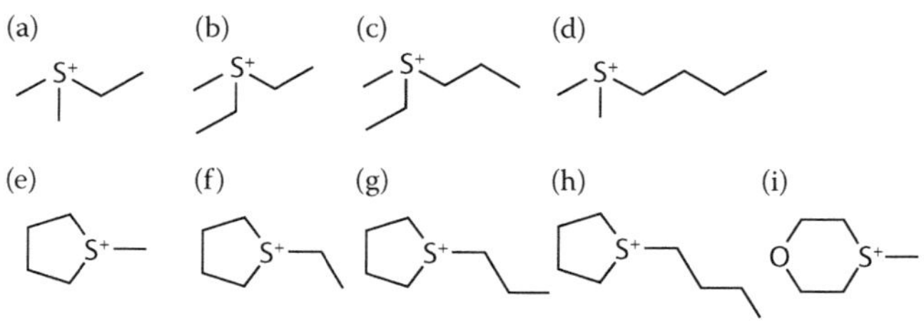

FIGURE 2.79 Chemical structures of cations used in Reference 706: (a) DMES (dimethylethylsulfonium), (b) DEMS (diethylethylsulfonium), (c) EMPS (ethylmethylpropylsulfonium), (d) BDMS (butyldimethylsulfonium), (e) MTT (1-methyltetrahydrothiophenium), (f) ETT (1-ethyltetrahydrothiophenium), (g) PTT (1-propyltetrahydrothiophenium), (h) BTT (1-butyltetrahydrothiophenium), and (i) MOT (1-methyl-[1,4]-thioxonium). (Reprinted from *Journal of Power Sources*, 195, Orita, A. et al., Application of sulfonium-, thiophenium-, and thioxonium-based salts as electric double-layer capacitor electrolytes, 6970–6976, Copyright 2010, with permission from Elsevier.)

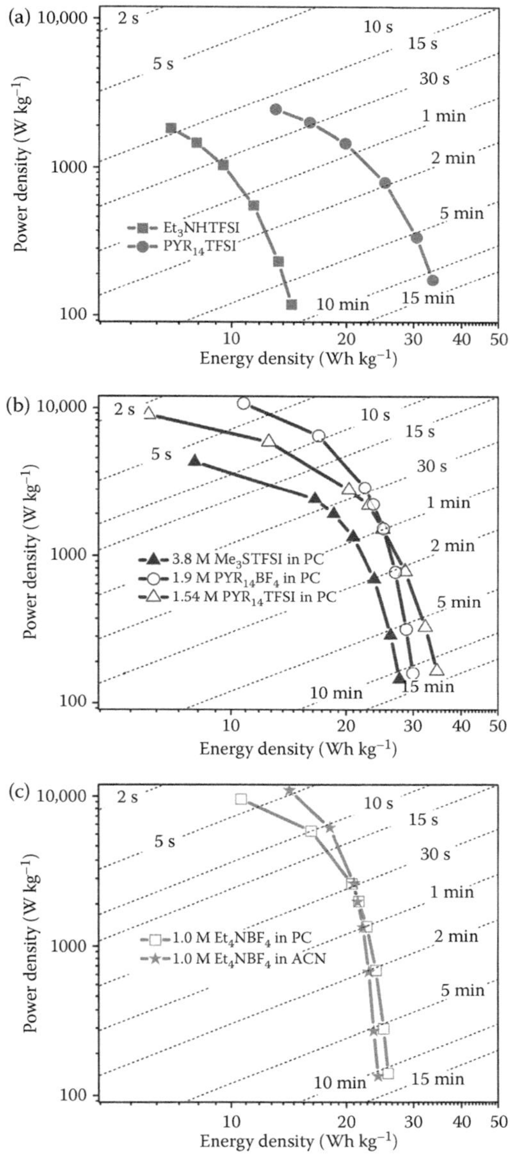

FIGURE 2.80 Ragone-like plots of EDLCs containing the electrolytes considered in Reference 707. The average values of energy and power density refer to the total AC loading in the cell. For this comparison current densities between 0.1 and 10 A g^{-1} (per total AC loading in the cell) were applied at 20°C. Times shown are the time constants of the EDLCs obtained by dividing the energy density by the power. (Reprinted from *Journal of Power Sources*, 250, Brandt, A., and A. Balducci, Theoretical and practical energy limitations of organic and ionic liquid-based electrolytes for high voltage electrochemical double layer capacitors, 343–351, Copyright 2014, with permission from Elsevier.)

to the piperidinium-based ILs, Lewandowski and Olejniczak [708] reported the ionic conductivity of 52 wt.% ACN mixed with *N*-methyl-*N*-propylpiperidinium bis(trifluoromethanesulfonyl)imide ([MePrPip][TFSI]) IL was out 40 mS cm^{-1}, which was much higher than that of the pure [MePrPip][TFSI] IL (1.5 mS cm^{-1} at 25°C). An electrochemically stable window of about 3.7 V was obtained on AC electrodes in the mixture electrolyte containing 52 wt.% ACN and [MePrPip][TFSI] IL.

Although protic ILs have attracted much less interest compared with aprotic ILs, they have also been investigated to be mixed with organic solvents [647,649,709,710]. Generally, compared to the aprotic ILs in organic solvents, a much lower operative cell voltage of 1.5–2.5 V was obtained in electrolyte mixtures containing organic solvents and protic ILs such as tributylphosphonium tetrafluoroborate ([Bu_3HP][BF_4]) [709], triethylammonium bis(tetrafluoromethylsulfonyl)amide ([Et_3NH][TFSA]) [647,649], and pyrrolidinium nitrate ([Pyrr][NO_3]) [710]. This is possibly the main reason why they have received limited interest so far. However, protic ILs have some advantages such as being generally easier and cheaper to synthesize compared with aprotic ILs [649,711,712]. Several studies found protic ILs mixed with organic solvents such as [Bu_3PH][BF_4] in GBL [709] and [Pyrr][NO_3] in GBL [710] provided a larger working temperature window from −40°C to 80°C for EDLCs. Coadou et al. [713] compared performances of trimethylammonium-based protic IL ([HN_{111}][TFSI]) and trimethylsulfonium-based aprotic ([S_{111}][TESI]) ILs, as mixtures with three organic solvents: GBL, PC, and ACN for AC-based EDLCs. It was found that for each solvent investigated under the same working conditions, no significant differences in the electrochemical performance were observed between the electrolytes based on the protic IL and aprotic IL. GBL-based electrolytes resulted in a better specific capacitance even at low temperature. [S_{111}][TESI] + GBL mixture gave specific capacitances up to 131 and 80 F g^{-1} at 25 and −30°C, respectively.

Theoretical and simulation work has also been done to investigate the projected behavior and performance of ILs mixed with organic solvents for EDLCs [517,714–717]. Using MD simulations, Feng et al. [517] found although the dielectric constant of bulk ACN was more than twice as high as that of pure [BMIM][BF_4] IL, the capacitance of the EDLs in the [BMIM][BF_4]/ACN mixture increased only slightly with the increasing mass fraction of ACN rising from 0% to 50%. Based on MD simulations, Shim et al. [715] found that for 1.1 M [EMIM][BF_4] in ACN, ACN played an important role by aligning its dipoles near the electrodes. Shim et al. [714] also investigated the EDLCs based on a single-sheet graphene electrode with pure [EMIM][BF_4] or [EMIM][BF_4]/ACN mixture electrolyte via MD simulations. The capacitance normalized to the electrode surface area was found to increase by 55%–60% in pure [EMIM][BF_4] compared to [EMIM][BF_4]–ACN mixture, which was explained by the strong screening of the electrode charge by IL ions in the former. Owing to the difference in the size and shape between cations and anions of ILs, the specific capacitance of the positively charged electrode was higher than that of the negatively charged electrode by more than 10% for both pure [EMIM][BF_4] and [EMIM][BF_4]/ACN mixture electrolyte. Merlet et al. [716] used MD simulations to study the effect of ACN solvent on the EDL structural and capacitive properties for EDLCs based on two different ILs ([BMIM][PF_6] and [BMIM][BF_4]). The salvation was found to reduce the ionic/molecular layering region near the graphite electrodes.

A larger dissymmetry of the capacitance between the positively and negatively charged electrodes was observed in pure ILs compared with the IL–ACN mixtures. Okada et al. [717] performed MD calculations for pure [EMIM][TFSI] and [EMIM][TFSI]–PC mixture confined in slit graphite pore with three pore sizes of 1.6, 2.8, and 4.0 nm. The layering of $[EMIM]^+$ was found to be interfered by PC solvent. The solvation effect of PC increased the diffusion coefficients of ILs in the bulk, however, this effect was drastically inhibited in the graphite nanopore.

2.5 SOLID- OR QUASI-SOLID-STATE ELECTROLYTES

Recently, solid-state or quasi-solid-state electrolytes have received increasing attention since they can greatly promote the development of ESs for portable electronics, wearable electronics, printable electronics, microelectronics, and especially flexible electronic devices.

The solid-state electrolytes can also be used as separators, and using solid-state electrolytes may simplify the fabrication and packaging processes of ESs. Solid electrolyte-based ESs can also reduce the leakage concern that related to the liquid electrolyte-based ESs. To data, various kinds of solid-state electrolytes have been developed for ESs. Most of them are polymer-based electrolytes, and inorganic solid materials such as ceramic electrolytes have attracted only very limited attention [718–721].

The polymer-based electrolytes for ESs can be further classified as the solid polymer electrolyte (SPEs, also called as dry polymer electrolytes), the gel polymer electrolyte (GPE), and the polyelectrolyte. Some studies regarded GPEs as quasi-solid-state electrolytes since they contain certain amount of liquid phase [722,723]. As shown in Figure 2.81a, the SPE consists of a polymer matrix such as poly(ethyl oxide) (PEO) and conducting salt such as LiCl, and there is no solvent such as water [2]. The transfer of ions from the salts through the polymer provides the ionic conductivity of the SPE. The GPE is composed of polymer matrix such as poly(vinyl alcohol) (PVA) and a conducting salt dissolved in the liquid solvent or an aqueous electrolyte such as H_2SO_4/H_2O. The polymer matrix is swollen by the solvent, which is different with the SPE. The electrolyte ions transport in the liquid solvent instead of the polymer phase. In the polyelectrolyte, the charged polymer chains provided the ionic conductivity. Each group of solid-state electrolytes has their advantages and shortcomings [2]. GPEs generally have the highest ionic conductivity among these types of solid-state electrolytes, due to the good ion transportation ability in the liquid phase. Owing to this reason, GPEs have been the most extensively studied electrolytes for solid electrolyte-based ESs [724,725].

As will be discussed below, these solid electrolytes have so far been used for EDLCs, pseudocapacitors, and hybrid ESs. Generally, when developing solid-state electrolytes for ESs, one should take into account a number of requirements such as high ionic conductivity, high electrochemical and thermal stability, good mechanical strength, and dimensional stability. Actually, it is extremely difficult for a solid-state electrolyte to meet all the requirements. For example, there is often a trade-off between high ionic conductivity and high electrochemical stability. This has stimulated considerable research to improve the overall performance of solid-state

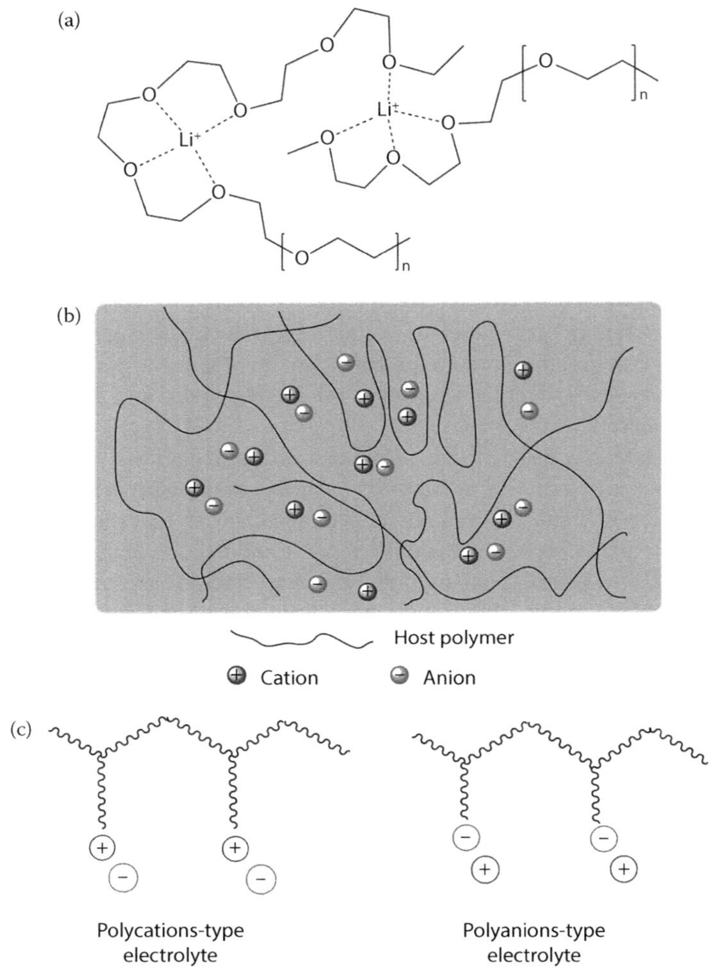

FIGURE 2.81 Schematic diagrams of (a) dry solid-state polymer electrolyte (e.g., PEO/Li^+), (b) gel polymer electrolyte, and (c) polyelectrolyte. (Zhong, C. et al. 2015. A review of electrolyte materials and compositions for electrochemical supercapacitors. *Chemical Society Reviews* 44:7484–7539. Reproduced by permission of The Royal Society of Chemistry.)

electrolytes for ESs. To date, there have already been several excellent and comprehensive reviews published in this area [725–730] and two of them have just been published recently in 2013–2014 [725,728]. Therefore, this book chapter focuses mainly on the latest developments reported in 2013–2015.

2.5.1 Gel Polymer Electrolytes for ESs

As mentioned previously, GPEs have been so far the most commonly used electrolytes for solid-state ESs due the high ionic conductivity. Various polymer hosts have been studied for GPEs, such as PVA, poly(-acrylic acid) (PAA), potassium polyacrylate (PAAK), PEO, poly(methylmethacrylate) (PMMA), poly(ether ether ketone) (PEEK), poly(vinylidene fluoride-co-hexafluoropropylene) (PVDF-HFP), and poly(acrylonitrile)-block-poly(ethylene glycol)-block-poly(acrylonitrile) (PAN-*b*-PEG-*b*-PAN). When water is used as the solvent, the GPE is also referred to as hydrogel polymer electrolyte. Water is trapped in the 3D polymer networks of GPE by the surface tension [730]. Apart from water, organic solvents, such as PC [731], EC and DMF [732] or their mixtures such as PC–EC [733,734], PC–EC–DMC [735], and PC–EC–nanoclay [736], have been investigated for GPEs. The proportion of polymer and solvent has a profound influence in the plasticization degree and thus affects the glass transition temperature of GPEs [730].

One of the most distinct advantages of developing GPEs as electrolytes is that it facilitates the development of ESs with interesting diverse structures and shapes for various applications. It is exciting to note that, for example, using PVA-based hydrogels, various solid-state ESs, such as flexible ESs [737–739], stretchable ESs [740,741], origami-type foldable ESs [742], flexible micro-ESs [743], printable micro-ESs [744–746], on chip micro-ESs [747], 3D micro-ESs [748], yarn ESs [749–751], wire or fiber-shaped ESs [134,752–755], transparent ESs [756–758], ultrathin ESs [759–761], weaveable ESs [762], paper-like ESs [763], and integrated ESs with other devices [764–766], have been developed. Considering the application of PVA-based hydrogels for flexible and stretchable ESs, Zhao et al. [767] investigated the mechanical property of the PVA/H_3PO_4 hydrogel electrolyte. It was found that the mechanical properties could be optimized by controlling the molecular weight of PVA and its weight ratio to H_3PO_4. The optimized PVA/H_3PO_4 hydrogel showed a tensile strength of 2 MPa and a Young's modulus of 1 MPa, and exhibited only 5% plastic deformation after 1000 stretching cycles applied with 100% strain.

2.5.1.1 Hydrogel Polymer Electrolyte

As a linear polymer, PVA has so far been the most frequently used polymer host among various types of polymers for hydrogels due to its easy preparation, high hydrophilicity, good film forming properties, nontoxic characteristic, and low cost [768,769]. Hydrogel electrolytes are generally prepared from PVA mixed with aqueous solutions such as strong acid (e.g., H_2SO_4 and H_3PO_3), strong alkaline (e.g., KOH), and neutral (e.g., LiCl) electrolytes. These hydrogels have been studied for different types of ESs such as EDLCs, pseudocapacitors, hybrid ESs, and asymmetric ESs.

2.5.1.1.1 Hydrogel Polymer Electrolytes for Carbon-Based Electrodes

When ESs use GPEs as electrolytes, it is important to properly design the interface between the electrode and the electrolyte, in order to make the performance of the ESs using GPEs comparable to those using aqueous counterparts. For example, previous studies showed ESs with GPEs suffered from limited rate capability [770–772], which might be attributed to a limited ion mobility or a large resistance that occurred at the interface between the electrode and the electrolyte [770,773]. For instance, for PVA/H_3PO_3-based ESs with SWCNTs on polyethylene-terephthalate (PET) substrates as electrodes, the specific capacitance decreased significantly from about 110 to about 36 F g^{-1} as the discharging current density increased from 1 to 30 A g^{-1}

[770]. In this study, it was also found the electrode thickness significantly affected the capacitance per geometric area (F cm^{-2}) of the PVA/H_3PO_3-based ESs [770]. As the electrode thickness increased, the capacitance increased linearly for the 1 M H_2SO_4 aqueous electrolyte whereas the capacitance value did not increase when the electrode thickness was higher than 15 μm for the PVA/H_3PO_3 hydrogel electrolyte (Figure 2.82). This was due to the restricted penetration of ions into porous carbon electrodes, and thus might limit the use of this hydrogel as electrolyte for high charge storage applications.

The optimization of the electrode/hydrogel electrolyte interface can be achieved through the coordinated development of the electrode materials and the hydrogel electrolyte. Electrode materials with various structures, such as highly porous graphene on carbon cloth [774], CNTs grown on carbon cloth [775], AC cloth [773,776], 3D graphene network [777,778], porous graphene ribbons [779], rGO–carbon black hybrid films [780], and graphene–porous carbon aerogels [781], have been investigated for high-performance ESs using PVA-based hydrogel electrolytes. The development of these novel carbon structures can promote the penetration of hydrogel electrolyte into the porous electrode and greatly improves the utilization degree of the active electrode materials. For example, in order to improve the rate capacity, Wang et al. [773] proposed a method to activate carbon cloth via chemical oxidation followed by a two-step reduction process in hydrazine and ammonia. Exfoliated carbon fibers were formed on the surface of carbon fibers by this method, which remarkably increased the effective surface area. As a result, the PVA/H_3PO_4-based ES with AC cloth electrodes showed a high capacitance retention of 81% as the current density increased from 1 to 20 mA cm^{-2}, and it exhibited an a real capacitance of about 23 mF cm^{-2} at a high current density of 20 mA cm^{-2}. In an effort to enhance the overall or a real specific capacitance of ESs using hydrogels,

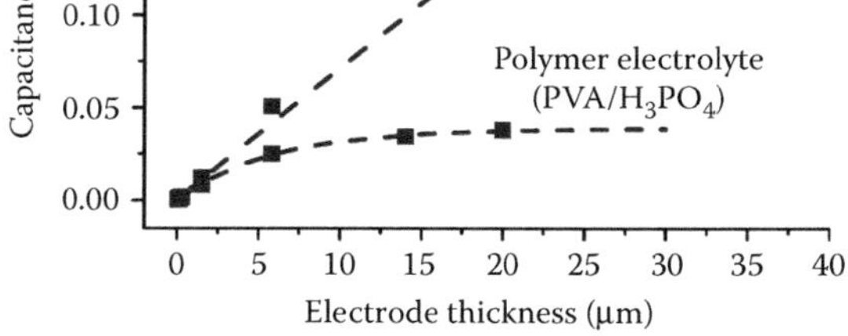

FIGURE 2.82 Thickness dependence of the capacitance per area for CNT films comparing a liquid (1 M H_2SO_4) and a gel electrolyte (PVA/H_3PO_4). (Reprinted with permission from Kaempgen, M. et al. Printable thin film supercapacitors using single-walled carbon nanotubes. *Nano Letters* 9:1872–1876. Copyright 2009 American Chemical Society.)

Xu et al. [778] prepared highly interconnected 3D graphene hydrogel films as electrodes for PVA/H_2SO_4-based flexible solid-state ESs. The PVA/H_2SO_4-based ESs with 120 μm thick graphene hydrogel electrodes exhibited a very high real capacitance of 372 mF cm^{-2}, a high specific capacitance of 186 F g^{-1} at 1 A g^{-1}, a good rate capability (70% retention at 20 A g^{-1}), and excellent cycling stability (8.4% capacitance decrease over 10,000 charging/discharging cycles at 10 A g^{-1}). The surface functionalization of 3D graphene hydrogels by incorporation of hydroquinones was found to further increase the specific capacitance (412 F g^{-1} at 1 A g^{-1}) of PVA/H_2SO_4-based ESs, due to the presence of pseudocapacitive hydroquinones [777].

Regarding the electrolyte, the performance (e.g., ionic conductivity, thermal, and environmental stability) can be optimized through the selection of the electrolyte solution for PVA, the modification of PVA, and the incorporation of additives into the hydrogels. Chen et al. [782] investigated six different PVA-based hydrogels (i.e., H_3PO_4, H_2SO_4, KOH, NaOH, KCl, and NaCl) as electrolytes for ESs using graphene electrodes. Based on the CV measurements, it was found that the use of PVA/H_3PO_4 hydrogel as the electrolyte resulted in the best capacitive performance of the ES among all the investigated hydrogel electrolytes (Figure 2.83). Although the specific capacitance of PVA/H_2SO_4-based ESs was higher than that of the PVA/H_3PO_4-based

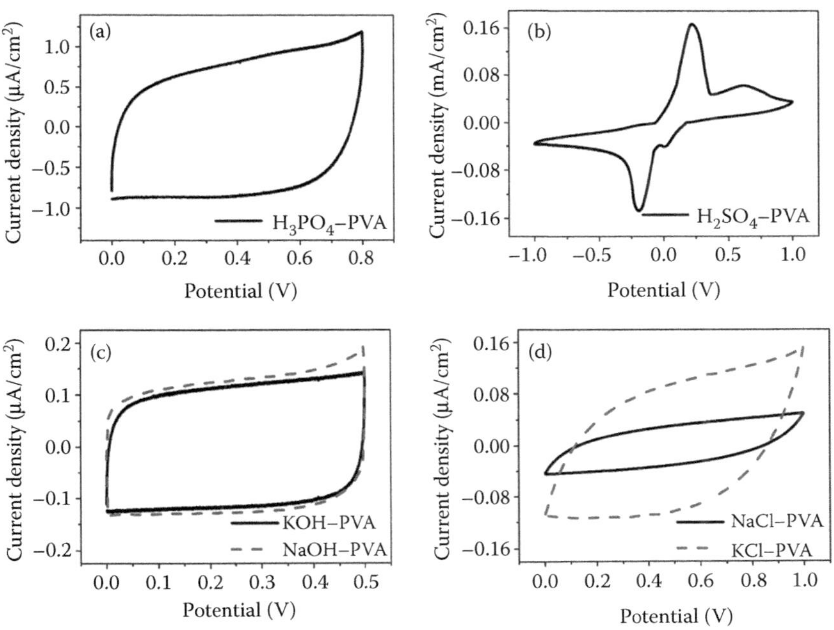

FIGURE 2.83 CV curves of as-prepared solid-state supercapacitors with different electrolytes at 100 mV s^{-1} scan rate. (a) H_3PO_4–PVA and (b) H_2SO_4–PVA electrolyte. (c) Base gel electrolytes: KOH (straight line) and NaOH (dashed line) and (d) neutral gel electrolytes: NaCl (straight line) and KCl (dashed line). (Chen, Q. et al. 2014. Effect of different gel electrolytes on graphene-based solid-state supercapacitors. *RSC Advances* 4:36253–36256. Reproduced by permission of The Royal Society of Chemistry.)

ESs due to the pseudocapacitance contribution in this study, the former had a poorer stability.

The PVA-based electrolyte generally suffered from the fluidity problem at high temperatures [769]. In order to improve the performance of PVA/H_2SO_4 hydrogels at higher temperature, Fei et al. [769] used the crosslinking agent (glutaraldehyde [GA]) to prepare cross-linked PVA/H_2SO_4 hydrogel electrolyte for ESs using graphene–carbon black nanoparticle film electrodes. The cycling stability of the resultant ESs was good, with the capacitance retention of 78.3% after 1000 cycles at 70°C. Various additives (fillers), such as SiO_2 [783], TiO_2 [784], Sb_2O_3 [785], graphene oxide (GO) [786], and hydroxyethylcellulose (HEC) [787], have also been added in PVA-based hydrogels with the aim to improve the performance. For example, Lim et al. [785] found the addition of Sb_2O_3 into PVA/$LiClO_4$ hydrogels increased the maximum operative potential region. AC-based EDLCs using PVA/$LiClO_4$ hydrogel with 6 wt.% Sb_2O_3 showed higher specific capacitance compared with those without Sb_2O_3. This was explained by considering that the Sb_2O_3 in the polymer system promoted the dissociation of lithium salt and acted as solid plasticizer. To improve the ionic conductivity, replacement of Sb_2O_3 with TiO_2 filler was also investigated by Lim et al. [784]. The introduction of TiO_2 into PVA/$LiClO_4$ was found to increase the ionic conductivity, and the highest ionic conductivity value of 1.3×10^{-4} S cm^{-1} at room temperature was obtained for the PVA/$LiClO_4$ electrolyte containing 8 wt.% of TiO_2. Differential scanning calorimetry analyses demonstrated the TiO_2 addition enhanced the flexibility of the polymer chain and favored the ion mobility. EDLCs using PVA/$LiClO_4$ containing TiO_2 exhibited higher specific capacitances compared with those without TiO_2. Huang et al. [786] added electrically insulating GO into boron cross-linked PVA/KOH hydrogels for EDLCs, and the GO served as an ionic conducting promoter. The synergetic effect of the GO sheets and the boron cross-links on the specific capacitance of the EDLCs was observed. GO containing PVA/KOH hydrogel electrolytes exhibited higher ionic conductivity compared with the KOH aqueous electrolyte. When the GO content was low, the ionic conductivity of the electrolyte increased with the increasing amount of GO at the low content of GO, but it decreased as the GO content further increased due to the aggregation of GO sheets that led to the blocking of ion channels (Figure 2.84). The ionic conductivity of boron cross-linked PVA/KOH hydrogels containing 20 wt.% GO was 0.1950 S cm^{-1}. The addition of 20 wt.% GO into the boron cross-linked PVA/KOH hydrogels resulted in the 29% increase of the specific capacitance of the EDLC compared to the EDLC using KOH aqueous electrolyte. The authors assumed the capacitance increase originated from the GO sheets at the electrode/electrolyte interface. Concerning the organic additives, Rosi et al. [787] investigated the effect of the addition of HEC (a gelatin/thickening agent) in the PVA–H_2SO_4 hydrogel electrolyte. Fourier transform infrared spectroscopy indicated the interaction of HEC with PVA/H_2SO_4. Different amounts (1, 2.5, and 5 wt.%) of HEC in the PVA/H_2SO_4 hydrogels were studied, and the addition of 1 wt.% HEC resulted in highest ionic conductivity of the electrolyte and the highest specific capacitance in the ESs.

When hydrogel electrolytes are used, it is also necessary to consider the environmental stability issue of the electrolytes. Gao et al. [788] found the ES with PVA/KOH

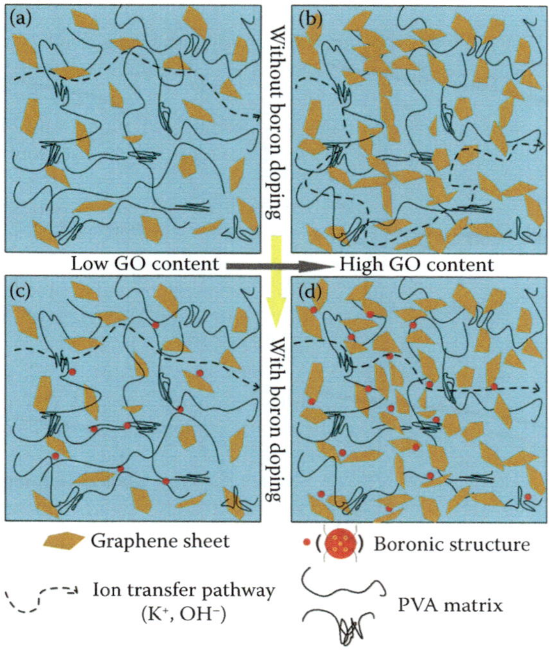

FIGURE 2.84 Illustration of ion transport mechanism in GO–PVA and GO–B–VA alkaline gel electrolytes with low and high GO content. (Reprinted from *Electrochimica Acta*, 132, Huang, Y. F. et al., Boron cross-linked graphene oxide/polyvinyl alcohol nanocomposite gel electrolyte for flexible solid-state electric double layer capacitor with high performance, 103–111, Copyright 2014, with permission from Elsevier.)

hydrogel electrolyte suffered a remarkable performance loss after shelf storage for 40–67 days. This was attributed to the dehydration of KOH/PVA. Therefore, Gao et al. [788] proposed the use of tetraethylammonium hydroxide (TEAOH) to replace KOH for PVA-based hydrogel electrolytes. The shelf life of the PVA/TEAOH-based ESs was much longer than that of the PVA/KOH-based ESs, owing to the higher water retention capability of TEAOH (Figure 2.85). Gao et al. [783] also found the water content in the PVA/silicotungstic acid (SiWA)–H_3PO4 hydrogel electrolyte at lower relative humidity (RH) could be stabilized by the addition of SiO_2 into the electrolyte and thus the environmental stability of the ESs could be improved. The increased environmental stability was attributed to the bond interaction between SiO_2 and water molecules.

2.5.1.1.2 Hydrogel Polymer Electrolytes for Pseudocapacitors and Hybrid ESs

As mentioned in Section 2.2 and similar to the research trend in the field of aqueous electrolytes, considerable work has focused on the development of hydrogel electrolytes for ESs with enhanced charge storage capacity, such as pseudocapacitors and hybrid ESs including asymmetric ESs. The electrochemical behavior of the pseudocapacitive electrode materials is directly influenced by the nature of the electrolytes.

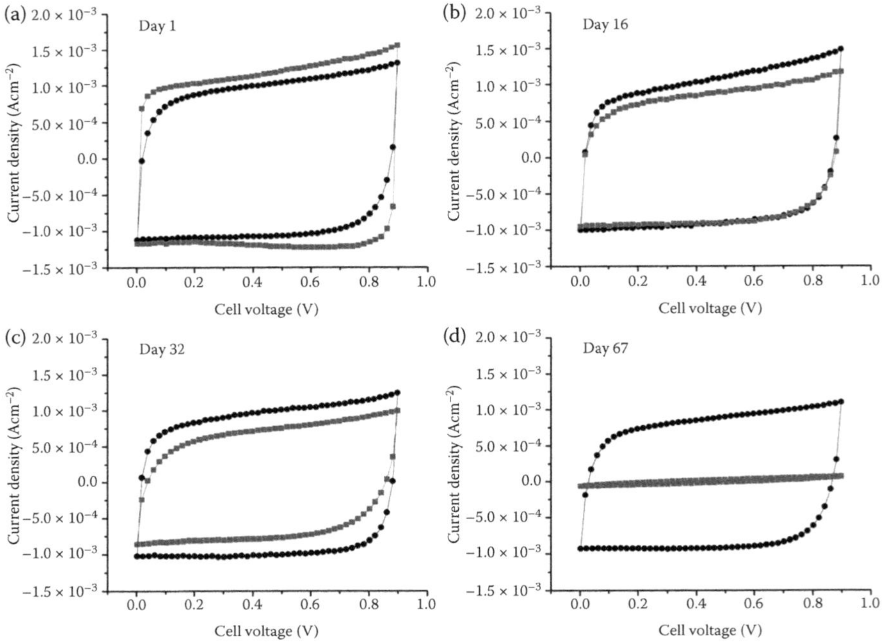

FIGURE 2.85 CVs of a solid graphite/TEAOH–PVA ES and a solid graphite/KOH–PVA ES at (a) day 1, (b) day 16, (c) day 32, and (d) day 67 (sweep rate = 1 V s^{-1}). (Gao, H., J. Li, and K. Lian. 2014. Alkaline quaternary ammonium hydroxides and their polymer electrolytes for electrochemical capacitors. *RSC Advances* 4:21332–21339. Reproduced by permission of The Royal Society of Chemistry.)

For ESs with acid PVA-based hydrogels such as PVA/H_2SO_4 and PVA/H_3PO_4, typically investigated pseudocapacitive electrode materials include transition metal oxides such as RuO_2 [789,790], manganese oxide [761,791,792], TiO_2 [793], $ZnCo_2O_4$ [794], and WO_3 [795], metal nitrides such as molybdenum nitride [796], sulfides such as MoS_2 [754], TiS_2 [754], and TaS_2 [754], selenides such as tin selenide [797], $GeSe_2$ [738], and $NbSe_2$ [754], and ECPs such as PANI [798–803], PPy [804], PEDOT [805,806], and polyselenophene (PSe) [807]. When alkaline PVA-based (mostly PVA/KOH) hydrogels are used as electrolytes for ESs, the most frequently investigated pseudocapacitive electrode material is $Ni(OH)_2$ [134,808–812], and other pseudocapacitive electrode materials such as $NiCo_2O_4$ [813,814], CoMn-LDH [815], MnO_2 [816], and $Mn_3(PO_4)_2 \cdot 3H_2O$ [817], have also been explored. In the case of ESs using neutral PVA-based hydrogel electrolytes such as PVA/LiCl [818–821], PVA/Na_2SO_4 [822], and PVA/$NaNO_3$ [823], typically studied pseudocapacitive electrode materials include MnO_2 [822–824], V_2O_5 [820,825], ZnO [819,826], $VOPO_4$ [827], vanadium nitride (VN) [828], KCu_7S_4 [829], and KCu_4Se_8 [820]. With the aim to enhance the intercalation/deintercalation kinetics for pseudocapacitive electrodes, PVA-based hydrogels containing Li salts such as PVA/LiCl [818–821,827], PVA/LiOH [815], and PVA/$LiNO_3$ [830], have been extensively studied.

Interestingly, the utilization of hydrogels as electrolytes has also been found to have a beneficial effect on improving the electrochemical stability of some pseudocapacitive electrode materials such as vanadium oxide [818] and VN [828]. For example, Wang et al. [818] found the vanadium oxide-based pseudocapacitor with PVA/LiCl hydrogel electrolyte showed a much higher cycling stability (more than 85% capacitance retention after 5000 cycles) than that with the 5 M LiCl aqueous electrolyte (Figure 2.86). This result occurred as the PVA/LiCl hydrogel electrolyte minimized the water content and provided a neutral pH medium, inhibiting the chemical dissolution of vanadium oxides in the aqueous electrolyte. In addition, the hydrogel electrolyte was able to serve as a supporting matrix and maintain the structure of the prepared vanadium oxide nanowire network.

It is also of significance to optimize the interface between the electrode and the PVA-based electrolyte. For example, to improve the contact between the electrode and the hydrogel electrolyte, Wang et al. [798] proposed to use hydrogel PANI materials instead of conventional solid materials as the electrodes for the ESs. For hydrogel PANI electrode materials, active materials were swollen with water. A high specific capacitance of 430 F g^{-1} was subsequently obtained for the PVA/H_2SO_4 electrolyte-based ESs using hydrogel PANI electrodes. To improve conductivity and the utilization factor of the electrode materials in the hydrogel electrolytes, the pseudocapacitive materials are often dispersed on high surface area carbonaceous supports (e.g., carbon fibers [815], carbon cloth [795,813,819], carbon paper [831], carbon nanowalls [790], graphene [761,832], and graphitic petal [803]) or combined with carbonaceous materials (e.g., CNTs [825,833–835] and graphene [754,807,836]) to make hybrid composites. In order to further enhance the performance of ESs using PVA-based hydrogel electrolytes, various properly designed structures, such

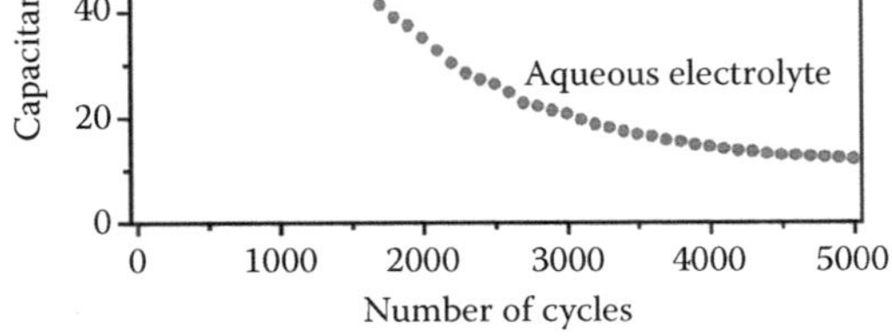

FIGURE 2.86 Comparison of cycling performance of vanadium oxide pseudocapacitors in a 5 M LiCl aqueous electrolyte and a LiCl/PVA gel electrolyte. The capacitance was calculated on the basis of the CV scans collected at a scan rate of 100 mV s^{-1}. (Reprinted with permission from Wang, G. M. et al. LiCl/PVA gel electrolyte stabilizes vanadium oxide nanowire electrodes for pseudocapacitors. *ACS Nano* 6:10296–10302. Copyright 2012 American Chemical Society.)

as ultrathin nanosheets [761,827], mesoporous nanostructures [810,835], hierarchical nanostructures [801,811], and 3D expressway-like architectures [808] have been developed.

Hydrogel electrolytes generally share similar disadvantages as aqueous electrolytes. For example, hydrogel electrolyte-based symmetric ESs had a limited working cell voltages and thus low energy density due to the presence of aqueous solvent that has low electrochemical stability. In an initiative to increase operative cell voltage, asymmetric ESs based on two different electrodes with complementary potential windows in hydrogel electrolytes, recently attracted considerable interest. Table 2.11 shows selected data for asymmetric ESs using PVA-based hydrogel electrolytes, from the recently published literature (2013–2014). Compared to the symmetric ESs using PVA hydrogel-based electrolytes, the operative cell voltage of these asymmetric ESs was significantly increased up to higher values such as 1.8 V. The enhanced operative cell voltage benefits the improved energy and power densities of ESs. Several studies involved the comparison of the performance of asymmetric ESs based on PVA-based hydrogel electrolytes with that of asymmetric ESs utilizing an aqueous electrolyte (without PVA). Compared to aqueous electrolytes, PVA-based hydrogel electrolytes inhibited the dissolution of some pseudocapacitive materials (e.g., MnO_2 [837] and VN [828]), and thereby improved cycling stability. Xu et al. [838] found both aqueous (3 M KOH) and PVA/KOH hydrogel electrolyte-based asymmetric ESs could be operated at 1.6 V, and the former delivered an energy density of 1.21 mWh cm^{-3} at a power density of 13.29 W cm^{-3} while the latter gave an energy density of 1.44 mWh cm^{-3} at a power density of 0.89 W cm^{-3}. The higher ESR of the hydrogel electrolyte-based asymmetric ESs was what limited the power density compared with the aqueous electrolyte-based asymmetric ESs. Besides, asymmetric PVA/LiCl hydrogel-based ESs using different carbon electrodes with different functional group was reported by Jin et al. [839], which showed a relatively high cell voltage of 1.6 V.

In addition to PVA, other polymers have also been studied as polymer hosts in hydrogels as electrolytes for ESs. For example, several studies demonstrated that poly(acrylate)- or poly(acrylic acid)-based hydrogels could be used as promising electrolytes for ESs. Due to the easy subtraction of their side chains in the aqueous solution, the use of these polymers can facilitate the proton conduction [844–849]. These hydrogel electrolytes can be used for AC-based EDLCs [849], pseudocapacitors based on RuO_2 [844], MnO_2 [847], ECPs [845], etc, and asymmetric ESs [846,848]. Since the nature of the electrolyte greatly affected the capacitive behavior of pseudocapacitive electrode material, it is expected that the performance of the resultant ES is strongly dependent on the property of the polymer host. Kim et al. [844] investigated different acrylic hydrogel electrolytes including PAA, PAAK, and poly(2-acrylamido-2-methyl-1-propanesulfonic acid) (PAMPS) for ESs using RuO_2 electrodes. It was found that the specific capacitance of the RuO_2 electrode in various electrolytes decreased in the following order: PAMPS/H_2O > 1 M H_2SO_4 (aqueous electrolyte) > PAA/H_2SO_4 > PAAK/H_2SO_4 > PAMPS/H_2SO_4. The side chain groups of PAMPS provided the best proton accommodation, contributing the highest capacitance of the RuO_2 electrode in the PAMPS/H_2O electrolyte. For ESs with PANI electrodes, Ko et al. [845] found that the PAA/H_2SO_4, PAAK/H_2SO_4, and

TABLE 2.11
Some Data from Selected Literature Published in 2013–2014 Concerning the Asymmetric ESs Using PVA-Based Hydrogel Electrolyte

Electrolyte	Negative Electrode	Positive Electrode	Capacitance ($F\ cm^{-3}$)	Cell Voltage (V)	Energy Density ($mWh\ cm^{-3}$)	Power Density ($mW\ cm^{-3}$)	Cycle Numbers	ESR (Ω)	References
PVA/LiCl	H-TiO_2@carbon NWs	H-TiO_2@MnO_2 NWs	0.70 (at 0.5 $mA\ cm^{-2}$)	1.8	0.3	230	5000 (91.2% retention)		[837]
PVA/LiCl	VO_x nanowire	VN nanowire	1.35 (at 0.5 $mA\ cm^{-2}$)	1.8	0.61 (at 0.5 $mA\ cm^{-2}$)	850 (at 5 $mA\ cm^{-2}$)	10,000 (95.3% retention)		[828]
PVA/LiCl	Reduced GO	MnO_2@ZnO nanorod array	0.52 $F\ cm^{-3}$ (at 10 $mV\ s^{-1}$)	1.8	0.234	133	5000 (98.5% retention)		[840]
PVA/KOH	Co_3O_4@RuO_2 nanosheet arrays	Co_9S_8 nanorod	4.28 $F\ cm^{-3}$ (at 2.5 $mA\ cm^{-2}$)	1.6	1.44	890	2000 (89% retention)		[838]
PVA/KOH	Graphene	$Ni_{11}(HPO_3)_8(OH)_6$	1.64 $F\ cm^{-3}$ (at 0.5 $mA\ cm^{-2}$)	1.4	0.45 (at 0.5 $mA\ cm^{-2}$)	33 (at 5 $mA\ cm^{-2}$)	10,000 (93.3% retention)		[841]
PVA/KOH	CNT	Layer-by-layer graphene/$Ni(OH)_2$/graphene	58.5	1.7	18 $Wh\ kg^{-1}$	850 $W\ kg^{-1}$	20,000		[808]
PVA/KOH	AC	Plate-like β-$Ni(OH)_2$	37.5 $F\ g^{-1}$ (at 2 mA)	1.4	9.8 $Wh\ kg^{-1}$	154 $W\ kg^{-1}$	2000 (76% retention)	~24	[842]

(Continued)

TABLE 2.11 (*Continued*)
Some Data from Selected Literature Published in 2013–2014 Concerning the Asymmetric ESs Using PVA-Based Hydrogel Electrolyte

Electrolyte	Negative Electrode	Positive Electrode	Capacitance ($F\ cm^{-3}$)	Cell Voltage (V)	Energy Density ($mWh\ cm^{-3}$)	Power Density ($mW\ cm^{-3}$)	Cycle Numbers	ESR (Ω)	References
PVA/KOH	Graphene	Co_3O_4 NWs on Ni fibers	2.1 (at 20 mA cm^{-3})	1.5	0.62	147	1000 (84% retention)	55.5	[764]
PVA/H_3PO_4	Functionalized carbon fiber thread	Carbon fiber thread @PANI		1.6	2	10,220 W cm^{-3} (maximum value)	10,000 (81% retention)		[753]
PVA/H_3PO_4	MnO_2@PPy on carbon cloth	AC on carbon cloth	1.41 F cm^{-2} (at 0.5 mA cm^{-2}); 261 F g^{-1}	1.8	8.69	12.35	1000 (98.6% retention)		[843]

Note: A@B, A is the core and B is the shell.

Abbreviations: NWs, nanowires; H-TiO_2, hydrogen-treated TiO_2; AC, activated carbon; PANI, polyaniline; PPy, polypyrrole; CNT, carbon nanotube; and ESR, equivalent series resistance.

PAMPS/H_2SO_4 electrolyte gave a higher specific capacitance of 400–500 F g^{-1} up to even a high scan rate of 1000 mV s^{-1} and also much better cycling stability compared with the PAMPS/H_2O electrolyte. This indicated the certain interaction between the H_2SO_4 solvent and acrylic polymers was beneficial for the PANI-based ESs compared with the H_2O-based PAMPS electrolyte. Nam et al. [847] found nanowire MnO_2-based ESs using the PAAK/KCl hydrogel electrolyte showed higher specific capacitance compared with those using the PAAK-co-polyacrylamide (PAAK-PAAM)/KCl hydrogel and 1 M KCl aqueous electrolyte.

Generally, the above-mentioned hydrogel electrolytes have limited thermal stability. In order to improve the thermal stability, Shimamoto et al. [850] proposed the use of the phosphosilicate gel electrolyte for the asymmetric ES. The negative and positive electrodes were AC and MnO_2–CNT composite, respectively. A large operating temperature range (–30°C to 100°C) was obtained for the assembled ESs. The ES using phosphosilicate gel electrolyte showed high capacitance retention after 20,000 cycles, with 83% at –30°C, 87% at 25°C, and 85% at 100°C, respectively. The specific capacitance and rate performance increased with the increasing temperature, due to the ion conductivity enhancement of the phosphosilicate gel as the temperature increased. In an effort to increase the operative voltage, Ramasamy et al. [851] investigated a gel electrolyte composed of polyvinylpyrrolidine (PVP) and Na_2SO_4 salt in an ethylene glycol (EG)–water mixture for AC-based EDLCs. The addition of 25% EG into the aqueous-based gels extended the operative cell potential limit to 1.75 V, and thereby increased the energy density.

As a proton-conducting polymer, Nafion® (a perfluorosulfonic acid polymer) has high ionic conductivity and has also been investigated as an electrolyte for different types of solid-state ESs [852–858]. It has been found that high scan rates could be achieved for ESs with Nafion electrolytes [854,855,859].

2.5.1.2 Organogel Electrolytes

In order to increase the operative cell voltage, GPEs containing organic solvents (plasticizers), also referred to organogel electrolytes, have been investigated as electrolytes for ESs. Various polymer matrixes, such as PEO [735,860], PMMA [861,862], polyvinylpyrrolidone (PVP), PEEK [863], and copolymers [722,732,864–866], have been used in organogel electrolytes. The utilization of PMMA-based GPEs as electrolytes is beneficial for achieving transparent ESs [862]. Copolymers that have different constituent units, such as poly(acrylonitrile)-block-poly(ethylene glycol)-block-poly(acrylonitrile) (PAN-*b*-PEG-*b*-PAN) have been proposed as promising polymeric hosts used in organogel electrolytes for ESs [732,864]. Hsueh et al. [864] found there was a synergistic effect between PAN and PEG in the GPE host swollen with DMF dissolving $LiClO_4$ for EDLCs. The PAN nitrile group was found to facilitate ion solvation and motion into the electrode materials, and the PEG ether group coordinated solvent molecules to promote polymer swelling and ion transport. As a result, through controlling the chain length ratio between AN and EG, the ionic conductivity of copolymer organogels could be improved. The ionic conductivity of the optimized copolymer organogel could achieve a maximum value of 1.1×10^{-2} S cm^{-1}, whereas $LiClO_4$/DMF organic liquid electrolyte provided limited ionic conductivity of the 1.6×10^{-3} S cm^{-1}. Due to the positive synergistic effect of the PAN and PEG blocks, copolymer organogel electrolyte-based EDLCs provided an energy density of 20 Wh kg^{-1} at 10 kW kg^{-1}, which was higher than organic liquid electrolyte-based EDLCs.

For organic solvents in the organogel electrolyte, typically investigated systems include PC, EC, DMC, or their mixtures [722,735,860,867], ACN [861,862], DMSO [863,865], and DMF [732,864]. The cell voltage of ESs using organogel electrolytes can be generally increased to 2.5–3 V [732,864–866,868]. As shown in Table 2.11, this operative cell voltage was much higher compared to both symmetric and asymmetric ESs using hydrogel electrolytes. This is one of the most significant advantages of organogel electrolyte-based ESs. As a result, most ESs using organogel electrolytes exhibited relatively high energy densities of 18–25 Wh kg^{-1} [735,860,861,864]. It is interesting to note Schroeder et al. [868] reported a high operative cell voltage of 4 V for the LIC with the methacrylate-based organogel electrolyte, and this greatly improved its energy density (Figure 2.87).

Concerning the conducting salt for organogel electrolytes, recently investigated salts include $LiClO_4$ [732,864,869], $LiPF_6$ [866], $TBAPF_6$ [861,862], NaTFSI [735], etc. The ionic conductivity of the electrolyte is strongly dependent on the type of the salt and the ratio between salt and the polymer host [869]. Besides, considering the

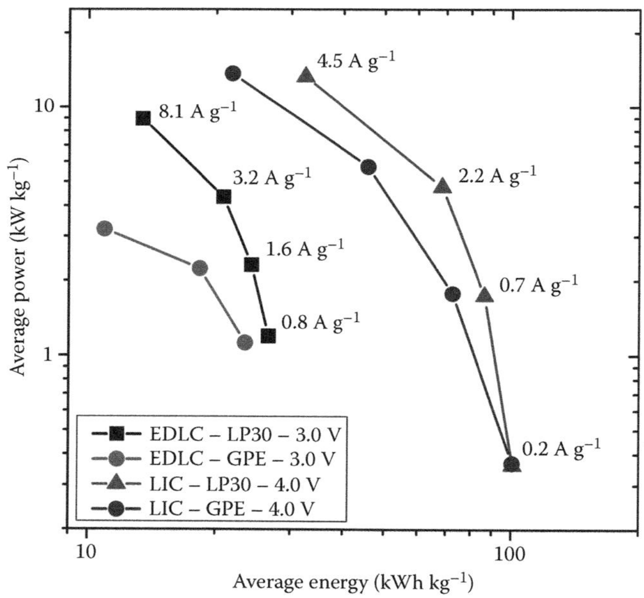

FIGURE 2.87 Ragone-like plot for two ESs and two LICs each containing either the liquid electrolyte LP30 or the methacrylate-based GPE. The average energy and power indicated in the figure is referred to the total active mass of the LIC. (The data points obtained at 0.2 and 4.5 A g^{-1} of the LIC measurements are added for better comparison). (Reproduced by permission of The Electrochemical Society from Schroeder, M. et al. 2013. *Journal of the Electrochemical Society* 160:A1753–A1758.)

abundance of magnesium (Mg), Mg salts have also been investigated for organogel electrolytes for ESs [860,865]. EDLCs using Mg-based organogel electrolyte such as PEO/MgTFSI [860] and PVDF-HFP/$Mg(CF_3SO_3)_2$ [865] showed encouraging performance in terms of high operative cell voltage such as 2.5 V reported by Ramasamy et al. [860]. Generally, the ionic conductivity of organogel electrolyte is relatively low. In order to improve the ionic conductivity, Jain and Tripathi [870] proposed the addition of SiO_2 particles into the PVDF-HFP/$Mg(ClO_4)_2$ – PC as electrolytes for EDLCs. A relatively high ionic conductivity of 5.4×10^{-3} S cm^{-1} at room temperature was obtained for this organogel electrolyte.

As mentioned in Section 2.3.2, when the cell voltage of organic electrolyte-based ESs is beyond the operating potential limits, the decomposition of organic electrolytes and the possible SEI formation on the electrodes occur. This seriously deteriorated the performance of ESs. The application of organogel electrolytes have been found to benefit the cycling stability of ESs especially at higher cell voltages, as reported by Chiou et al. [866]. In this study, CNT-based ESs with a PVDF-HFP/$LiPF_6$ organogel electrolyte showed a higher stability compared with those with $LiPF_6$ organic liquid electrolytes when the cell voltage was enlarged to 3.5 or 4 V (Figure 2.88). The authors attributed this to the presence of the PVDF-HFP polymer, which slowed down the electrolyte decomposition kinetics.

2.5.1.3 IL-Based Solid-State Electrolytes

Taking into account the better electrochemical stability and higher thermal stability of ILs, IL-based GPEs (also referred to ionogels) recently have attracted great interest. Ionogels are composed of IL and suitable polymer host. Ionogel electrolytes have been frequently investigated for EDLCs based on electrode materials such as AC [871], porous carbon [872], and CNT [873–876]. Ionogels can also be used as electrolytes for pseudocapacitive electrode materials such as MnO_2 [877], PEDOT [878], and composite materials such as PMeT combined with hydrous RuO_2 [879] and 3,4-ethylenedioxythiophene):poly (styrene sulfonate) (PEDOT:PSS) combined with hydrous ruthenium oxide [880]. With the use of ionogel electrolytes, it is possible to fabricate different types of ESs such as flexible ESs [877], stretchable ESs [881], micro-ESs [872], flexible micro-ESs [873], and stretchable micro-ESs [874,876].

The electrochemical stability and ionic conductivity of ionogels are dependent on the nature of ILs and host polymers, and the interactions between them. Liew et al. [882] found the incorporation of [BMIM][Cl] IL decreased the glass transition temperature of PVA-based electrolytes down to sub-ambient value. A polymer electrolyte with 50 wt.% of [BMIM][Cl] showed the maximum ionic conductivity of 7.31 ± 0.01 mS cm^{-1} at 120°C. It is also important to consider the polymer host to improve the performance of ESs using IL-based GPEs. A wide variety of polymer hosts, such as PEO [883], PVA [882], PMMA [874,881], poly(ethylene glycol) diacrylate [876], PVA–PVP [880], and PVDF-HFP [733,875,877,878,884–886], have been studied for ionogel electrolytes. Several studies found the chitosan host had a positive effect on the performance of EDLCs using ionogels [887,888]. Chitosan is a linear polysaccharide composed of D-glucosamine and *N*-acetyl-D-glucosamine, which is an environment-friendly material and abundant in nature [888]. Yamagata et al. [888] reported the EDLC using chitosan/[EMIM][BF_4] gel electrolyte showed

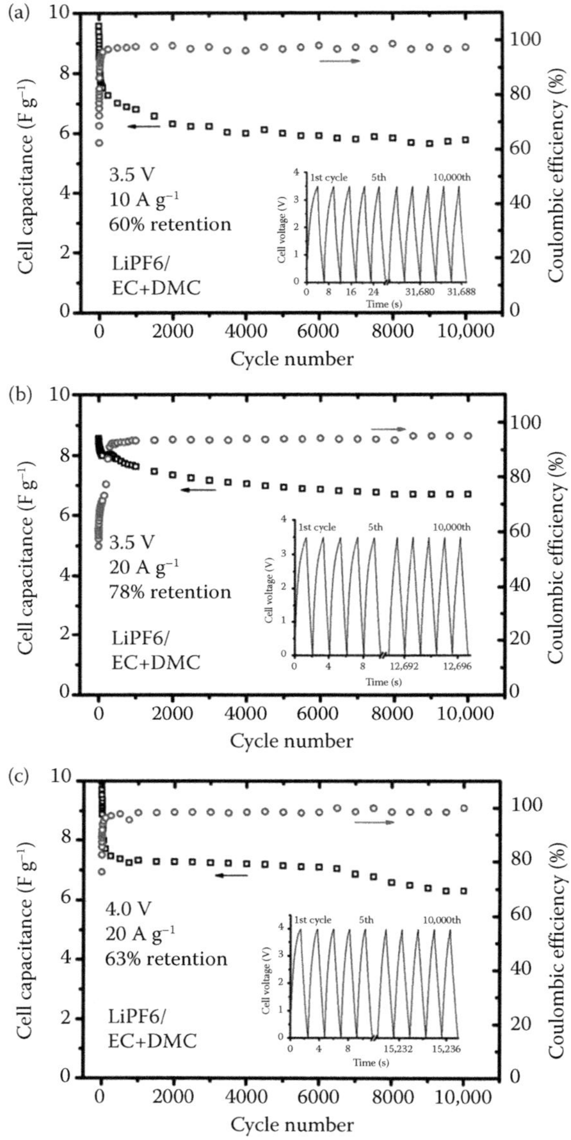

FIGURE 2.88 Cycle stability of the $LiPF_6$ capacitor at (a) 10 A g^{-1} and 3.5 V; (b) 20 A g^{-1} and 3.5 V; and (c) 20 A g^{-1} and 4.0 V. (Chiou, Y. D. et al. 2013. Cycle stability of the electrochemical capacitors patterned with vertically aligned carbon nanotubes in an $LiPF_6$-based electrolyte. *Nanoscale* 5:8122–8129. Reproduced by permission of The Royal Society of Chemistry.)

enhanced discharge capacitance and rate performance compared to that using liquid [EMIM][BF_4] IL electrolyte. This was probably due to the decrease of the electrode/electrolyte interfacial resistance contributed by the high affinity of chitosan for the AC electrode, as indicated by the EIS measurements.

To take advantage of the high chemical and thermal stability of PVDF while improving its flexibility, Yang et al. [889] blended amorphous polyvinyl acetate (PVAc) with PVDF as polymer host for [BMIM][BF_4] IL. The PVDF–PVAc with 50 wt.% [BMIM][BF_4] possessed good flexibility, and showed the ionic conductivity of 2.42×10^{-3} S cm^{-1} at room temperature, as well as exhibited high thermal stability up to about 300°C. AC-based EDLCs with this ionogel electrolyte exhibited a cell voltage of 3 V, with the specific capacitance of 93.3 F g^{-1} at 200 mA g^{-1}, and the capacitance loss was lower than 10% after 5000 charging/discharging cycles. With the use of poly(styrene-block-ethylene oxide-block-styrene) (PS–PEO–PS) copolymer as a matrix for [EMIM][NTf_2] IL, the assembled CNT-based EDLC exhibited good mechanical integrity and its performance was maintained after 200 bending cycles [890]. The hydrophobic PS block was assumed to provide good adhesion to the hydrophobic CNT electrode via van der Waals interaction, while the PEO block could retain hydrophilic IL in the polymeric matrix [890].

As mentioned in Section 2.4, the performance of ESs is strongly dependent on the properties of ILs. Therefore, it is also very important to rationally select and design the IL component in ionogels. For example, some ILs with low viscosity such as [EMIM][SCN] [891], [EMIM][FAP] [884], and [EMIM][TCB] [875] have been proposed. The reported PVDF-HFP/[EMIM][TCB] and PAN/[EMIM][SCN] ionogels showed ionic conductivities of about 9×10^{-3} S cm^{-1} [875] and 1×10^{-3} S cm^{-1} [891] at room temperature, respectively. However, these ionogel electrolytes resulted in a limited cell voltage and thus led to relatively low specific energy densities of EDLCs (~3.5–4.7 Wh kg^{-1}) [875,891]. Ketabi and Lian [883] found that the ionic conductivity of the PEO/$EMIHSO_4$ electrolyte could be increased through the addition of SiO_2 into the electrolyte. Suleman et al. [886] added plastic crystalline materials such as succinonitrile (SN) into IL-based GPEs. One interesting feature of the SN-based GPE is that it may accommodate the mechanical stress owing to the plastic character of SN [886].

The property of IL-based GPEs was also affected by the preparation methods [892,893]. Liu et al. [893] used [BMIM][Cl], chitosan, and hydroxyethyl methacrylate (HEMA) to prepare ionogel electrolytes by a self-triggered UV photopolymerization. This method did not involve the use of additional initiators and crosslinkers (Figure 2.89). EDLCs with this ionogel electrolyte exhibited good performance stability at 0.5 A g^{-1} during 2000 cycles under normal and bent conditions, and they can work at a high temperature of 100°C. It is interesting to note that this ionogel electrolyte also had self-recovering ability, due to the crosslinking interaction by hydrogen bonds [892,893].

Apart from the IL-based GPEs (ionogels), other new IL-based polymer electrolytes have also been developed for ESs [871,894]. Ayalneh Tiruye et al. [871] prepared a binary blend of a polymeric IL: [PYR_{14}][TFSI] and poly(diallyldimethylammonium) bis(trifluoromethanesulfonyl)imide ([pDADMA][TFSI]) (6:4 by mass ratio).

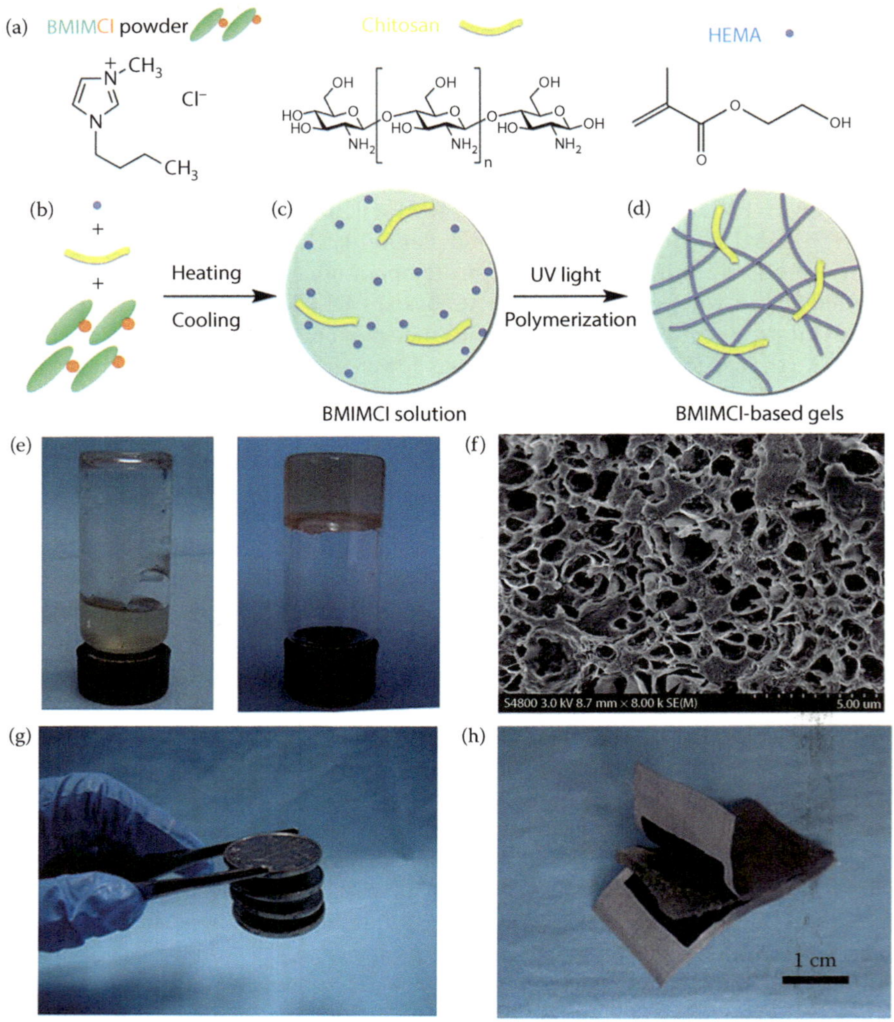

FIGURE 2.89 Preparation process, optical images, and SEM pictures of BMIMCl-based gels. (a) The molecular structures of the chemicals: BMIMCl, chitosan, and HEMA. (b) The chitosan solution prepared by blending chitosan with BMIMCl at 110°C for 6 h. (c) The precursor obtained by mixing chitosan solution and HEMA. (d) The UV photo-crosslinking of the as-prepared mixture. (e) Free flowing liquid-like precursor. (f) SEM image of all-solid-state BMIMCl-based gels. (g) Viscous demo of BMIMCl/chitosan/poly (hydroxyethyl methacrylate) (BCH)-GPE with coins (about 6 g per coin). (h) The sandwiched structure of a symmetrical cell of supercapacitors with an ionogel film as the electrolyte and activated charcoal as the electrode. (Liu, X. H. et al. 2014. Tough BMIMCl-based ionogels exhibiting excellent and adjustable performance in high-temperature supercapacitors. *Journal of Materials Chemistry A* 2:11569–11573. Reproduced by permission of The Royal Society of Chemistry.)

Due to the higher electrochemical stability of pyrrolidinium-based IL, a large operative cell voltage of 3.5 V could be obtained for EDLC using this electrolyte. As a result, a high energy density of 32 Wh kg^{-1} could be achieved for this EDLC. Taniki et al. [894] reported the use of ionic plastic crystal of *N,N*-dimethylpyrrolidinium fluorohydrogenate ($DMPyr(FH)_2F$) as the solid electrolyte for AC-based ESs. CV measurements showed typical EDL capacitive behavior within the narrow potential regions, whereas the redox reactions involving fluorohydrogenate anions occur on both electrodes within the wide potential regions.

2.5.1.4 Environmental-Friendly Gel Polymer Electrolytes

Due to environmental concerns, there has been growing interest in the development of environmental-friendly polymer hosts made from natural sources such as corn starch [895,896] and chitosan [734,897] or biodegradable materials such as poly(epsilon-caprolactone) [898] for solid-state electrolytes of ESs. Since the mechanical strength of natural polymers such as starch film was generally low, several studies reported the use of blended materials made from chitosan and starch [897] or chitosan and PEG [734].

2.5.1.5 Structural Electrolytes

Taking into account the use of solid-state ESs in applications involving vibration (e.g., electric vehicles), the development of load-bearing solid-state electrolytes (structural electrolytes) has received growing interest [899–902]. Westover et al. [902] used PEO/[EMIM][BF_4] infiltrated into nanoporous silicon to assemble load-bearing solid-state ESs. This ES could maintain the energy density of nearly 10 Wh kg^{-1} under the tensile stress at 300 kPa and 80 g vibratory accelerations.

2.5.2 Inorganic Solid-State Electrolytes for ESs

There are some studies focusing on inorganic solid-state electrolytes for ESs [718,719,903], but so far they have attracted very limited interest compared to solid-state electrolytes based on polymers. Despite the lack of flexibility compared with polymer-based solid electrolytes, inorganic solid electrolytes are usually mechanically robust and thermally stable and may also be considered as potential electrolytes for solid-state ESs [2]. Owing to its relatively high Li-ion conductivity, Francisco et al. [718] used $Li_2S–P_2S_5$ glass ceramic as both electrolyte and separator for solid-state ESs. A specific capacitance of nearly 10 F g^{-1} was obtained for the resultant ES. Ulihin et al. [719] employed $LiClO_4–Al_2O_3$ composites as solid electrolyte for symmetric and asymmetric ESs. Manganese oxides were mixed with the lithium–manganese–nickel spinel to serve as electrode materials. The assembled ESs could work at high temperatures up to 150–200°C. Unfortunately, this ES exhibited low specific capacitance value of 29 F g^{-1} at 0.05 A g^{-1} under 150°C. The ES's capacity decreased by 10% during five charging/discharging cycles, probably due to the loss of the contact between the composite solid electrolyte and electrode materials. Zhang et al. [903] and Gao et al. [904] used GO as the solid electrolyte for ESs. Zhang et al. [903] presented all-solid-state layer-by-layer rGO//GO//rGO capacitor. Comparable to conventional ESs, this capacitor exhibited a fairly high specific capacitance up

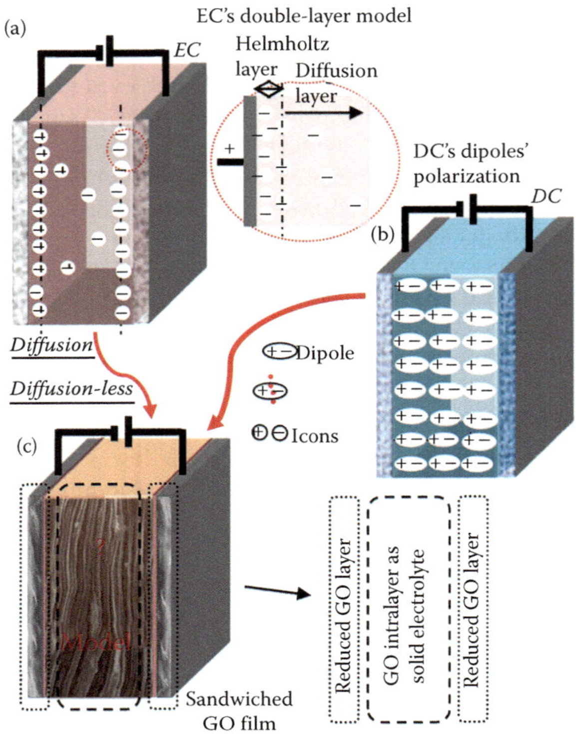

FIGURE 2.90 Schematics of charge storage mechanisms. (a) The EC, which stores energy through ionic diffusion under electrostatic attraction to form one or two atomic layers of compact electrolytic ions followed with a relatively uncondensed diffusion layer. (b) The DC, where dipoles in dielectric polarize under external potential field. (c) The GO-capacitor with the sandwiched structure of rGO/GO/rGO that behaves abnormally. (Reprinted with permission from Zhang, Q. et al., Anomalous capacitive behaviors of graphene oxide based solid-state supercapacitors. *Nano Letters* 14:1938–1943. Copyright 2014 American Chemical Society.)

to about 0.86 mF cm^{-2}. However, no detectable ionic diffusion was observed for this capacitor, which is similar to dielectric capacitors. This study implied a new capacitive mechanism beyond the conventional understanding on ESs and dielectric capacitors. The authors proposed a new capacitive model that combined the two distinct charge storage mechanisms of ESs and dielectric capacitors to explain the anomalous capacitive behavior of this capacitor (Figure 2.90).

2.6 REDOX-ACTIVE ELECTROLYTES

Recently, redox-active electrolytes have been proposed to be used for ESs to provide addition pseudocapacitive contribution, which utilized the faradaic reactions originated from the electrolyte to store charge [905–907]. Therefore, both the electrode materials and electrolytes contribute to the overall capacitance [908].

2.6.1 Redox-Active Aqueous Electrolytes

2.6.1.1 Redox-Active Aqueous Electrolytes for Carbon Electrodes

One typical redox pair example is iodide/iodine, and Lota and Frackowiak [909] applied the iodide/iodine redox pair for redox-active aqueous electrolytes of carbon-based ESs. An extremely high capacitance value of 1840 F g^{-1} was obtained for the carbon positive electrode in 1 M KI electrolyte. The following reactions (Equations 2.12 through 2.15) occur in this electrolyte, contributing to the pseudo-capacitance [909]:

$$3I^{-1} \leftrightarrow I_3^{-1} + 2e^- \tag{2.12}$$

$$2I^{-1} \leftrightarrow I_2 + 2e^- \tag{2.13}$$

$$2I_3^{-1} \leftrightarrow 3I_2 + 2e^- \tag{2.14}$$

$$I_2 + 6H_2O \leftrightarrow 2IO_3^{-1} + 12H^+ + 10e^- \tag{2.15}$$

However, only the positive electrode could provide such high capacitance value within a narrow potential range (Figure 2.91). A specific capacitance of 125 F g^{-1} at

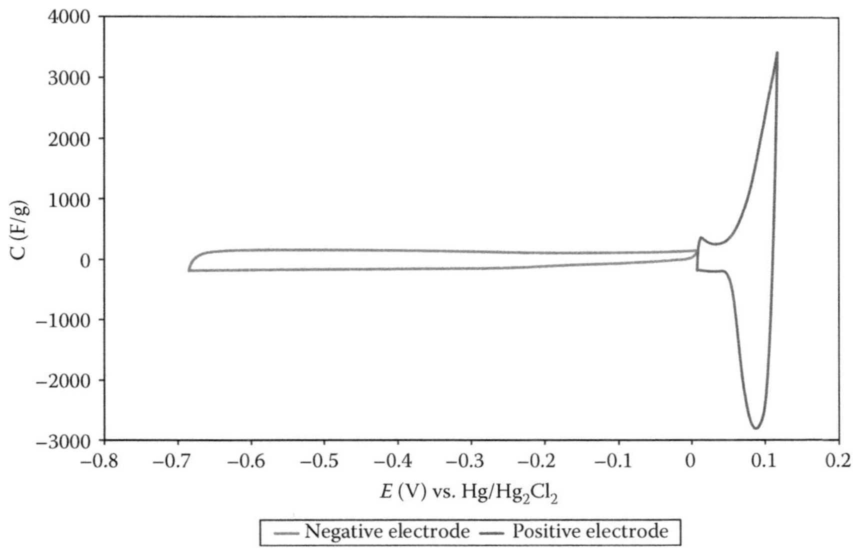

FIGURE 2.91 Voltammetry curve at 5 mV s^{-1} scan rate for carbon AC1 in 1 M KI solution. Potentials of both electrodes are measured separately versus saturated calomel electrode. (Reprinted from *Electrochemistry Communications*, 11, Lota, G., and E. Frackowiak, Striking capacitance of carbon/iodide interface, 87–90, Copyright 2009, with permission from Elsevier.)

50 A g^{-1} was obtained for two-electrode ES based on KI electrolyte. Senthilkumar et al. [910] found the addition of 0.08 M KI into the 1 M H_2SO_4 electrolyte led to nearly two-fold increase of the specific capacitance and energy density of the AC-based ES compared to the 1 M H_2SO_4 without KI. In comparison, the 0.08 M KI + 1 M Na_2SO_4 and 0.08 M KBr + 1 M H_2SO_4 electrolytes resulted in a lower specific capacitance and energy density.

As shown in Figure 2.92, the capacitive behavior of the electrode strongly depends on the nature of the alkali counter-ion of the iodides [911]. The capacitance of the positive electrode increased as the van der Waals radius of the alkali metal ion increased, following the order: LiI (300 F g^{-1}) < NaI (492 F g^{-1}) < KI (1078 F g^{-1}) < RbI (2272 F g^{-1}) (all with 1 M concentration) [911].

The overall ES capacitance is mainly determined by the electrode with low capacitance value. Therefore, in order to improve the capacitance of the negative electrode in the iodide-based electrolyte, Frackowiak et al. [912] applied 1 M $VOSO_4$ aqueous electrolyte for the negative electrode and 1 M KI aqueous electrolyte for the positive electrode. The following faradaic reactions (Equations 2.16 through 2.19) occurred at the negative electrode in $VOSO_4$ electrolyte, contributing to the high pseudocapacitance from the negative electrode [912]:

$$VOH^{2+} + H^+ + e^- \leftrightarrow V^{2+} + H_2O \tag{2.16}$$

$$[H_2V_{10}O_{28}]^{4-} + 54H^+ + 30e^- \leftrightarrow 10V^{2+} + 28H_2O \tag{2.17}$$

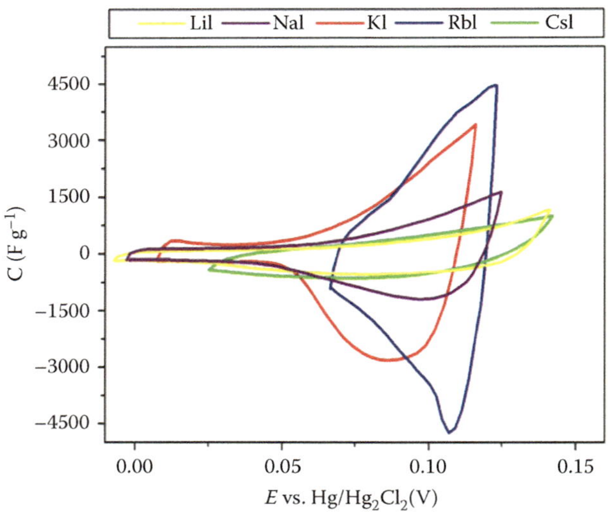

FIGURE 2.92 Capacitance values of the positive carbon electrode estimated from the voltammetry characteristics at 5 mV s^{-1} in the different alkali metal iodide solutions. (Reprinted from *Electrochemistry Communications*, 13, Lota, G., K. Fic, and E. Frackowiak, Alkali metal iodide/carbon interface as a source of pseudocapacitance, 38–41, Copyright 2011, with permission from Elsevier.)

$$[H_2V_{10}O_{28}]^{4-} + 44H^+ + 20e^- \leftrightarrow 10VOH^{2+} + 18H_2O \quad (2.18)$$

$$HV_2O_7^{3-} + 9H^+ + 6e^- \leftrightarrow 2VO + 5H_2O \quad (2.19)$$

As shown in Figure 2.93, these two different electrolytes were separated by a glassy paper and a Nafion 117 membrane so as to avoid the mixing of ions from different electrolytes [912]. A high capacitance value of 500 F g^{-1} at 0.5 A g^{-1} could be obtained for this ES at a cell voltage of 1 V, and the energy density is of about 20 Wh kg^{-1}. The addition of $VOSO_4$ into 1 M H_2SO_4 was also found to increase the specific capacitance and energy density, as well as decrease the ESR of the AC-based ESs [913].

Owing to their high proton conductivity and multiple rapid electron transfer processes, heteropoly acids, such as PTA and silicotungstic acid (SiWA), have also been proposed as attractive redox mediators for redox-active electrolytes [914–917]. For example, Suárez-Guevara et al. [915] found the phosphotungstic acid ($H_3PW_{12}O_{40}$, PW_{12}) electrolyte could provide pseudocapacitance. More interestingly, it could increase the working cell voltage up to 1.6 V for AC-based ESs. The improved cell voltage was attributed to the high overpotential of PW_{12} toward H_2

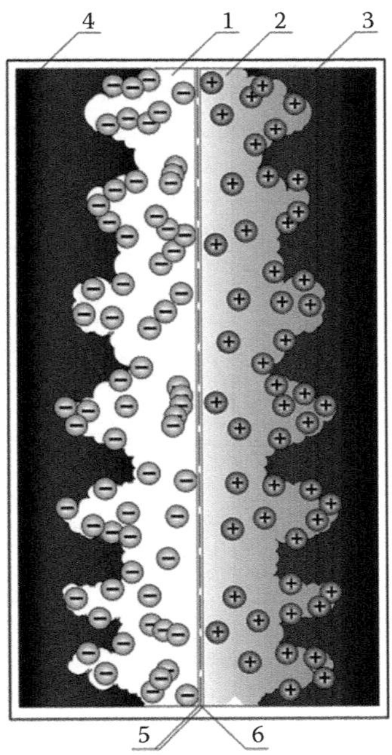

FIGURE 2.93 Capacitor scheme: (1) KI solution, (2) $VOSO_4$ solution, (3) negative electrode, (4) positive electrode, (5) glassy paper separators, and (6) Nafion 117 membrane. (Frackowiak, E. et al.: Electrochemistry serving people and nature: High-energy ecocapacitors based on redox-active electrolytes. *ChemSusChem*. 2012. 5. 1181–1185. Copyright Wiley-VCH GmbH & Co. KGaA. Reproduced with permission.)

generation. Besides, the redox pair of metal ions can also be used for redox-active electrolytes of ESs [918,919]. Mai et al. [918] utilized Cu^{2+} reduction with the surface oxygen group on carbons to increase the specific capacitance of the ES. CV measurements showed an approximately 10 times increase in the capacitance of the electrode in 0.09 M $CuCl_2$ + 1 M HNO_3 electrolyte compared to the 1 M H_2SO_4 electrolyte (Figure 2.94), owing to the pseudocapacitance provided from the following reaction [918]:

$$CuCl_2 + {>}C{=}O^{+} e^{-} \rightleftharpoons CuClCO_{\text{surface layer}} + Cl^{-} \qquad (2.20)$$

Tanahashi [919] reported an increase of about 300% in the maximum specific capacitance of the carbonaceous electrode in the $AgNO_3 + H_2SO_4$ electrolyte compared with H_2SO_4 electrolyte, which was attributed to the pseudocapacitance related with the redox reaction of Ag^+/Ag redox couple. However, the capacitance of the ES using $AgNO_3$-containing electrolyte decreased by 16% after only 1020 cycles, due to the deposition of Ag particles on the surface of carbonaceous electrodes. It is worth noticing that the self-discharge of ESs may be a problem for ESs using electrolytes containing metal ions.

Furthermore, organic redox mediators, such as hydroquinone (HQ) [920–923], methylene blue (MB) [924], indigo carmine [925], *p*-phenylenediamine (PPD) [926,927], *m*-phenylenediamine (MPD) [928], lignosulfonates [929], sulfonated polyaniline (SPAni) [930], and humic acids [931], have also been investigated. Table 2.12 shows some typical aqueous electrolytes with organic redox-active molecules, the corresponding redox processes, and the performance of the assembled

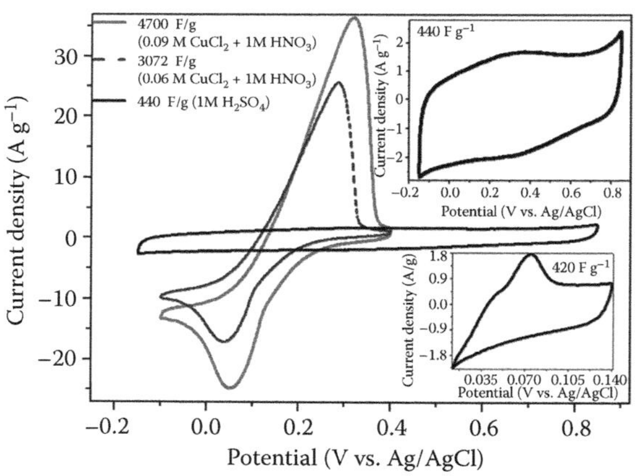

FIGURE 2.94 Enhanced capacitance and charge kinetics. Cyclic voltammogram at a scan rate of 5 mV s^{-1} in an acidic aqueous solution of the electrolyte $CuCl_2$. The top inset is the magnified curve of the cyclic voltammogram in a 1 M H_2SO_4 solution, and the bottom inset shows the curve in 1 M HNO_3 and 0.12 M $CuSO_4$. (Reprinted by permission from Macmillan Publishers Ltd. *Nature Communications* Mai, L. Q. et al., Synergistic interaction between redox-active electrolyte and binder-free functionalized carbon for ultrahigh supercapacitor performance. 4:1–7, copyright 2013.)

TABLE 2.12
Some Data from the Selected Literature Concerning the Redox-Active Electrolyte-Based ESs

Redox Mediator	Redox Reactions	Supporting Electrolyte	Capacitance without Redox Mediator ($F\ g^{-1}$)	Capacitance with Redox Mediator ($F\ g^{-1}$)	Cycling Stability	References
HQ	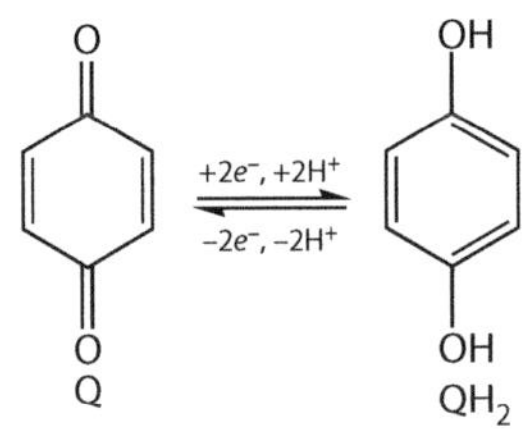	1 M H_2SO_4	72 (at 2.65 mA cm^{-2})	220 (at 2.65 mA cm^{-2})	–	[921]
IC	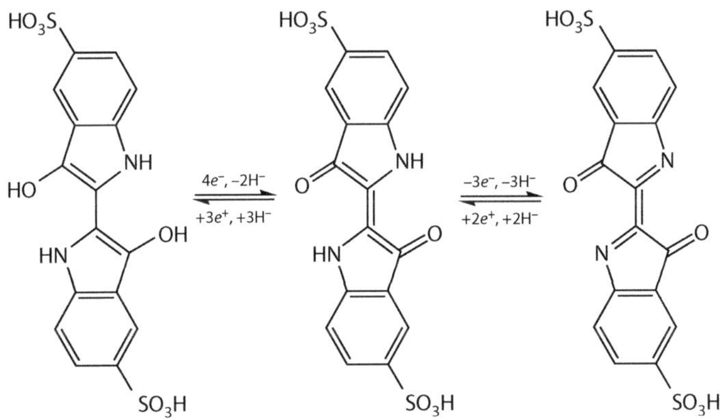	1 M H_2SO_4	17 (at 41 mA g^{-1})	50 (at 41 mA g^{-1})	70% capacitance retention after 10,000 cycles	[925]

(Continued)

TABLE 2.12 (*Continued*)
Some Data from the Selected Literature Concerning the Redox-Active Electrolyte-Based ESs

Redox Mediator	Redox Reactions	Supporting Electrolyte	Capacitance without Redox Mediator (F g^{-1})	Capacitance with Redox Mediator (F g^{-1})	Cycling Stability	References
MB	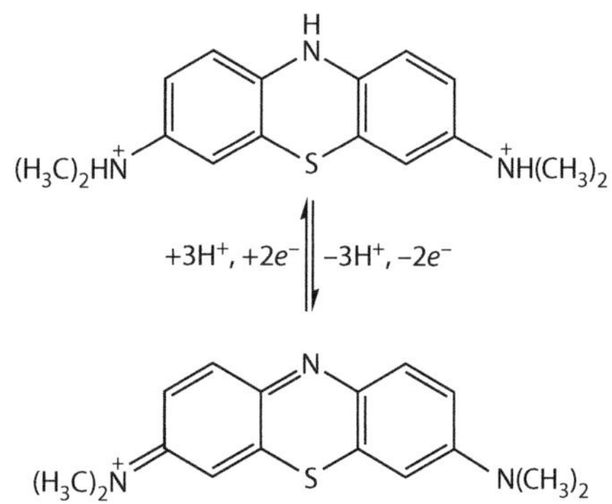	1 M H_2SO_4	5	23	88% capacitance retention after 6000 cycles	[924]

(*Continued*)

TABLE 2.12 (*Continued*)
Some Data from the Selected Literature Concerning the Redox-Active Electrolyte-Based ESs

Redox Mediator	Redox Reactions	Supporting Electrolyte	Capacitance without Redox Mediator (F g^{-1})	Capacitance with Redox Mediator (F g^{-1})	Cycling Stability	References
Sulfonated PANI	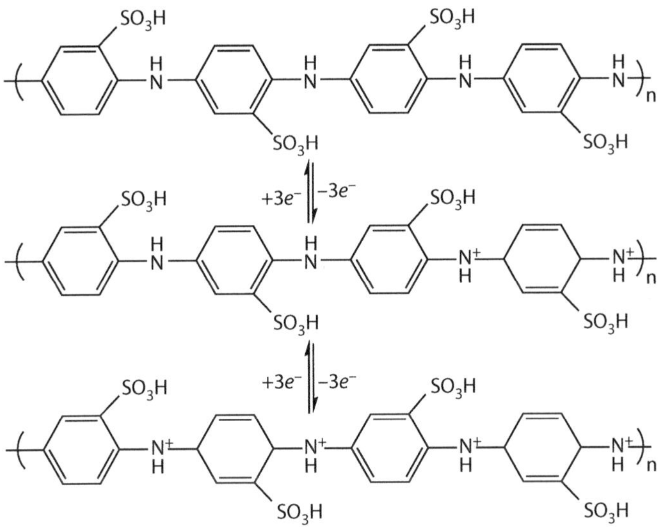	4 M H_2SO_4	47.5 (at 5.22 A g^{-1})	57 (at 5.22 A g^{-1})	99% capacitance retention after 1000 cycles	[930]

(Continued)

TABLE 2.12 (*Continued*)
Some Data from the Selected Literature Concerning the Redox-Active Electrolyte-Based ESs

Redox Mediator	Redox Reactions	Supporting Electrolyte	Capacitance without Redox Mediator ($F\ g^{-1}$)	Capacitance with Redox Mediator ($F\ g^{-1}$)	Cycling Stability	References
PPD	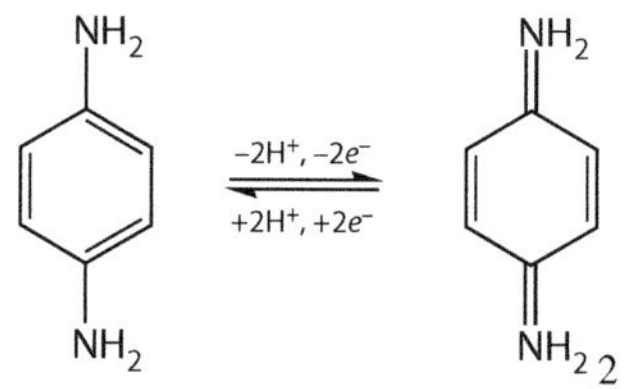	2 M KOH	144.037 (at 1 A g^{-1}) (electrode capacitance)	605.225 (at 1 A g^{-1}) (electrode capacitance)	94.53% capacitance retention after 4000 cycles	[926]
MPD		2 M KOH	36.43 (at 0.5 A g^{-1})	78.01 (at 0.5 A g^{-1})	90.68% capacitance retention after 10,000 cycles	[928]

Source: Adapted from Zhong, C. et al. 2015. *Chemical Society Reviews* 44:7484–7539.
Abbreviations: HQ, hydroquinone; IC, indigo carmine; MB, methylene blue; PANI, polyaniline; PPD, *p*-phenylenediamine; and MPD, *m*-phenylenediamine.

carbon-based ESs [2]. It can be seen that the specific capacitance value of the carbon-based ESs can be generally increased by about two to four times after adding these organic redox additives.

For example, Roldán et al. [920] found that the specific capacitance of the carbon-based ESs was significantly improved at least twice after the addition of hydroquinone (HQ) into the H_2SO_4 electrolyte. This research investigated four different types of carbon-based materials (MWCNTs, carbon aerogel, and AC activated by KOH or NaOH). It was found that the enhanced degree of the specific capacitance also depended on the type and activation process of the carbon-based materials (Figure 2.95). Chemically AC material by KOH exhibited the greatest increase in the specific capacitance value, which trebled to 901 F g^{-1} at 2.65 mA cm^{-2}. A maximum energy density of 31.3 Wh kg^{-1} was achieved for the ES using electrolyte containing HQ with the best AC electrodes. However, further work is necessary to improve the cycling stability of this ES (65% decrease in the specific capacitance value after 4000 cycles). In another study, Roldán et al. [921] found that the energy storage mechanism of both carbon electrodes of the device was also changed upon the addition of HQ into the H_2SO_4 electrolyte. A battery-type behavior was observed for the positive electrode, which was attributed to the development of the quinone/hydroquinone redox reactions on the electrode surface. Besides the EDL charge storage mechanism, a pseudocapacitive hydrogen electrosorption process occurred at the negative electrode due to the asymmetric voltage distribution between the two electrodes after the introduction of HQ. The specific capacitance of the positive electrode drastically increased to 5017 F g^{-1} while that of the negative electrode increased moderately from 290 to 477 F g^{-1}. Wang et al. [922] found that there existed an optimum concentration of HQ in the 1 M H_2SO_4 aqueous electrolyte for ESs. Too high HQ concentration resulted in the decrease of the ES's performance and the optimum concentration of HQ in this work was 0.075 M.

Noticeably, Chen et al. [923] reported that ESs using the HQ-based redox-active electrolyte suffered from a much faster self-discharge process, which was due to the transfer of redox-active species in the electrolytes between positive and negative electrodes. The authors also proposed two strategies to address this issue: (a) the selection of a redox-active electrolyte that could be reversibly converted into insoluble species during the charging/discharging process and (b) the use of a suitable separator membrane such as Nafion.

2.6.1.2 Redox-Active Aqueous Electrolytes for Pseudocapacitive Electrodes

Redox-active electrolytes can also be used for pseudocapacitors and hybrid ESs based on pseudocapacitive electrode materials such as RuO_2 [916,917], MnO_2 [155,932], $Co(OH)_2$ [933], Bi_2O_3 [934], CuS [935], and ECPs [936–939]. Different types of redox-active electrolytes, such as KI [934], heteropoly acids [916,917], HQ [936–940], PPD [155], and $K_3Fe(CN)_6$ [154,932,933,941,942], have been investigated for these pseudocapacitors and hybrid ESs. The specific capacitance can be significantly increased by the introduction of redox mediators into electrolytes, which increases the energy and power densities of the ESs using pseudocapacitive electrode materials. Several studies found that the use of redox-active electrolytes resulted in the lower cycling stability especially during the beginning cycles [936–938] whereas other studies showed the similar cycling stability compared to aqueous

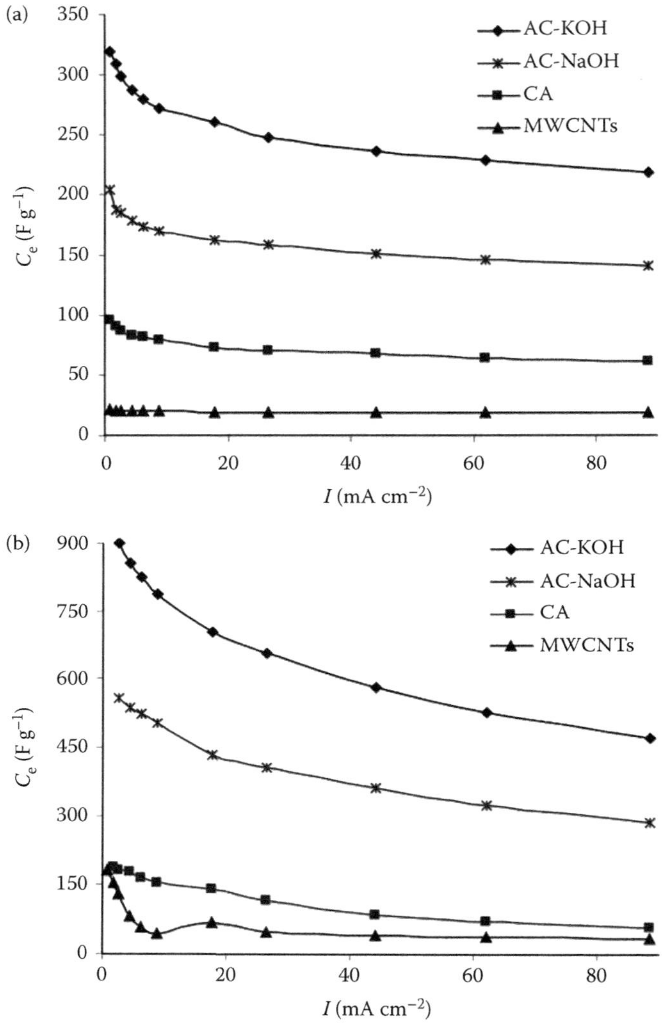

FIGURE 2.95 Variation of specific capacitance with current density in: (a) H_2SO_4 and (b) HQ/H_2SO_4 (1 V). (Roldán, S. et al.: Towards a further generation of high-energy carbon-based capacitors by using redox-active electrolytes. *Angewandte Chemie International Edition*. 2011. 50. 1699–1701. Copyright Wiley-VCH Verlag GmbH & Co. KGaA. Reproduced with permission.)

electrolytes without redox mediators [939,940]. Furthermore, besides increasing the specific capacitance via the pseudocapacitance contribution from the reduction/oxidation of redox-active species in electrolytes, researchers found that the redox-active electrolytes could also compensate (promote) the relatively slow faradaic reactions that occurred on pseudocapacitive electrode materials through supplying the electron buffer source [932,933,942]. For example, Su et al. [942] assumed that the redox reaction of $Fe(CN)_6^{3-}/Fe(CN)_6^{4-}$ served as "electron shuttle" during the

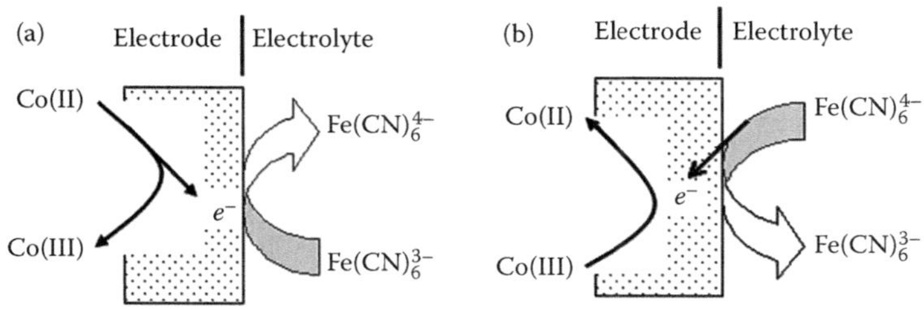

FIGURE 2.96 Illustrations of the roles of hexacyanoferrate (II) and (III) in the courses of charge (a) and discharge (b) for the Co–Al LDH electrode. (Su, L. H. et al. 2009. Improvement of the capacitive performances for Co–Al layered double hydroxide by adding hexacyanoferrate into the electrolyte. *Physical Chemistry Chemical Physics* 11:2195–2202. Reproduced by permission of The Royal Society of Chemistry.)

charging/discharging processes of the Co–Al LDH, accelerating the transition of Co(II)/Co(III) (Figure 2.96 and Equations 2.21 and 2.22). Similar effects have also been observed for the $Co(OH)_2$/graphene [933] and MnO_2 [932] electrodes in the $K_3Fe(CN)_6$ or $K_4Fe(CN)_6$ contained aqueous electrolyte. Zhao et al. [933] found that as the $K_3Fe(CN)_6$ concentration in 1 M KOH increased, both the specific capacitance and coulombic efficiency for the $Co(OH)_2$/graphene electrode increased, however, the rate performance decreased.

$$\mathrm{Co(II)} - e^- \rightarrow \mathrm{Co(III)} \tag{2.21}$$

$$\mathrm{Fe(CN)_6^{3-}} + e^- \rightarrow \mathrm{Fe(CN)_6^{4-}} \tag{2.22}$$

2.6.2 Redox-Active Nonaqueous Electrolytes

To increase the cell voltage and thus to improve the energy density, the application of redox mediators have also been extended to the nonaqueous electrolyte including organic and IL-based electrolytes [943–946]. Ionica-Bousquet et al. [943] found that adding redox-active polyfluorododecaborate cluster ions ($[B_{12}F_xH_{12-x}]^{2-}$) into PC–DMC mixture solvent (2:1 by volume) could provide the additional pseudocapacitance to the total capacitance of carbon-based ESs. It is interesting to note that this electrolyte could also give the protection of ESs from the overcharge. For the IL-based electrolytes, Yamazaki et al. [946] reported that the ES using [EMIM][BF_4] IL containing 1 M [EMIM][Br] as the electrolyte exhibited much higher specific capacitance compared to other ESs without bromide, due to redox reactions of bromide. This redox-active IL-based ES also showed excellent cycling stability with 98% capacitance retention after 10,000 cycles. Similarly, Tooming et al. [945] found that the specific capacitance of the ES increased by nearly 50% after adding 5 wt.% 1-ethyl-3-methylimidazolium iodide ([EMIM][I]) to the [EMIM][BF_4] IL electrolyte. ES based on the electrolyte consisting of 5 wt.% [EMIM][I] and [EMIM][BF_4] could deliver a high specific energy density of 36.7 Wh kg^{-1}, with minor increase in the characteristic relaxation time constant (from 1.45 to 1.48 s). Taniki et al. [944]

found the specific capacitance of ESs could be greatly increased though the use of *N*-ethyl-*N*-methylpyrrolidinium fluorohydrogenate (EMPyr(FH)$_{2.3}$F) IL electrolyte. The positive and negative electrodes had different faradaic reactions. With the cell voltage of 2.5 V at 238 mA g^{-1}, the EDL capacitance and redox capacitance of the positive electrode was 140 and 150 F g^{-1}, respectively; the corresponding capacitances of the negative electrode was 130 and 116 F g^{-1}.

2.6.3 Redox-Active Solid-State Electrolytes

This redox-mediator enhanced strategy has also been proven effective to be used in solid-state electrolytes for ESs. To date, investigated redox-active additives for solid-state electrolytes are almost the same as those used in the liquid electrolytes, such as NaI [947–949], KI [723,950], $K_3Fe(CN)_6$ [908,951], Na_2MoO_4 [952], organic redox mediators such as hydroquinone [953], PPD [954], *p*-benzenediol [955], and MB [956], mixture of redox mediators such as KI–$VOSO_4$ [723]. In addition, most of the redox-active mediators are used in GPEs based on polymer hosts such as PEO [908,949,957], PVA [723,950,952–956], and Nafion [947,948,951]. Similar to the results obtained in the aqueous and organic electrolytes, these redox-active GPEs have been shown to result in a much higher specific capacitance of ESs compared to the GPEs without redox-active species. As shown in Figure 2.97, Fan et al. [723] reported the use of PVA/H_2SO_4 GPE containing KI–$VOSO_4$ could increase the specific capacitance of the AC-based ESs up to 1232.8 F g^{-1}. This ES exhibited an energy density of 25.4 Wh kg^{-1}, which was nearly eight times larger than that of the ES without redox mediators. It can also be seen from Figure 2.97 that the energy density of ESs based on GPE containing mixtures of redox mediators was higher that with one redox additive. Senthilkumar et al. [952] found that the addition of Na_2MoO_4 into the PVA/H_2SO_4 GPE resulted in a 57.2% increase in the

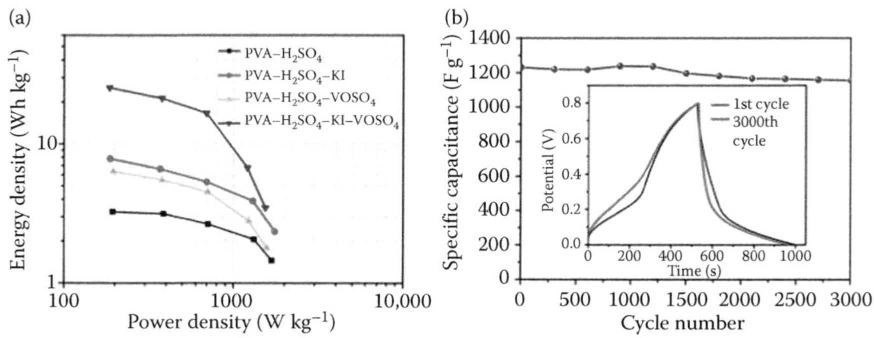

FIGURE 2.97 (a) Ragone plots of EDLCs with PVA–H_2SO_4, PVA–H_2SO_4–KI, PVA–H_2SO_4–$VOSO_4$, and PVA–H_2SO_4–KI–$VOSO_4$ GPEs. (b) Cyclic performance of EDLC with PVA–H_2SO_4–KI–$VOSO_4$ GPE at a current density of 0.5 A g^{-1}; the inset shows the GCD curves of the EDLC at the 1st and 3000th cycles. (Fan, L. Q. et al. 2014. Improving the energy density of quasi-solid-state electric double-layer capacitors by introducing redox additives into gel polymer electrolytes. *Journal of Materials Chemistry A* 2:9011–9014. Reproduced by permission of The Royal Society of Chemistry.)

specific capacitance (648 F g^{-1}) and energy density (14.4 Wh kg^{-1}) of AC-based ESs compared to the PVA/H_2SO_4 GPE without Na_2MoO_4. The redox reactions related with Mo(VI)/Mo(V) contributed to the improved performance. ESs using PVA/H_2SO_4 GPE containing Na_2MoO_4 showed a slightly lower cycling stability (about 93% capacitance retention) compared with the PVA/H_2SO_4 GPE (98% capacitance retention) after 3000 cycles.

In order to improve the ES's working cell voltage, Zhou et al. [951] tested PEO/$LiClO_4$–ACN organogel containing NaI/I_2 or $K_3Fe(CN)_6$/$K_4Fe(CN)_6$ as an electrolyte for ESs. ESs with electrolytes containing NaI/I_2 and $K_3Fe(CN)_6$/$K_4Fe(CN)_6$ showed specific capacitances of 209.0 and 138.8 F g^{-1}, respectively; and they gave specific energy densities of 49.1 Wh kg^{-1} at 1.6 kW kg^{-1} and 33.6 Wh kg^{-1} at 1.3 kW kg^{-1}, respectively. This suggested that the NaI/I_2 mediator resulted in a better improvement effect than the $K_3Fe(CN)_6$/$K_4Fe(CN)_6$ mediator in the authors' work. Using *in situ* XAS combined with CV measurements, Mansour et al. [957] pointed out that the redox couple of I^-/I_2 proceeded via the formation of I_3^- ions as an intermediate step. In addition, it is interesting to note several research groups found that the ionic conductivity of the GPE could be significantly increased by the addition of a redox mediator into the GPE [908,948,949,956]. This improvement effect is more obvious for the GPE using an organic solvent such as ACN to dissolve the conducting salt [908,948,949,956]. For instance, Zhou et al. [948] found that the ionic conductivity of PEO/$LiClO_4$–ACN organogel increased by 32 folds after the addition of NaI–I_2 redox mediators. Ma et al. [908] demonstrated that the addition of $K_3Fe(CN)_6$ into PVA/KOH hydrogel remarkably decreased the ESR of the ES from 19.23 to 4.42 Ω cm^2.

2.7 SUMMARY

In this chapter, several important aspects of electrolytes for ESs were discussed. The main concepts covered include: (1) fundamental role of electrolytes and possible interactions between electrolytes and electrode materials in the charge storage/release processes of ESs; (2) general requirements of electrolytes for ESs including EDLCs, pseudocapacitors, and hybrid ESs; (3) effects of electrolyte properties (e.g., electrochemical stability, ionic conductivity, viscosity, and impurities) on the performances (e.g., capacitance, energy and power densities, and cycling stability) of different kinds of ESs; (4) the importance of making rational design choices that meet the energy, power, safety, and durability needs of ESs; and (5) an overview of choices of electrolyte materials and compositions to improve ES performance.

REFERENCES

1. Burke, A., and M. Miller. 2011. The power capability of ultracapacitors and lithium batteries for electric and hybrid vehicle applications. *Journal of Power Sources* 196:514–522.
2. Zhong, C., Y. D. Deng, W. B. Hu, J. L. Qiao, L. Zhang, and J. J. Zhang. 2015. A review of electrolyte materials and compositions for electrochemical supercapacitors. *Chemical Society Reviews* 44:7484–7539.
3. Lide, D. 2009. *CRC Handbook of Chemistry and Physics*, 89th edn. Boca Raton, FL: CRC Press.

4. Davies, A., and A. Yu. 2011. Material advancements in supercapacitors: From activated carbon to carbon nanotube and graphene. *Canadian Journal of Chemical Engineering* 89:1342–1357.
5. Burke, A. 2007. R&D considerations for the performance and application of electrochemical capacitors. *Electrochimica Acta* 53:1083–1091.
6. Yu, A., V. Chabot, and J. Zhang. 2013. *Electrochemical Supercapacitors for Energy Storage and Delivery: Fundamentals and Applications*. Boca Raton, FL: CRC Press.
7. Conway, B. E., and W. G. Pell. 2003. Double-layer and pseudocapacitance types of electrochemical capacitors and their applications to the development of hybrid devices. *Journal of Solid State Electrochemistry* 7:637–644.
8. Pandolfo, A. G., and A. F. Hollenkamp. 2006. Carbon properties and their role in supercapacitors. *Journal of Power Sources* 157:11–27.
9. Speight, J. G. 2005. *Lange's Handbook of Chemistry*, 16th edn. New York: McGraw-Hill.
10. Volkov, A. G., S. Paula, and D. W. Deamer. 1997. Two mechanisms of permeation of small neutral molecules and hydrated ions across phospholipid bilayers. *Bioelectrochemistry and Bioenergetics* 42:153–160.
11. Nightingale, E. R. 1959. Phenomenological theory of ion solvation. Effective radii of hydrated ions. *The Journal of Physical Chemistry* 63:1381–1387.
12. Kiriukhin, M. Y., and K. D. Collins. 2002. Dynamic hydration numbers for biologically important ions. *Biophysical Chemistry* 99:155–168.
13. Yan, J., Q. Wang, T. Wei, and Z. J. Fan. 2014. Recent advances in design and fabrication of electrochemical supercapacitors with high energy densities. *Advanced Energy Materials* 4:1300816.
14. Béguin, F., V. Presser, A. Balducci, and E. Frackowiak. 2014. Carbons and electrolytes for advanced supercapacitors. *Advanced Materials* 26:2219–2251.
15. Zhai, Y. P., Y. Q. Dou, D. Y. Zhao, P. F. Fulvio, R. T. Mayes, and S. Dai. 2011. Carbon materials for chemical capacitive energy storage. *Advanced Materials* 23:4828–4850.
16. Zhang, L. L., and X. S. Zhao. 2009. Carbon-based materials as supercapacitor electrodes. *Chemical Society Reviews* 38:2520–2531.
17. Lee, S. W., B. M. Gallant, H. R. Byon, P. T. Hammond, and Y. Shao-Horn. 2011. Nanostructured carbon-based electrodes: Bridging the gap between thin-film lithium-ion batteries and electrochemical capacitors. *Energy & Environmental Science* 4:1972–1985.
18. Mun, Y., C. Jo, T. Hyeon et al. 2013. Simple synthesis of hierarchically structured partially graphitized carbon by emulsion/block-copolymer co-template method for high power supercapacitors. *Carbon* 64:391–402.
19. Lee, J. S., S. I. Kim, J. C. Yoon, and J. H. Jang. 2013. Chemical vapor deposition of mesoporous graphene nanoballs for supercapacitor. *ACS Nano* 7:6047–6055.
20. Lauševića, Z., P. Y. Apel, J. B. Krstićc, and I. V. Blonskaya. 2013. Porous carbon thin films for electrochemical capacitors. *Carbon* 64:456–463.
21. Kunowsky, M., A. Garcia-Gomez, V. Barranco et al. 2014. Dense carbon monoliths for supercapacitors with outstanding volumetric capacitances. *Carbon* 68:553–562.
22. Kang, D. Y., and J. H. Moon. 2014. Carbon nanotube balls and their application in supercapacitors. *ACS Applied Materials & Interfaces* 6:706–711.
23. Jin, Z., X. D. Yan, Y. H. Yu, and G. J. Zhao. 2014. Sustainable activated carbon fibers from liquefied wood with controllable porosity for high-performance supercapacitors. *Journal of Materials Chemistry A* 2:11706–11715.
24. Fedorovskaya, E. O., L. G. Bulusheva, A. G. Kurenya et al. 2014. Supercapacitor performance of vertically aligned multiwall carbon nanotubes produced by aerosol-assisted CCVD method. *Electrochimica Acta* 139:165–172.
25. Chen, Q., Y. Hu, C. G. Hu et al. 2014. Graphene quantum dots–three-dimensional graphene composites for high-performance supercapacitors. *Physical Chemistry Chemical Physics* 16:19307–19313.

26. Torad, N. L., R. R. Salunkhe, Y. Q. Li et al. 2014. Electric double-layer capacitors based on highly graphitized nanoporous carbons derived from ZIF-67. *Chemistry – A European Journal* 20:7895–7900.
27. Estevez, L., R. Dua, N. Bhandari, A. Ramanujapuram, P. Wang, and E. P. Giannelis. 2013. A facile approach for the synthesis of monolithic hierarchical porous carbons—High performance materials for amine based CO_2 capture and supercapacitor electrode. *Energy & Environmental Science* 6:1785–1790.
28. Jiménez-Cordero, D., F. Heras, M. A. Gilarranz, and E. Raymundo-Piñero. 2014. Grape seed carbons for studying the influence of texture on supercapacitor behaviour in aqueous electrolytes. *Carbon* 71:127–138.
29. Yang, Z. Y., Y. F. Zhao, Q. Q. Xiao et al. 2014. Controllable growth of CNTs on graphene as high-performance electrode material for supercapacitors. *ACS Applied Materials & Interfaces* 6:8497–8504.
30. Wang, Q., J. Yan, Y. B. Wang et al. 2014. Three-dimensional flower-like and hierarchical porous carbon materials as high-rate performance electrodes for supercapacitors. *Carbon* 67:119–127.
31. Wang, H. J., X. X. Sun, Z. H. Liu, and Z. B. Lei. 2014. Creation of nanopores on graphene planes with MgO template for preparing high-performance supercapacitor electrodes. *Nanoscale* 6:6577–6584.
32. Saha, D., Y. C. Li, Z. H. Bi et al. 2014. Studies on supercapacitor electrode material from activated lignin-derived mesoporous carbon. *Langmuir* 30:900–910.
33. Mao, B. S., Z. H. Wen, Z. Bo, J. B. Chang, X. K. Huang, and J. H. Chen. 2014. Hierarchical nanohybrids with porous CNT-networks decorated crumpled graphene balls for supercapacitors. *ACS Applied Materials & Interfaces* 6:9881–9889.
34. Ma, X. M., L. H. Gan, M. X. Liu et al. 2014. Mesoporous size controllable carbon microspheres and their electrochemical performances for supercapacitor electrodes. *Journal of Materials Chemistry A* 2:8407–8415.
35. Ma, J., T. Xue, and X. Qin. 2014. Sugar-derived carbon/graphene composite materials as electrodes for supercapacitors. *Electrochimica Acta* 115:566–572.
36. Hsu, Y. H., C. C. Lai, C. L. Ho, and C. T. Lo. 2014. Preparation of interconnected carbon nanofibers as electrodes for supercapacitors. *Electrochimica Acta* 127:369–376.
37. Fan, Y., X. Yang, B. Zhu, P. F. Liu, and H. T. Lu. 2014. Micro-mesoporous carbon spheres derived from carrageenan as electrode material for supercapacitors. *Journal of Power Sources* 268:584–590.
38. Wang, Y. F., R. L. Xiong, L. H. Dong, and A. G. Hu. 2014. Synthesis of carbon nanomembranes through cross-linking of phenyl self-assembled monolayers for electrode materials in supercapacitors. *Journal of Materials Chemistry A* 2:5212–5217.
39. Abbas, Q., D. Pajak, E. Frąckowiak, and F. Béguin. 2014. Effect of binder on the performance of carbon/carbon symmetric capacitors in salt aqueous electrolyte. *Electrochimica Acta* 140:132–138.
40. Ban, S., X. Jing, H. J. Zhou, L. Zhang, and J. J. Zhang. 2014. Experimental and modeling study on charge storage/transfer mechanism of graphene-based supercapacitors. *Journal of Power Sources* 268:604–609.
41. Ramasamy, C., J. P. del Val, and M. Anderson. 2014. An analysis of ethylene glycol-aqueous based electrolyte system for supercapacitor applications. *Journal of Power Sources* 248:370–377.
42. Saghafi, M., F. Mahboubi, S. Mohajerzadeh, M. Fathi, and R. Holze. 2014. Longitudinal unzipping of carbon nanotubes and their electrochemical performance in supercapacitors. *Current Applied Physics* 14:1335–1343.
43. Shi, K., M. Ren, and I. Zhitomirsky. 2014. Activated carbon-coated carbon nanotubes for energy storage in supercapacitors and capacitive water purification. *ACS Sustainable Chemistry & Engineering* 2:1289–1298.

44. Wang, S. G., Z. H. Ren, J. P. Li, Y. Q. Ren, L. Zhao, and J. Yu. 2014. Cotton-based hollow carbon fibers with high specific surface area prepared by ammonia etching for supercapacitor application. *RSC Advances* 4:31300–31307.
45. Tsai, W. Y., P. C. Gao, B. Daffos et al. 2013. Ordered mesoporous silicon carbide-derived carbon for high-power supercapacitors. *Electrochemistry Communications* 34:109–112.
46. Thomberg, T., T. Tooming, T. Romann, R. Palm, A. Jänes, and E. Lust. 2013. High power density supercapacitors based on the carbon dioxide activated D-glucose derived carbon electrodes and acetonitrile electrolyte. *Journal of the Electrochemical Society* 160:A1834–A1841.
47. Sevilla, M., and A. B. Fuertes. 2014. Direct synthesis of highly porous interconnected carbon nanosheets and their application as high-performance supercapacitors. *ACS Nano* 8:5069–5078.
48. Qian, W. J., F. X. Sun, Y. H. Xu et al. 2014. Human hair-derived carbon flakes for electrochemical supercapacitors. *Energy & Environmental Science* 7:379–386.
49. Liang, Y., F. X. Liang, H. Zhong, Z. H. Li, R. W. Fu, and D. C. Wu. 2013. An advanced carbonaceous porous network for high-performance organic electrolyte supercapacitors. *Journal of Materials Chemistry A* 1:7000–7005.
50. Karthikeyan, K., S. Amaresh, S. N. Lee et al. 2014. Construction of high-energy-density supercapacitors from pine-cone-derived high-surface-area carbons. *ChemSusChem* 7:1435–1442.
51. Jung, N., S. Kwon, D. Lee et al. 2013. Synthesis of chemically bonded graphene/carbon nanotube composites and their application in large volumetric capacitance supercapacitors. *Advanced Materials* 25:6854–6858.
52. Bhattacharjya, D., and J. S. Yu. 2014. Activated carbon made from cow dung as electrode material for electrochemical double layer capacitor. *Journal of Power Sources* 262:224–231.
53. Bhattacharjya, D., M. S. Kim, T. S. Bae, and J. S. Yu. 2013. High performance supercapacitor prepared from hollow mesoporous carbon capsules with hierarchical nanoarchitecture. *Journal of Power Sources* 244:799–805.
54. Lee, J. H., N. Park, B. G. Kim et al. 2013. Restacking-inhibited 3D reduced graphene oxide for high performance supercapacitor electrodes. *ACS Nano* 7:9366–9374.
55. Han, X. G., M. R. Funk, F. Shen et al. 2014. Scalable holey graphene synthesis and dense electrode fabrication toward high-performance ultracapacitors. *ACS Nano* 8:8255–8265.
56. Kong, C. Y., W. Z. Qian, C. Zheng, Y. T. Yu, C. J. Cui, and F. Wei. 2013. Raising the performance of a 4 V supercapacitor based on an $EMIBF_4$-single walled carbon nanotube nanofluid electrolyte. *Chemical Communications* 49:10727–10729.
57. Lei, Z. B., Z. H. Liu, H. J. Wang, X. X. Sun, L. Lu, and X. S. Zhao. 2013. A high-energy-density supercapacitor with graphene–CMK-5 as the electrode and ionic liquid as the electrolyte. *Journal of Materials Chemistry A* 1:2313–2321.
58. Li, X. J., W. Xing, J. Zhou et al. 2014. Excellent capacitive performance of three-dimensional hierarchical porous graphene/carbon composite with a superhigh surface area. *Chemistry – A European Journal* 20:13314–13320.
59. Peng, C. X., Z. B. Wen, Y. Qin et al. 2014. Three-dimensional graphitized carbon nanovesicles for high-performance supercapacitors based on ionic liquids. *ChemSusChem* 7:777–784.
60. Tsai, W. Y., R. Lin, S. Murali et al. 2013. Outstanding performance of activated graphene based supercapacitors in ionic liquid electrolyte from −50°C to 80°C. *Nano Energy* 2:403–411.
61. Vu, A., X. Y. Li, J. Phillips et al. 2013. Three-dimensionally ordered mesoporous (3DOm) carbon materials as electrodes for electrochemical double-layer capacitors with ionic liquid electrolytes. *Chemistry of Materials* 25:4137–4148.

62. Wang, H., Z. W. Xu, A. Kohandehghan et al. 2013. Interconnected carbon nanosheets derived from hemp for ultrafast supercapacitors with high energy. *ACS Nano* 7:5131–5141.
63. Zheng, C., X. F. Zhou, H. L. Cao, G. H. Wang, and Z. P. Liu. 2014. Edge-enriched porous graphene nanoribbons for high energy density supercapacitors. *Journal of Materials Chemistry A* 2:7484–7490.
64. Zhong, H., F. Xu, Z. Li, R. Fu, and D. Wu. 2013. High-energy supercapacitors based on hierarchical porous carbon with an ultrahigh ion-accessible surface area in ionic liquid electrolytes. *Nanoscale* 5:4678–4682.
65. Zhou, Y., M. Ghaffari, M. Lin et al. 2013. High volumetric electrochemical performance of ultra-high density aligned carbon nanotube supercapacitors with controlled nanomorphology. *Electrochimica Acta* 111:608–613.
66. Kang, J. H., S. H. Jayaram, J. Rawlins, and J. Wen. 2014. Characterization of thermal behaviors of electrochemical double layer capacitors (EDLCs) with aqueous and organic electrolytes. *Electrochimica Acta* 144:1–11.
67. Niu, J., W. G. Pell, and B. E. Conway. 2006. Requirements for performance characterization of C double-layer supercapacitors: Applications to a high specific-area C-cloth material. *Journal of Power Sources* 156:725–740.
68. Soneda, Y., M. Toyoda, Y. Tani et al. 2004. Electrochemical behavior of exfoliated carbon fibers in H_2SO_4 electrolyte with different concentrations. *Journal of Physics and Chemistry of Solids* 65:219–222.
69. Conway, B. E. 1999. *Electrochemical Supercapacitors: Scientific Fundamentals and Technological Applications*. New York: Kluwer Academic/Plenum.
70. Oda, H., A. Yamashita, S. Minoura, M. Okamoto, and T. Morimoto. 2006. Modification of the oxygen-containing functional group on activated carbon fiber in electrodes of an electric double-layer capacitor. *Journal of Power Sources* 158:1510–1516.
71. Hulicova-Jurcakova, D., E. Fiset, G. Q. M. Lu, and T. J. Bandosz. 2012. Changes in surface chemistry of carbon materials upon electrochemical measurements and their effects on capacitance in acidic and neutral electrolytes. *ChemSusChem* 5:2188–2199.
72. Yan, X. D., Y. H. Yu, and X. P. Yang. 2014. Effects of electrolytes on the capacitive behavior of nitrogen/phosphorus co-doped nonporous carbon nanofibers: An insight into the role of phosphorus groups. *RSC Advances* 4:24986–24990.
73. Hulicova-Jurcakova, D., M. Seredych, G. Q. Lu, and T. J. Bandosz. 2009. Combined effect of nitrogen- and oxygen-containing functional groups of microporous activated carbon on its electrochemical performance in supercapacitors. *Advanced Functional Materials* 19:438–447.
74. Hulicova-Jurcakova, D., A. M. Puziy, O. I. Poddubnaya, F. Suarez-Garcia, J. M. D. Tascon, and G. Q. Lu. 2009. Highly stable performance of supercapacitors from phosphorus-enriched carbons. *Journal of the American Chemical Society* 131:5026–5027.
75. Hulicova, D., M. Kodama, and H. Hatori. 2006. Electrochemical performance of nitrogen-enriched carbons in aqueous and non-aqueous supercapacitors. *Chemistry of Materials* 18:2318–2326.
76. Pognon, G., T. Brousse, L. Demarconnay, D. Bélanger. 2011. Performance and stability of electrochemical capacitor based on anthraquinone modified activated carbon. *Journal of Power Sources* 196:4117–4122.
77. Andreas, H. A., and B. E. Conway. 2006. Examination of the double-layer capacitance of an high specific-area C-cloth electrode as titrated from acidic to alkaline pHs. *Electrochimica Acta* 51:6510–6520.
78. Calvo, E. G., N. Rey-Raap, A. Arenillas, and J. A. Menéndez. 2014. The effect of the carbon surface chemistry and electrolyte pH on the energy storage of supercapacitors. *RSC Advances* 4:32398–32404.

79. Ruiz, V., C. Blanco, E. Raymundo-Piñero, V. Khomenko, F. Béguin, and R. Santamaría. 2007. Effects of thermal treatment of activated carbon on the electrochemical behaviour in supercapacitors. *Electrochimica Acta* 52:4969–4973.
80. Singh, K., M. Seredych, E. R. Castellon, and T. J. Bandosz. 2014. Effect of visible-light exposure and electrolyte oxygen content on the capacitance of sulfur-doped carbon. *ChemElectroChem* 1:565–572.
81. Oickle, A. M., and H. A. Andreas. 2011. Examination of water electrolysis and oxygen reduction as self-discharge mechanisms for carbon-based, aqueous electrolyte electrochemical capacitors. *Journal of Physical Chemistry C* 115:4283–4288.
82. Jana, M., P. Khanra, N. C. Murmu, P. Samanta, J. H. Lee, and T. Kuila. 2014. Covalent surface modification of chemically derived graphene and its application as supercapacitor electrode material. *Physical Chemistry Chemical Physics* 16:7618–7626.
83. Ruiz, V., R. Santamaría, M. Granda, and C. Blanco. 2009. Long-term cycling of carbon-based supercapacitors in aqueous media. *Electrochimica Acta* 54:4481–4486.
84. Ruiz, V., C. Blanco, M. Granda, and R. Santamaría. 2008. Enhanced life-cycle supercapacitors by thermal treatment of mesophase-derived activated carbons. *Electrochimica Acta* 54:305–310.
85. Huang, C. C., T. Sun, and D. Hulicova-Jurcakova. 2013. Wide electrochemical window of supercapacitors from coffee bean-derived phosphorus-rich carbons. *ChemSusChem* 6:2330–2339.
86. Toyoda, M., Y. Tani, and Y. Soneda. 2004. Exfoliated carbon fibers as an electrode for electric double layer capacitors in a 1 mol/dm^3 H_2SO_4 electrolyte. *Carbon* 42:2833–2837.
87. Soneda, Y, M. Toyoda, K Hashiya et al. 2003. Huge electrochemical capacitance of exfoliated carbon fibers. *Carbon* 41:2680–2682.
88. Vol'fkovich, Y. M., A. A. Mikhalin, D. A. Bograchev, and V. E. Sosenkin. 2012. Carbon electrodes with high pseudocapacitance for supercapacitors. *Russian Journal of Electrochemistry* 48:424–433.
89. Wang, G., L. Zhang, and J. Zhang. 2012. A review of electrode materials for electrochemical supercapacitors. *Chemical Society Reviews* 41:797–828.
90. Augustyn, V., P. Simon, and B. Dunn. 2014. Pseudocapacitive oxide materials for high-rate electrochemical energy storage. *Energy & Environmental Science* 7:1597–1614.
91. Liu, X. R., and P. G. Pickup. 2010. Ru oxide/carbon fabric composites for supercapacitors. *Journal of Solid State Electrochemistry* 14:231–240.
92. Mendoza-Sánchez, B., T. Brousse, C. Ramirez-Castro, V. Nicolosi, and P. S. Grant. 2013. An investigation of nanostructured thin film α-MoO_3 based supercapacitor electrodes in an aqueous electrolyte. *Electrochimica Acta* 91:253–260.
93. Purushothaman, K. K., M. Cuba, and G. Muralidharan. 2012. Supercapacitor behavior of alpha-$MnMoO_4$ nanorods on different electrolytes. *Materials Research Bulletin* 47:3348–3351.
94. Lu, M., F. Béguin, and E. Frąckowiak. 2013. *Supercapacitors: Materials, Systems and Applications*. Weinheim, Germany: John Wiley & Sons.
95. Long, J. W., D. Bélanger, T. Brousse, W. Sugimoto, M. B. Sassin, and O. Crosnier. 2011. Asymmetric electrochemical capacitors—Stretching the limits of aqueous electrolytes. *MRS Bulletin* 36:513–522.
96. Perret, P., Z. Khani, T. Brousse, D. Bélanger, and D. Guay. 2011. Carbon/PbO_2 asymmetric electrochemical capacitor based on methanesulfonic acid electrolyte. *Electrochimica Acta* 56:8122–8128.
97. Algharaibeh, Z., X. R. Liu, and P. G. Pickup. 2009. An asymmetric anthraquinone-modified carbon/ruthenium oxide supercapacitor. *Journal of Power Sources* 187:640–643.

98. Salinas-Torres, D., J. M. Sieben, D. Lozano-Castelló, D. Cazorla-Amorós, and E. Morallón. 2013. Asymmetric hybrid capacitors based on activated carbon and activated carbon fibre-PANI electrodes. *Electrochimica Acta* 89:326–333.
99. Khomenko, V., E. Raymundo-Piñero, and F. Béguin. 2010. A new type of high energy asymmetric capacitor with nanoporous carbon electrodes in aqueous electrolyte. *Journal of Power Sources* 195:4234–4241.
100. Linden, D., and T. B. Reddy. 2002. *Handbook of Batteries*, 3rd edn. New York: McGraw-Hill.
101. Zheng, J. P. 2003. The limitations of energy density of battery/double-layer capacitor asymmetric cells. *Journal of the Electrochemical Society* 150:A484–A492.
102. Yu, N. F., and L. J. Gao. 2009. Electrodeposited PbO_2 thin film on Ti electrode for application in hybrid supercapacitor. *Electrochemistry Communications* 11:220–222.
103. Perret, P., T. Brousse, D. Bélanger, and D. Guay. 2009. Electrochemical template synthesis of ordered lead dioxide nanowires. *Journal of the Electrochemical Society* 156:A645–A651.
104. Suppes, G. M., C. G. Cameron, and M. S. Freund. 2010. A polypyrrole/phosphomolybdic acid|poly (3,4-ethylenedioxythiophene)/phosphotungstic acid asymmetric supercapacitor. *Journal of the Electrochemical Society* 157:A1030–A1034.
105. Shen, J., C. Yang, X. Li, and G. Wang. 2013. High-performance asymmetric supercapacitor based on nanoarchitectured polyaniline/graphene/carbon nanotube and activated graphene electrodes. *ACS Applied Materials & Interfaces* 5:8467–8476.
106. Hou, Y., L. Y. Chen, L. Zhang et al. 2013. Ultrahigh capacitance of nanoporous metal enhanced conductive polymer pseudocapacitors. *Journal of Power Sources* 225:304–310.
107. Andreas, H. A., K. Lussier, and A. M. Oickle. 2009. Effect of Fe-contamination on rate of self-discharge in carbon-based aqueous electrochemical capacitors. *Journal of Power Sources* 187:275–283.
108. Kazaryan, S. A., S. V. Litvinenko, and G. G. Kharisov. 2008. Self-discharge of heterogeneous electrochemical supercapacitor of $PbO_2|H_2SO_4|C$ related to manganese and titanium ions. *Journal of the Electrochemical Society* 155:A464–A473.
109. He, X. J., Y. J. Geng, J. S. Qiu, M. D. Zheng, S. A. Long, and X. Y. Zhang. 2010. Effect of activation time on the properties of activated carbons prepared by microwave-assisted activation for electric double layer capacitors. *Carbon* 48:1662–1669.
110. He, X. J., Y. J. Geng, J. S. Qiu, M. D. Zheng, X. Y. Zhang, and H. F. Shui. 2010. Influence of KOH/coke mass ratio on properties of activated carbons made by microwave-assisted activation for electric double-layer capacitors. *Energy & Fuels* 24:3603–3609.
111. Stepniak, I., and A. Ciszewski. 2010. New design of electric double layer capacitors with aqueous LiOH electrolyte as alternative to capacitor with KOH solution. *Journal of Power Sources* 195:2564–2569.
112. Xu, J., R. Zhang, P. Chen, and S. Ge. 2014. Effects of adding ethanol to KOH electrolyte on electrochemical performance of titanium carbide-derived carbon. *Journal of Power Sources* 246:132–140.
113. Xu, J., R. Zhang, C. Wu, Y. Zhao, X. Ye, and S. Ge. 2014. Electrochemical performance of graphitized carbide-derived-carbon with hierarchical micro- and meso-pores in alkaline electrolyte. *Carbon* 74:226–236.
114. Xu, B., S. Hou, G. Cao, F. Wu, and Y. Yang. 2012. Sustainable nitrogen-doped porous carbon with high surface areas prepared from gelatin for supercapacitors. *Journal of Materials Chemistry* 22:19088–19093.
115. Si, W., J. Zhou, S. Zhang, S. Li, W. Xing, and S. Zhuo. 2013. Tunable N-doped or dual N, S-doped activated hydrothermal carbons derived from human hair and glucose for supercapacitor applications. *Electrochimica Acta* 107:397–405.

116. Chen, X. Y., C. Chen, Z. J. Zhang, D. H. Xie, X. Deng, and J. W. Liu. 2013. Nitrogen-doped porous carbon for supercapacitor with long-term electrochemical stability. *Journal of Power Sources* 230:50–58.
117. Lang, J. W., X. B. Yan, W. W. Liu, R. T. Wang, and Q. J. Xue. 2012. Influence of nitric acid modification of ordered mesoporous carbon materials on their capacitive performances in different aqueous electrolytes. *Journal of Power Sources* 204:220–229.
118. Wang, D. W., F. Li, L. C. Yin et al. 2012. Nitrogen-doped carbon monolith for alkaline supercapacitors and understanding nitrogen-induced redox transitions. *Chemistry – A European Journal* 18:5345–5351.
119. Rennie, A. J. R., and P. J. Hall. 2013. Nitrogen-enriched carbon electrodes in electrochemical capacitors: Investigating accessible porosity using CM-SANS. *Physical Chemistry Chemical Physics* 15:16774–16778.
120. Wang, C. L., L. Sun, Y. Zhou, P. Wan, X. Zhang, and J. S. Qiu. 2013. P/N co-doped microporous carbons from H_3PO_4-doped polyaniline by *in situ* activation for supercapacitors. *Carbon* 59:537–546.
121. Jurewicz, K., E. Frackowiak, and F. Béguin. 2004. Towards the mechanism of electrochemical hydrogen storage in nanostructured carbon materials. *Applied Physics A – Materials Science & Processing* 78:981–987.
122. Béguin, F., M. Friebe, K. Jurewicz, C. Vix-Guterl, J. Dentzer, and E. Frackowiak. 2006. State of hydrogen electrochemically stored using nanoporous carbons as negative electrode materials in an aqueous medium. *Carbon* 44:2392–2398.
123. Qu, D. Y. 2008. Mechanism for electrochemical hydrogen insertion in carbonaceous materials. *Journal of Power Sources* 179:310–316.
124. Nam, K. W., and K. B. Kim. 2002. A study of the preparation of NiO_x electrode via electrochemical route for supercapacitor applications and their charge storage mechanism. *Journal of the Electrochemical Society* 149:A346–A354.
125. Yu, D. B., J. F. Yao, L. Qiu et al. 2014. *In situ* growth of Co_3O_4 nanoparticles on α-MnO_2 nanotubes: A new hybrid for high-performance supercapacitors. *Journal of Materials Chemistry A* 2:8465–8471.
126. Feng, C., J. F. Zhang, Y. He et al. 2015. Sub-3 nm Co_3O_4 nanofilms with enhanced supercapacitor properties. *ACS Nano* 9:1730–1739.
127. Wang, D., Q. Wang, and T. Wang. 2011. Morphology-controllable synthesis of cobalt oxalates and their conversion to mesoporous Co_3O_4 nanostructures for application in supercapacitors. *Inorganic Chemistry* 50:6482–6492.
128. Palmas, S., F. Ferrara, A. Vacca, M. Mascia, and A. M. Polcaro. 2007. Behavior of cobalt oxide electrodes during oxidative processes in alkaline medium. *Electrochimica Acta* 53:400–406.
129. Yuan, A., and Q. Zhang. 2006. A novel hybrid manganese dioxide/activated carbon supercapacitor using lithium hydroxide electrolyte. *Electrochemistry Communications* 8:1173–1178.
130. Yuan, A. B., M. Zhou, X. L. Wang, Z. H. Sun, and Y. Q. Wang. 2008. Synthesis and characterization of nanostructured manganese dioxide used as positive electrode material for electrochemical capacitor with lithium hydroxide electrolyte. *Chinese Journal of Chemistry* 26:65–69.
131. Naveen, A. N., and S. Selladurai. 2014. Investigation on physiochemical properties of Mn substituted spinel cobalt oxide for supercapacitor applications. *Electrochimica Acta* 125:404–414.
132. Li, X., J. Shao, J. Li, L. Zhang, Q. Qu, and H. Zheng. 2013. Ordered mesoporous MoO_2 as a high-performance anode material for aqueous supercapacitors. *Journal of Power Sources* 237:80–83.
133. Yang, Y., L. Li, G. D. Ruan et al. 2014. Hydrothermally formed three-dimensional nanoporous $Ni(OH)_2$ thin-film supercapacitors. *ACS Nano* 8:9622–9628.

134. Dong, X. L., Z. Y. Guo, Y. F. Song et al. 2014. Flexible and wire-shaped micro-supercapacitor based on $Ni(OH)_2$-nanowire and ordered mesoporous carbon electrodes. *Advanced Functional Materials* 24:3405–3412.
135. Yan, H. J., J. W. Bai, J. Wang et al. 2013. Graphene homogeneously anchored with $Ni(OH)_2$ nanoparticles as advanced supercapacitor electrodes. *CrystEngComm* 15:10007–10015.
136. Wang, L., Z. H. Dong, Z. G. Wang, F. X. Zhang, and J. Jin. 2013. Layered α-$Co(OH)_2$ nanocones as electrode materials for pseudocapacitors: Understanding the effect of interlayer space on electrochemical activity. *Advanced Functional Materials* 23:2758–2764.
137. Yu, J. W., H. Z. Wan, J. J. Jiang et al. 2014. Activation mechanism study of dandelion-like Co_9S_8 nanotubes in supercapacitors. *Journal of the Electrochemical Society* 161: A996–A1000.
138. Rakhi, R. B., N. A. Alhebshi, D. H. Anjum, and H. N. Alshareef. 2014. Nanostructured cobalt sulfide-on-fiber with tunable morphology as electrodes for asymmetric hybrid supercapacitors. *Journal of Materials Chemistry A* 2:16190–16198.
139. Ramachandran, R., M. Saranya, C. Santhosh et al. 2014. Co_9S_8 nanoflakes on graphene (Co_9S_8/G) nanocomposites for high performance supercapacitors. *RSC Advances* 4:21151–21162.
140. Wang, H. L., Y. Y. Liang, T. Mirfakhrai, Z. Chen, H. S. Casalongue, and H. J. Dai. 2011. Advanced asymmetrical supercapacitors based on graphene hybrid materials. *Nano Research* 4:729–736.
141. Apostolova, R. D., Y. A. Tkachenko, O. V. Kolomoyets, and E. M. Shembel. 2012. Thin-layer electrolytic nickel hydroxide $Ni(OH)_2$ in an electrochemical capacitor. *Surface Engineering and Applied Electrochemistry* 48:170–174.
142. Joseph, J., R. Rajagopalan, S. S. Anoop et al. 2014. Shape tailored $Ni_3(NO_3)_2(OH)_4$ nano-flakes simulating 3-D bouquet-like structures for supercapacitors: Exploring the effect of electrolytes on stability and performance. *RSC Advances* 4:39378–39385.
143. Li, L. L., S. J. Peng, Y. L. Cheah et al. 2013. Electrospun porous $NiCo_2O_4$ nanotubes as advanced electrodes for electrochemical capacitors. *Chemistry – A European Journal* 19:5892–5898.
144. Choi, D., G. E. Blomgren, and P. N. Kumta. 2006. Fast and reversible surface redox reaction in nanocrystalline vanadium nitride supercapacitors. *Advanced Materials* 18:1178–1182.
145. Lucio-Porto, R., S. Bouhtiyya, J. F. Pierson et al. 2014. VN thin films as electrode materials for electrochemical capacitors. *Electrochimica Acta* 141:203–211.
146. Apostolova, R. D., O. V. Kolomoyets, Y. A. Tkachenko, and E. M. Shembel. 2012. Electrolytic nickel sulfide in a model electrochemical capacitor. *Russian Journal of Applied Chemistry* 85:612–615.
147. Hou, L. R., C. Z. Yuan, D. K. Li et al. 2011. Electrochemically induced transformation of NiS nanoparticles into $Ni(OH)_2$ in KOH aqueous solution toward electrochemical capacitors. *Electrochimica Acta* 56:7454–7459.
148. Yuan, C. Z., B. Gao, L. H. Su, L. Chen, and X. G. Zhang. 2009. Electrochemically induced phase transformation and charge-storage mechanism of amorphous CoS_x nanoparticles prepared by interface-hydrothermal method. *Journal of the Electrochemical Society* 156:A199–A203.
149. Mefford, J. T., W. G. Hardin, S. Dai, K. P. Johnston, and K. J. Stevenson. 2014. Anion charge storage through oxygen intercalation in $LaMnO_3$ perovskite pseudocapacitor electrodes. *Nature Materials* 13:726–732.
150. Wang, L., X. H. Liu, X. Wang, X. J. Yang, and L. D. Lu. 2011. Electrochemical capacitance study on Co_3O_4 nanowires for super capacitors application. *Journal of Materials Science: Materials in Electronics* 22:601–606.

151. Tian, Y., J. W. Yan, R. Xue, and B. L. Yi. 2011. Influence of electrolyte concentration and temperature on the capacitance of activated carbon. *Acta Physico—Chimica Sinica* 27:479–485.
152. Patil, U. M., R. R. Salunkhe, K. V. Gurav, and C. D. Lokhande. 2008. Chemically deposited nanocrystalline NiO thin films for supercapacitor application. *Applied Surface Science* 255:2603–2607.
153. Zhang, Y. Q., L. Li, S. J. Shi et al. 2014. Synthesis of porous Co_3O_4 nanoflake array and its temperature behavior as pseudo-capacitor electrode. *Journal of Power Sources* 256:200–205.
154. Su, L. H., Z. W. Song, L. H. Lu, and G. L. Pang. 2013. Improving the low-temperature capacitance of CoNiAl three-component layered double hydroxide in a redox electrolyte. *Materials Research Bulletin* 48:3636–3639.
155. Su, L., L. Gong, H. Lü, and Q. Xü. 2014. Enhanced low-temperature capacitance of MnO_2 nanorods in a redox-active electrolyte. *Journal of Power Sources* 248:212–217.
156. Wang, X. L., A. B. Yuan, and Y. Q. Wang. 2007. Supercapacitive behaviors and their temperature dependence of sol–gel synthesized nanostructured manganese dioxide in lithium hydroxide electrolyte. *Journal of Power Sources* 172:1007–1011.
157. Misnon, I. I., R. A. Aziz, N. K. M. Zain, B. Vidhyadharan, S. G. Krishnan, and R. Jose. 2014. High performance MnO_2 nanoflower electrode and the relationship between solvated ion size and specific capacitance in highly conductive electrolytes. *Materials Research Bulletin* 57:221–230.
158. Reddy, R. N., and R. G. Reddy. 2003. Sol–gel MnO_2 as an electrode material for electrochemical capacitors. *Journal of Power Sources* 124:330–337.
159. Fic, K., G. Lota, M. Meller, and E. Frackowiak. 2012. Novel insight into neutral medium as electrolyte for high-voltage supercapacitors. *Energy & Environmental Science* 5:5842–5850.
160. Wang, R. Y., Q. Li, L. L. Cheng et al. 2014. Electrochemical properties of manganese ferrite-based supercapacitors in aqueous electrolyte: The effect of ionic radius. *Colloids and Surfaces A: Physicochemical and Engineering Aspects* 457:94–99.
161. Su, L., L. Gong, and J. Gao. 2012. The supercapacitive performances of $Co(OH)_2$/$Ni(OH)_2$ composites in lithium hydroxide solution: Selection of electrolyte and effect of weight ratio. *Journal of Power Sources* 209:141–146.
162. Inamdar, A. I., Y. Kim, S. M. Pawar, J. H. Kim, H. Im, and H. Kim. 2011. Chemically grown, porous, nickel oxide thin-film for electrochemical supercapacitors. *Journal of Power Sources* 196:2393–2397.
163. Wang, X. J., Z. Z. Yan, H. Pang et al. 2013. $NH_4CoPO_4 \cdot H_2O$ microflowers and porous $Co_2P_2O_7$ microflowers: Effective electrochemical supercapacitor behavior in different alkaline electrolytes. *International Journal of Electrochemical Science* 8:3768–3785.
164. Nithya, V. D., R. K. Selvan, D. Kalpana, L. Vasylechko, and C. Sanjeeviraja. 2013. Synthesis of Bi_2WO_6 nanoparticles and its electrochemical properties in different electrolytes for pseudocapacitor electrodes. *Electrochimica Acta* 109:720–731.
165. Wang, C., Y. Zhou, L. Sun, P. Wan, X. Zhang, and J. Qiu. 2013. Sustainable synthesis of phosphorus- and nitrogen-co-doped porous carbons with tunable surface properties for supercapacitors. *Journal of Power Sources* 239:81–88.
166. Bichat, M. P., E. Raymundo-Piñero, and F. Béguin. 2010. High voltage supercapacitor built with seaweed carbons in neutral aqueous electrolyte. *Carbon* 48:4351–4361.
167. Yuan, C., L. Hou, D. Li, Y. Zhang, S. Xiong, and X. Zhang. 2013. Unusual electrochemical behavior of Ru–Cr binary oxide-based aqueous symmetric supercapacitors in KOH solution. *Electrochimica Acta* 88:654–658.
168. Park, J. H., O. O. Park, K. H. Shin, C. S. Jin, and J. H. Kim. 2002. An electrochemical capacitor based on a $Ni(OH)_2$/activated carbon composite electrode. *Electrochemical and Solid-State Letters* 5:H7–H10.

169. Wang, Y. G., Z. D. Wang, and Y. Y. Xia. 2005. An asymmetric supercapacitor using RuO_2/TiO_2 nanotube composite and activated carbon electrodes. *Electrochimica Acta* 50:5641–5646.
170. Aida, T., I. Murayama, K. Yamada, and M. Morita. 2007. High-energy-density hybrid electrochemical capacitor using graphitizable carbon activated with KOH for positive electrode. *Journal of Power Sources* 166:462–470.
171. Huang, J. C., P. P. Xu, D. X. Cao et al. 2014. Asymmetric supercapacitors based on β-$Ni(OH)_2$ nanosheets and activated carbon with high energy density. *Journal of Power Sources* 246:371–376.
172. Long, C. L., T. Wei, J. Yan, L. L. Jiang, and Z. J. Fan. 2013. Supercapacitors based on graphene-supported iron nanosheets as negative electrode materials. *ACS Nano* 7:11325–11332.
173. Kong, L. B., M. Liu, J. W. Lang, Y. C. Luo, and L. Kang. 2009. Asymmetric supercapacitor based on loose-packed cobalt hydroxide nanoflake materials and activated carbon. *Journal of the Electrochemical Society* 156:A1000–A1004.
174. Xie, L. J., J. F. Wu, C. M. Chen et al. 2013. A novel asymmetric supercapacitor with an activated carbon cathode and a reduced graphene oxide–cobalt oxide nanocomposite anode. *Journal of Power Sources* 242:148–156.
175. Zhang, C., L. Xie, W. Song, J. Wang, G. Sun, and K. Li. 2013. Electrochemical performance of asymmetric supercapacitor based on Co_3O_4/AC materials. *Journal of Electroanalytical Chemistry* 706:1–6.
176. Vidyadharan, B., R. Abd Aziz, I. I. Misnon, II et al. 2014. High energy and power density asymmetric supercapacitors using electrospun cobalt oxide nanowire anode. *Journal of Power Sources* 270:526–535.
177. Dai, C. S., P. Y. Chien, J. Y. Lin et al. 2013. Hierarchically structured Ni_3S_2/carbon nanotube composites as high performance cathode materials for asymmetric supercapacitors. *ACS Applied Materials & Interfaces* 5:12168–12174.
178. Zhang, W. F., C. Ma, J. H. Fang et al. 2013. Asymmetric electrochemical capacitors with high energy and power density based on graphene/CoAl-LDH and activated carbon electrodes. *RSC Advances* 3:2483–2490.
179. Yuan, C. Z., B. Gao, and X. G. Zhang. 2007. Electrochemical capacitance of NiO/$Ru_{0.35}V_{0.65}O_2$ asymmetric electrochemical capacitor. *Journal of Power Sources* 173:606–612.
180. Tang, C. H., X. Yin, and H. Gong. 2013. Superior performance asymmetric supercapacitors based on a directly grown commercial mass 3D Co_3O_4@$Ni(OH)_2$ core–shell electrode. *ACS Applied Materials & Interfaces* 5:10574–10582.
181. Vaquero, S., J. Palma, M. Anderson, and R. Marcilla. 2013. Mass-balancing of electrodes as a strategy to widen the operating voltage window of carbon/carbon supercapacitors in neutral aqueous electrolytes. *International Journal of Electrochemical Science* 8:10293–10307.
182. Demarconnay, L., E. Raymundo-Piñero, and F. Béguin. 2010. A symmetric carbon/carbon supercapacitor operating at 1.6 V by using a neutral aqueous solution. *Electrochemistry Communications* 12:1275–1278.
183. Wang, Y., J. Cao, Y. Zhou, J. H. Ouyang, D. Jia, and L. Guo. 2012. Ball-milled graphite as an electrode material for high voltage supercapacitor in neutral aqueous electrolyte. *Journal of the Electrochemical Society* 159:A579–A583.
184. Zhao, L., Y. Qiu, J. Yu, X. Deng, C. Dai, and X. Bai. 2013. Carbon nanofibers with radially grown graphene sheets derived from electrospinning for aqueous supercapacitors with high working voltage and energy density. *Nanoscale* 5:4902–4909.
185. Gao, Q., L. Demarconnay, E. Raymundo-Piñero, and F. Béguin. 2012. Exploring the large voltage range of carbon/carbon supercapacitors in aqueous lithium sulfate electrolyte. *Energy & Environmental Science* 5:9611–9617.

186. Ratajczak, P., K. Jurewicz, and F. Béguin. 2014. Factors contributing to ageing of high voltage carbon/carbon supercapacitors in salt aqueous electrolyte. *Journal of Applied Electrochemistry* 44:475–480.
187. Ratajczak, P., K. Jurewicz, P. Skowron, Q. Abbas, and F. Béguin. 2014. Effect of accelerated ageing on the performance of high voltage carbon/carbon electrochemical capacitors in salt aqueous electrolyte. *Electrochimica Acta* 130:344–350.
188. Tsay, K. C., L. Zhang, and J. J. Zhang. 2012. Effects of electrode layer composition/thickness and electrolyte concentration on both specific capacitance and energy density of supercapacitor. *Electrochimica Acta* 60:428–436.
189. Menzel, J., K. Fic, M. Meller, and E. Frackowiak. 2014. The effect of halide ion concentration on capacitor performance. *Journal of Applied Electrochemistry* 44:439–445.
190. Yin, J., C. Zheng, L. Qi, and H. Y. Wang. 2011. Concentrated $NaClO_4$ aqueous solutions as promising electrolytes for electric double-layer capacitors. *Journal of Power Sources* 196:4080–4087.
191. Chae, J. H., and G. Z. Chen. 2014. Influences of ions and temperature on performance of carbon nano-particulates in supercapacitors with neutral aqueous electrolytes. *Particuology* 15:9–17.
192. Hsieh, C. T., S. M. Hsu, J. Y. Lin, and H. Teng. 2011. Electrochemical capacitors based on graphene oxide sheets using different aqueous electrolytes. *Journal of Physical Chemistry C* 115:12367–12374.
193. Qu, Q. T., B. Wang, L. C. Yang, Y. Shi, S. Tian, and Y. P. Wu. 2008. Study on electrochemical performance of activated carbon in aqueous Li_2SO_4, Na_2SO_4 and K_2SO_4 electrolytes. *Electrochemistry Communications* 10:1652–1655.
194. Chandrasekaran, R., M. Koh, A. Yamauchi, and M. Ishikawa. 2009. Electrochemical study on aqueous magnesium nitrate electrolyte system for EDLC applications. *Electrochemistry* 77:51–55.
195. Gao, H., A. Virya, and K. Lian. 2014. Monovalent silicotungstate salts as electrolytes for electrochemical supercapacitors. *Electrochimica Acta* 138:240–246.
196. Yamazaki, S., K. Obata, Y. Okuhama, Y. Matsuda, and M. Ishikawa. 2007. Activated carbon/DNA composite electrodes for electric double layer capacitors with neutral aqueous electrolytes. *Electrochemistry* 75:592–594.
197. Yin, J., L. Qi, and H. Y. Wang. 2013. Anti-freezing aqueous electrolytes for electric double-layer capacitors. *Electrochimica Acta* 88:208–216.
198. Lee, H. Y., and J. B. Goodenough. 1999. Supercapacitor behavior with KCl electrolyte. *Journal of Solid State Chemistry* 144:220–223.
199. Simon, P., and Y. Gogotsi. 2008. Materials for electrochemical capacitors. *Nature Materials* 7:845–854.
200. Toupin, M., T. Brousse, and D. Bélanger. 2004. Charge storage mechanism of MnO_2 electrode used in aqueous electrochemical capacitor. *Chemistry of Materials* 16:3184–3190.
201. Kuo, S. L., and N. L. Wu. 2006. Investigation of pseudocapacitive charge-storage reaction of $MnO_2 \cdot nH_2O$ supercapacitors in aqueous electrolytes. *Journal of the Electrochemical Society* 153:A1317–A1324.
202. Zheng, L., Y. Xu, D. Jin, and Y. Xie. 2010. Well-aligned molybdenum oxide nanorods on metal substrates: Solution-based synthesis and their electrochemical capacitor application. *Journal of Materials Chemistry* 20:7135–7143.
203. Gao, F. L., L. J. Zhang, and S. M. Huang. 2010. Fabrication horizontal aligned MoO_2/single-walled carbon nanotube nanowires for electrochemical supercapacitor. *Materials Letters* 64:537–540.
204. Xu, C. J., B. H. Li, H. D. Du, F. Y. Kang, and Y. Q. Zeng. 2008. Supercapacitive studies on amorphous MnO_2 in mild solutions. *Journal of Power Sources* 184:691–694.

205. Xu, C. J., C. G. Wei, B. H. Li, F. Y. Kang, and Z. C. Guan. 2011. Charge storage mechanism of manganese dioxide for capacitor application: Effect of the mild electrolytes containing alkaline and alkaline-earth metal cations. *Journal of Power Sources* 196:7854–7859.
206. Wen, S., J. W. Lee, I. H. Yeo, J. Park, and S. Mho. 2004. The role of cations of the electrolyte for the pseudocapacitive behavior of metal oxide electrodes, MnO_2 and RuO_2. *Electrochimica Acta* 50:849–855.
207. Li, J., Q. M. Yang, and I. Zhitornirsky. 2008. Nickel foam-based manganese dioxide–carbon nanotube composite electrodes for electrochemical supercapacitors. *Journal of Power Sources* 185:1569–1574.
208. Li, S. H., L. Qi, L. H. Lu, and H. Y. Wang. 2012. Facile preparation and performance of mesoporous manganese oxide for supercapacitors utilizing neutral aqueous electrolytes. *RSC Advances* 2:3298–3308.
209. Liang, C. H., and C. S. Hwang. 2010. Effects of temperatures and cations of electrolyte on the capacitive characteristics of the manganese oxide deposited by hydrothermal electrochemical method. *Journal of Alloys and Compounds* 500:102–107.
210. Wang, Y. Q., A. B. Yuan, and X. L. Wang. 2008. Pseudocapacitive behaviors of nanostructured manganese dioxide/carbon nanotubes composite electrodes in mild aqueous electrolytes: Effects of electrolytes and current collectors. *Journal of Solid State Electrochemistry* 12:1101–1107.
211. Jeong, Y. U., and A. Manthiram. 2002. Nanocrystalline manganese oxides for electrochemical capacitors with neutral electrolytes. *Journal of the Electrochemical Society* 149:A1419–A1422.
212. Shao, J., X. Li, Q. Qu, and Y. Wu. 2013. Study on different power and cycling performance of crystalline $K_xMnO_2 \cdot nH_2O$ as cathode material for supercapacitors in Li_2SO_4, Na_2SO_4, and K_2SO_4 aqueous electrolytes. *Journal of Power Sources* 223:56–61.
213. Wu, M. Q., L. P. Zhang, J. H. Gao et al. 2008. Effects of thickness and electrolytes on the capacitive characteristics of anodically deposited hydrous manganese oxide coatings. *Journal of Electroanalytical Chemistry* 613:125–130.
214. Qu, Q., P. Zhang, B. Wang et al. 2009. Electrochemical performance of MnO_2 nanorods in neutral aqueous electrolytes as a cathode for asymmetric supercapacitors. *Journal of Physical Chemistry C* 113:14020–14027.
215. Ji, C. C., M. W. Xu, S. J. Bao, C. J. Cai, R. Y. Wang, and D. Z. Jia. 2013. Effect of alkaline and alkaline-earth cations on the supercapacitor performance of MnO_2 with various crystallographic structures. *Journal of Solid State Electrochemistry* 17:1357–1368.
216. Xu, C. J., H. D. Du, B. H. Li, F. Y. Kang, and Y. Q. Zeng. 2009. Capacitive behavior and charge storage mechanism of manganese dioxide in aqueous solution containing bivalent cations. *Journal of the Electrochemical Society* 156:A73–A78.
217. Munaiah, Y., B. G. Sundara Raj, T. Prem Kumar, and P. Ragupathy. 2013. Facile synthesis of hollow sphere amorphous MnO_2: The formation mechanism, morphology and effect of a bivalent cation-containing electrolyte on its supercapacitive behavior. *Journal of Materials Chemistry A* 1:4300–4306.
218. Boisset, A., L. Athouël, J. Jacquemin, P. Porion, T. Brousse, and M. Anouti. 2013. Comparative performances of birnessite and cryptomelane MnO_2 as electrode material in neutral aqueous lithium salt for supercapacitor application. *Journal of Physical Chemistry C* 117:7408–7422.
219. Jiang, Y. M., Z. Li, J. L. Xu, and J. Li. 2012. Fast morphological evolution and enhanced capacitive behavior of MnO_2 at high operating temperature. *ECS Electrochemistry Letters* 1:A47–A49.
220. Wang, J. G., Y. Yang, Z. H. Huang, and F. Kang. 2013. Effect of temperature on the pseudo-capacitive behavior of freestanding MnO_2@carbon nanofibers composites electrodes in mild electrolyte. *Journal of Power Sources* 224:86–92.

221. Yan, J., E. Khoo, A. Sumboja, and P. S. Lee. 2010. Facile coating of manganese oxide on tin oxide nanowires with high-performance capacitive behavior. *ACS Nano* 4:4247–4255.
222. Wang, J. G., Y. Yang, Z. H. Huang, and F. Kang. 2012. Synthesis and electrochemical performance of MnO_2/CNTs-embedded carbon nanofibers nanocomposites for supercapacitors. *Electrochimica Acta* 75:213–219.
223. Komaba, S., T. Tsuchikawa, M. Tomita, N. Yabuuchi, and A. Ogata. 2013. Efficient electrolyte additives of phosphate, carbonate, and borate to improve redox capacitor performance of manganese oxide electrodes. *Journal of the Electrochemical Society* 160:A1952–A1961.
224. Wang, B., P. Guo, H. Bi et al. 2013. Electrocapacitive properties of $MnFe_2O_4$ electrodes in aqueous $LiNO_3$ electrolyte with surfactants. *International Journal of Electrochemical Science* 8:8966–8977.
225. Ataherian, F., and N. L. Wu. 2011. 1.2 Volt manganese oxide symmetric supercapacitor. *Electrochemistry Communications* 13:1264–1267.
226. Athouël, L., F. Moser, R. Dugas, O. Crosnier, D. Bélanger, and T. Brousse. 2008. Variation of the MnO_2 birnessite structure upon charge/discharge in an electrochemical supercapacitor electrode in aqueous Na_2SO_4 electrolyte. *Journal of Physical Chemistry C* 112:7270–7277.
227. Tao, X. Y., J. Du, Y. Sun et al. 2013. Exploring the energy storage mechanism of high performance MnO_2 electrochemical capacitor electrodes: An *in situ* atomic force microscopy study in aqueous electrolyte. *Advanced Functional Materials* 23:4745–4751.
228. Wang, S. Y., and N. L. Wu. 2003. Operating characteristics of aqueous magnetite electrochemical capacitors. *Journal of Applied Electrochemistry* 33:345–348.
229. Shivakumara, S., T. R. Penki, and N. Munichandraiah. 2014. Preparation and electrochemical performance of porous hematite (α-Fe_2O_3) nanostructures as supercapacitor electrode material. *Journal of Solid State Electrochemistry* 18:1057–1066.
230. Lin, Y., X. Y. Wang, G. Qian, and J. J. Watkins. 2014. Additive-driven self-assembly of well-ordered mesoporous carbon/iron oxide nanoparticle composites for supercapacitors. *Chemistry of Materials* 26:2128–2137.
231. Liu, Y., Y. Jiao, Z. L. Zhang, F. Y. Qu, A. Umar, and X. Wu. 2014. Hierarchical SnO_2 nanostructures made of intermingled ultrathin nanosheets for environmental remediation, smart gas sensor, and supercapacitor applications. *ACS Applied Materials & Interfaces* 6:2174–2184.
232. Qu, Q. T., L. L. Liu, Y. P. Wu, and R. Holze. 2013. Electrochemical behavior of $V_2O_5 \cdot 0.6H_2O$ nanoribbons in neutral aqueous electrolyte solution. *Electrochimica Acta* 96:8–12.
233. Prakash, A., and D. Bahadur. 2014. The role of ionic electrolytes on capacitive performance of ZnO-reduced graphene oxide nanohybrids with thermally tunable morphologies. *ACS Applied Materials & Interfaces* 6:1394–1405.
234. Yang, Q., S. K. Pang, and K. C. Yung. 2014. Study of PEDOT–PSS in carbon nanotube/conducting polymer composites as supercapacitor electrodes in aqueous solution. *Journal of Electroanalytical Chemistry* 728:140–147.
235. Ammam, M., and J. Fransaer. 2012. Performance of SLS/MWCNTs/PANI capacitor electrodes in a physiological electrolyte and in serum. *Chemical Communications* 48:2036–2038.
236. Reddy, R. N., and R. G. Reddy. 2006. Porous structured vanadium oxide electrode material for electrochemical capacitors. *Journal of Power Sources* 156:700–704.
237. Wu, N. L., S. Y. Wang, C. Y. Han, D. S. Wu, and L. R. Shiue. 2003. Electrochemical capacitor of magnetite in aqueous electrolytes. *Journal of Power Sources* 113:173–178.
238. Xia, H., Y. Shirley Meng, G. Yuan, C. Cui, and L. Lu. 2012. A symmetric RuO_2/RuO_2 supercapacitor operating at 1.6 V by using a neutral aqueous electrolyte. *Electrochemical and Solid-State Letters* 15:A60–A63.

239. Sun, W., R. Zheng, and X. Chen. 2010. Symmetric redox supercapacitor based on micro-fabrication with three-dimensional polypyrrole electrodes. *Journal of Power Sources* 195:7120–7125.
240. Hong, M. S., S. H. Lee, and S. W. Kim. 2002. Use of KCl aqueous electrolyte for 2 V manganese oxide/activated carbon hybrid capacitor. *Electrochemical and Solid-State Letters* 5:A227–A230.
241. Brousse, T., M. Toupin, and D. Bélanger. 2004. A hybrid activated carbon–manganese dioxide capacitor using a mild aqueous electrolyte. *Journal of the Electrochemical Society* 151:A614–A622.
242. Deng, L. J., Z. P. Hao, J. F. Wang et al. 2013. Preparation and capacitance of graphene/multiwall carbon nanotubes/MnO_2 hybrid material for high-performance asymmetrical electrochemical capacitor. *Electrochimica Acta* 89:191–198.
243. Zhang, G. N., L. J. Ren, L. J. Deng, J. F. Wang, L. P. Kang, and Z. H. H. Liu. 2014. Graphene–MnO_2 nanocomposite for high-performance asymmetrical electrochemical capacitor. *Materials Research Bulletin* 49:577–583.
244. Santos-Peña, J., O. Crosnier, and T. Brousse. 2010. Nanosized α-$LiFeO_2$ as electrochemical supercapacitor electrode in neutral sulfate electrolytes. *Electrochimica Acta* 55:7511–7515.
245. Ng, K. C., S. Zhang, and G. Z. Chen. 2008. An asymmetrical supercapacitor based on CNTs/SnO_2 and CNTs/MnO_2 nanocomposites working at 1.7 V in aqueous electrolyte. *ECS Transactions* 16:153–162.
246. Lu, L. H., Y. D. Zhu, F. J. Li, W. Zhuang, K. Y. Chan, and X. H. Lu. 2010. Carbon titania mesoporous composite whisker as stable supercapacitor electrode material. *Journal of Materials Chemistry* 20:7645–7651.
247. Nagaraju, D. H., Q. Wang, P. Beaujuge, and H. N. Alshareef. 2014. Two-dimensional heterostructures of V_2O_5 and reduced graphene oxide as electrodes for high energy density asymmetric supercapacitors. *Journal of Materials Chemistry A* 2:17146–17152.
248. Mak, W. F., G. Wee, V. Aravindan, N. Gupta, S. G. Mhaisalkar, and S. Madhavi. 2012. High-energy density asymmetric supercapacitor based on electrospun vanadium pentoxide and polyaniline nanofibers in aqueous electrolyte. *Journal of the Electrochemical Society* 159:A1481–A1488.
249. Xia, H., and C. Huo. 2011. Electrochemical properties of MnO_2/CNT nanocomposite in neutral aqueous electrolyte as cathode material for asymmetric supercapacitors. *International Journal of Smart and Nano Materials* 2:283–291.
250. Deng, L. J., G. Zhu, J. F. Wang et al. 2011. Graphene–MnO_2 and graphene asymmetrical electrochemical capacitor with a high energy density in aqueous electrolyte. *Journal of Power Sources* 196:10782–10787.
251. Lekitima, J. N., K. I. Ozoemena, C. J. Jafta et al. 2013. High-performance aqueous asymmetric electrochemical capacitors based on graphene oxide/cobalt((II))-tetrapyrazinoporphyrazine hybrids. *Journal of Materials Chemistry A* 1:2821–2826.
252. Bélanger, D., T. Brousse, and J. W. Long. 2008. Manganese oxides: Battery materials make the leap to electrochemical capacitors. *The Electrochemical Society Interface* Spring 49–52.
253. Calvo, E. G., F. Lufrano, A. Arenillas, A. Brigandi, J. A. Menéndez, and P. Staiti. 2014. Effect of unequal load of carbon xerogel in electrodes on the electrochemical performance of asymmetric supercapacitors. *Journal of Applied Electrochemistry* 44:481–489.
254. Chen, Y. C., Y. G. Lin, Y. K. Hsu, S. C. Yen, K. H. Chen, and L. C. Chen. 2014. Novel iron oxyhydroxide lepidocrocite nanosheet as ultrahigh power density anode material for asymmetric supercapacitors. *Small* 10:3803–3810.
255. Chou, T. C., R. A. Doong, C. C. Hu, B. S. Zhang, and D. S. Su. 2014. Hierarchically porous carbon with manganese oxides as highly efficient electrode for asymmetric supercapacitors. *ChemSusChem* 7:841–847.

256. Fan, L. Q., G. J. Liu, J. H. Wu, L. Liu, J. M. Lin, and Y. L. Wei. 2014. Asymmetric supercapacitor based on graphene oxide/polypyrrole composite and activated carbon electrodes. *Electrochimica Acta* 137:26–33.
257. Huang, M., Y. X. Zhang, F. Li et al. 2014. Merging of Kirkendall growth and Ostwald ripening: CuO@MnO_2 core–shell architectures for asymmetric supercapacitors. *Scientific Reports* 4:4518.
258. Kim, M., and J. Kim. 2014. Development of high power and energy density microsphere silicon carbide–MnO_2 nanoneedles and thermally oxidized activated carbon asymmetric electrochemical supercapacitors. *Physical Chemistry Chemical Physics* 16:11323–11336.
259. Wang, C. H., H. C. Hsu, and J. H. Hu. 2014. High-energy asymmetric supercapacitor based on petal-shaped MnO_2 nanosheet and carbon nanotube-embedded polyacrylonitrile-based carbon nanofiber working at 2 V in aqueous neutral electrolyte. *Journal of Power Sources* 249:1–8.
260. Wang, F. X., S. Y. Xiao, Y. S. Zhu et al. 2014. Spinel $LiMn_2O_4$ nanohybrid as high capacitance positive electrode material for supercapacitors. *Journal of Power Sources* 246:19–23.
261. Wu, S. S., W. F. Chen, and L. F. Yan. 2014. Fabrication of a 3D MnO_2/graphene hydrogel for high-performance asymmetric supercapacitors. *Journal of Materials Chemistry A* 2:2765–2772.
262. Yang, J., L. F. Lian, H. C. Ruan, F. Y. Xie, and M. D. Wei. 2014. Nanostructured porous MnO_2 on Ni foam substrate with a high mass loading via a CV electrodeposition route for supercapacitor application. *Electrochimica Acta* 136:189–194.
263. Zhang, B. H., Y. Liu, Z. Chang et al. 2014. Nanowire $Na_{0.35}MnO_2$ from a hydrothermal method as a cathode material for aqueous asymmetric supercapacitors. *Journal of Power Sources* 253:98–103.
264. Zhang, Y. X., M. Kuang, X. D. Hao et al. 2014. Rational design of hierarchically porous birnessite-type manganese dioxides nanosheets on different one-dimensional titania-based nanowires for high performance supercapacitors. *Journal of Power Sources* 270:675–683.
265. Zhao, F. P., Y. Y. Wang, X. N. Xu et al. 2014. Cobalt hexacyanoferrate nanoparticles as a high-rate and ultra-stable supercapacitor electrode material. *ACS Applied Materials & Interfaces* 6:11007–11012.
266. Mosqueda, H. A., O. Crosnier, L. Athouël et al. 2010. Electrolytes for hybrid carbon–MnO_2 electrochemical capacitors. *Electrochimica Acta* 55:7479–7483.
267. Chen, L. M., Q. Y. Lai, Y. J. Hao, J. H. Huang, and X. Y. Ji. 2008. Pseudo-capacitive properties of $LiCoO_2$/AC electrochemical capacitor in various aqueous electrolytes. *Ionics* 14:441–447.
268. Xu, C. J., H. D. Du, B. H. Li, F. Y. Kang, and Y. Q. Zeng. 2009. Asymmetric activated carbon–manganese dioxide capacitors in mild aqueous electrolytes containing alkaline-earth cations. *Journal of the Electrochemical Society* 156:A435–A441.
269. Tomko, T., R. Rajagopalan, M. Lanagan, and H. C. Foley. 2011. High energy density capacitor using coal tar pitch derived nanoporous carbon/MnO_2 electrodes in aqueous electrolytes. *Journal of Power Sources* 196:2380–2386.
270. Makino, S., Y. Shinohara, T. Ban et al. 2012. 4 V class aqueous hybrid electrochemical capacitor with battery-like capacity. *RSC Advances* 2:12144–12147.
271. Shimizu, W., S. Makino, K. Takahashi, N. Imanishi, and W. Sugimoto. 2013. Development of a 4.2 V aqueous hybrid electrochemical capacitor based on MnO_2 positive and protected Li negative electrodes. *Journal of Power Sources* 241:572–577.
272. Chae, J. H., and G. Z. Chen. 2012. 1.9 V aqueous carbon–carbon supercapacitors with unequal electrode capacitances. *Electrochimica Acta* 86:248–254.
273. Zheng, C., L. Qi, M. Yoshio, and H. Y. Wang. 2010. Cooperation of micro- and mesoporous carbon electrode materials in electric double-layer capacitors. *Journal of Power Sources* 195:4406–4409.

274. Torchala, K., K. Kierzek, and J. Machnikowski. 2012. Capacitance behavior of KOH activated mesocarbon microbeads in different aqueous electrolytes. *Electrochimica Acta* 86:260–267.
275. Eskusson, J., A. Jänes, A. Kikas, L. Matisen, and E. Lust. 2011. Physical and electrochemical characteristics of supercapacitors based on carbide derived carbon electrodes in aqueous electrolytes. *Journal of Power Sources* 196:4109–4116.
276. Hu, C. C., and C. C. Wang. 2004. Effects of electrolytes and electrochemical pretreatments on the capacitive characteristics of activated carbon fabrics for supercapacitors. *Journal of Power Sources* 125:299–308.
277. Wu, F. C., R. L. Tseng, C. C. Hu, and C. C. Wang. 2005. Effects of pore structure and electrolyte on the capacitive characteristics of steam- and KOH-activated carbons for supercapacitors. *Journal of Power Sources* 144:302–309.
278. Wu, F. C., R. L. Tseng, C. C. Hu, and C. C. Wang. 2006. The capacitive characteristics of activated carbons—Comparisons of the activation methods on the pore structure and effects of the pore structure and electrolyte on the capacitive performance. *Journal of Power Sources* 159:1532–1542.
279. Wu, H., X. Wang, L. Jiang et al. 2013. The effects of electrolyte on the supercapacitive performance of activated calcium carbide-derived carbon. *Journal of Power Sources* 226:202–209.
280. Zhang, X., X. Wang, L. Jiang, H. Wu, C. Wu, and J. Su. 2012. Effect of aqueous electrolytes on the electrochemical behaviors of supercapacitors based on hierarchically porous carbons. *Journal of Power Sources* 216:290–296.
281. Vaquero, S., R. Díaz, M. Anderson, J. Palma, and R. Marcilla. 2012. Insights into the influence of pore size distribution and surface functionalities in the behaviour of carbon supercapacitors. *Electrochimica Acta* 86:241–247.
282. Sun, X., X. Zhang, H. Zhang, D. Zhang, and Y. Ma. 2012. A comparative study of activated carbon-based symmetric supercapacitors in Li_2SO_4 and KOH aqueous electrolytes. *Journal of Solid State Electrochemistry* 16:2597–2603.
283. Fic, K., G. Lota, and E. Frackowiak. 2010. Electrochemical properties of supercapacitors operating in aqueous electrolyte with surfactants. *Electrochimica Acta* 55:7484–7488.
284. Wu, C., X. Wang, B. Ju et al. 2013. Supercapacitive behaviors of the nitrogen-enriched activated mesocarbon microbead in aqueous electrolytes. *Journal of Solid State Electrochemistry* 17:1693–1700.
285. Fedorov, F. S., J. Linnemann, K. Tschulik, L. Giebeler, M. Uhlemann, and A. Gebert. 2013. Capacitance performance of cobalt hydroxide-based capacitors with utilization of near-neutral electrolytes. *Electrochimica Acta* 90:166–170.
286. Huang, Q. H., X. Y. Wang, and J. Li. 2006. Characterization and performance of hydrous manganese oxide prepared by electrochemical method and its application for supercapacitors. *Electrochimica Acta* 52:1758–1762.
287. Sankar, K. V., D. Kalpana, and R. K. Selvan. 2012. Electrochemical properties of microwave-assisted reflux-synthesized Mn_3O_4 nanoparticles in different electrolytes for supercapacitor applications. *Journal of Applied Electrochemistry* 42:463–470.
288. Sarpoushi, M. R., M. Nasibi, M. A. Golozar, M. R. Shishesaz, M. R. Borhani, and S. Noroozi. 2014. Electrochemical investigation of graphene/cerium oxide nanoparticles as an electrode material for supercapacitors. *Materials Science in Semiconductor Processing* 26:374–378.
289. Zhang, J., J. Ma, L. L. Zhang, P. Guo, J. Jiang, and X. S. Zhao. 2010. Template synthesis of tubular ruthenium oxides for supercapacitor applications. *Journal of Physical Chemistry C* 114:13608–13613.
290. Zolfaghari, A., F. Ataherian, M. Ghaemi, and A. Gholami. 2007. Capacitive behavior of nanostructured MnO_2 prepared by sonochemistry method. *Electrochimica Acta* 52:2806–2814.

291. Wang, S. Y., K. C. Ho, S. L. Kuo, and N. L. Wu. 2006. Investigation on capacitance mechanisms of Fe_3O_4 electrochemical capacitors. *Journal of the Electrochemical Society* 153:A75–A80.
292. Kartachova, O., A. M. Glushenkov, Y. Chen, H. Zhang, X. J. Dai, and Y. Chen. 2012. Electrochemical capacitance of mesoporous tungsten oxynitride in aqueous electrolytes. *Journal of Power Sources* 220:298–305.
293. Sun, G. W., D. H. Long, X. J. Liu et al. 2011. Asymmetric capacitance response from the chemical characteristics of activated carbons in KOH electrolyte. *Journal of Electroanalytical Chemistry* 659:161–167.
294. Hulicova, D., J. Yamashita, Y. Soneda, H. Hatori, and M. Kodama. 2005. Supercapacitors prepared from melamine-based carbon. *Chemistry of Materials* 17:1241–1247.
295. Xiao, Y. H., Y. B. Cao, Y. Y. Gong et al. 2014. Electrolyte and composition effects on the performances of asymmetric supercapacitors constructed with Mn_3O_4 nanoparticles-graphene nanocomposites. *Journal of Power Sources* 246:926–933.
296. Barthel, J., H. J. Gores, G. Schmeer, and R. Wechter. 1983. *Non-Aqueous Electrolyte Solutions in Chemistry and Modern Technology*. Berlin Heidelberg: Springer.
297. Coetzee, J. E. 1982. *Recommended Methods for Purification of Solvents and Tests for Impurities*. New York: Pergamon Press, Commission on Electroanalytical Chemistry International Union of Pure and Applied Chemistry.
298. Ue, M., K. Ida, and S. Mori. 1994. Electrochemical properties of organic liquid electrolytes based on quaternary onium salts for electrical double-layer capacitors. *Journal of the Electrochemical Society* 141:2989–2996.
299. Aldrich Chemical Company. 1993. *Catalog Handbook of Fine Chemicals*. Milwaukee, USA: Aldrich Chemical Co.
300. Kelly, M. J., and W. R. Heineman. 1988. Electrochemical studies in 3-methyl-2-oxazolidone. *Journal of Electroanalytical Chemistry* 248:441–446.
301. Rosenfarb, J., and J. A. Caruso. 1976. Conductance behavior of electrolytes in 1, 3-dimethylethyleneurea and 1,3-dimethylpropyleneurea at 25°C. *Journal of Solution Chemistry* 5:345–354.
302. Rosenfarb, J., M. Martin, C. Prakash, and J. A. Caruso. 1976. Conductance behavior of electrolytes in 3-methylsulfolane at 25°C. *Journal of Solution Chemistry* 5:311–318.
303. Olivares-Marín, M., J. A. Fernández, M. J. Lázaro et al. 2009. Cherry stones as precursor of activated carbons for supercapacitors. *Materials Chemistry and Physics* 114:323–327.
304. Vol'fkovich, Y. M., V. M. Mazin, and N. A. Urisson. 1998. Operation of double-layer capacitors based on carbon materials. *Russian Journal of Electrochemistry* 34:740–746.
305. Dobele, G., T. Dizhbite, M. V. Gil, A. Volperts, and T. A. Centeno. 2012. Production of nanoporous carbons from wood processing wastes and their use in supercapacitors and CO_2 capture. *Biomass and Bioenergy* 46:145–154.
306. Puthusseri, D., V. Aravindan, S. Madhavi, and S. Ogale. 2014. 3D micro-porous conducting carbon beehive by single step polymer carbonization for high performance supercapacitors: The magic of *in situ* porogen formation. *Energy & Environmental Science* 7:728–735.
307. Fiset, E., J. S. Bae, T. E. Rufford, S. Bhatia, G. Q. Lu, and D. Hulicova-Jurcakova. 2014. Effects of structural properties of silicon carbide-derived carbons on their electrochemical double-layer capacitance in aqueous and organic electrolytes. *Journal of Solid State Electrochemistry* 18:703–711.
308. Wang, H. L., Q. M. Gao, and J. Hu. 2010. Preparation of porous doped carbons and the high performance in electrochemical capacitors. *Microporous and Mesoporous Materials* 131:89–96.
309. Barranco, V., M. A. Lillo-Rodenas, A. Linares-Solano et al. 2010. Amorphous carbon nanofibers and their activated carbon nanofibers as supercapacitor electrodes. *Journal of Physical Chemistry C* 114:10302–10307.

310. Wang, K., Y. Wang, Y. Wang, E. Hosono, and H. Zhou. 2009. Mesoporous carbon nanofibers for supercapacitor application. *Journal of Physical Chemistry C* 113:1093–1097.
311. McDonough, J. K., A. I. Frolov, V. Presser et al. 2012. Influence of the structure of carbon onions on their electrochemical performance in supercapacitor electrodes. *Carbon* 50:3298–3309.
312. Qie, L., W. Chen, H. Xu et al. 2013. Synthesis of functionalized 3D hierarchical porous carbon for high-performance supercapacitors. *Energy & Environmental Science* 6:2497–2504.
313. Zheng, Z., and Q. Gao. 2011. Hierarchical porous carbons prepared by an easy one-step carbonization and activation of phenol–formaldehyde resins with high performance for supercapacitors. *Journal of Power Sources* 196:1615–1619.
314. Xu, B., S. Hou, G. Cao, M. Chu, and Y. Yang. 2013. Easy synthesis of a high surface area, hierarchical porous carbon for high-performance supercapacitors. *RSC Advances* 3:17500–17506.
315. Barranco, V., V. Celorrio, M. J. Lazaro, and J. M. Rojo. 2012. Carbon nanocoils as unusual electrode materials for supercapacitors. *Journal of the Electrochemical Society* 159:A464–A469.
316. Yang, C. M., Y. J. Kim, M. Endo et al. 2007. Nanowindow-regulated specific capacitance of supercapacitor electrodes of single-wall carbon nanohorns. *Journal of the American Chemical Society* 129:20–21.
317. Sun, D., X. Yan, J. Lang, and Q. Xue. 2013. High performance supercapacitor electrode based on graphene paper via flame-induced reduction of graphene oxide paper. *Journal of Power Sources* 222:52–58.
318. Zheng, C., X. F. Zhou, H. L. Cao, G. H. Wang, and Z. P. Liu. 2014. Synthesis of porous graphene/activated carbon composite with high packing density and large specific surface area for supercapacitor electrode material. *Journal of Power Sources* 258:290–296.
319. Centeno, T. A., M. Hahn, J. A. Fernández, R. Kötz, and F. Stoeckli. 2007. Correlation between capacitances of porous carbons in acidic and aprotic EDLC electrolytes. *Electrochemistry Communications* 9:1242–1246.
320. Centeno, T. A., and F. Stoeckli. 2011. Surface-related capacitance of microporous carbons in aqueous and organic electrolytes. *Electrochimica Acta* 56:7334–7339.
321. Fernández, J. A., M. Arulepp, J. Leis, F. Stoeckli, and T. A. Centeno. 2008. EDLC performance of carbide-derived carbons in aprotic and acidic electrolytes. *Electrochimica Acta* 53:7111–7116.
322. Mladenov, M., K. Alexandrova, N. V. Petrov et al. 2013. Synthesis and electrochemical properties of activated carbons and $Li_4Ti_5O_{12}$ as electrode materials for supercapacitors. *Journal of Solid State Electrochemistry* 17:2101–2108.
323. Xu, B., S. Hou, H. Duan, G. Cao, M. Chu, and Y. Yang. 2013. Ultramicroporous carbon as electrode material for supercapacitors. *Journal of Power Sources* 228:193–197.
324. Kim, Y. J., Y. Horie, S. Ozaki et al. 2004. Correlation between the pore and solvated ion size on capacitance uptake of PVDC-based carbons. *Carbon* 42:1491–1500.
325. Sun, G., W. Song, X. Liu, D. Long, W. Qiao, and L. Ling. 2011. Capacitive matching of pore size and ion size in the negative and positive electrodes for supercapacitors. *Electrochimica Acta* 56:9248–9256.
326. Liu, C. L., W. S. Dong, G. P. Cao et al. 2008. Some effects of textural properties of carbon fibers from phenolic resins on double-layer capacitance in aprotic electrolyte. *Journal of the Electrochemical Society* 155:F124–F131.
327. Salitra, G., A. Soffer, L. Eliad, Y. Cohen, and D. Aurbach. 2000. Carbon electrodes for double-layer capacitors—I. Relations between ion and pore dimensions. *Journal of the Electrochemical Society* 147:2486–2493.

328. Chmiola, J., G. Yushin, Y. Gogotsi, C. Portet, P. Simon, and P. L. Taberna. 2006. Anomalous increase in carbon capacitance at pore sizes less than 1 nanometer. *Science* 313:1760–1763.
329. Vix-Guterl, C., E. Frackowiak, K. Jurewicz, M. Friebe, J. Parmentier, and F. Béguin. 2005. Electrochemical energy storage in ordered porous carbon materials. *Carbon* 43:1293–1302.
330. Decaux, C., C. M. Ghimbeu, M. Dahbi et al. 2014. Influence of electrolyte ion–solvent interactions on the performances of supercapacitors porous carbon electrodes. *Journal of Power Sources* 263:130–140.
331. Lin, R., P. L. Taberna, J. Chmiola, D. Guay, Y. Gogotsi, and P. Simon. 2009. Microelectrode study of pore size, ion size, and solvent effects on the charge/discharge behavior of microporous carbons for electrical double-layer capacitors. *Journal of the Electrochemical Society* 156:A7–A12.
332. Segalini, J., B. Daffos, P. L. Taberna, Y. Gogotsi, and P. Simon. 2010. Qualitative electrochemical impedance spectroscopy study of ion transport into sub-nanometer carbon pores in electrochemical double layer capacitor electrodes. *Electrochimica Acta* 55:7489–7494.
333. Jiang, D. E., and J. Z. Wu. 2013. Microscopic insights into the electrochemical behavior of nonaqueous electrolytes in electric double-layer capacitors. *Journal of Physical Chemistry Letters* 4:1260–1267.
334. Feng, G., J. Huang, B. G. Sumpter, V. Meunier, and R. Qiao. 2010. Structure and dynamics of electrical double layers in organic electrolytes. *Physical Chemistry Chemical Physics* 12:5468–5479.
335. Takana, A., T. Iiyama, T. Ohba, S. Ozeki, K. Urita, T. Fujimori, H. Kanoh, and K. Kaneko. 2010. Effect of a quaternary ammonium salt on propylene carbonate structure in slit-shape carbon nanopores. *Journal of the American Chemical Society* 132:2112–2113.
336. Fukano, M., T. Fujimori, J. Ségalini et al. 2013. Vertically oriented propylene carbonate molecules and tetraethyl ammonium ions in carbon slit pores. *Journal of Physical Chemistry C* 117:5752–5757.
337. Ohba, T., and K. Kaneko. 2013. Competition of desolvation and stabilization of organic electrolytes in extremely narrow nanopores. *Journal of Physical Chemistry C* 117:17092–17098.
338. Jiang, D. E., Z. H. Jin, D. Henderson, and J. Z. Wu. 2012. Solvent effect on the pore-size dependence of an organic electrolyte supercapacitor. *Journal of Physical Chemistry Letters* 3:1727–1731.
339. Wang, H. N., J. Varghese, and L. Pilon. 2011. Simulation of electric double layer capacitors with mesoporous electrodes: Effects of morphology and electrolyte permittivity. *Electrochimica Acta* 56:6189–6197.
340. Yang, L., B. H. Fishbine, A. Migliori, and L. R. Pratt. 2009. Molecular simulation of electric double-layer capacitors based on carbon nanotube forests. *Journal of the American Chemical Society* 131:12373–12376.
341. Lee, S. I., K. Saito, K. Kanehashi et al. 2006. ^{11}B NMR study of the BF_4 anion in activated carbons at various stages of charge of EDLCs in organic electrolyte. *Carbon* 44:2578–2586.
342. Wang, H., A. C. Forse, J. M. Griffin et al. 2013. *In situ* NMR spectroscopy of supercapacitors: Insight into the charge storage mechanism. *Journal of the American Chemical Society* 135:18968–18980.
343. Wang, H., T. K. J. Koster, N. M. Trease et al. 2011. Real-time NMR studies of electrochemical double-layer capacitors. *Journal of the American Chemical Society* 133:19270–19273.

344. Levi, M. D., G. Salitra, N. Levy, D. Aurbach, and J. Maier. 2009. Application of a quartz-crystal microbalance to measure ionic fluxes in microporous carbons for energy storage. *Nature Materials* 8:872–875.
345. Deschamps, M., E. Gilbert, P. Azais et al. 2013. Exploring electrolyte organization in supercapacitor electrodes with solid-state NMR. *Nature Materials* 12:351–358.
346. Forse, A. C., J. M. Griffin, H. Wang et al. 2013. Nuclear magnetic resonance study of ion adsorption on microporous carbide-derived carbon. *Physical Chemistry Chemical Physics* 15:7722–7730.
347. Borchardt, L., M. Oschatz, S. Paasch, S. Kaskel, and E. Brunner. 2013. Interaction of electrolyte molecules with carbon materials of well-defined porosity: Characterization by solid-state NMR spectroscopy. *Physical Chemistry Chemical Physics* 15:15177–15184.
348. Boukhalfa, S., D. Gordon, L. L. He et al. 2014. *In situ* small angle neutron scattering revealing ion sorption in microporous carbon electrical double layer capacitors. *ACS Nano* 8:2495–2503.
349. Bonhomme, F., J. C. Lassègues, and L. Servant. 2001. Raman spectroelectrochemistry of a carbon supercapacitor. *Journal of the Electrochemical Society* 148:E450–E458.
350. Ilott, A. J., N. M. Trease, C. P. Grey, and A. Jerschow. 2014. Multinuclear *in situ* magnetic resonance imaging of electrochemical double-layer capacitors. *Nature Communications* 5:4536.
351. Blanc, F., M. Leskes, and C. P. Grey. 2013. *In situ* solid-state NMR spectroscopy of electrochemical cells: Batteries, supercapacitors, and fuel cells. *Accounts of Chemical Research* 46:1952–1963.
352. Harris, R. K., T. V. Thompson, P. R. Norman, and C. Pottage. 1996. Adsorption competition onto activated carbon, studied by magic-angle spinning NMR. *Journal of the Chemical Society—Faraday Transactions* 92:2615–2618.
353. Dickinson, L. M., R. K. Harris, J. A. Shaw, M. Chinn, and P. R. Norman. 2000. Oxygen-17 and deuterium NMR investigation into the adsorption of water on activated carbon. *Magnetic Resonance in Chemistry* 38:918–924.
354. Harris, R. K., T. V. Thompson, P. R. Norman, and C. Pottage. 1999. Phosphorus-31 NMR studies of adsorption onto activated carbon. *Carbon* 37:1425–1430.
355. Maeshima, H., H. Moriwake, A. Kuwabara, and C. A. J. Fisher. 2010. Quantitative evaluation of electrochemical potential windows of electrolytes for electric double-layer capacitors using *ab initio* calculations. *Journal of the Electrochemical Society* 157:A696–A701.
356. Maeshima, H., H. Moriwake, A. Kuwabara, C. A. J. Fisher, and I. Tanaka. 2014. An improved method for quantitatively predicting the electrochemical stabilities of organic liquid electrolytes using *ab initio* calculations. *Journal of the Electrochemical Society* 161:G7–G14.
357. Kurzweil, P., and M. Chwistek. 2008. Electrochemical stability of organic electrolytes in supercapacitors: Spectroscopy and gas analysis of decomposition products. *Journal of Power Sources* 176:555–567.
358. Ricketts, B. W., and C. Ton-That. 2000. Self-discharge of carbon-based supercapacitors with organic electrolytes. *Journal of Power Sources* 89:64–69.
359. Farahmandi, C. J., J. M. Dispennette, E. Blank, and A. C. Kolb. 1998. Multi-electrode double layer capacitor. WO9815962 for Maxwell Tech.
360. Hardwick, L. J., M. Hahn, P. Ruch et al. 2006. An *in situ* Raman study of the intercalation of supercapacitor-type electrolyte into microcrystalline graphite. *Electrochimica Acta* 52:675–680.
361. Ruch, P. W., M. Hahn, F. Rosciano et al. 2007. *In situ* X-ray diffraction of the intercalation of $(C_2H_5)_4N^+$ and BF_4^- into graphite from acetonitrile and propylene carbonate based supercapacitor electrolytes. *Electrochimica Acta* 53:1074–1082.

362. Hahn, M., O. Barbieri, F. P. Campana, R. Kötz, and R. Gallay. 2006. Carbon based double layer capacitors with aprotic electrolyte solutions: The possible role of intercalation/insertion processes. *Applied Physics A: Materials Science & Processing* 82:633–638.
363. Foelske, A., O. Barbieri, M. Hahn, and R. Kötz. 2006. An X-ray photoelectron spectroscopy study of hydrous ruthenium oxide powders with various water contents for supercapacitors. *Electrochemical and Solid-State Letters* 9:A268–A272.
364. Campana, F. P., M. Hahn, A. Foelske, P. Ruch, R. Kötz, and H. Siegenthaler. 2006. Intercalation into and film formation on pyrolytic graphite in a supercapacitor-type electrolyte $(C_2H_5)_4NBF_4$/propylene carbonate. *Electrochemistry Communications* 8:1363–1368.
365. Hahn, M., R. Kötz, R. Gallay, and A. Siggel. 2006. Pressure evolution in propylene carbonate based electrochemical double layer capacitors. *Electrochimica Acta* 52:1709–1712.
366. Kötz, R., M. Hahn, and R. Gallay. 2006. Temperature behavior and impedance fundamentals of supercapacitors. *Journal of Power Sources* 154:550–555.
367. Barbieri, O., M. Hahn, A. Foelske, and R. Kötz. 2006. Effect of electronic resistance and water content on the performance of RuO_2 for supercapacitors. *Journal of the Electrochemical Society* 153:A2049–A2054.
368. Foelske-Schmitz, A., P. W. Ruch, and R. Kötz. 2010. Ion intercalation into HOPG in supercapacitor electrolyte—An X-ray photoelectron spectroscopy study. *Journal of Electron Spectroscopy and Related Phenomena* 182:57–62.
369. Galiński, M., A. Lewandowski, and I. Stępniak. 2006. Ionic liquids as electrolytes. *Electrochimica Acta* 51:5567–5580.
370. Gualous, H., D. Bouquain, A. Berthon, and J. M. Kauffmann. 2003. Experimental study of supercapacitor serial resistance and capacitance variations with temperature. *Journal of Power Sources* 123:86–93.
371. Brandon, E. J., W. C. West, M. C. Smart, L. D. Whitcanack, and G. A. Plett. 2007. Extending the low temperature operational limit of double-layer capacitors. *Journal of Power Sources* 170:225–232.
372. Zhang, Q., J. Rong, D. Ma, and B. Wei. 2011. The governing self-discharge processes in activated carbon fabric-based supercapacitors with different organic electrolytes. *Energy & Environmental Science* 4:2152–2159.
373. Hanlon, D., C. Backes, T. M. Higgins et al. 2014. Production of molybdenum trioxide nanosheets by liquid exfoliation and their application in high-performance supercapacitors. *Chemistry of Materials* 26:1751–1763.
374. Chen, H. Y., G. Wee, R. Al-Oweini et al. 2014. A polyoxovanadate as an advanced electrode material for supercapacitors. *ChemPhysChem* 15:2162–2169.
375. Lim, E., H. Kim, C. Jo et al. 2014. Advanced hybrid supercapacitor based on a mesoporous niobium pentoxide/carbon as high-performance anode. *ACS Nano* 8:8968–8978.
376. Yang, Y., Z. W. Peng, G. N. Wang et al. 2014. Three-dimensional thin film for lithium-ion batteries and supercapacitors. *ACS Nano* 8:7279–7287.
377. Huang, X. D., B. Sun, S. Q. Chen, and G. X. Wang. 2014. Self-assembling synthesis of free-standing nanoporous graphene-transition-metal oxide flexible electrodes for high-performance lithium-ion batteries and supercapacitors. *Chemistry – An Asian Journal* 9:206–211.
378. Cho, M. Y., M. H. Kim, H. K. Kim, K. B. Kim, J. R. Yoon, and K. C. Roh. 2014. Electrochemical performance of hybrid supercapacitor fabricated using multi-structured activated carbon. *Electrochemistry Communications* 47:5–8.
379. Pan, D. Y., H. Huang, X. Y. Wang et al. 2014. *C*-axis preferentially oriented and fully activated TiO_2 nanotube arrays for lithium ion batteries and supercapacitors. *Journal of Materials Chemistry A* 2:11454–11464.

380. Qu, W. H., F. Han, A. H. Lu, C. Xing, M. Qiao, and W. C. Li. 2014. Combination of a SnO_2–C hybrid anode and a tubular mesoporous carbon cathode in a high energy density non-aqueous lithium ion capacitor: Preparation and characterisation. *Journal of Materials Chemistry A* 2:6549–6557.
381. Cai, Y., B. T. Zhao, J. Wang, and Z. P. Shao. 2014. Non-aqueous hybrid supercapacitors fabricated with mesoporous TiO_2 microspheres and activated carbon electrodes with superior performance. *Journal of Power Sources* 253:80–89.
382. Sun, X. Z., X. Zhang, H. T. Zhang, N. S. Xu, K. Wang, and Y. W. Ma. 2014. High performance lithium-ion hybrid capacitors with pre-lithiated hard carbon anodes and bifunctional cathode electrodes. *Journal of Power Sources* 270:318–325.
383. Kaiyappan, K., S. Amaresh, and Y. S. Lee. 2014. $LiMnBO_3$ nanobeads as an innovative anode material for high power lithium ion capacitor applications. *ACS Applied Materials & Interfaces* 6:11357–11367.
384. Aravindan, V., J. Sundaramurthy, A. Jain et al. 2014. Unveiling $TiNb_2O_7$ as an insertion anode for lithium ion capacitors with high energy and power density. *ChemSusChem* 7:1858–1863.
385. Amaresh, S., K. Karthikeyan, I. C. Jang, and Y. S. Lee. 2014. Single-step microwave mediated synthesis of the CoS_2 anode material for high rate hybrid supercapacitors. *Journal of Materials Chemistry A* 2:11099–11106.
386. Zhang, F., T. F. Zhang, X. Yang et al. 2013. A high-performance supercapacitor-battery hybrid energy storage device based on graphene-enhanced electrode materials with ultrahigh energy density. *Energy & Environmental Science* 6:1623–1632.
387. Kalubarme, R. S., H. S. Jadhav, and C. J. Park. 2013. Electrochemical characteristics of two-dimensional nano-structured MnO_2 for symmetric supercapacitor. *Electrochimica Acta* 87:457–465.
388. Wee, G., H. Z. Soh, Y. L. Cheah, S. G. Mhaisalkar, and M. Srinivasan. 2010. Synthesis and electrochemical properties of electrospun V_2O_5 nanofibers as supercapacitor electrodes. *Journal of Materials Chemistry* 20:6720–6725.
389. Bonso, J. S., A. Rahy, S. D. Perera et al. 2012. Exfoliated graphite nanoplatelets–V_2O_5 nanotube composite electrodes for supercapacitors. *Journal of Power Sources* 203:227–232.
390. Yiğit, D., M. Güllü, T. Yumak, and A. Sınağ. 2014. Heterostructured poly(3,6-dithien-2-yl-9H-carbazol-9-yl acetic acid)/TiO_2 nanoparticles composite redox-active materials as both anode and cathode for high-performance symmetric supercapacitor applications. *Journal of Materials Chemistry A* 2:6512–6524.
391. Nam, K. W., C. W. Lee, X. Q. Yang, B. W. Cho, W. S. Yoon, and K. B. Kim. 2009. Electrodeposited manganese oxides on three-dimensional carbon nanotube substrate: Supercapacitive behaviour in aqueous and organic electrolytes. *Journal of Power Sources* 188:323–331.
392. Ryu, K. S., Y. G. Lee, Y. S. Hong et al. 2004. Poly(ethylenedioxythiophene) (PEDOT) as polymer electrode in redox supercapacitor. *Electrochimica Acta* 50:843–847.
393. Cho, S. H., J. S. Joo, B. R. Jung, T. M. Ha, and J. Y. Lee. 2009. PET fabric/poly(3,4-ethylenedioxythiophene) composite as polymer electrode in redox supercapacitor. *Macromolecular Research* 17:746–749.
394. Zhang, H., J. Wang, Y. Chen, Z. Wang, and S. Wang. 2013. Long-term cycling stability of polyaniline on graphite electrodes used for supercapacitors. *Electrochimica Acta* 105:69–74.
395. Zhou, S., H. Zhang, X. Wang, J. Li, and F. Wang. 2013. Sandwich nanocomposites of polyaniline embedded between graphene layers and multi-walled carbon nanotubes for cycle-stable electrode materials of organic supercapacitors. *RSC Advances* 3:1797–1807.
396. Li, L., D. C. Loveday, D. S. K. Mudigonda, and J. P. Ferraris. 2002. Effect of electrolytes on performance of electrochemical capacitors based on poly[3-(3,4-difluorophenyl) thiophene]. *Journal of the Electrochemical Society* 149:A1201–A1207.

397. Ertas, M., R. M. Walczak, R. K. Das, A. G. Rinzler, and J. R. Reynolds. 2012. Supercapacitors based on polymeric dioxypyrroles and single walled carbon nanotubes. *Chemistry of Materials* 24:433–443.
398. Arsov, Lj. D., W. Plieth, and G. Koßmehl. 1998. Electrochemical and Raman spectroscopic study of polyaniline; Influence of the potential on the degradation of polyaniline. *Journal of Solid State Electrochemistry* 2:355–361.
399. Planes, G. A., J. L. Rodriguez, M. C. Miras, G. Garcia, E. Pastor, and C. A. Barbero. 2010. Spectroscopic evidence for intermediate species formed during aniline polymerization and polyaniline degradation. *Physical Chemistry Chemical Physics* 12:10584–10593.
400. Amatucci, G. G., F. Badway, A. Du Pasquier, and T. Zheng. 2001. An asymmetric hybrid nonaqueous energy storage cell. *Journal of the Electrochemical Society* 148:A930–A939.
401. Zheng, C., M. Yoshio, L. Qi, and H. Y. Wang. 2014. A 4 V-electrochemical capacitor using electrode and electrolyte materials free of metals. *Journal of Power Sources* 260:19–26.
402. Khomenko, V., E. Raymundo-Piñero, and F. Béguin. 2008. High-energy density graphite/AC capacitor in organic electrolyte. *Journal of Power Sources* 177:643–651.
403. Wang, L., T. Morishita, M. Toyoda, and M. Inagaki. 2007. Asymmetric electric double layer capacitors using carbon electrodes with different pore size distributions. *Electrochimica Acta* 53:882–886.
404. Wang, H. Q., Z. S. Li, Y. G. Huang, Q. Y. Li, and X. Y. Wang. 2010. A novel hybrid supercapacitor based on spherical activated carbon and spherical MnO_2 in a non-aqueous electrolyte. *Journal of Materials Chemistry* 20:3883–3889.
405. Wang, D. W., H. T. Fang, F. Li et al. 2008. Aligned titania nanotubes as an intercalation anode material for hybrid electrochemical energy storage. *Advanced Functional Materials* 18:3787–3793.
406. Naoi, K., S. Ishimoto, Y. Isobe, and S. Aoyagi. 2010. High-rate nano-crystalline $Li_4Ti_5O_{12}$ attached on carbon nano-fibers for hybrid supercapacitors. *Journal of Power Sources* 195:6250–6254.
407. Ni, J. F., L. X. Yang, H. B. Wang, and L. J. Gao. 2012. A high-performance hybrid supercapacitor with $Li_4Ti_5O_{12}$-C nano-composite prepared by *in situ* and *ex situ* carbon modification. *Journal of Solid State Electrochemistry* 16:2791–2796.
408. Cheng, L., H. J. Liu, J. J. Zhang, H. M. Xiong, and Y. Y. Xia. 2006. Nanosized $Li_4Ti_5O_{12}$ prepared by molten salt method as an electrode material for hybrid electrochemical supercapacitors. *Journal of the Electrochemical Society* 153:A1472–A1477.
409. Sivaraman, P., A. R. Bhattacharrya, S. P. Mishra, A. P. Thakur, K. Shashidhara, and A. B. Samui. 2013. Asymmetric supercapacitor containing poly(3-methyl thiophene)-multiwalled carbon nanotubes nanocomposites and activated carbon. *Electrochimica Acta* 94:182–191.
410. Zou, W. Y., W. Wang, B. L. He, M. L. Sun, and Y. S. Yin. 2010. Supercapacitive properties of hybrid films of manganese dioxide and polyaniline based on active carbon in organic electrolyte. *Journal of Power Sources* 195:7489–7493.
411. Ferraris, J. P., M. M. Eissa, I. D. Brotherston, and D. C. Loveday. 1998. Performance evaluation of poly 3-(phenylthiophene) derivatives as active materials for electrochemical capacitor applications. *Chemistry of Materials* 10:3528–3535.
412. Fusalba, F., P. Gouérec, D. Villers, and D. Bélanger. 2001. Electrochemical characterization of polyaniline in nonaqueous electrolyte and its evaluation as electrode material for electrochemical supercapacitors. *Journal of the Electrochemical Society* 148:A1–A6.
413. Güllü, M., and D. Yiğit. 2012. A novel asymmetric pseudocapacitor based on poly(5,12-dihydrothieno [3′,4′:2,3][1,4]dioxocino[6,7-b] quinoxaline) coated graphite anode and poly(ethylenedioxythiophene) coated graphite cathode. *Synthetic Metals* 162:1434–1442.

414. Schroeder, M., S. Menne, J. Segalini et al. 2014. Considerations about the influence of the structural and electrochemical properties of carbonaceous materials on the behavior of lithium-ion capacitors. *Journal of Power Sources* 266:250–258.
415. Smith, P. H., T. N. Tran, T. L. Jiang, and J. Chung. 2013. Lithium-ion capacitors: Electrochemical performance and thermal behavior. *Journal of Power Sources* 243:982–992.
416. Pasquier, A. D., I. Plitz, J. Gural, F. Badway, and G. G. Amatucci. 2004. Power-ion battery: Bridging the gap between Li-ion and supercapacitor chemistries. *Journal of Power Sources* 136:160–170.
417. Lee, C. H., F. Xu, and C. Jung. 2014. Influence of the electrolyte distribution near the micropores of the activated carbon (AC) electrode on high rate performance of high voltage capacitors. *Electrochimica Acta* 131:240–244.
418. Xu, K. 2004. Nonaqueous liquid electrolytes for lithium-based rechargeable batteries. *Chemical Reviews* 104:4303–4417.
419. Omar, N., M. Daowd, O. Hegazy et al. 2012. Assessment of lithium-ion capacitor for using in battery electric vehicle and hybrid electric vehicle applications. *Electrochimica Acta* 86:305–315.
420. Karthikeyan, K., S. Amaresh, S. N. Lee, J. Y. An, and Y. S. Lee. 2014. High-power lithium-ion capacitor using $LiMnBO_3$-nanobead anode and polyaniline-nanofiber cathode with excellent cycle life. *ChemSusChem* 7:2310–2316.
421. Zhang, J., Z. Q. Shi, and C. Y. Wang. 2014. Effect of pre-lithiation degrees of mesocarbon microbeads anode on the electrochemical performance of lithium-ion capacitors. *Electrochimica Acta* 125:22–28.
422. Bhuvanalogini, G., N. Murugananthem, V. Shobana, and A. Subramania. 2014. Preparation, characterization, and evaluation of $LiNi_{0.4}Co_{0.6}O_2$ nanofibers for supercapacitor applications. *Journal of Solid State Electrochemistry* 18:2387–2392.
423. Puthusseri, D., V. Aravindan, S. Madhavi, and S. Ogale. 2014. Improving the energy density of Li-ion capacitors using polymer-derived porous carbons as cathode. *Electrochimica Acta* 130:766–770.
424. Karthikeyan, K., S. H. Kim, K. J. Kim, S. N. Lee, and Y. S. Lee. 2013. Low cost, eco-friendly layered $Li_{1.2}(Mn_{0.32}Ni_{0.32}Fe_{0.16})O_2$ nanoparticles for hybrid supercapacitor applications. *Electrochimica Acta* 109:595–601.
425. Lei, Y., Z. H. Huang, Y. Yang et al. 2013. Porous mesocarbon microbeads with graphitic shells: Constructing a high-rate, high-capacity cathode for hybrid supercapacitor. *Scientific Reports* 3:2477.
426. Perera, S. D., M. Rudolph, R. G. Mariano et al. 2013. Manganese oxide nanorod-graphene/vanadium oxide nanowire-graphene binder-free paper electrodes for metal oxide hybrid supercapacitors. *Nano Energy* 2:966–975.
427. Karthikeyan, K., S. Amaresh, V. Aravindan, and Y. S. Lee. 2013. Microwave assisted green synthesis of MgO–carbon nanotube composites as electrode material for high power and energy density supercapacitors. *Journal of Materials Chemistry A* 1:4105–4111.
428. Simon, P., and Y. Gogotsi. 2013. Capacitive energy storage in nanostructured carbon-electrolyte systems. *Accounts of Chemical Research* 46:1094–1103.
429. Ue, M. 2007. Chemical capacitors and quaternary ammonium salts. *Electrochemistry* 75:565–572.
430. Naoi, K. 2010. 'Nanohybrid capacitor': The next generation electrochemical capacitors. *Fuel Cells* 10:825–833.
431. Ruch, P. W., D. Cericola, A. Foelske, R. Kötz, and A. Wokaun. 2010. A comparison of the aging of electrochemical double layer capacitors with acetonitrile and propylene carbonate-based electrolytes at elevated voltages. *Electrochimica Acta* 55:2352–2357.
432. Ruch, P. W., D. Cericola, A. Foelske-Schmitz, R. Kötz, and A. Wokaun. 2010. Aging of electrochemical double layer capacitors with acetonitrile-based electrolyte at elevated voltages. *Electrochimica Acta* 55:4412–4420.

433. Liu, P., M. Verbrugge, and S. Soukiazian. 2006. Influence of temperature and electrolyte on the performance of activated-carbon supercapacitors. *Journal of Power Sources* 156:712–718.
434. Lust, E., A. Jänes, and M. Arulepp. 2004. Influence of solvent nature on the electrochemical parameters of electrical double layer capacitors. *Journal of Electroanalytical Chemistry* 562:33–42.
435. Arulepp, M., L. Permann, J. Leis et al. 2004. Influence of the solvent properties on the characteristics of a double layer capacitor. *Journal of Power Sources* 133:320–328.
436. Ding, M. S., K. Xu, J. P. Zheng, and T. R. Jow. 2004. γ-Butyrolactone–acetonitrile solution of triethylmethylammonium tetrafluoroborate as an electrolyte for double-layer capacitors. *Journal of Power Sources* 138:340–350.
437. Ue, M., M. Takeda, M. Takehara, and S. Mori. 1997. Electrochemical properties of quaternary ammonium salts for electrochemical capacitors. *Journal of the Electrochemical Society* 144:2684–2688.
438. Chiba, K., T. Ueda, Y. Yamaguchi, Y. Oki, F. Saiki, and K. Naoi. 2011. Electrolyte systems for high withstand voltage and durability II. Alkylated cyclic carbonates for electric double-layer capacitors. *Journal of the Electrochemical Society* 158:A1320–A1327.
439. Nakajima, T., and H. Groult (Eds). 2006. *Fluorinated Materials for Energy Conversion*. Oxford: Elsevier.
440. Francke, R., D. Cericola, R. Kötz, D. Weingarth, and S. R. Waldvogel. 2012. Novel electrolytes for electrochemical double layer capacitors based on 1,1,1,3,3,3-hexafluoropropan-2-ol. *Electrochimica Acta* 62:372–380.
441. Nanbu, N., K. Suzuki, N. Yagi et al. 2007. Use of fluoroethylene carbonate as solvent for electric double-layer capacitors. *Electrochemistry* 75:607–610.
442. Hagiyama, K., K. Suzuki, M. Ohtake et al. 2008. Physical properties of substituted 1,3-dioxolan-2-ones. *Chemistry Letters* 37:210–211.
443. Riddick, J. A., W. B. Bunger, and T. K. Sakano. 1986. *Organic Solvents: Physical Properties and Methods of Purification*, 4th edn. New York: Wiley-Interscience.
444. Suzuki, K., M. Shin-Ya, Y. Ono et al. 2007. Physical and electrochemical properties of fluoroacetonitrile and its application to electric double-layer capacitors. *Electrochemistry* 75:611–614.
445. Chiba, K., T. Ueda, Y. Yamaguchi, Y. Oki, F. Shimodate, and K. Naoi. 2011. Electrolyte systems for high withstand voltage and durability I. Linear sulfones for electric double-layer capacitors. *Journal of the Electrochemical Society* 158:A872–A882.
446. Brandt, A., P. Isken, A. Lex-Balducci, and A. Balducci. 2012. Adiponitrile-based electrochemical double layer capacitor. *Journal of Power Sources* 204:213–219.
447. Brandt, A., and A. Balducci. 2012. The Influence of pore structure and surface groups on the performance of high voltage electrochemical double layer capacitors containing adiponitrile-based electrolyte. *Journal of the Electrochemical Society* 159:A2053–A2059.
448. Jänes, A., and E. Lust. 2005. Organic carbonate-organic ester-based non-aqueous electrolytes for electrical double layer capacitors. *Electrochemistry Communications* 7:510–514.
449. Nambu, N., D. Kobayashi, and Y. Sasaki. 2013. Physical and electrolytic properties of different cyclic carbonates as solvents for electric double-layer capacitors. *Electrochemistry* 81:814–816.
450. Tian, S. F., L. Qi, M. Yoshio, and H. Y. Wang. 2014. Tetramethylammonium difluoro(oxalato)borate dissolved in ethylene/propylene carbonates as electrolytes for electrochemical capacitors. *Journal of Power Sources* 256:404–409.
451. Jänes, A., T. Thomberg, J. Eskusson, and E. Lust. 2013. Fluoroethylene carbonate as co-solvent for propylene carbonate based electrical double layer capacitors. *Journal of the Electrochemical Society* 160:A1025–A1030.

452. Kim, M., I. J. Kim, S. Yang, and S. Kim. 2014. Fluoroethylene carbonate addition effect on electrochemical properties of mixed carbonate-based organic electrolyte solution for a capacitor. *Bulletin of the Korean Chemical Society* 35:466–470.
453. Jänes, A., J. Eskusson, T. Thomberg, and E. Lust. 2014. Supercapacitors based on propylene carbonate with small addition of different sulfur containing organic solvents. *Journal of the Electrochemical Society* 161:A1284–A1290.
454. Laheäär, A., H. Kurig, A. Jänes, and E. Lust. 2009. $LiPF_6$ based ethylene carbonate–dimethyl carbonate electrolyte for high power density electrical double layer capacitor. *Electrochimica Acta* 54:4587–4594.
455. Jänes, A., and E. Lust. 2006. Use of organic esters as co-solvents for electrical double layer capacitors with low temperature performance. *Journal of Electroanalytical Chemistry* 588:285–295.
456. Perricone, E., M. Chamas, J. C. Leprêtre et al. 2013. Safe and performant electrolytes for supercapacitor. Investigation of esters/carbonate mixtures. *Journal of Power Sources* 239:217–224.
457. Perricone, E., M. Chamas, L. Cointeaux et al. 2013. Investigation of methoxypropionitrile as co-solvent for ethylene carbonate based electrolyte in supercapacitors. A safe and wide temperature range electrolyte. *Electrochimica Acta* 93:1–7.
458. Iwama, E., P. L. Taberna, P. Azais, L. Brégeon, and P. Simon. 2012. Characterization of commercial supercapacitors for low temperature applications. *Journal of Power Sources* 219:235–239.
459. West, W. C., M. C. Smart, E. J. Brandon, L. D. Whitcanack, and G. A. Plett. 2008. Double-layer capacitor electrolytes using 1,3-dioxolane for low temperature operation. *Journal of the Electrochemical Society* 155:A716–A720.
460. Korenblit, Y., A. Kajdos, W. C. West et al. 2012. *In situ* studies of ion transport in microporous supercapacitor electrodes at ultralow temperatures. *Advanced Functional Materials* 22:1655–1662.
461. Baek, B., and C. Jung. 2010. Enhancement of the Li^+ ion transfer reaction at the $LiCoO_2$ interface by 1,3,5-trifluorobenzene. *Electrochimica Acta* 55:3307–3311.
462. Zheng, J. P. 2005. Theoretical energy density for electrochemical capacitors with intercalation electrodes. *Journal of the Electrochemical Society* 152:A1864–A1869.
463. Zheng, J. P., and T. R. Jow. 1997. The effect of salt concentration in electrolytes on the maximum energy storage for double layer capacitors. *Journal of the Electrochemical Society* 144:2417–2420.
464. House, H. O., E. Feng, and N. P. Peet. 1971. Comparison of various tetraalkylammonium salts as supporting electrolytes in organic electrochemical reactions. *The Journal of Organic Chemistry* 36:2371–2375.
465. Rousseau, K., G. C. Farrington, and D. Dolphin. 1972. Tetraalkylammonium trifluoromethanesulfonates as supporting electrolytes. *The Journal of Organic Chemistry* 37:3968–3971.
466. Ue, M., A. Murakami, and S. Nakamura. 2002. Anodic stability of several anions examined by *ab initio* molecular orbital and density functional theories. *Journal of the Electrochemical Society* 149:A1572–A1577.
467. Xu, K., M. S. Ding, and T. R. Jow. 2001. Quaternary onium salts as nonaqueous electrolytes for electrochemical capacitors. *Journal of the Electrochemical Society* 148:A267–A274.
468. Koh, A. R., B. Hwang, K. C. Roh, and K. Kim. 2014. The effect of the ionic size of small quaternary ammonium BF_4 salts on electrochemical double layer capacitors. *Physical Chemistry Chemical Physics* 16:15146–15151.
469. Jänes, A., H. Kurig, T. Romann, and E. Lust. 2010. Novel doubly charged cation based electrolytes for non-aqueous supercapacitors. *Electrochemistry Communications* 12:535–539.

470. Zheng, C., M. Yoshio, L. Qi, and H. Y. Wang. 2012. A divalent quaternary alkyl ammonium salt as the electrolyte for high-energy electric double-layer capacitors. *Journal of Power Sources* 220:169–172.
471. Yokoyama, Y., N. Shimosaka, H. Matsumoto, M. Yoshio, and T. Ishiharaa. 2008. Effects of supporting electrolyte on the storage capacity of hybrid capacitors using graphitic and activated carbon. *Electrochemical and Solid-State Letters* 11:A72–A75.
472. Wang, H. Y., and M. Yoshio. 2012. Feasibility of quaternary alkyl ammonium-intercalated graphite as negative electrode materials in electrochemical capacitors. *Journal of Power Sources* 200:108–112.
473. Nambu, N., R. Takahashi, M. Takehara, M. Ue, and Y. Sasaki. 2013. Electrolytic characteristics of fluoroethylene carbonate for electric double-layer capacitors at high concentrations of electrolyte. *Electrochemistry* 81:817–819.
474. Chiba, K., T. Ueda, and H. Yamamoto. 2007. Performance of electrolyte composed of spiro-type quaternary ammonium salt and electric double-layer capacitor using it. *Electrochemistry* 75:664–667.
475. Chiba, K., T. Ueda, and H. Yamamoto. 2007. Highly conductive electrolytic solution for electric double-layer capacitor using dimethylcarbonate and spiro-type quaternary ammonium salt. *Electrochemistry* 75:668–671.
476. Yu, X. W., D. B. Ruan, C. C. Wu, J. Wang, and Z. Q. Shi. 2014. Spiro-(1,1′)-bipyrrolidinium tetrafluoroborate salt as high voltage electrolyte for electric double layer capacitors. *Journal of Power Sources* 265:309–316.
477. Zheng, C., J. C. Gao, M. Yoshio, L. Qi, and H. Y. Wang. 2013. Non-porous activated mesophase carbon microbeads as a negative electrode material for asymmetric electrochemical capacitors. *Journal of Power Sources* 231:29–33.
478. Kurig, H., A. Jänes, and E. Lust. 2010. Substituted phosphonium cation based electrolytes for nonaqueous electrical double-layer capacitors. *Journal of Materials Research* 25:1447–1450.
479. Ebina, T., H. Uno, S. Ishizawa, N. Nanbu, and Y. Sasaki. 2005. Use of tetraethylammonium bis(oxalato)borate as electrolyte for electrical double-layer capacitors. *Chemistry Letters* 34:1014–1015.
480. Waldvogel, S. R., I. M. Malkowsky, U. Griesbach et al. 2009. Novel fluorine-free electrolyte system for supercapacitors. *Electrochemistry Communications* 11:1237–1241.
481. Nanbu, N., T. Ebina, H. Uno, S. Ishizawa, and Y. Sasaki. 2006. Physical and electrochemical properties of quaternary ammonium bis(oxalato)borates and their application to electric double-layer capacitors. *Electrochimica Acta* 52:1763–1770.
482. Nambu, N., R. Takahashi, K. Suzuki, and Y. Sasaki. 2013. Electrolytic properties of tetramethylammonium compound in highly concentrated solutions and its application to electric double-layer capacitors. *Electrochemistry* 81:811–813.
483. Nanbu, N., K. Suzuki, T. Ebina, H. Uno, and Y. Sasaki. 2007. Tetramethyl ammonium difluoro(oxalato)borate as novel electrolyte for electric double-layer capacitors. *Electrochemistry* 75:615–618.
484. Lai, Y. Q., X. J. Chen, Z. A. Zhang, J. Li, and Y. X. Liu. 2011. Tetraethylammonium difluoro(oxalato)borate as electrolyte salt for electrochemical double-layer capacitors. *Electrochimica Acta* 56:6426–6430.
485. Francke, R., D. Cericola, R. Kötz, G. Schnakenburg, and S. R. Waldvogel. 2011. Bis(2,2′-biphenoxy)borates for electrochemical double-layer capacitor electrolytes. *Chemistry – A European Journal* 17:3082–3085.
486. Kim, I. T., M. Egashira, N. Yoshimoto, and M. Morita. 2012. Combination of alkali-treated soft carbon and activated carbon fiber electrodes for asymmetric electric double-layer capacitor. *Electrochemistry* 80:415–420.

487. Laheäär, A., A. Jänes, and E. Lust. 2011. Electrochemical properties of carbide-derived carbon electrodes in non-aqueous electrolytes based on different Li-salts. *Electrochimica Acta* 56:9048–9055.
488. Li, Q., X. X. Zuo, J. S. Liu, X. Xiao, D. Shu, and J. M. Nan. 2011. The preparation and properties of a novel electrolyte of electrochemical double layer capacitors based on $LiPF_6$ and acetamide. *Electrochimica Acta* 58:330–335.
489. Lota, G., M. Walkowiak, G. Schroeder, and B. Gierczyk. 2013. Electrochemical behavior of tripodand-type silanes as electrolyte solvents for electrochemical capacitors. *International Journal of Electrochemical Science* 8:9222–9230.
490. Laheäär, A., A. Jänes, and E. Lust. 2012. Lithium bis(oxalato)borate as an electrolyte for micromesoporous carbide-derived carbon based supercapacitors. *Journal of Electroanalytical Chemistry* 669:67–72.
491. Ohta, T., I. T. Kim, M. Egashira, N. Yoshimoto, and M. Morita. 2012. Effects of electrolyte composition on the electrochemical activation of alkali-treated soft carbon as an electric double-layer capacitor electrode. *Journal of Power Sources* 198:408–415.
492. Chen, W., R. B. Rakhi, and H. N. Alshareef. 2012. High energy density supercapacitors using macroporous kitchen sponges. *Journal of Materials Chemistry* 22:14394–14402.
493. Laheäär, A., A. Jänes, and E. Lust. 2012. $NaClO_4$ and $NaPF_6$ as potential non-aqueous electrolyte salts for electrical double layer capacitor application. *Electrochimica Acta* 82:309–313.
494. Väli, R., A. Laheäär, A. Jänes, and E. Lust. 2014. Characteristics of non-aqueous quaternary solvent mixture and Na-salts based supercapacitor electrolytes in a wide temperature range. *Electrochimica Acta* 121:294–300.
495. Chandrasekaran, R., M. Koh, A. Yamauchi, and M. Ishikawa. 2010. Electrochemical cell studies based on non-aqueous magnesium electrolyte for electric double layer capacitor applications. *Journal of Power Sources* 195:662–666.
496. Laheäär, A., A. Jänes, and E. Lust. 2014. Cesium carborane as an unconventional non-aqueous electrolyte salt for electrochemical capacitors. *Electrochimica Acta* 125:482–487.
497. Rogers, R. D., and G. A. Voth. 2007. Ionic liquids. *Accounts of Chemical Research* 40:1077–1078.
498. Armand, M., F. Endres, D. R. MacFarlane, H. Ohno, and B. Scrosati. 2009. Ionic-liquid materials for the electrochemical challenges of the future. *Nature Materials* 8:621–629.
499. Del Pópolo, M. G., and G. A. Voth. 2004. On the structure and dynamics of ionic liquids. *Journal of Physical Chemistry B* 108:1744–1752.
500. Marsh, K. N., J. A. Boxall, and R. Lichtenthaler. 2004. Room temperature ionic liquids and their mixtures—A review. *Fluid Phase Equilibria* 219:93–98.
501. Brandt, A., S. Pohlmann, A. Varzi, A. Balducci, and S. Passerini. 2013. Ionic liquids in supercapacitors. *MRS Bulletin* 38:554–559.
502. Plechkova, N. V., and K. R. Seddon. 2008. Applications of ionic liquids in the chemical industry. *Chemical Society Reviews* 37:123–150.
503. Wasserscheid, P., and T. Welton. 2002. *Ionic Liquids in Synthesis*. Weinheim, Germany: Wiley-VCH Verlag GmbH & Co. KGaA.
504. Freemantle, M. 1998. Designer solvents – Ionic liquids may boost clean technology development. *Chemical & Engineering News Archive* 76:32–37.
505. Lewandowski, A., and M. Galinski. 2007. Practical and theoretical limits for electrochemical double-layer capacitors. *Journal of Power Sources* 173:822–828.
506. Lewandowski, A., A. Olejniczak, M. Galinski, and I. Stepniak. 2010. Performance of carbon–carbon supercapacitors based on organic, aqueous and ionic liquid electrolytes. *Journal of Power Sources* 195:5814–5819.
507. Borges, R. S., A. L. M. Reddy, M. T. F. Rodrigues et al. 2013. Supercapacitor operating at 200 degrees Celsius. *Scientific Reports* 3:2572.

508. Balducci, A., R. Dugas, P. L. Taberna et al. 2007. High temperature carbon–carbon supercapacitor using ionic liquid as electrolyte. *Journal of Power Sources* 165:922–927.
509. Fletcher, S. I., F. B. Sillars, R. C. Carter et al. 2010. The effects of temperature on the performance of electrochemical double layer capacitors. *Journal of Power Sources* 195:7484–7488.
510. Huddleston, J. G., A. E. Visser, W. M. Reichert, H. D. Willauer, G. A. Broker, and R. D. Rogers. 2001. Characterization and comparison of hydrophilic and hydrophobic room temperature ionic liquids incorporating the imidazolium cation. *Green Chemistry* 3:156–164.
511. Zhou, Z. B., H. Matsumoto, and K. Tatsumi. 2004. Low-melting, low-viscous, hydrophobic ionic liquids: 1-alkyl (alkyl ether)-3-methylimidazolium perfluoroalkyltrifluoroborate. *Chemistry – A European Journal* 10:6581–6591.
512. Chen, Y., X. Zhang, D. Zhang, P. Yu, and Y. Ma. 2011. High performance supercapacitors based on reduced graphene oxide in aqueous and ionic liquid electrolytes. *Carbon* 49:573–580.
513. Kurig, H., A. Jänes, and E. Lust. 2010. Electrochemical characteristics of carbide-derived carbon vertical bar 1-ethyl-3-methylimidazolium tetrafluoroborate supercapacitor cells. *Journal of the Electrochemical Society* 157:A272–A279.
514. Xu, B., F. Wu, R. J. Chen et al. 2008. Highly mesoporous and high surface area carbon: A high capacitance electrode material for EDLCs with various electrolytes. *Electrochemistry Communications* 10:795–797.
515. Frackowiak, E. 2007. Carbon materials for supercapacitor application. *Physical Chemistry Chemical Physics* 9:1774–1785.
516. Merlet, C., B. Rotenberg, P. A. Madden et al. 2012. On the molecular origin of supercapacitance in nanoporous carbon electrodes. *Nature Materials* 11:306–310.
517. Feng, G., J. Huang, B. G. Sumpter, V. Meunier, and R. Qiao. 2011. A "counter-charge layer in generalized solvents" framework for electrical double layers in neat and hybrid ionic liquid electrolytes. *Physical Chemistry Chemical Physics* 13:14723–14734.
518. Fedorov, M. V., and A. A. Kornyshev. 2014. Ionic liquids at electrified interfaces. *Chemical Reviews* 114:2978–3036.
519. Merlet, C., B. Rotenberg, P. A. Madden, and M. Salanne. 2013. Computer simulations of ionic liquids at electrochemical interfaces. *Physical Chemistry Chemical Physics* 15:15781–15792.
520. Burt, R., G. Birkett, and X. S. Zhao. 2014. A review of molecular modelling of electric double layer capacitors. *Physical Chemistry Chemical Physics* 16:6519–6538.
521. Feng, G., S. Li, V. Presser, and P. T. Cummings. 2013. Molecular insights into carbon supercapacitors based on room-temperature ionic liquids. *Journal of Physical Chemistry Letters* 4:3367–3376.
522. Baldelli, S. 2008. Surface structure at the ionic liquid-electrified metal interface. *Accounts of Chemical Research* 41:421–431.
523. Zhou, H., M. Rouha, G. Feng et al. 2012. Nanoscale perturbations of room temperature ionic liquid structure at charged and uncharged interfaces. *ACS Nano* 6:9818–9827.
524. Mezger, M., H. Schröder, H. Reichert et al. 2008. Molecular layering of fluorinated ionic liquids at a charged sapphire (0001) surface. *Science* 322:424–428.
525. Feng, G. A., R. Qiao, J. S. Huang, S. Dai, B. G. Sumpter, and V. Meunier. 2011. The importance of ion size and electrode curvature on electrical double layers in ionic liquids. *Physical Chemistry Chemical Physics* 13:1152–1161.
526. Wang, S., S. Li, Z. Cao, and T. Y. Yan. 2010. Molecular dynamic simulations of ionic liquids at graphite surface. *Journal of Physical Chemistry C* 114:990–995.
527. Feng, G., J. S. Zhang, and R. Qiao. 2009. Microstructure and capacitance of the electrical double layers at the interface of ionic. *Journal of Physical Chemistry C* 113:4549–4559.

528. Jiang, D. E., D. Meng, and J. Wu. 2011. Density functional theory for differential capacitance of planar electric double layers in ionic liquids. *Chemical Physics Letters* 504:153–158.
529. Wu, J. Z., T. Jiang, D. E. Jiang, Z. H. Jin, and D. Henderson. 2011. A classical density functional theory for interfacial layering of ionic liquids. *Soft Matter* 7:11222–11231.
530. Henderson, D., and J. Z. Wu. 2012. Electrochemical properties of the double layer of an ionic liquid using a dimer model electrolyte and density functional theory. *Journal of Physical Chemistry B* 116:2520–2525.
531. Pinilla, C., M. G. Del Pópolo, J. Kohanoff, and R. M. Lynden-Bell. 2007. Polarization relaxation in an ionic liquid confined between electrified walls. *Journal of Physical Chemistry B* 111:4877–4884.
532. Kornyshev, A. A. 2007. Double-layer in ionic liquids: Paradigm change? *Journal of Physical Chemistry B* 111:5545–5557.
533. Fedorov, M. V., and A. A. Kornyshev. 2008. Towards understanding the structure and capacitance of electrical double layer in ionic liquids. *Electrochimica Acta* 53:6835–6840.
534. Fedorov, M. V., and A. A. Kornyshev. 2008. Ionic liquid near a charged wall structure and capacitance of electrical double layer. *Journal of Physical Chemistry B* 112:11868–11872.
535. Smith, A. M., K. R. J. Lovelock, N. N. Gosvami et al. 2013. Monolayer to bilayer structural transition in confined pyrrolidinium-based ionic liquids. *Journal of Physical Chemistry Letters* 4:378–382.
536. Smith, A. M., K. R. J. Lovelock, and S. Perkin. 2013. Monolayer and bilayer structures in ionic liquids and their mixtures confined to nano-films. *Faraday Discussions* 167:279–292.
537. Perkin, S., L. Crowhurst, H. Niedermeyer, T. Welton, A. M. Smith, and N. N. Gosvami. 2011. Self-assembly in the electrical double layer of ionic liquids. *Chemical Communications* 47:6572–6574.
538. Perkin, S. 2012. Ionic liquids in confined geometries. *Physical Chemistry Chemical Physics* 14:5052–5062.
539. Black, J. M., D. Walters, A. Labuda et al. 2013. Bias-dependent molecular-level structure of electrical double layer in ionic liquid on graphite. *Nano Letters* 13:5954–5960.
540. Lockett, V., R. Sedev, J. Ralston, M. Horne, and T. Rodopoulos. 2008. Differential capacitance of the electrical double layer in imidazolium-based ionic liquids: Influence of potential, cation size, and temperature. *Journal of Physical Chemistry C* 112:7486–7495.
541. Alam, M. T., M. M. Islam, T. Okajima, and T. Ohsaka. 2007. Measurements of differential capacitance in room temperature ionic liquid at mercury, glassy carbon and gold electrode interfaces. *Electrochemistry Communications* 9:2370–2374.
542. Kornyshev, A. A. 2013. The simplest model of charge storage in single file metallic nanopores. *Faraday Discussions* 164:117–133.
543. Kilic, M. Z., M. Z. Bazant, and A. Ajdari. 2007. Steric effects in the dynamics of electrolytes at large applied voltages. I. Double-layer charging. *Physical Review E* 75:021502.
544. Georgi, N., A. A. Kornyshev, and M. V. Fedorov. 2010. The anatomy of the double layer and capacitance in ionic liquids with anisotropic ions: Electrostriction vs. lattice saturation. *Journal of Electroanalytical Chemistry* 649:261–267.
545. Fedorov, M. V., N. Georgi, and A. A. Kornyshev. 2010. Double layer in ionic liquids: The nature of the camel shape of capacitance. *Electrochemistry Communications* 12:296–299.
546. Henderson, D., S. Lamperski, Z. H. Jin, and J. Z. Wu. 2011. Density functional study of the electric double layer formed by a high density electrolyte. *Journal of Physical Chemistry B* 115:12911–12914.

547. Breitsprecher, K., P. Košovan, and C. Holm. 2014. Coarse-grained simulations of an ionic liquid-based capacitor: I. Density, ion size, and valency effects. *Journal of Physics: Condensed Matter* 26:284108.
548. Breitsprecher, K., P. Košovan, and C. Holm. 2014. Coarse-grained simulations of an ionic liquid-based capacitor: II. Asymmetry in ion shape and charge localization. *Journal of Physics Condensed Matter* 26:284114.
549. Merlet, C., M. Salanne, B. Rotenberg, and P. A. Madden. 2011. Imidazolium ionic liquid interfaces with vapor and graphite: Interfacial tension and capacitance from coarse-grained molecular simulations. *Journal of Physical Chemistry C* 115:16613–16618.
550. Bedrov, D., J. Vatamanu, and Z. Z. Hu. 2015. Ionic liquids at charged surfaces: Insight from molecular simulations. *Journal of Non-Crystalline Solids* 407:339–348.
551. Vatamanu, J., L. L. Cao, O. Borodin, D. Bedrov, and G. D. Smith. 2011. On the influence of surface topography on the electric double layer structure and differential capacitance of graphite/ionic liquid interfaces. *Journal of Physical Chemistry Letters* 2:2267–2272.
552. Li, S., K. L. Van Aken, J. K. McDonough, G. Feng, Y. Gogotsi, and P. T. Cummings. 2014. The electrical double layer of dicationic ionic liquids at onion-like carbon surface. *Journal of Physical Chemistry C* 118:3901–3909.
553. Costa, R., C. M. Pereira, and F. Silva. 2010. Double layer in room temperature ionic liquids: Influence of temperature and ionic size on the differential capacitance and electrocapillary curves. *Physical Chemistry Chemical Physics* 12:11125–11132.
554. Alam, M. T., J. Masud, M. M. Islam, T. Okajima, and T. Ohsaka. 2011. Differential capacitance at Au(111) in 1-alkyl-3-methylimidazolium tetrafluoroborate based room-temperature ionic liquids. *Journal of Physical Chemistry C* 115:19797–19804.
555. Su, Y. Z., J. W. Yan, M. G. Li, M. Zhang, and B. W. Mao. 2013. Electric double layer of Au(100)/imidazolium-based ionic liquids interface: Effect of cation size. *Journal of Physical Chemistry C* 117:205–212.
556. Alam, M. T., M. M. Islam, T. Okajima, and T. Ohsaka. 2007. Measurements of differential capacitance at mercury/room-temperature ionic liquids interfaces. *Journal of Physical Chemistry C* 111:18326–18333.
557. Vatamanu, J., O. Borodin, D. Bedrov, and G. D. Smith. 2012. Molecular dynamics simulation study of the interfacial structure and differential capacitance of alkylimidazolium bis (trifluoromethanesulfonyl) imide [$C_n$mim][TFSI] ionic liquids at graphite electrodes. *Journal of Physical Chemistry C* 116:7940–7951.
558. Lockett, V., M. Horne, R. Sedev, T. Rodopoulos, and J. Ralston. 2010. Differential capacitance of the double layer at the electrode/ionic liquids interface. *Physical Chemistry Chemical Physics* 12:12499–12512.
559. Hu, Z. Z., J. Vatamanu, O. Borodin, and D. Bedrov. 2013. A molecular dynamics simulation study of the electric double layer and capacitance of [BMIM][PF_6] and [BMIM][BF_4] room temperature ionic liquids near charged surfaces. *Physical Chemistry Chemical Physics* 15:14234–14247.
560. Lauw, Y., M. D. Horne, T. Rodopoulos, A. Nelson, and F. A. M. Leermakers. 2010. Electrical double-layer capacitance in room temperature ionic liquids: Ion-size and specific adsorption effects. *Journal of Physical Chemistry B* 114:11149–11154.
561. Si, X. J., S. Li, Y. L. Wang, S. H. Ye, and T. Y. Yan. 2012. Effects of specific adsorption on the differential capacitance of imidazolium-based ionic liquid electrolytes. *ChemPhysChem* 13:1671–1676.
562. Li, S., G. Feng, and P. T. Cummings. 2014. Interfaces of dicationic ionic liquids and graphene: A molecular dynamics simulation study. *Journal of Physics: Condensed Matter* 26:284106.
563. Silva, F., C. Gomes, M. Figueiredo, R. Costa, A. Martins, and C. M. Pereira. 2008. The electrical double layer at the [BMIM][PF6] ionic liquid/electrode interface – Effect of temperature on the differential capacitance. *Journal of Electroanalytical Chemistry* 622:153–160.

564. Reszko-Zygmunt, J., S. Sokołowski, D. Henderson, and D. Boda. 2005. Temperature dependence of the double layer capacitance for the restricted primitive model of an electrolyte solution from a density functional approach. *The Journal of Chemical Physics* 122:084504.
565. Vatamanu, J., O. Borodin, and G. D. Smith. 2010. Molecular insights into the potential and temperature dependences of the differential capacitance of a room-temperature ionic liquid at graphite electrodes. *Journal of the American Chemical Society* 132:14825–14833.
566. Wakai, C., A. Oleinikova, M. Ott, and H. Weingärtner. 2005. How polar are ionic liquids? Determination of the static dielectric constant of an imidazolium-based ionic liquid by microwave dielectric spectroscopy. *Journal of Physical Chemistry B* 109:17028–17030.
567. Feng, G., D. E. Jiang, and P. T. Cummings. 2012. Curvature effect on the capacitance of electric double layers at ionic liquid/onion-like carbon interfaces. *Journal of Chemical Theory and Computation* 8:1058–1063.
568. Paek, E., A. J. Pak, and G. S. Hwang. 2013. Curvature effects on the interfacial capacitance of carbon nanotubes in an ionic liquid. *Journal of Physical Chemistry C* 117:23539–23546.
569. Xing, L. D., J. Vatamanu, G. D. Smith, and D. Bedrov. 2012. Nanopatterning of electrode surfaces as a potential route to improve the energy density of electric double-layer capacitors: Insight from molecular simulations. *Journal of Physical Chemistry Letters* 3:1124–1129.
570. Drüschler, M., N. Borisenko, J. Wallauer et al. 2012. New insights into the interface between a single-crystalline metal electrode and an extremely pure ionic liquid: Slow interfacial processes and the influence of temperature on interfacial dynamics. *Physical Chemistry Chemical Physics* 14:5090–5099.
571. Liu, X. H., Y. N. Han, and T. Y. Yan. 2014. Temperature effects on the capacitance of an imidazolium-based ionic liquid on a graphite electrode: A molecular dynamics simulation. *ChemPhysChem* 15:2503–2509.
572. Vatamanu, J., O. Borodin, and G. D. Smith. 2011. Molecular simulations of the electric double layer structure, differential capacitance, and charging kinetics for *N*-methyl-*N*-propylpyrrolidinium bis (fluorosulfonyl) imide at graphite electrodes. *Journal of Physical Chemistry B* 115:3073–3084.
573. Kislenko, S. A., R. H. Amirov, and I. S. Samoylov. 2010. Influence of temperature on the structure and dynamics of the [BMIM][PF_6] ionic liquid/graphite interface. *Physical Chemistry Chemical Physics* 12:11245–11250.
574. Chen, R. J., Z. Y. He, L. Li, F. Wu, B. Xu, and M. Xie. 2012. Pore size effect of carbon electrodes on the electrochemical double-layer capacitance in LiTFSI/2-oxazolidinone complex electrolyte. *Journal of Physical Chemistry C* 116:2594–2599.
575. Lazzari, M., M. Mastragostino, A. G. Pandolfo, V. Ruiz, and F. Soavi. 2011. Role of carbon porosity and ion size in the development of ionic liquid based supercapacitors. *Journal of the Electrochemical Society* 158:A22–A25.
576. Pohlmann, S., B. Lobato, T. A. Centeno, and A. Balducci. 2013. The influence of pore size and surface area of activated carbons on the performance of ionic liquid based supercapacitors. *Physical Chemistry Chemical Physics* 15:17287–17294.
577. Ania, C. O., J. Pernak, F. Stefaniak, E. Raymundo-Piñero, and F. Béguin. 2006. Solvent-free ionic liquids as *in situ* probes for assessing the effect of ion size on the performance of electrical double layer capacitors. *Carbon* 44:3126–3130.
578. Largeot, C., P. L. Taberna, Y. Gogotsi, and P. Simon. 2011. Microporous carbon-based electrical double layer capacitor operating at high temperature in ionic liquid electrolyte. *Electrochemical and Solid-State Letters* 14:A174–A176.
579. Ania, C. O., J. Pernak, F. Stefaniak, E. Raymundo-Piñero, and F. Béguin. 2009. Polarization-induced distortion of ions in the pores of carbon electrodes for electrochemical capacitors. *Carbon* 47:3158–3166.

580. Sillars, F. B., S. I. Fletcher, M. Mirzaeian, and P. J. Hall. 2011. Effect of activated carbon xerogel pore size on the capacitance performance of ionic liquid electrolytes. *Energy & Environmental Science* 4:695–706.
581. Jiang, D. E., Z. Jin, and J. Wu. 2011. Oscillation of capacitance inside nanopores. *Nano Letters* 11:5373–5377.
582. Xing, L. D., J. Vatamanu, O. Borodin, and D. Bedrov. 2013. On the atomistic nature of capacitance enhancement generated by ionic liquid electrolyte confined in subnanometer pores. *Journal of Physical Chemistry Letters* 4:132–140.
583. Rajput, N. N., J. Monk, R. Singh, and F. R. Hung. 2012. On the influence of pore size and pore loading on structural and dynamical heterogeneities of an ionic liquid confined in a slit nanopore. *Journal of Physical Chemistry C* 116:5169–5181.
584. Feng, G., S. Li, J. S. Atchison, V. Presser, and P. T. Cummings. 2013. Molecular insights into carbon nanotube supercapacitors: Capacitance independent of voltage and temperature. *Journal of Physical Chemistry C* 117:9178–9186.
585. Jover, J. F., R. Lugo, H. Toulhoat, P. Simon, and T. de Bruin. 2014. Screening methodology for the efficient pairing of ionic liquids and carbonaceous electrodes applied to electric energy storage. *Journal of Physical Chemistry C* 118:864–872.
586. Shim, Y., and H. J. Kim. 2010. Nanoporous carbon supercapacitors in an ionic liquid: A computer simulation study. *ACS Nano* 4:2345–2355.
587. Largeot, C., C. Portet, J. Chmiola, P. L. Taberna, Y. Gogotsi, and P. Simon. 2008. Relation between the ion size and pore size for an electric double-layer capacitor. *Journal of the American Chemical Society* 130:2730–2731.
588. Feng, G., and P. T. Cummings. 2011. Supercapacitor capacitance exhibits oscillatory behavior as a function of nanopore size. *Journal of Physical Chemistry Letters* 2:2859–2864.
589. Wu, P., J. S. Huang, V. Meunier, B. G. Sumpter, and R. Qiao. 2011. Complex capacitance scaling in ionic liquids-filled nanopores. *ACS Nano* 5:9044–9051.
590. Vatamanu, J., Z. Hu, D. Bedrov, C. Perez, and Y. Gogotsi. 2013. Increasing energy storage in electrochemical capacitors with ionic liquid electrolytes and nanostructured carbon electrodes. *Journal of Physical Chemistry Letters* 4:2829–2837.
591. Shilina, Y., M. D. Levi, V. Dargel et al. 2013. Ion size to pore width ratio as a factor that determines the electrochemical stability window of activated carbon electrodes. *Journal of the Electrochemical Society* 160:A629–A635.
592. Rajput, N. N., J. Monk, and F. R. Hung. 2012. Structure and dynamics of an ionic liquid confined inside a charged slit graphitic nanopore. *Journal of Physical Chemistry C* 116:14504–14513.
593. Singh, R., J. Monk, and F. R. Hung. 2010. A computational study of the behavior of the ionic liquid [BMIM][PF_6] confined inside multiwalled carbon nanotubes. *Journal of Physical Chemistry C* 114:15478–15485.
594. Pak, A. J., E. Paek, and G. S. Hwang. 2014. Impact of graphene edges on enhancing the performance of electrochemical double layer capacitors. *Journal of Physical Chemistry C* 118:21770–21777.
595. Paek, E., A. J. Pak, and G. S. Hwang. 2014. Large capacitance enhancement induced by metal-doping in graphene-based supercapacitors: A first-principles-based assessment. *ACS Applied Materials & Interfaces* 6:12168–12176.
596. Paek, E., A. J. Pak, K. E. Kweon, and G. S. Hwang. 2013. On the origin of the enhanced supercapacitor performance of nitrogen-doped graphene. *Journal of Physical Chemistry C* 117:5610–5616.
597. Pak, A. J., E. Paekw, and G. S. Hwang. 2013. Relative contributions of quantum and double layer capacitance to the supercapacitor performance of carbon nanotubes in an ionic liquid. *Physical Chemistry Chemical Physics* 15:19741–19747.
598. Kondrat, S., P. Wu, R. Qiao, and A. A. Kornyshev. 2014. Accelerating charging dynamics in subnanometre pores. *Nature Materials* 13:387–393.

599. Huang, H. C., C. W. Huang, C. T. Hsieh, and H. Teng. 2014. Electric double layer capacitors of high volumetric energy based on ionic liquids and hierarchical-pore carbon. *Journal of Materials Chemistry A* 2:14963–14972.
600. Zhou, J., W. Xing, S. P. Zhuo, and Y. Zhao. 2011. Capacitive performance of ordered mesoporous carbons with tunable porous texture in ionic liquid electrolytes. *Solid State Sciences* 13:2000–2006.
601. Richey, F. W., B. Dyatkin, Y. Gogotsi, and Y. A. Elabd. 2013. Ion dynamics in porous carbon electrodes in supercapacitors using *in situ* infrared spectroelectrochemistry. *Journal of the American Chemical Society* 135:12818–12826.
602. Richey, F. W., C. Tran, V. Kalra, and Y. A. Elabd. 2014. Ionic liquid dynamics in nanoporous carbon nanofibers in supercapacitors measured with in operando infrared spectroelectrochernistry. *Journal of Physical Chemistry C* 118:21846–21855.
603. Péan, C., C. Merlet, B. Rotenberg et al. 2014. On the dynamics of charging in nanoporous carbon-based supercapacitors. *ACS Nano* 8:1576–1583.
604. Jiang, J., D. P. Cao, D. E. Jiang, and J. Z. Wu. 2014. Kinetic charging inversion in ionic liquid electric double layers. *Journal of Physical Chemistry Letters* 5:2195–2200.
605. Van Aken, K. L., J. K. McDonough, S. Li et al. 2014. Effect of cation on diffusion coefficient of ionic liquids at onion-like carbon electrodes. *Journal of Physics Condensed Matter* 26:284104.
606. Pinkert, K., M. Oschatz, L. Borchardt et al. 2014. Role of surface functional groups in ordered mesoporous carbide-derived carbon/ionic liquid electrolyte double-layer capacitor interfaces. *ACS Applied Materials & Interfaces* 6:2922–2928.
607. Tamilarasan, P., and S. Ramaprabhu. 2014. Ionic liquid-functionalized partially exfoliated multiwalled carbon nanotubes for high-performance supercapacitors. *Journal of Materials Chemistry A* 2:14054–14063.
608. Cho, S. D., J. K. Im, H. K. Kim, H. S. Kim, and H. S. Park. 2012. Functionalization of reduced graphene oxides by redox-active ionic liquids for energy storage. *Chemical Communications* 48:6381–6383.
609. Trigueiro, J. P. C., R. L. Lavall, and G. G. Silva. 2014. Supercapacitors based on modified graphene electrodes with poly(ionic liquid). *Journal of Power Sources* 256:264–273.
610. Mao, L., Y. Li, C. Y. Chi, H. S. O. Chan, and J. S. Wu. 2014. Conjugated polyfluorene imidazolium ionic liquids intercalated reduced graphene oxide for high performance supercapacitor electrodes. *Nano Energy* 6:119–128.
611. Kim, T. Y., H. W. Lee, M. Stoller et al. 2011. High-performance supercapacitors based on poly(ionic liquid)-modified graphene electrodes. *ACS Nano* 5:436–442.
612. Kunze, M., E. Paillard, S. Jeong et al. 2011. Inhibition of self-aggregation in ionic liquid electrolytes for high-energy electrochemical devices. *Journal of Physical Chemistry C* 115:19431–19436.
613. Kunze, M., S. Jeong, E. Paillard, M. Schönhoff, M. Winter, and S. Passerini. 2011. New insights to self-aggregation in ionic liquid electrolytes for high-energy electrochemical devices. *Advanced Energy Materials* 1:274–281.
614. Anouti, M., P. Porion, C. Brigouleix, H. Galiano, and D. Lemordant. 2010. Transport properties in two pyrrolidinium-based protic ionic liquids as determined by conductivity, viscosity and NMR self-diffusion measurements. *Fluid Phase Equilibria* 299:229–237.
615. Ue, M., M. Takeda, T. Takahashi, and M. Takehara. 2002. Ionic liquids with low melting points and their application to double-layer capacitor electrolytes. *Electrochemical and Solid-State Letters* 5:A119–A121.
616. Ue, M., M. Takeda, A. Toriumi, A. Kominato, R. Hagiwara, and Y. Ito. 2003. Application of low-viscosity ionic liquid to the electrolyte of double-layer capacitors. *Journal of the Electrochemical Society* 150:A499–A502.

617. Sun, G. H., K. X. Li, and C. G. Sun. 2006. Application of 1-ethyl-3-methylimidazolium thiocyanate to the electrolyte of electrochemical double layer capacitors. *Journal of Power Sources* 162:1444–1450.
618. Handa, N., T. Sugimoto, M. Yamagata, M. Kikuta, M. Kono, and M. Ishikawa. 2008. A neat ionic liquid electrolyte based on FSI anion for electric double layer capacitor. *Journal of Power Sources* 185:1585–1588.
619. Matsumoto, K., and R. Hagiwara. 2010. Electrochemical properties of the ionic liquid 1-ethyl-3-methylimidazolium difluorophosphate as an electrolyte for electric double-layer capacitors. *Journal of the Electrochemical Society* 157:A578–A581.
620. Pandey, G. P., and S. A. Hashmi. 2013. Studies on electrical double layer capacitor with a low-viscosity ionic liquid 1-ethyl-3-methylimidazolium tetracyanoborate as electrolyte. *Bulletin of Materials Science* 36:729–733.
621. Kurig, H., M. Vestli, K. Tõnurist, A. Jänes, and E. Lust. 2012. Influence of room temperature ionic liquid anion chemical composition and electrical charge delocalization on the supercapacitor properties. *Journal of the Electrochemical Society* 159:A944–A951.
622. Shi, M., S. Kou, and X. Yan. 2014. Engineering the electrochemical capacitive properties of graphene sheets in ionic-liquid electrolytes by correct selection of anions. *ChemSusChem* 7:3053–3062.
623. Nagao, Y., Y. Nakayama, H. Oda, and M. Ishikawa. 2007. Activation of an ionic liquid electrolyte for electric double layer capacitors by addition of $BaTiO_3$ to carbon electrodes. *Journal of Power Sources* 166:595–598.
624. Murayama, I., N. Yoshimoto, M. Egashira, M. Morita, Y. Kobayashi, and M. Ishikawa. 2005. Characteristics of electric double layer capacitors with an ionic liquid electrolyte containing Li ion. *Electrochemistry* 73:600–602.
625. Matsumoto, H., H. Kageyama, and Y. Miyazaki. 2003. Effect of ionic additives on the limiting cathodic potential of EMI-based room temperature ionic liquids. *Electrochemistry* 71:1058–1060.
626. Fukuda, Y., R. Tanaka, and M. Tshikawa. 2007. Beneficial effects of a Li salt on electrode behavior in an ionic liquid for electric double layer capacitors. *Electrochemistry* 75:589–591.
627. Borges, R. S., H. Ribeiro, R. L. Lavall, and G. G. Silva. 2012. Temperature stable supercapacitors based on ionic liquid and mixed functionalized carbon nanomaterials. *Journal of Solid State Electrochemistry* 16:3573–3580.
628. Bettini, L. G., M. Galluzzi, A. Podestà, P. Milani, and P. Piseri. 2013. Planar thin film supercapacitor based on cluster-assembled nanostructured carbon and ionic liquid electrolyte. *Carbon* 59:212–220.
629. Lazzari, M., F. Soavi, and M. Mastragostino. 2008. High voltage, asymmetric EDLCs based on xerogel carbon and hydrophobic IL electrolytes. *Journal of Power Sources* 178:490–496.
630. Lazzari, M., F. Soavi, and M. Mastragostino. 2009. Dynamic pulse power and energy of ionic-liquid-based supercapacitor for HEV application. *Journal of the Electrochemical Society* 156:A661–A666.
631. Arbizzani, C., M. Biso, D. Cericola, M. Lazzari, F. Soavi, and M. Mastragostino. 2008. Safe, high-energy supercapacitors based on solvent-free ionic liquid electrolytes. *Journal of Power Sources* 185:1575–1579.
632. Sato, T., G. Masuda, and K. Takagi. 2004. Electrochemical properties of novel ionic liquids for electric double layer capacitor applications. *Electrochimica Acta* 49:3603–3611.
633. Kim, Y. J., Y. Matsuzawa, S. Ozaki et al. 2005. High energy-density capacitor based on ammonium salt type ionic liquids and their mixing effect by propylene carbonate. *Journal of the Electrochemical Society* 152:A710–A715.

634. Senda, A., K. Matsumoto, T. Nohira, and R. Hagiwara. 2010. Effects of the cationic structures of fluorohydrogenate ionic liquid electrolytes on the electric double layer capacitance. *Journal of Power Sources* 195:4414–4417.
635. Anouti, M., L. Timperman, M. el hilali, A. Boisset, and H. Galiano. 2012. Sulfonium bis(trifluorosulfonimide) plastic crystal ionic liquid as an electrolyte at elevated temperature for high-energy supercapacitors. *Journal of Physical Chemistry C* 116:9412–9418.
636. Zhou, Z. B., H. Matsumoto, and K. Tatsumi. 2006. Cyclic quaternary ammonium ionic liquids with perfluoroalkyltrifluoroborates: Synthesis, characterization, and properties. *Chemistry* 12:2196–2212.
637. Mastragostino, M., and F. Soavi. 2009. Electrochemical capacitors ionic liquid electrolytes. In *Encyclopedia of Electrochemical Power Sources*, C. K. Dyer, P. T. Moseley, Z. Ogumi, D. A. J. Rand, B. Scrosati, and J. Garche (Eds), 649–657, Amsterdam, the Netherlands: Elsevier B.V.
638. Pohlmann, S., T. Olyschläger, P. Goodrich, J. Alvarez Vicente, J. Jacquemin, and A. Balducci. 2014. Azepanium-based ionic liquids as green electrolytes for high voltage supercapacitors. *Journal of Power Sources* 273:931–936.
639. Sillars, F. B., S. I. Fletcher, M. Mirzaeian, and P. J. Hall. 2012. Variation of electrochemical capacitor performance with room temperature ionic liquid electrolyte viscosity and ion size. *Physical Chemistry Chemical Physics* 14:6094–6100.
640. Rennie, A. J. R., N. Sanchez-Ramirez, R. M. Torresi, and P. J. Hall. 2013. Ether-bond-containing ionic liquids as supercapacitor electrolytes. *Journal of Physical Chemistry Letters* 4:2970–2974.
641. Cho, W. J., C. G. Yeom, B. C. Kim, K. M. Kim, J. M. Ko, and K. H. Yu. 2013. Supercapacitive properties of activated carbon electrode in organic electrolytes containing single- and double-cationic liquid salts. *Electrochimica Acta* 89:807–813.
642. Ye, Y. S., J. Rick, and B. J. Hwang. 2013. Ionic liquid polymer electrolytes. *Journal of Materials Chemistry A* 1:2719–2743.
643. Seddon, K. R., A. Stark, and M. J. Torres. 2000. Influence of chloride, water, and organic solvents on the physical properties of ionic liquids. *Pure and Applied Chemistry* 72:2275–2287.
644. Barrosse-Antle, L. E., A. M. Bond, R. G. Compton, A. M. O'Mahony, E. I. Rogers, and D. S. Silvester. 2010. Voltammetry in room temperature ionic liquids: Comparisons and contrasts with conventional electrochemical solvents. *Chemistry – An Asian Journal* 5:202–230.
645. Winoto, H. P., S. Agarwal, J. K. Im, M. Cheong, and J. S. Lee. 2012. Ionic liquids containing 1,1-dicyano-1-acetylmethanide anion as potential electrolytes. *Bulletin of the Korean Chemical Society* 33:2999–3003.
646. Demarconnay, L., E. G. Calvo, L. Timperman et al. 2013. Optimizing the performance of supercapacitors based on carbon electrodes and protic ionic liquids as electrolytes. *Electrochimica Acta* 108:361–368.
647. Timperman, L., P. Skowron, A. Boisset et al. 2012. Triethylammonium bis-(tetrafluoromethylsulfonyl)amide protic ionic liquid as an electrolyte for electrical double-layer capacitors. *Physical Chemistry Chemical Physics* 14:8199–8207.
648. Timperman, L., F. Béguin, E. Frackowiak, and M. Anouti. 2014. Comparative study of two protic ionic liquids as electrolyte for electrical double-layer capacitors. *Journal of the Electrochemical Society* 161:A228–A238.
649. Brandt, A., J. Pires, M. Anouti, and A. Balducci. 2013. An investigation about the cycling stability of supercapacitors containing protic ionic liquids as electrolyte components. *Electrochimica Acta* 108:226–231.
650. Higashiya, S., T. S. Devarajan, M. V. Rane-Fondacaro, and P. Haldar. 2012. Eutectic mixtures of ionic liquids electrolytes for electric double layer capacitors. *ECS Transactions* 41:103–109.

651. Lin, R. Y., P. L. Taberna, S. Fantini et al. 2011. Capacitive energy storage from −50 to 100°C using an ionic liquid electrolyte. *Journal of Physical Chemistry Letters* 2:2396–2401.
652. Kunze, M., S. Jeong, G. B. Appetecchi, M. Schönhoff, M. Winter, and S. Passerini. 2012. Mixtures of ionic liquids for low temperature electrolytes. *Electrochimica Acta* 82:69–74.
653. Huang, P. H., D. Pech, R. Y. Lin et al. 2013. On-chip micro-supercapacitors for operation in a wide temperature range. *Electrochemistry Communications* 36:53–56.
654. Palm, R., H. Kurig, K. Tõnurist, A. Jänes, and E. Lust. 2012. Is the mixture of 1-ethyl-3-methylimidazolium tetrafluoroborate and 1-butyl-3-methylimidazolium tetrafluoroborate applicable as electrolyte in electrical double layer capacitors? *Electrochemistry Communications* 22:203–206.
655. Li, S., G. Feng, P. F. Fulvio et al. 2012. Molecular dynamics simulation study of the capacitive performance of a binary mixture of ionic liquids near an onion-like carbon electrode. *Journal of Physical Chemistry Letters* 3:2465–2469.
656. Mysyk, R., E. Raymundo-Piñero, M. Anouti, D. Lemordant, and F. Béguin. 2010. Pseudo-capacitance of nanoporous carbons in pyrrolidinium-based protic ionic liquids. *Electrochemistry Communications* 12:414–417.
657. Rochefort, D., and A. L. Pont. 2006. Pseudocapacitive behaviour of RuO_2 in a proton exchange ionic liquid. *Electrochemistry Communications* 8:1539–1543.
658. Mayrand-Provencher, L., and D. Rochefort. 2009. Influence of the conductivity and viscosity of protic ionic liquids electrolytes on the pseudocapacitance of RuO_2 electrodes. *Journal of Physical Chemistry C* 113:1632–1639.
659. Mayrand-Provencher, L., S. X. Lin, D. Lazzerini, and D. Rochefort. 2010. Pyridinium-based protic ionic liquids as electrolytes for RuO_2 electrochemical capacitors. *Journal of Power Sources* 195:5114–5121.
660. Richey, F. W., and Y. A. Elabd. 2013. *In situ* spectroscopic measurements of individual cation and anion dynamics in a RuO_2 electrochemical capacitor. *Journal of the Electrochemical Society* 160:A862–A868.
661. Chang, J. K., M. T. Lee, C. W. Cheng, W. T. Tsai, M. J. Deng, and I. W. Sun. 2009. Evaluation of ionic liquid electrolytes for use in manganese oxide supercapacitors. *Electrochemical and Solid-State Letters* 12:A19–A22.
662. Chang, J. K., M. T. Lee, W. T. Tsai, M. J. Deng, and I. W. Sun. 2009. X-ray photoelectron spectroscopy and *in situ* X-ray absorption spectroscopy studies on reversible insertion/desertion of dicyanamide anions into/from manganese oxide in ionic liquid. *Chemistry of Materials* 21:2688–2695.
663. Chang, J. K., M. T. Lee, W. T. Tsai, M. J. Deng, H. F. Cheng, and I. W. Sun. 2009. Pseudocapacitive mechanism of manganese oxide in 1-ethyl-3-methytimidazolium thiocyanate ionic liquid electrolyte studied using X-ray photoelectron spectroscopy. *Langmuir* 25:11955–11960.
664. Lee, M. T., W. T. Tsai, M. J. Deng, H. F. Cheng, I. W. Sun, and J. K. Chang. 2010. Pseudocapacitance of MnO_2 originates from reversible insertion/desertion of thiocyanate anions studied using *in situ* X-ray absorption spectroscopy in ionic liquid electrolyte. *Journal of Power Sources* 195:919–922.
665. Li, Y. S., I. W. Sun, J. K. Chang, C. J. Su, and M. T. Lee. 2012. Doped butylmethylpyrrolidinium–dicyanamide ionic liquid as an electrolyte for MnO_2 supercapacitors. *Journal of Materials Chemistry* 22:6274–6279.
666. Benedetti, T. M., V. R. Gonçales, S. I. C. de Torresi, and R. M. Torresi. 2013. In search of an appropriate ionic liquid as electrolyte for macroporous manganese oxide film electrochemistry. *Journal of Power Sources* 239:1–8.
667. Lee, M. T., Y. S. Li, I. W. Sun, and J. K. Chang. 2014. Pseudocapacitive behavior of manganese oxide in lithium-ion-doped butylmethylpyrrolidinium–dicyanamide ionic liquid investigated using *in situ* X-ray absorption spectroscopy. *Journal of Power Sources* 246:269–276.

668. Lee, M. T., W. T. Tsai, H. F. Cheng, I. W. Sun, and J. K. Chang. 2013. Effect of electrolyte temperature on the pseudo-capacitive behavior of manganese oxide in *N*-butyl-*N*-methylpyrrolidinium–dicyanamide ionic liquid. *Journal of Power Sources* 233:28–33.
669. Ruiz, C. A. C., D. Bélanger, and D. Rochefort. 2013. Electrochemical and spectroelectrochemical evidence of redox transitions involving protons in thin MnO_2 electrodes in protic ionic liquids. *Journal of Physical Chemistry C* 117:20397–20405.
670. Fechler, N., G. A. Tiruye, R. Marcilla, and M. Antonietti. 2014. Vanadium nitride Tuning of pore structure and particle size through salt templating and its influence on supercapacitance in ionic liquid media. *RSC Advances* 4:26981–26989.
671. Kobayashi, N., T. Sakumoto, S. Mori et al. 2013. Investigation on capacitive behaviors of porous Ni electrodes in ionic liquids. *Electrochimica Acta* 105:455–461.
672. Fabregat-Santiago, F., H. Randriamahazaka, A. Zaban, J. Garcia-Cañadas, G. Garcia-Belmonte, and J. Bisquert. 2006. Chemical capacitance of nanoporous-nanocrystalline TiO_2 in a room temperature ionic liquid. *Physical Chemistry Chemical Physics* 8:1827–1833.
673. Shaikh, J. S., R. C. Pawar, R. S. Devan et al. 2011. Synthesis and characterization of Ru doped CuO thin films for supercapacitor based on Bronsted acidic ionic liquid. *Electrochimica Acta* 56:2127–2134.
674. Sun, S. X., J. W. Lang, R. T. Wang, L. B. Kong, X. C. Li, and X. B. Yan. 2014. Identifying pseudocapacitance of Fe_2O_3 in an ionic liquid and its application in asymmetric supercapacitors. *Journal of Materials Chemistry A* 2:14550–14556.
675. Arbizzani, C., F. Soavi, and M. Mastragostino. 2006. A novel galvanostatic polymerization for high specific capacitance poly(3-methylthiophene) in ionic liquid. *Journal of Power Sources* 162:735–737.
676. Chen, W., R. B. Rakhi, and H. N. Alshareef. 2013. Facile synthesis of polyaniline nanotubes using reactive oxide templates for high energy density pseudocapacitors. *Journal of Materials Chemistry A* 1:3315–3324.
677. Wang, K., J. Huang, and Z. Wei. 2010. Conducting polyaniline nanowire arrays for high performance supercapacitors. *Journal of Physical Chemistry C* 114:8062–8067.
678. Zhang, X., D. Zhao, Y. Zhao et al. 2013. High performance asymmetric supercapacitor based on MnO_2 electrode in ionic liquid electrolyte. *Journal of Materials Chemistry A* 1:3706–3712.
679. Balducci, A., U. Bardi, S. Caporali, M. Mastragostino, and F. Soavi. 2004. Ionic liquids for hybrid supercapacitors. *Electrochemistry Communications* 6:566–570.
680. Balducci, A., W. A. Henderson, M. Mastragostino, S. Passerini, P. Simon, and F. Soavi. 2005. Cycling stability of a hybrid activated carbon//poly(3-methylthiophene) supercapacitor with *N*-butyl-*N*-methylpyrrolidinium bis(trifluoromethanesulfonyl) imide ionic liquid as electrolyte. *Electrochimica Acta* 50:2233–2237.
681. Balducci, A., F. Soavi, and M. Mastragostino. 2006. The use of ionic liquids as solvent-free green electrolytes for hybrid supercapacitors. *Applied Physics A – Materials Science & Processing* 82:627–632.
682. Arbizzani, C., S. Beninati, M. Lazzari, F. Soavi, and M. Mastragostino. 2007. Electrode materials for ionic liquid-based supercapacitors. *Journal of Power Sources* 174:648–652.
683. McEwen, A. B., S. F. McDevitt, and V. R. Koch. 1997. Nonaqueous electrolytes for electrochemical capacitors: Imidazolium cations and inorganic fluorides with organic carbonates. *Journal of the Electrochemical Society* 144:L84–L86.
684. Orita, A., K. Kamijima, and M. Yoshida. 2010. Allyl-functionalized ionic liquids as electrolytes for electric double-layer capacitors. *Journal of Power Sources* 195:7471–7479.
685. Lin, R., P. Huang, J. Segalini et al. 2009. Solvent effect on the ion adsorption from ionic liquid electrolyte into sub-nanometer carbon pores. *Electrochimica Acta* 54:7025–7032.

686. Mysyk, R., V. Ruiz, E. Raymundo-Piñero, R. Santamaria, and F. Béguin. 2010. Capacitance evolution of electrochemical capacitors with tailored nanoporous electrodes in pure and dissolved ionic liquids. *Fuel Cells* 10:834–839.
687. Chen, X. Y., Y. Y. He, H. Song, and Z. J. Zhang. 2014. Structure and electrochemical performance of highly nanoporous carbons from benzoate-metal complexes by a template carbonization method for supercapacitor application. *Carbon* 72:410–420.
688. Zhang, Z. J., D. H. Xie, P. Cui, and X. Y. Chen. 2014. Conversion of a zinc salicylate complex into porous carbons through a template carbonization process as a superior electrode material for supercapacitors. *RSC Advances* 4:6664–6671.
689. Chen, X. Y., C. Chen, Z. J. Zhang, and D. H. Xie. 2013. Gelatin-derived nitrogen-doped porous carbon *via* a dual-template carbonization method for high performance supercapacitors. *Journal of Materials Chemistry A* 1:10903–10911.
690. Ghamouss, F., A. Brugere, and J. Jacquemin. 2014. Physicochemical investigation of adiponitrile-based electrolytes for electrical double layer capacitor. *Journal of Physical Chemistry C* 118:14107–14123.
691. Palm, R., H. Kurig, K. Tõnurist, A. Jänes, and E. Lust. 2013. Influence of different organic solvent additives on 1-ethyl-3-methylimidazolium tetrafluoroborate electrolyte based electrical double layer capacitors. *Journal of the Electrochemical Society* 160:A1741–A1745.
692. Liu, W. W., X. B. Yan, J. W. Lang, and Q. J. Xue. 2012. Effects of concentration and temperature of $EMIMBF_4$/acetonitrile electrolyte on the supercapacitive behavior of graphene nanosheets. *Journal of Materials Chemistry* 22:8853–8861.
693. Liu, W. W., X. B. Yan, J. W. Lang, and Q. J. Xue. 2011. Electrochemical behavior of graphene nanosheets in alkylimidazolium tetrafluoroborate ionic liquid electrolytes: Influences of organic solvents and the alkyl chains. *Journal of Materials Chemistry* 21:13205–13212.
694. Kumar, R., H. J. Kim, S. Park, A. Srivastava, and I. K. Oh. 2014. Graphene-wrapped and cobalt oxide-intercalated hybrid for extremely durable super-capacitor with ultrahigh energy and power densities. *Carbon* 79:192–202.
695. Hantel, M. M., A. Platek, T. Kaspar, R. Nesper, A. Wokaun, and R. Kötz. 2013. Investigation of diluted ionic liquid 1-ethyl-3-methyl-imidazolium tetrafluoroborate electrolytes for intercalation-like electrodes used in supercapacitors. *Electrochimica Acta* 110:234–239.
696. Ruiz, V., T. Huynh, S. R. Sivakkumar, and A. G. Pandolfo. 2012. Ionic liquid-solvent mixtures as supercapacitor electrolytes for extreme temperature operation. *RSC Advances* 2:5591–5598.
697. Pohlmann, S., R. S. Kuhnel, T. A. Centeno, and A. Balducci. 2014. The influence of anion–cation combinations on the physicochemical properties of advanced electrolytes for supercapacitors and the capacitance of activated carbons. *ChemElectroChem* 1:1301–1311.
698. Pohlmann, S., and A. Balducci. 2013. A new conducting salt for high voltage propylene carbonate-based electrochemical double layer capacitors. *Electrochimica Acta* 110:221–227.
699. Krause, A., and A. Balducci. 2011. High voltage electrochemical double layer capacitor containing mixtures of ionic liquids and organic carbonate as electrolytes. *Electrochemistry Communications* 13:814–817.
700. Frackowiak, E. 2006. Supercapacitors based on carbon materials and ionic liquids. *Journal of the Brazilian Chemical Society* 17:1074–1082.
701. Frackowiak, E., G. Lota, and J. Pernak. 2005. Room-temperature phosphonium ionic liquids for supercapacitor application. *Applied Physics Letters* 86:1–3.
702. Abdallah, T., D. Lemordant, and B. Claude-Montigny. 2012. Are room temperature ionic liquids able to improve the safety of supercapacitors organic electrolytes without degrading the performances? *Journal of Power Sources* 201:353–359.

703. Brandt, A., C. Ramirez-Castro, M. Anouti, and A. Balducci. 2013. An investigation about the use of mixtures of sulfonium-based ionic liquids and propylene carbonate as electrolytes for supercapacitors. *Journal of Materials Chemistry A* 1:12669–12678.
704. Liu, W. W., X. B. Yan, J. W. Lang, J. B. Pu, and Q. J. Xue. 2013. Supercapacitors based on graphene nanosheets using different non-aqueous electrolytes. *New Journal of Chemistry* 37:2186–2195.
705. Yuyama, K., G. Masuda, H. Yoshida, and T. Sato. 2006. Ionic liquids containing the tetrafluoroborate anion have the best performance and stability for electric double layer capacitor applications. *Journal of Power Sources* 162:1401–1408.
706. Orita, A., K. Kamijima, M. Yoshida, and L. Yang. 2010. Application of sulfonium-, thiophenium-, and thioxonium-based salts as electric double-layer capacitor electrolytes. *Journal of Power Sources* 195:6970–6976.
707. Brandt, A., and A. Balducci. 2014. Theoretical and practical energy limitations of organic and ionic liquid-based electrolytes for high voltage electrochemical double layer capacitors. *Journal of Power Sources* 250:343–351.
708. Lewandowski, A., and A. Olejniczak. 2007. *N*-methyl-*N*-propylpiperidinium bis-(trifluoromethanesulphonyl) imide as an electrolyte for carbon-based double-layer capacitors. *Journal of Power Sources* 172:487–492.
709. Timperman, L., H. Galiano, D. Lemordant, and M. Anouti. 2011. Phosphonium-based protic ionic liquid as electrolyte for carbon-based supercapacitors. *Electrochemistry Communications* 13:1112–1115.
710. Anouti, M., and L. Timperman. 2013. A pyrrolidinium nitrate protic ionic liquid-based electrolyte for very low-temperature electrical double-layer capacitors. *Physical Chemistry Chemical Physics* 15:6539–6548.
711. Xu, W., and C. A. Angell. 2003. Solvent-free electrolytes with aqueous solution-like conductivities. *Science* 302:422–425.
712. Yoshizawa, M., W. Xu, and C. A. Angell. 2003. Ionic liquids by proton transfer: Vapor pressure, conductivity, and the relevance of ΔpK_a from aqueous solutions. *Journal of the American Chemical Society* 125:15411–15419.
713. Coadou, E., L. Timperman, J. Jacquemin, H. Galiano, C. Hardacre, and M. Anouti. 2013. Comparative study on performances of trimethyl-sulfonium and trimethyl-ammonium based ionic liquids in molecular solvents as electrolyte for electrochemical double layer capacitors. *Journal of Physical Chemistry C* 117:10315–10325.
714. Shim, Y., Y. Jung, and H. J. Kim. 2011. Graphene-based supercapacitors: A computer simulation study. *Journal of Physical Chemistry C* 115:23574–23583.
715. Shim, Y., H. J. Kim, and Y. Jung. 2012. Graphene-based supercapacitors in the parallel-plate electrode configuration: Ionic liquids versus organic electrolytes. *Faraday Discussions* 154:249–263.
716. Merlet, C., M. Salanne, B. Rotenberg, and P. A. Madden. 2013. Influence of solvation on the structural and capacitive properties of electrical double layer capacitors. *Electrochimica Acta* 101:262–271.
717. Okada, Y., T. Ito, T. Minamikawa, H. Kamisuki, S. Higai, and K. Shiratsuyu. 2013. Molecular dynamics study of ionic liquids in graphite nanopores. *Electrochemistry* 81:808–810.
718. Francisco, B. E., C. M. Jones, S. H. Lee, and C. R. Stoldt. 2012. Nanostructured all-solid-state supercapacitor based on $Li_2S–P_2S_5$ glass-ceramic electrolyte. *Applied Physics Letters* 100:103902.
719. Ulihin, A. S., Y. G. Mateyshina, and N. F. Uvarov. 2013. All-solid-state asymmetric supercapacitors with solid composite electrolytes. *Solid State Ionics* 251:62–65.
720. Yoon, Y. S., W. I. Cho, J. H. Lim, and D. J. Choi. 2001. Solid-state thin-film supercapacitor with ruthenium oxide and solid electrolyte thin films. *Journal of Power Sources* 101:126–129.

721. Kim, H. K., S. H. Cho, Y. W. Ok, T. Y. Seong, and Y. S. Yoon. 2003. All solid-state rechargeable thin-film microsupercapacitor fabricated with tungsten cosputtered ruthenium oxide electrodes. *Journal of Vacuum Science and Technology B: Microelectronics and Nanometer Structures* 21:949–952.
722. Łatoszyńska, A. A., G. Z. Zukowska, I. A. Rutkowska et al. 2014. Non-aqueous gel polymer electrolyte with phosphoric acid ester and its application for quasi solid-state supercapacitors. *Journal of Power Sources* 274:1147–1154.
723. Fan, L. Q., J. Zhong, J. H. Wu, J. M. Lin, and Y. F. Huang. 2014. Improving the energy density of quasi-solid-state electric double-layer capacitors by introducing redox additives into gel polymer electrolytes. *Journal of Materials Chemistry A* 2:9011–9014.
724. Verma, M. L., M. Minakshi, and N. K. Singh. 2014. Synthesis and characterization of solid polymer electrolyte based on activated carbon for solid state capacitor. *Electrochimica Acta* 137:497–503.
725. Fang, X., and D. Yao. 2013. An overview of solid-like electrolytes for supercapacitors. *Proceedings of the ASME 2013 International Mechanical Engineering Congress and Exposition*, San Diego, California.
726. Agrawal, R. C., and G. P. Pandey. 2008. Solid polymer electrolytes: Materials designing and all-solid-state battery applications: An overview. *Journal of Physics D: Applied Physics* 41:223001.
727. Samui, A. B., and P. Sivaraman. 2010. Solid polymer electrolytes for supercapacitors. In *Polymer Electrolytes: Fundamentals and Applications*, C. Sequeira and D. Santos (Eds), 431–470, Cambridge, UK: Woodhead Publishing Limited.
728. Gao, H., and K. Lian. 2014. Proton-conducting polymer electrolytes and their applications in solid supercapacitors: A review. *RSC Advances* 4:33091–33113.
729. Varshney, P. K., and S. Gupta. 2011. Natural polymer-based electrolytes for electrochemical devices: A review. *Ionics* 17:479–483.
730. Choudhury, N. A., S. Sampath, and A. K. Shukla. 2009. Hydrogel-polymer electrolytes for electrochemical capacitors: An overview. *Energy & Environmental Science* 2:55–67.
731. Duay, J., E. Gillette, R. Liu, and S. B. Lee. 2012. Highly flexible pseudocapacitor based on freestanding heterogeneous MnO_2/conductive polymer nanowire arrays. *Physical Chemistry Chemical Physics* 14:3329–3337.
732. Huang, C. W., C. A. Wu, S. S. Hou, P. L. Kuo, C. T. Hsieh, and H. S. Teng. 2012. Gel electrolyte derived from poly(ethylene glycol) blending poly(acrylonitrile) applicable to roll-to-roll assembly of electric double layer capacitors. *Advanced Functional Materials* 22:4677–4685.
733. Kumar, Y., G. P. Pandey, and S. A. Hashmi. 2012. Gel polymer electrolyte based electrical double layer capacitors: Comparative study with multiwalled carbon nanotubes and activated carbon electrodes. *Journal of Physical Chemistry C* 116:26118–26127.
734. Sudhakar, Y. N., M. Selvakumar, and D. K. Bhat. 2013. $LiClO_4$-doped plasticized chitosan and poly(ethylene glycol) blend as biodegradable polymer electrolyte for supercapacitors. *Ionics* 19:277–285.
735. Ramasamy, C., J. P. del Vel, and M. Anderson. 2014. An activated carbon supercapacitor analysis by using a gel electrolyte of sodium salt-polyethylene oxide in an organic mixture solvent. *Journal of Solid State Electrochemistry* 18:2217–2223.
736. Sivaraman, P., A. Thakur, R. K. Kushwaha, D. Ratna, and A. B. Samui. 2006. Poly(3-methyl thiophene)-activated carbon hybrid supercapacitor based on gel polymer electrolyte. *Electrochemical and Solid-State Letters* 9:A435–A438.
737. Chen, L. F., Z. H. Huang, H. W. Liang, W. T. Yao, Z. Y. Yu, and S. H. Yu. 2013. Flexible all-solid-state high-power supercapacitor fabricated with nitrogen-doped carbon nanofiber electrode material derived from bacterial cellulose. *Energy & Environmental Science* 6:3331–3338.

738. Wang, X. F., B. Liu, Q. F. Wang et al. 2013. Three-dimensional hierarchical $GeSe_2$ nanostructures for high performance flexible all-solid-state supercapacitors. *Advanced Materials* 25:1479–1486.
739. El-Kady, M. F., V. Strong, S. Dubin, and R. B. Kaner. 2012. Laser scribing of high-performance and flexible graphene-based electrochemical capacitors. *Science* 335: 1326–1330.
740. Meng, Y. N., Y. Zhao, C. G. Hu et al. 2013. All-graphene core-sheath microfibers for all-solid-state, stretchable fibriform supercapacitors and wearable electronic textiles. *Advanced Materials* 25:2326–2331.
741. Niu, Z. Q., H. B. Dong, B. W. Zhu et al. 2013. Highly stretchable, integrated supercapacitors based on single-walled carbon nanotube films with continuous reticulate architecture. *Advanced Materials* 25:1058–1064.
742. Nam, I., G. P. Kim, S. Park, J. W. Han, and J. Yi. 2014. All-solid-state, origami-type foldable supercapacitor chips with integrated series circuit analogues. *Energy & Environmental Science* 7:1095–1102.
743. Sun, G., J. An, C. K. Chua, H. Pang, J. Zhang, and P. Chen. 2015. Layer-by-layer printing of laminated graphene-based interdigitated microelectrodes for flexible planar micro-supercapacitors. *Electrochemistry Communications* 51:33–36.
744. Liu, S. Y., J. Xie, H. B. Li et al. 2014. Nitrogen-doped reduced graphene oxide for high-performance flexible all-solid-state micro-supercapacitors. *Journal of Materials Chemistry A* 2:18125–18131.
745. Zhao, C., C. Y. Wang, R. Gorkin, S. Beirne, K. W. Shu, and G. G. Wallace. 2014. Three dimensional (3D) printed electrodes for interdigitated supercapacitors. *Electrochemistry Communications* 41:20–23.
746. Wang, Y., Y. M. Shi, C. X. Zhao, J. I. Wong, X. W. Sun, and H. Y. Yang. 2014. Printed all-solid flexible microsupercapacitors: Towards the general route for high energy storage devices. *Nanotechnology* 25:094010.
747. Si, W. P., C. L. Yan, Y. Chen, S. Oswald, L. Y. Han, and O. G. Schmidt. 2013. On chip, all solid-state and flexible micro-supercapacitors with high performance based on MnO_x/Au multilayers. *Energy & Environmental Science* 6:3218–3223.
748. Meng, C. Z., J. Maeng, S. W. M. John, and P. P. Irazoqui. 2014. Ultrasmall integrated 3D micro-supercapacitors solve energy storage for miniature devices. *Advanced Energy Materials* 4:1301269.
749. Zang, X. B., Q. Chen, P. X. Li et al. 2014. Highly flexible and adaptable, all-solid-state supercapacitors based on graphene woven-fabric film electrodes. *Small* 10:2583–2588.
750. Kou, L., T. Q. Huang, B. N. Zheng et al. 2014. Coaxial wet-spun yarn supercapacitors for high-energy density and safe wearable electronics. *Nature Communications* 5:3754.
751. Meng, Q. H., H. P. Wu, Y. N. Meng, K. Xie, Z. X. Wei, and Z. X. Guo. 2014. High-performance all-carbon yarn micro-supercapacitor for an integrated energy system. *Advanced Materials* 26:4100–4106.
752. Sun, H., X. You, J. Deng et al. 2014. Novel graphene/carbon nanotube composite fibers for efficient wire-shaped miniature energy devices. *Advanced Materials* 26:2868–2873.
753. Jin, H. Y., L. M. Zhou, C. L. Mak, H. T. Huang, W. M. Tang, and H. L. W. Chan. 2015. High-performance fiber-shaped supercapacitors using carbon fiber thread (CFT)@ polyanilne and functionalized CFT electrodes for wearable/stretchable electronics. *Nano Energy* 11:662–670.
754. Sun, G., J. Liu, X. Zhang et al. 2014. Fabrication of ultralong hybrid microfibers from nanosheets of reduced graphene oxide and transition-metal dichalcogenides and their application as supercapacitors. *Angewandte Chemie International Edition* 53:12576–12580.

755. Yu, D., K. Goh, Q. Zhang et al. 2014. Controlled functionalization of carbonaceous fibers for asymmetric solid-state micro-supercapacitors with high volumetric energy density. *Advanced Materials* 26:6790–6797.
756. Chen, T., H. S. Peng, M. Durstock, and L. M. Dai. 2014. High-performance transparent and stretchable all-solid supercapacitors based on highly aligned carbon nanotube sheets. *Scientific Reports* 4:3612.
757. Gao, Y., Y. S. Zhou, W. Xiong et al. 2013. Transparent, flexible, and solid-state supercapacitors based on graphene electrodes. *APL Materials* 1:012101.
758. Nam, I., S. Park, G. P. Kim, J. Park, and J. Yi. 2013. Transparent and ultra-bendable all-solid-state supercapacitors without percolation problems. *Chemical Science* 4: 1663–1667.
759. Niu, Z. Q., L. Zhang, L. L. Liu, B. W. Zhu, H. B. Dong, and X. D. Chen. 2013. All-solid-state flexible ultrathin micro-supercapacitors based on graphene. *Advanced Materials* 25:4035–4042.
760. Meng, F. H., and Y. Ding. 2011. Sub-micrometer-thick all-solid-state supercapacitors with high power and energy densities. *Advanced Materials* 23:4098–4102.
761. Peng, L., X. Peng, B. Liu, C. Wu, Y. Xie, and G. Yu. 2013. Ultrathin two-dimensional MnO2/graphene hybrid nanostructures for high-performance, flexible planar supercapacitors. *Nano Letters* 13:2151–2157.
762. Ren, J., W. Y. Bai, G. Z. Guan, Y. Zhang, and H. S. Peng. 2013. Flexible and weaveable capacitor wire based on a carbon nanocomposite fiber. *Advanced Materials* 25:5965–5970.
763. Yuan, L., B. Yao, B. Hu, K. Huo, W. Chen, and J. Zhou. 2013. Polypyrrole-coated paper for flexible solid-state energy storage. *Energy & Environmental Science* 6:470–476.
764. Wang, X. F., B. Liu, R. Liu et al. 2014. Fiber-based flexible all-solid-state asymmetric supercapacitors for integrated photodetecting system. *Angewandte Chemie International Edition* 53:1849–1853.
765. Wang, K., H. P. Wu, Y. N. Meng, Y. J. Zhang, and Z. X. Wei. 2012. Integrated energy storage and electrochromic function in one flexible device: An energy storage smart window. *Energy & Environmental Science* 5:8384–8389.
766. Zhang, Z. T., X. L. Chen, P. N. Chen et al. 2014. Integrated polymer solar cell and electrochemical supercapacitor in a flexible and stable fiber format. *Advanced Materials* 26:466–470.
767. Zhao, C., C. Wang, Z. Yue, K. Shu, and G. G. Wallace. 2013. Intrinsically stretchable supercapacitors composed of polypyrrole electrodes and highly stretchable gel electrolyte. *ACS Applied Materials & Interfaces* 5:9008–9014.
768. Kim, S. Y., H. S. Shin, Y. M. Lee, and C. N. Jeong. 1999. Properties of electroresponsive poly(vinyl alcohol)/poly(acrylic acid) IPN hydrogels under an electric stimulus. *Journal of Applied Polymer Science* 73:1675–1683.
769. Fei, H. J., C. Y. Yang, H. Bao, and G. C. Wang. 2014. Flexible all-solid-state supercapacitors based on graphene/carbon black nanoparticle film electrodes and cross-linked poly(vinyl alcohol)-H_2SO_4 porous gel electrolytes. *Journal of Power Sources* 266:488–495.
770. Kaempgen, M., C. K. Chan, J. Ma, Y. Cui, and G. Gruner. 2009. Printable thin film supercapacitors using single-walled carbon nanotubes. *Nano Letters* 9:1872–1876.
771. Yuan, L., X. H. Lu, X. Xiao et al. 2012. Flexible solid-state supercapacitors based on carbon nanoparticles/MnO_2 nanorods hybrid structure. *ACS Nano* 6:656–661.
772. Yuan, L. Y., X. Xiao, T. P. Ding et al. 2012. Paper-based supercapacitors for self-powered nanosystems. *Angewandte Chemie International Edition* 51:4934–4938.
773. Wang, G. M., H. Y. Wang, X. H. Lu et al. 2014. Solid-state supercapacitor based on activated carbon cloths exhibits excellent rate capability. *Advanced Materials* 26:2676–2682.

774. Wang, S., B. Pei, X. Zhao, and R. A. W. Dryfe. 2013. Highly porous graphene on carbon cloth as advanced electrodes for flexible all-solid-state supercapacitors. *Nano Energy* 2:530–536.
775. Zhou, C., and J. P. Liu. 2014. Carbon nanotube network film directly grown on carbon cloth for high-performance solid-state flexible supercapacitors. *Nanotechnology* 25:035402.
776. Jiang, S. L., T. L. Shi, X. B. Zhan et al. 2014. High-performance all-solid-state flexible supercapacitors based on two-step activated carbon cloth. *Journal of Power Sources* 272:16–23.
777. Xu, Y., Z. Lin, X. Huang, Y. Wang, Y. Huang, and X. Duan. 2013. Functionalized graphene hydrogel-based high-performance supercapacitors. *Advanced Materials* 25:5779–5784.
778. Xu, Y. X., Z. Y. Lin, X. Q. Huang, Y. Liu, Y. Huang, and X. F. Duan. 2013. Flexible solid-state supercapacitors based on three-dimensional graphene hydrogel films. *ACS Nano* 7:4042–4049.
779. Huang, G. J., C. Y. Hou, Y. L. Shao et al. 2015. High-performance all-solid-state yarn supercapacitors based on porous graphene ribbons. *Nano Energy* 12:26–32.
780. Wang, Y., J. Chen, J. Cao et al. 2014. Graphene/carbon black hybrid film for flexible and high rate performance supercapacitor. *Journal of Power Sources* 271:269–277.
781. Ju, H. F., W. L. Song, and L. Z. Fan. 2014. Rational design of graphene/porous carbon aerogels for high-performance flexible all-solid-state supercapacitors. *Journal of Materials Chemistry A* 2:10895–10903.
782. Chen, Q., X. M. Li, X. B. Zang et al. 2014. Effect of different gel electrolytes on graphene-based solid-state supercapacitors. *RSC Advances* 4:36253–36256.
783. Gao, H., and K. Lian. 2013. Effect of SiO_2 on silicotungstic acid–H_3PO_4–poly(vinyl alcohol) electrolyte for electrochemical supercapacitors. *Journal of the Electrochemical Society* 160:A505–A510.
784. Lim, C. S., K. H. Teoh, C. W. Liew, and S. Ramesh. 2014. Capacitive behavior studies on electrical double layer capacitor using poly (vinyl alcohol)-lithium perchlorate based polymer electrolyte incorporated with TiO_2. *Materials Chemistry and Physics* 143:661–667.
785. Lim, C. S., K. H. Teoh, C. W. Liew, and S. Ramesh. 2014. Electric double layer capacitor based on activated carbon electrode and biodegradable composite polymer electrolyte. *Ionics* 20:251–258.
786. Huang, Y. F., P. F. Wu, M. Q. Zhang, W. H. Ruan, and E. P. Giannelis. 2014. Boron cross-linked graphene oxide/polyvinyl alcohol nanocomposite gel electrolyte for flexible solid-state electric double layer capacitor with high performance. *Electrochimica Acta* 132:103–111.
787. Rosi, M., F. Iskandar, M. Abdullah, and Khairurrijal. 2014. Hydrogel-polymer electrolytes based on polyvinyl alcohol and hydroxyethylcellulose for supercapacitor applications. *International Journal of Electrochemical Science* 9:4251–4256.
788. Gao, H., J. Li, and K. Lian. 2014. Alkaline quaternary ammonium hydroxides and their polymer electrolytes for electrochemical capacitors. *RSC Advances* 4:21332–21339.
789. Dubal, D. P., G. S. Gund, R. Holze, H. S. Jadhav, C. D. Lokhande, and C. J. Park. 2013. Solution-based binder-free synthetic approach of RuO_2 thin films for all solid state supercapacitors. *Electrochimica Acta* 103:103–109.
790. Dinh, T. M., A. Achour, S. Vizireanu et al. 2014. Hydrous RuO_2/carbon nanowalls hierarchical structures for all-solid-state ultrahigh-energy-density micro-supercapacitors. *Nano Energy* 10:288–294.
791. Dubal, D. P., and R. Holze. 2013. All-solid-state flexible thin film supercapacitor based on Mn_3O_4 stacked nanosheets with gel electrolyte. *Energy* 51:407–412.
792. Yu, M. H., T. Zhai, X. H. Lu et al. 2013. Manganese dioxide nanorod arrays on carbon fabric for flexible solid-state supercapacitors. *Journal of Power Sources* 239:64–71.

793. Zheng, H. M., T. Zhai, M. H. Yu et al. 2013. TiO_2@C core–shell nanowires for high-performance and flexible solid-state supercapacitors. *Journal of Materials Chemistry C* 1:225–229.
794. Liu, B., D. S. Tan, X. F. Wang, D. Chen, and G. Z. Shen. 2013. Flexible, planar-integrated, all-solid-state fiber supercapacitors with an enhanced distributed-capacitance effect. *Small* 9:1998–2004.
795. Gao, L. N., X. F. Wang, Z. Xie et al. 2013. High-performance energy-storage devices based on WO_3 nanowire arrays/carbon cloth integrated electrodes. *Journal of Materials Chemistry A* 1:7167–7173.
796. Gao, H., Y. J. Ting, N. P. Kherani, and K. Lian. 2013. Ultra-high-rate all-solid pseudocapacitive electrochemical capacitors. *Journal of Power Sources* 222:301–304.
797. Zhang, C. L., H. H. Yin, M. Han et al. 2014. Two-dimensional tin selenide nanostructures for flexible all-solid-state supercapacitors. *ACS Nano* 8:3761–3770.
798. Wang, K., X. Zhang, C. Li et al. 2014. Flexible solid-state supercapacitors based on a conducting polymer hydrogel with enhanced electrochemical performance. *Journal of Materials Chemistry A* 2:19726–19732.
799. Hu, H., K. Zhang, S. Li, S. Ji, and C. Ye. 2014. Flexible, in-plane, and all-solid-state micro-supercapacitors based on printed interdigital Au/polyaniline network hybrid electrodes on a chip. *Journal of Materials Chemistry A* 2:20916–20922.
800. Zhang, K., H. Hu, W. Yao, and C. Ye. 2015. Flexible and all-solid-state supercapacitors with long-time stability constructed on PET/Au/polyaniline hybrid electrodes. *Journal of Materials Chemistry A* 3:617–623.
801. Liu, L. L., Z. Q. Niu, L. Zhang, W. Y. Zhou, X. D. Chen, and S. S. Xie. 2014. Nanostructured graphene composite papers for highly flexible and foldable supercapacitors. *Advanced Materials* 26:4855–4862.
802. Wee, B. H., and J. D. Hong. 2014. Multilayered poly(*p*-phenylenevinylene)/reduced graphene oxide film: An efficient organic current collector in an all-plastic supercapacitor. *Langmuir* 30:5267–5275.
803. Xiong, G. P., C. Z. Meng, R. G. Reifenberger, P. P. Irazoqui, and T. S. Fisher. 2014. Graphitic petal electrodes for all-solid-state flexible supercapacitors. *Advanced Energy Materials* 4:1300515.
804. Huang, Y., J. Y. Tao, W. J. Meng et al. 2015. Super-high rate stretchable polypyrrole-based supercapacitors with excellent cycling stability. *Nano Energy* 11:518–525.
805. Anothumakkool, B., A. T. A. Torris, S. N. Bhange, M. V. Badiger, and S. Kurungot. 2014. Electrodeposited polyethylenedioxythiophene with infiltrated gel electrolyte interface: A close contest of an all-solid-state supercapacitor with its liquid-state counterpart. *Nanoscale* 6:5944–5952.
806. Kurra, N., J. Park, and H. N. Alshareef. 2014. A conducting polymer nucleation scheme for efficient solid-state supercapacitors on paper. *Journal of Materials Chemistry A* 2:17058–17065.
807. Park, J. W., S. J. Park, O. S. Kwon et al. 2014. *In situ* synthesis of graphene/polyselenophene nanohybrid materials as highly flexible energy storage electrodes. *Chemistry of Materials* 26:2354–2360.
808. Li, M., Z. Tang, M. Leng, and J. M. Xue. 2014. Flexible solid-state supercapacitor based on graphene-based hybrid films. *Advanced Functional Materials* 24:7495–7502.
809. Zhou, W. J., X. J. Liu, Y. H. Sang et al. 2014. Enhanced performance of layered titanate nanowire-based supercapacitor electrodes by nickel ion exchange. *ACS Applied Materials & Interfaces* 6:4578–4586.
810. Kurra, N., N. A. Alhebshi, and H. N. Alshareef. 2014. Microfabricated pseudocapacitors using $Ni(OH)_2$ electrodes exhibit remarkable volumetric capacitance and energy density. *Advanced Energy Materials* 4:1401303.

811. Hao, C. X., F. S. Wen, J. Y. Xiang et al. 2014. Controlled incorporation of $Ni(OH)_2$ nanoplates into flowerlike MoS_2 nanosheets for flexible all-solid-state supercapacitors. *Advanced Functional Materials* 24:6700–6707.
812. Xie, J., X. Sun, N. Zhang, K. Xu, M. Zhou, and Y. Xie. 2013. Layer-by-layer β-$Ni(OH)_2$/graphene nanohybrids for ultraflexible all-solid-state thin-film supercapacitors with high electrochemical performance. *Nano Energy* 2:65–74.
813. Yang, W. L., Z. Gao, J. Ma, X. M. Zhang, J. Wang, and J. Y. Liu. 2014. Hierarchical $NiCo_2O_4$@NiO core–shell hetero-structured nanowire arrays on carbon cloth for a high-performance flexible all-solid-state electrochemical capacitor. *Journal of Materials Chemistry A* 2:1448–1457.
814. Wang, Q. F., X. F. Wang, B. Liu et al. 2013. $NiCo_2O_4$ nanowire arrays supported on Ni foam for high-performance flexible all-solid-state supercapacitors. *Journal of Materials Chemistry A* 1:2468–2473.
815. Zhao, J., J. Chen, S. Xu et al. 2013. CoMn-layered double hydroxide nanowalls supported on carbon fibers for high-performance flexible energy storage devices. *Journal of Materials Chemistry A* 1:8836–8843.
816. Xie, Y. B., and X. Q. Fang. 2014. Electrochemical flexible supercapacitor based on manganese dioxide–titanium nitride nanotube hybrid. *Electrochimica Acta* 120:273–283.
817. Yang, C., L. Dong, Z. X. Chen, and H. B. Lu. 2014. High-performance all-solid-state supercapacitor based on the assembly of graphene and manganese(II) phosphate nanosheets. *Journal of Physical Chemistry C* 118:18884–18891.
818. Wang, G. M., X. H. Lu, Y. C. Ling et al. 2012. LiCl/PVA gel electrolyte stabilizes vanadium oxide nanowire electrodes for pseudocapacitors. *ACS Nano* 6:10296–10302.
819. Yang, P., X. Xiao, Y. Li et al. 2013. Hydrogenated ZnO core–shell nanocables for flexible supercapacitors and self-powered systems. *ACS Nano* 7:2617–2626.
820. Zhang, K. Y., H. Chen, X. Wang et al. 2014. Synthesis and structure determination of potassium copper selenide nanowires and solid-state supercapacitor application. *Journal of Power Sources* 268:522–532.
821. Dai, S. G., Y. Xi, C. G. Hu, X. L. Yue, L. Cheng, and G. Wang. 2014. MnO_2@KCu_7S_4 NWs hybrid compositions for high-power all-solid-state supercapacitor. *Journal of Power Sources* 274:477–482.
822. Kang, Y. J., H. Chung, and W. Kim. 2013. 1.8-V flexible supercapacitors with asymmetric configuration based on manganese oxide, carbon nanotubes, and a gel electrolyte. *Synthetic Metals* 166:40–44.
823. Sawangphruk, M., P. Srimuk, P. Chiochan, A. Krittayavathananon, S. Luanwuthi, and J. Limtrakul. 2013. High-performance supercapacitor of manganese oxide/reduced graphene oxide nanocomposite coated on flexible carbon fiber paper. *Carbon* 60:109–116.
824. Xu, H., X. Hu, Y. Sun, H. Yang, X. Liu, and Y. Huang. 2015. Flexible fiber-shaped supercapacitors based on hierarchically nanostructured composite electrodes. *Nano Research* 8:1148–1158.
825. Kim, D., J. Yun, G. Lee, and J. S. Ha. 2014. Fabrication of high performance flexible micro-supercapacitor arrays with hybrid electrodes of MWNT/V_2O_5 nanowires integrated with a SnO_2 nanowire UV sensor. *Nanoscale* 6:12034–12041.
826. Li, Z. J., Z. H. Zhou, G. Q. Yun, K. Shi, X. W. Lv, and B. C. Yang. 2013. High-performance solid-state supercapacitors based on graphene–ZnO hybrid nanocomposites. *Nanoscale Research Letters* 8:473.
827. Wu, C. Z., X. L. Lu, L. L. Peng et al. 2013. Two-dimensional vanadyl phosphate ultrathin nanosheets for high energy density and flexible pseudocapacitors. *Nature Communications* 4:2431.

828. Lu, X. H., M. H. Yu, T. Zhai et al. 2013. High energy density asymmetric quasi-solid-state supercapacitor based on porous vanadium nitride nanowire anode. *Nano Letters* 13:2628–2633.
829. Dai, S. G., Y. Xi, C. G. Hu et al. 2013. KCu_7S_4 nanowires and the Mn/KCu_7S_4 nanostructure for solid-state supercapacitors. *Journal of Materials Chemistry A* 1:15530–15534.
830. Qian, T., N. Xu, J. Zhou et al. 2015. Interconnected three-dimensional V_2O_5/polypyrrole network nanostructures for high performance solid-state supercapacitors. *Journal of Materials Chemistry A* 3:488–493.
831. Anothumakkool, B., Torris A. T. A., S. N. Bhange, S. M. Unni, M. V. Badiger, and S. Kurungot. 2013. Design of a high performance thin all-solid-state supercapacitor mimicking the active interface of its liquid-state counterpart. *ACS Applied Materials & Interfaces* 5:13397–13404.
832. Li, X. M., T. S. Zhao, Q. Chen et al. 2013. Flexible all solid-state supercapacitors based on chemical vapor deposition derived graphene fibers. *Physical Chemistry Chemical Physics* 15:17752–17757.
833. Meng, Q. H., K. Wang, W. Guo, J. Fang, Z. X. Wei, and X. L. She. 2014. Thread-like supercapacitors based on one-step spun nanocomposite yarns. *Small* 10:3187–3193.
834. Ge, D. T., L. L. Yang, L. Fan et al. 2015. Foldable supercapacitors from triple networks of macroporous cellulose fibers, single-walled carbon nanotubes and polyaniline nanoribbons. *Nano Energy* 11:568–578.
835. Xiao, X., X. Peng, H. Y. Jin et al. 2013. Freestanding mesoporous VN/CNT hybrid electrodes for flexible all-solid-state supercapacitors. *Advanced Materials* 25:5091–5097.
836. Bao, J., X. Zhang, L. Bai et al. 2014. All-solid-state flexible thin-film supercapacitors with high electrochemical performance based on a two-dimensional $V_2O_5 \cdot H_2O$/graphene composite. *Journal of Materials Chemistry A* 2:10876–10881.
837. Lu, X. H., M. H. Yu, G. M. Wang et al. 2013. H-TiO_2@MnO_2//H-TiO_2@C core–shell nanowires for high performance and flexible asymmetric supercapacitors. *Advanced Materials* 25:267–272.
838. Xu, J., Q. Wang, X. Wang et al. 2013. Flexible asymmetric supercapacitors based upon Co_9S_8 nanorod//Co_3O_4@RuO_2 nanosheet arrays on carbon cloth. *ACS Nano* 7:5453–5462.
839. Jin, H. Y., Z. H. Peng, W. M. Tang, and H. L. W. Chan. 2014. Controllable functionalized carbon fabric for high-performance all-carbon-based supercapacitors. *RSC Advances* 4:33022–33028.
840. Wang, Z. L., Z. L. Zhu, J. H. Qiu, and S. H. Yang. 2014. High performance flexible solid-state asymmetric supercapacitors from MnO_2/ZnO core–shell nanorods//specially reduced graphene oxide. *Journal of Materials Chemistry C* 2:1331–1336.
841. Gao, Y. P., J. H. Zhao, Z. Run, G. Q. Zhang, and H. Pang. 2014. Microporous $Ni_{11}(HPO_3)_8(OH)_6$ nanocrystals for high-performance flexible asymmetric all solid-state supercapacitors. *Dalton Transactions* 43:17000–17005.
842. Senthilkumar, S. T., and R. K. Selvan. 2014. Fabrication and performance studies of a cable-type flexible asymmetric supercapacitor. *Physical Chemistry Chemical Physics* 16:15692–15698.
843. Tao, J. Y., N. S. Liu, L. Y. Li, and Y. H. Gao. 2014. Hierarchical nanostructures of polypyrrole@MnO_2 composite electrodes for high performance solid-state asymmetric supercapacitors. *Nanoscale* 6:2922–2928.
844. Kim, K. M., J. H. Nam, Y. G. Lee, W. I. Cho, and J. M. Ko. 2013. Supercapacitive properties of electrodeposited RuO_2 electrode in acrylic gel polymer electrolytes. *Current Applied Physics* 13:1702–1706.
845. Ko, J. M., J. H. Nam, J. H. Won, and K. M. Kim. 2014. Supercapacitive properties of electrodeposited polyaniline electrode in acrylic gel polymer electrolytes. *Synthetic Metals* 189:152–156.

846. Liu, M. K., W. W. Tjiu, J. S. Pan, C. Zhang, W. Gao, and T. X. Liu. 2014. One-step synthesis of graphene nanoribbon–MnO_2 hybrids and their all-solid-state asymmetric supercapacitors. *Nanoscale* 6:4233–4242.
847. Nam, H. S., N. L. Wu, K. T. Lee et al. 2012. Electrochemical capacitances of a nanowire-structured MnO_2 in polyacrylate-based gel electrolytes. *Journal of the Electrochemical Society* 159:A899–A903.
848. Yang, C. Y., J. L. Shen, C. Y. Wang, H. J. Fei, H. Bao, and G. C. Wang. 2014. All-solid-state asymmetric supercapacitor based on reduced graphene oxide/carbon nanotube and carbon fiber paper/polypyrrole electrodes. *Journal of Materials Chemistry A* 2:1458–1464.
849. Kim, K. M., M. Latifatu, Y. G. Lee, J. M. Ko, J. H. Kim, and W. I. Cho. 2014. Effect of ceramic filler-containing polymer hydrogel electrolytes coated on the polyolefin separator on the electrochemical properties of activated carbon supercapacitor. *Journal of Electroceramics* 32:146–153.
850. Shimamoto, K., K. Tadanaga, and M. Tatsumisago. 2013. All-solid-state electrochemical capacitors using MnO_2/carbon nanotube composite electrode. *Electrochimica Acta* 109:651–655.
851. Ramasamy, C., J. P. del Val, and M. Anderson. 2014. An electrochemical cell study on polyvinylpyrrolidine aqueous gel with glycol addition for capacitor applications. *Electrochimica Acta* 135:181–186.
852. Gnerlich, M., E. Pomerantseva, K. Gregorczyk, D. Ketchum, G. Rubloff, and R. Ghodssi. 2013. Solid flexible electrochemical supercapacitor using Tobacco mosaic virus nanostructures and ALD ruthenium oxide. *Journal of Micromechanics and Microengineering* 23:114014.
853. Shimamoto, K., K. Tadanaga, and M. Tatsumisago. 2014. All-solid-state electrochemical capacitors using MnO_2 electrode/SiO_2–Nafion electrolyte composite prepared by the sol–gel process. *Journal of Power Sources* 248:396–399.
854. Subramaniam, C. K., and G. Boopalan. 2014. Study of storage capacity in various carbon/graphene-based solid-state supercapacitors. *Applied Physics A – Materials Science & Processing* 116:887–891.
855. Subramaniam, C. K., K. K. Cheralathan, G. Velayutham, and S. Bollepalli. 2012. Study of carbon based solid state EDLCs at high sweep rates. *ECS Transactions* 41:37–49.
856. Zhao, J. W., J. Chen, S. M. Xu et al. 2014. Hierarchical NiMn layered double hydroxide/carbon nanotubes architecture with superb energy density for flexible supercapacitors. *Advanced Functional Materials* 24:2938–2946.
857. Huang, C., and P. S. Grant. 2013. One-step spray processing of high power all-solid-state supercapacitors. *Scientific Reports* 3:2393.
858. Cole, D. P., A. L. M. Reddy, M. G. Hahm et al. 2014. Electromechanical properties of polymer electrolyte-based stretchable supercapacitors. *Advanced Energy Materials* 4:1300844.
859. Gao, H., and K. Lian. 2011. High rate all-solid electrochemical capacitors using proton conducting polymer electrolytes. *Journal of Power Sources* 196:8855–8857.
860. Ramasamy, C., J. Palma, and M. Anderson. 2014. A 3-V electrochemical capacitor study based on a magnesium polymer gel electrolyte by three different carbon materials. *Journal of Solid State Electrochemistry* 18:2903–2911.
861. DiCarmine, P. M., T. B. Schon, T. M. McCormick, P. P. Klein, and D. S. Seferos. 2014. Donor-acceptor polymers for electrochemical supercapacitors: Synthesis, testing, and theory. *Journal of Physical Chemistry C* 118:8295–8307.
862. Yuksel, R., Z. Sarioba, A. Cirpan, P. Hiralal, and H. E. Unalan. 2014. Transparent and flexible supercapacitors with single walled carbon nanotube thin film electrodes. *ACS Applied Materials & Interfaces* 6:15434–15439.
863. Chiu, K. F., and S. H. Su. 2013. Lithiated and sulphonated poly(ether ether ketone) solid state electrolyte films for supercapacitors. *Thin Solid Films* 544:144–147.

864. Hsueh, M. F., C. W. Huang, C. A. Wu, P. L. Kuo, and H. Teng. 2013. The synergistic effect of nitrile and ether functionalities for gel electrolytes used in supercapacitors. *Journal of Physical Chemistry C* 117:16751–16758.
865. Syahidah, S. N., and S. R. Majid. 2013. Super-capacitive electro-chemical performance of polymer blend gel polymer electrolyte (GPE) in carbon-based electrical double-layer capacitors. *Electrochimica Acta* 112:678–685.
866. Chiou, Y. D., D. S. Tsai, H. H. Lam, C. H. Chang, K. Y. Lee, and Y. S. Huang. 2013. Cycle stability of the electrochemical capacitors patterned with vertically aligned carbon nanotubes in an $LiPF_6$-based electrolyte. *Nanoscale* 5:8122–8129.
867. Jain, A., and S. K. Tripathi. 2013. Converting eucalyptus leaves into mesoporous carbon for its application in quasi solid-state supercapacitors. *Journal of Solid State Electrochemistry* 17:2545–2550.
868. Schroeder, M., P. Isken, M. Winter, S. Passerini, A. Lex-Balducci, and A. Balducci. 2013. An investigation on the use of a methacrylate-based gel polymer electrolyte in high power devices. *Journal of the Electrochemical Society* 160:A1753–A1758.
869. Rodríguez, J., E. Navarrete, E. A. Dalchiele, L. Sánchez, J. R. Ramos-Barrado, and F. Martín. 2013. Polyvinylpyrrolidone–$LiClO_4$ solid polymer electrolyte and its application in transparent thin film supercapacitors. *Journal of Power Sources* 237:270–276.
870. Jain, A., and S. K. Tripathi. 2013. Experimental studies on high-performance supercapacitor based on nanogel polymer electrolyte with treated activated charcoal. *Ionics* 19:549–557.
871. Tiruye, G. A., D. Muñoz-Torrero, J. Palma, M. Anderson, and R. Marcilla. 2015. All-solid state supercapacitors operating at 3.5 V by using ionic liquid based polymer electrolytes. *Journal of Power Sources* 279:472–480.
872. Wang, S., B. Hsia, C. Carraro, and R. Maboudian. 2014. High-performance all solid-state micro-supercapacitor based on patterned photoresist-derived porous carbon electrodes and an ionogel electrolyte. *Journal of Materials Chemistry A* 2:7997–8002.
873. Hsia, B., J. Marschewski, S. Wang et al. 2014. Highly flexible, all solid-state micro-supercapacitors from vertically aligned carbon nanotubes. *Nanotechnology* 25:055401.
874. Kim, D., G. Shin, Y. J. Kang, W. Kim, and J. S. Ha. 2013. Fabrication of a stretchable solid-state micro-supercapacitor array. *ACS Nano* 7:7975–7982.
875. Pandey, G. P., and S. A. Hashmi. 2013. Ionic liquid 1-ethyl-3-methylimidazolium tetracyanoborate-based gel polymer electrolyte for electrochemical capacitors. *Journal of Materials Chemistry A* 1:3372–3378.
876. Lim, Y., J. Yoon, J. Yun et al. 2014. Biaxially stretchable, integrated array of high performance microsupercapacitors. *ACS Nano* 8:11639–11650.
877. Gao, Y., Y. S. Zhou, M. Qian et al. 2013. High-performance flexible solid-state supercapacitors based on MnO_2-decorated nanocarbon electrodes. *RSC Advances* 3:20613–20618.
878. Pandey, G. P., A. C. Rastogi, and C. R. Westgate. 2014. All-solid-state supercapacitors with poly(3,4-ethylenedioxythiophene)-coated carbon fiber paper electrodes and ionic liquid gel polymer electrolyte. *Journal of Power Sources* 245:857–865.
879. Hashmi, S. S. A. 2014. Quasi-solid-state pseudocapacitors using proton-conducting gel polymer electrolyte and poly(3-methyl thiophene)–ruthenium oxide composite electrodes. *Journal of Solid State Electrochemistry* 18:465–475.
880. Sellam, and S. A. Hashmi. 2013. High rate performance of flexible pseudocapacitors fabricated using ionic-liquid-based proton conducting polymer electrolyte with poly(3, 4-ethylenedioxythiophene):poly(styrene sulfonate) and its hydrous ruthenium oxide composite electrodes. *ACS Applied Materials & Interfaces* 5:3875–3883.

881. Tamilarasan, P., and S. Ramaprabhu. 2014. Stretchable supercapacitors based on highly stretchable ionic liquid incorporated polymer electrolyte. *Materials Chemistry and Physics* 148:48–56.
882. Liew, C. W., S. Ramesh, and A. K. Arof. 2014. Good prospect of ionic liquid based-poly(vinyl alcohol) polymer electrolytes for supercapacitors with excellent electrical, electrochemical and thermal properties. *International Journal of Hydrogen Energy* 39:2953–2963.
883. Ketabi, S., and K. Lian. 2013. Effect of SiO_2 on conductivity and structural properties of PEO–$EMIHSO_4$ polymer electrolyte and enabled solid electrochemical capacitors. *Electrochimica Acta* 103:174–178.
884. Pandey, G. P., and S. A. Hashmi. 2013. Performance of solid-state supercapacitors with ionic liquid 1-ethyl-3-methylimidazolium tris(pentafluoroethyl) trifluorophosphate based gel polymer electrolyte and modified MWCNT electrodes. *Electrochimica Acta* 105:333–341.
885. Pandey, G. P., and S. A. Hashmi. 2013. Solid-state supercapacitors with ionic liquid based gel polymer electrolyte: Effect of lithium salt addition. *Journal of Power Sources* 243:211–218.
886. Suleman, M., Y. Kumar, and S. A. Hashmi. 2013. Structural and electrochemical properties of succinonitrile-based gel polymer electrolytes: Role of ionic liquid addition. *Journal of Physical Chemistry B* 117:7436–7443.
887. Soeda, K., M. Yamagata, S. Yamazaki, and M. Ishikawa. 2013. Application of chitosan-based gel electrolytes with ionic liquids for high-performance and safe electric double layer capacitors. *Electrochemistry* 81:867–872.
888. Yamagata, M., K. Soeda, S. Ikebe, S. Yamazaki, and M. Ishikawa. 2013. Chitosan-based gel electrolyte containing an ionic liquid for high-performance nonaqueous supercapacitors. *Electrochimica Acta* 100:275–280.
889. Yang, L., J. Hu, G. Lei, and H. Liu. 2014. Ionic liquid-gelled polyvinylidene fluoride/polyvinyl acetate polymer electrolyte for solid supercapacitor. *Chemical Engineering Journal* 258:320–326.
890. Kang, Y. J., S. J. Chun, S. S. Lee et al. 2012. All-solid-state flexible supercapacitors fabricated with bacterial nanocellulose papers, carbon nanotubes, and triblock-copolymer ion gels. *ACS Nano* 6:6400–6406.
891. Pandey, G. P., A. C. Rastogi, and C. R. Westgate. 2013. Polyacrylonitrile and 1-ethyl-3-methylimidazolium thiocyanate based gel polymer electrolyte for solid-state supercapacitors with graphene electrodes. *Electrochemical Capacitors* 50:145–151.
892. Liu, X. H., D. B. Wu, H. L. Wang, and Q. G. Wang. 2014. Self-recovering tough gel electrolyte with adjustable supercapacitor performance. *Advanced Materials* 26:4370–4375.
893. Liu, X. H., Z. B. Wen, D. B. Wu, H. L. Wang, J. H. Yang, and Q. G. Wang. 2014. Tough BMIMCl-based ionogels exhibiting excellent and adjustable performance in high-temperature supercapacitors. *Journal of Materials Chemistry A* 2:11569–11573.
894. Taniki, R., K. Matsumoto, T. Nohira, and R. Hagiwara. 2014. All solid-state electrochemical capacitors using *N,N*-dimethylpyrrolidinium fluorohydrogenate as ionic plastic crystal electrolyte. *Journal of Power Sources* 245:758–763.
895. Shukur, M. F., R. Ithnin, and M. F. Z. Kadir. 2014. Electrical characterization of corn starch-LiOAc electrolytes and application in electrochemical double layer capacitor. *Electrochimica Acta* 136:204–216.
896. Liew, C. W., and S. Ramesh. 2014. Comparing triflate and hexafluorophosphate anions of ionic liquids in polymer electrolytes for supercapacitor applications. *Materials* 7:4019–4033.
897. Sudhakar, Y. N., and M. Selvakumar. 2012. Lithium perchlorate doped plasticized chitosan and starch blend as biodegradable polymer electrolyte for supercapacitors. *Electrochimica Acta* 78:398–405.

898. Woo, H. J., C. W. Liew, S. R. Majid, and A. K. Arof. 2014. Poly(ϵ-caprolactone)-based polymer electrolyte for electrical double-layer capacitors. *High Performance Polymers* 26:637–640.
899. Shirshova, N., A. Bismarck, S. Carreyette et al. 2013. Structural supercapacitor electrolytes based on bicontinuous ionic liquid-epoxy resin systems. *Journal of Materials Chemistry A* 1:15300–15309.
900. Tamilarasan, P., and S. Ramaprabhu. 2013. Graphene based all-solid-state supercapacitors with ionic liquid incorporated polyacrylonitrile electrolyte. *Energy* 51:374–381.
901. Javaid, A., K. K. C. Ho, A. Bismarck, M. S. P. Shaffer, J. H. G. Steinke, and E. S. Greenhalgh. 2014. Multifunctional structural supercapacitors for electrical energy storage applications. *Journal of Composite Materials* 48:1409–1416.
902. Westover, A. S., J. W. Tian, S. Bernath et al. 2014. A multifunctional load-bearing solid-state supercapacitor. *Nano Letters* 14:3197–3202.
903. Zhang, Q., K. Scrafford, M. T. Li et al. 2014. Anomalous capacitive behaviors of graphene oxide based solid-state supercapacitors. *Nano Letters* 14:1938–1943.
904. Gao, W., N. Singh, L. Song et al. 2011. Direct laser writing of micro-supercapacitors on hydrated graphite oxide films. *Nature Nanotechnology* 6:496–500.
905. Fic, K., E. Frackowiak, and F. Béguin. 2012. Unusual energy enhancement in carbon-based electrochemical capacitors. *Journal of Materials Chemistry* 22:24213–24223.
906. Senthilkumar, S. T., R. K. Selvan, and J. S. Melo. 2013. Redox additive/active electrolytes: A novel approach to enhance the performance of supercapacitors. *Journal of Materials Chemistry A* 1:12386–12394.
907. Fic, K., M. Meller, and E. Frackowiak. 2014. Strategies for enhancing the performance of carbon/carbon supercapacitors in aqueous electrolytes. *Electrochimica Acta* 128:210–217.
908. Ma, G. F., J. J. Li, K. J. Sun, H. Peng, J. J. Mu, and Z. Q. Lei. 2014. High performance solid-state supercapacitor with PVA–KOH–$K_3[Fe(CN)_6]$ gel polymer as electrolyte and separator. *Journal of Power Sources* 256:281–287.
909. Lota, G., and E. Frackowiak. 2009. Striking capacitance of carbon/iodide interface. *Electrochemistry Communications* 11:87–90.
910. Senthilkumar, S. T., R. K. Selvan, Y. S. Lee, and J. S. Melo. 2013. Electric double layer capacitor and its improved specific capacitance using redox additive electrolyte. *Journal of Materials Chemistry A* 1:1086–1095.
911. Lota, G., K. Fic, and E. Frackowiak. 2011. Alkali metal iodide/carbon interface as a source of pseudocapacitance. *Electrochemistry Communications* 13:38–41.
912. Frackowiak, E., K. Fic, M. Meller, and G. Lota. 2012. Electrochemistry serving people and nature: High-energy ecocapacitors based on redox-active electrolytes. *ChemSusChem* 5:1181–1185.
913. Senthilkumar, S. T., R. K. Selvan, N. Ponpandian, J. S. Melo, and Y. S. Lee. 2013. Improved performance of electric double layer capacitor using redox additive (VO^{2+}/VO_2^+) aqueous electrolyte. *Journal of Materials Chemistry A* 1:7913–7919.
914. Lian, K., and C. M. Li. 2008. Heteropoly acid electrolytes for double-layer capacitors and pseudocapacitors. *Electrochemical and Solid-State Letters* 11:A158–A162.
915. Suárez-Guevara, J., V. Ruiz, and P. Gomez-Romero. 2014. Hybrid energy storage: High voltage aqueous supercapacitors based on activated carbon–phosphotungstate hybrid materials. *Journal of Materials Chemistry A* 2:1014–1021.
916. Lian, K., and C. M. Li. 2009. Asymmetrical electrochemical capacitors using heteropoly acid electrolytes. *Electrochemical and Solid-State Letters* 12:A10–A12.
917. Tian, Q. F., and K. Lian. 2010. *In situ* characterization of heteropolyacid based electrochemical capacitors. *Electrochemical and Solid-State Letters* 13:A4–A6.

918. Mai, L. Q., A. Minhas-Khan, X. Tian et al. 2013. Synergistic interaction between redox-active electrolyte and binder-free functionalized carbon for ultrahigh supercapacitor performance. *Nature Communications* 4:1–7.
919. Tanahashi, I. 2005. Capacitance enhancement of activated carbon fiber cloth electrodes in electrochemical capacitors with a mixed aqueous solution of H_2SO_4 and $AgNO_3$. *Electrochemical and Solid-State Letters* 8:A627–A629.
920. Roldán, S., C. Blanco, M. Granda, R. Menéndez, and R. Santamaría. 2011. Towards a further generation of high-energy carbon-based capacitors by using redox-active electrolytes. *Angewandte Chemie International Edition* 50:1699–1701.
921. Roldán, S., M. Granda, R. Menéndez, R. Santamaría, and C. Blanco. 2011. Mechanisms of energy storage in carbon-based supercapacitors modified with a quinoid redox-active electrolyte. *Journal of Physical Chemistry C* 115:17606–17611.
922. Wang, G. P., R. F. Liang, L. X. Liu, and B. H. Zhong. 2014. Improving the specific capacitance of carbon nanotubes-based supercapacitors by combining introducing functional groups on carbon nanotubes with using redox-active electrolyte. *Electrochimica Acta* 115:183–188.
923. Chen, L. B., H. Bai, Z. F. Huang, and L. Li. 2014. Mechanism investigation and suppression of self-discharge in active electrolyte enhanced supercapacitors. *Energy & Environmental Science* 7:1750–1759.
924. Roldán, S., M. Granda, R. Menéndez, R. Santamaría, and C. Blanco. 2012. Supercapacitor modified with methylene blue as redox active electrolyte. *Electrochimica Acta* 83:241–246.
925. Roldán, S., Z. González, C. Blanco, M. Granda, R. Menéndez, and R. Santamaría. 2011. Redox-active electrolyte for carbon nanotube-based electric double layer capacitors. *Electrochimica Acta* 56:3401–3405.
926. Wu, J., H. Yu, L. Fan, G. Luo, J. Lin, and M. Huang. 2012. A simple and high-effective electrolyte mediated with *p*-phenylenediamine for supercapacitor. *Journal of Materials Chemistry* 22:19025–19030.
927. Yu, H., J. Wu, J. Lin et al. 2013. A reversible redox strategy for SWCNT-based supercapacitors using a high-performance electrolyte. *ChemPhysChem* 14:394–399.
928. Yu, H., L. Fan, J. Wu et al. 2012. Redox-active alkaline electrolyte for carbon-based supercapacitor with pseudocapacitive performance and excellent cyclability. *RSC Advances* 2:6736–6740.
929. Lota, G., and G. Milczarek. 2011. The effect of lignosulfonates as electrolyte additives on the electrochemical performance of supercapacitors. *Electrochemistry Communications* 13:470–473.
930. Chen, L. B., Y. R. Chen, J. F. Wu, J. W. Wang, H. Bai, and L. Li. 2014. Electrochemical supercapacitor with polymeric active electrolyte. *Journal of Materials Chemistry A* 2:10526–10531.
931. Wasiński, K., M. Walkowiak, and G. Lota. 2014. Humic acids as pseudocapacitive electrolyte additive for electrochemical double layer capacitors. *Journal of Power Sources* 255:230–234.
932. Maiti, S., A. Pramanik, and S. Mahanty. 2014. Interconnected network of MnO2 nanowires with a "cocoonlike" morphology: Redox couple-mediated performance enhancement in symmetric aqueous supercapacitor. *ACS Applied Materials & Interfaces* 6:10754–10762.
933. Zhao, C. M., W. T. Zheng, X. Wang, H. B. Zhang, X. Q. Cui, and H. X. Wang. 2013. Ultrahigh capacitive performance from both Co(OH)2/graphene electrode and K3Fe(CN)6 electrolyte. *Scientific Reports* 3:2986.
934. Senthilkumar, S. T., R. Kalai Selvan, M. Ulaganathan, and J. S. Melo. 2014. Fabrication of Bi_2O_3||AC asymmetric supercapacitor with redox additive aqueous electrolyte and its improved electrochemical performances. *Electrochimica Acta* 115:518–524.

935. Qian, L., X. Q. Tian, L. Yang, J. F. Mao, H. Y. Yuan, and D. Xiao. 2013. High specific capacitance of CuS nanotubes in redox active polysulfide electrolyte. *RSC Advances* 3:1703–1708.
936. Chen, W., R. B. Rakhi, and H. N. Alshareef. 2013. Capacitance enhancement of polyaniline coated curved-graphene supercapacitors in a redox-active electrolyte. *Nanoscale* 5:4134–4138.
937. Chen, W., R. B. Rakhi, and H. N. Alshareef. 2013. Morphology-dependent enhancement of the pseudocapacitance of template-guided tunable polyaniline nanostructures. *Journal of Physical Chemistry C* 117:15009–15019.
938. Chen, W., C. Xia, R. B. Rakhi, and H. N. Alshareerf. 2014. A general approach toward enhancement of pseudocapacitive performance of conducting polymers by redox-active electrolytes. *Journal of Power Sources* 267:521–526.
939. Xie, H., Y. H. Zhu, Y. H. Wu, Z. L. Wu, and E. H. Liu. 2014. The effect of hydroquinone as an electrolyte additive on electrochemical performance of the polyaniline supercapacitor. *Materials Research Bulletin* 50:303–306.
940. Zhu, Y. H., E. H. Liu, Z. Y. Luo et al. 2014. A hydroquinone redox electrolyte for polyaniline/SnO_2 supercapacitors. *Electrochimica Acta* 118:106–111.
941. Chen, K. F., S. Y. Song, and D. F. Xue. 2014. An ionic aqueous pseudocapacitor system: Electroactive ions in both a salt electrode and redox electrolyte. *RSC Advances* 4:23338–23343.
942. Su, L. H., X. G. Zhang, C. H. Mi, B. Gao, and Y. Liu. 2009. Improvement of the capacitive performances for Co–Al layered double hydroxide by adding hexacyanoferrate into the electrolyte. *Physical Chemistry Chemical Physics* 11:2195–2202.
943. Ionica-Bousquet, C. M., W. J. Casteel Jr, R. M. Pearlstein et al. 2010. Polyfluorinated boron cluster-$[B_{12}F_{11}H]^{2-}$-based electrolytes for supercapacitors: Overcharge protection. *Electrochemistry Communications* 12:636–639.
944. Taniki, R., K. Matsumoto, T. Nohira, and R. Hagiwara. 2013. Evaluation of double-layer and redox capacitances of activated carbon electrodes in *N*-ethyl-*N*-methylpyrrolidinium fluorohydrogenate ionic liquid. *Journal of the Electrochemical Society* 160:A734–A738.
945. Tooming, T., T. Thomberg, L. Siinor, K. Tõnurist, A. Jänes, and E. Lust. 2014. A type high capacitance supercapacitor based on mixed room temperature ionic liquids containing specifically adsorbed iodide anions. *Journal of the Electrochemical Society* 161:A222–A227.
946. Yamazaki, S., T. Ito, M. Yamagata, and M. Ishikawa. 2012. Non-aqueous electrochemical capacitor utilizing electrolytic redox reactions of bromide species in ionic liquid. *Electrochimica Acta* 86:294–297.
947. Zhou, X., J. Zhou, Y. Yin, and A. N. Mansour. 2009. Development and study of mediator-enhanced polymer electrolyte supercapacitors. *ECS Transactions* 19:9–22.
948. Zhou, J., Y. Yin, A. N. Mansour, and X. Zhou. 2011. Experimental studies of mediator-enhanced polymer electrolyte supercapacitors. *Electrochemical and Solid-State Letters* 14:A25–A28.
949. Yin, Y., J. Zhou, A. N. Mansour, and X. Zhou. 2011. Effect of NaI/I_2 mediators on properties of PEO/$LiAlO_2$ based all-solid-state supercapacitors. *Journal of Power Sources* 196:5997–6002.
950. Yu, H., J. Wu, L. Fan et al. 2011. Improvement of the performance for quasi-solid-state supercapacitor by using PVA–KOH–KI polymer gel electrolyte. *Electrochimica Acta* 56:6881–6886.
951. Zhou, J., J. Cai, S. Cai, X. Zhou, and A. N. Mansour. 2011. Development of all-solid-state mediator-enhanced supercapacitors with polyvinylidene fluoride/lithium trifluoromethanesulfonate separators. *Journal of Power Sources* 196:10479–10483.

952. Senthilkumar, S. T., R. K. Selvan, J. S. Melo, and C. Sanjeeviraja. 2013. High performance solid-state electric double layer capacitor from redox mediated gel polymer electrolyte and renewable tamarind fruit shell derived porous carbon. *ACS Applied Materials & Interfaces* 5:10541–10550.
953. Senthilkumar, S. T., R. K. Selvan, N. Ponpandian, and J. S. Melo. 2012. Redox additive aqueous polymer gel electrolyte for an electric double layer capacitor. *RSC Advances* 2:8937–8940.
954. Ma, G. F., E. K. Feng, K. J. Sun, H. Peng, J. J. Li, and Z. Q. Lei. 2014. A novel and high-effective redox-mediated gel polymer electrolyte for supercapacitor. *Electrochimica Acta* 135:461–466.
955. Yu, H., J. Wu, L. Fan et al. 2012. A novel redox-mediated gel polymer electrolyte for high-performance supercapacitor. *Journal of Power Sources* 198:402–407.
956. Yu, F. D., M. L. Huang, J. H. Wu et al. 2014. A redox-mediator-doped gel polymer electrolyte applied in quasi-solid-state supercapacitors. *Journal of Applied Polymer Science* 131:39784.
957. Mansour, A. N., J. J. Zhou, and X. Y. Zhou. 2014. X-ray absorption spectroscopic study of sodium iodide and iodine mediators in a solid-state supercapacitor. *Journal of Power Sources* 245:270–276.

3 Compatibility of Electrolytes with Inactive Components of Electrochemical Supercapacitors

Cheng Zhong, Daoming Sun, Yida Deng, Wenbin Hu, Jinli Qiao, and Jiujun Zhang

CONTENTS

3.1 INTRODUCTION

As fully discussed in Chapter 2, the electrolyte has complex interactions with the electrode materials (active components) of electrochemical supercapacitors (ESs), which play an important role in the performance of ESs. Besides the active component of ESs, the compatibility or possible interaction between the electrolyte and inactive components such as current collectors, binders, and separators should also be considered. For example, the possible corrosion of current collectors in certain electrolytes could reduce the operative cell voltage and decrease the lifetime of ESs. Besides, the transfer of electrolyte ions across the separator could affect the equivalent series resistance (ESR) and the power performance of the ES. Therefore, the inactive components of ESs should be compatible with the electrolytes and electrode materials.

This chapter will provide a discussion of the compatibility issue or the possible interaction between electrolytes and inactive components of ESs including current collectors, binders, and separators.

3.2 CURRENT COLLECTORS

The chemical stability and electrochemical stability of current collectors in certain electrolytes have a profound influence on their performance characteristics such as the operating cell voltage, reliability, and lifetime of ESs. Besides, the contact between the active-electrode materials and the current collector may contribute a lot to the ESR of ESs in some cases. Furthermore, the loading amount and the utilization degree of active-electrode materials are also strongly dependent on the morphology of the current collectors.

For ESs using a strong acid electrolyte such as 1 M H_2SO_4, some metallic materials with high-corrosion resistance such as gold (Au) are traditionally used [1,2]. To save the cost of the current collectors, other materials, such as indium tin oxide (ITO) [3,4], carbon-based materials [5–7], and electrically conductive polymers (ECPs) [8], have been investigated as materials to fabricate the current collectors used in strong acid electrolyte-based ESs. The use of these recently developed current collectors may have other advantages. For instance, because of the relatively good conductivity and high transparency of ITO, transparent ESs can be fabricated with the use of ITO current collectors [3]. Noticeably, ITO may suffer from corrosion problems in some electrolytes. Folcher et al. [9] found ITO corroded in HCl solution at a potential of 0.9 V versus a saturated calomel electrode (SCE). This result was confirmed by quartz crystal microbalance (QCM), characterization of corrosion morphology, and electrochemical measurements. As the concentration of the HCl electrolyte increased from 0.04 to 1 M, QCM measurements indicated that the corrosion rate increased drastically from ≤5 to 3400 ng s^{-1} cm^{-2}. Obvious dissolution of the ITO electrode was observed after a potential scan from 0.33 to 1.33 V versus SCE in 0.1 M HCl (Figure 3.1). An intergranular corrosion attack of ITO was found by transmission electron microscopy images. The authors proposed that the corrosion mechanism of ITO was related to the electrochemical formation of Cl° and OH° radical species that

FIGURE 3.1 Scanning electron microscopy (SEM) image of an ITO layer immersed in a 0.1-M HCl solution after a potential scan from 0.3 to 1.2 V (SCE). (Reprinted from Anodic corrosion of indium tin oxide films induced by the electrochemical oxidation of chlorides, 301, Folcher, G. et al., *Thin Solid Films*, 242–248, Copyright 1997, with permission from Elsevier.)

destroyed the In–O surface bonds. In addition, using current collectors with carbon or ECP materials is beneficial for fabricating flexible ESs [5,8]. Furthermore, the use of free-standing electrodes such as the carbon-based composite paper for ESs has the benefit that it can be used without the use of additional current collectors [10].

In the case of ESs with alkaline electrolytes, nickel (Ni) current collectors have been extensively used owing to their good chemical and electrochemical stabilities in alkaline electrolytes, and the relatively low cost [11–17]. In some cases, Ni foam is used as the current collector to increase the total surface area, as well as the loading amount and utilization of the active-electrode materials [11,12,14,15]. However, it should be noted that the use of an Ni current collector may contribute to the additional pseudocapacitance due to the presence of Ni hydroxide/oxide on the surface, which can bring about an exaggerated capacitance value concerning the active-electrode materials, especially when the loading amount of the electrode material is small [18]. Therefore, Xing et al. [18] suggested the use of an electrochemically inert current collector such as titanium (Ti) and platinum (Pt) foil to characterize the true capacitive properties of active-electrode materials such as Ni and cobalt (Co) oxides/hydroxides. Apart from pure Ni, other types of metallic materials, such as stainless steel [19,20], Inconel 600 Ni-based alloy [21], and nanoporous Au [22], have also been explored to be used as current collectors in ESs using alkaline electrolytes. Owing to their good conductivity, high chemical, electrochemical, and thermal stabilities, lightweight, high flexibility, and good mechanical strength, carbon-based materials, such as carbon fibers [23], carbon fiber paper [24], carbon cloth [25], carbon nanotubes (CNTs) [26], and ultrathin-graphite foam [27], have all recently received growing interest.

Compared with strong acid and alkaline electrolytes, neutral aqueous electrolytes generally have much lower corrosiveness. Therefore, a wide variety of current-collector materials have been investigated for neutral aqueous electrolyte-based ESs. Some cited materials for current collectors are Ni [28–30], Ti [31], copper (Cu) [32], Ti oxynitride [33], stainless steel [34–38], modified TiO_2[39], ITO [40], CNT [41,42], etc. It is interesting to note that Wang et al. [43] found the type and morphology of the current collectors (Ti mesh and Ni foam) to be affecting the performance of ESs greatly. The nature of the current collector has a profound influence on the impedance behavior of the MnO_2–CNTs composite electrode (Figure 3.2), as well as the galvanic charge/discharge profiles (Figure 3.3). Figure 3.3 shows the specific capacitance of ES using Ni foam as current collectors had the highest value, suggesting that the Ni foam exhibited a better current-collection property than the Ti mesh.

With the aim to enhance the performance of ESs such as lowering the contact resistance between the electrode materials and the current collector, it has been proposed to directly grow the electrode materials on the current collector [44–47]. Du and Pan [45] prepared multiwalled carbon nanotube (MWCNT) thin films on the Ni foil current collector by the electrophoretic deposition technique. The obtained ES showed near-ideal rectangular cyclic voltammograms even at a very high scan rate of 1000 mV s^{-1}, a high specific power density of more than 20 kW kg^{-1}, and an excellent frequency response. Miller et al. [46] reported that vertically oriented graphene nanosheets grown directly on metal current collectors (Figure 3.4) could minimize both electronic and ionic resistance, with resistance–capacitance (RC) time constants of less than 200 μs. This is beneficial for achieving ES with high-power performance.

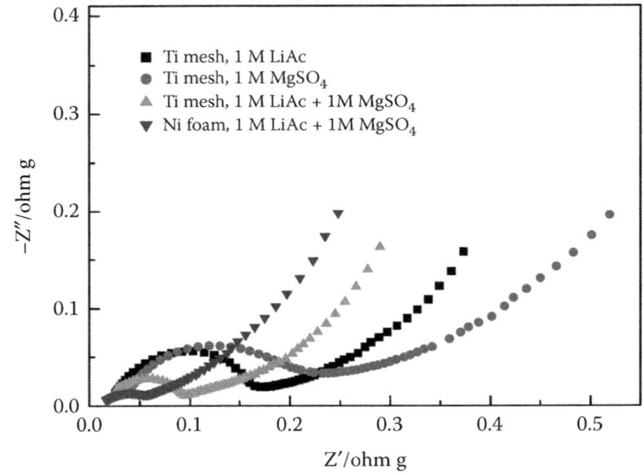

FIGURE 3.2 Nyquist plots of AC impedance of the nano-MnO_2/CNTs composite electrodes with different current collectors in different electrolytes. (With kind permission from Springer Science+Business Media: Pseudocapacitive behaviors of nanostructured manganese dioxide/carbon nanotubes composite electrodes in mild aqueous electrolytes: Effects of electrolytes and current collectors, *Journal of Solid State Electrochemistry*, 12, 2008, 1101–1107, Wang, Y. Q., A. B. Yuan, and X. L. Wang.)

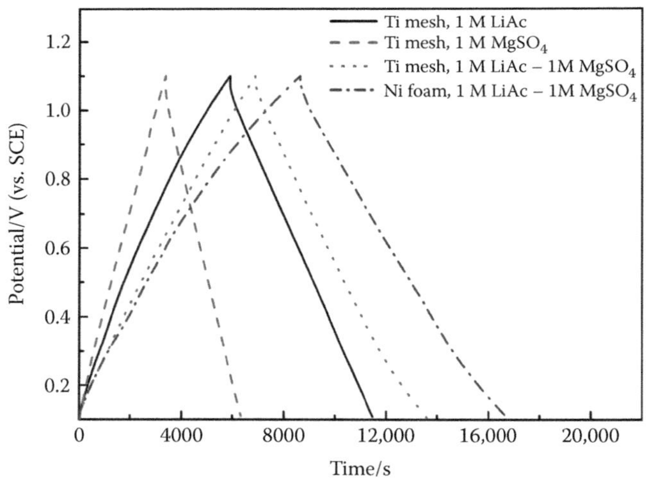

FIGURE 3.3 Galvanostatic charge/discharge profiles of the nano-MnO_2/CNTs composite electrodes with different current collectors in different electrolytes at a current rate of 25 mA g^{-1}. (With kind permission from Springer Science+Business Media: Pseudocapacitive behaviors of nanostructured manganese dioxide/carbon nanotubes composite electrodes in mild aqueous electrolytes: Effects of electrolytes and current collectors, Journal of Solid State Electrochemistry, 12, 2008, 1101–1107, Wang, Y. Q., A. B. Yuan, and X. L. Wang.)

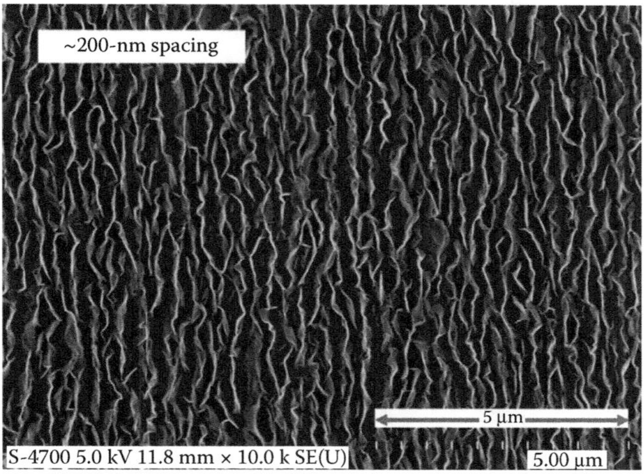

FIGURE 3.4 Vertically oriented and aligned graphene nanosheets. Alignment was achieved using a static direct current (dc) electric field. (Reprinted from Graphene electric double layer capacitor with ultra-high-power performance, 56, Miller, J. R., R. A. Outlaw, and B. C. Holloway, *Electrochimica Acta*, 10443–10449, Copyright 2011, with permission from Elsevier.)

Regarding ESs with organic electrolytes, aluminum (Al) has been the most commonly used material for current collectors [48–57]. The degradation of Al current collectors was found to be related to the performance aging of the organic electrolyte-based ESs [55,56]. It has been found by Jänes et al. [56] that during the aging tests of a microporous carbide-derived carbon-based ES with 1.5 M $(C_2H_5)_3CH_3NBF_4$/ACN electrolyte, partial dissolution of Al current collectors from the positive electrode and the deposition of Al on the negative electrode occurred (Figure 3.5). Besides, Väli et al. [52] found that the electrochemical stability of the Al current collector was strongly dependent on the nature of the conducting salt in the organic solvent. The operative cell voltages of ESs were 3.4, 3.2, and 2.5 V for $NaPF_6$, $NaClO_4$, and NaFSI electrolytes, respectively. NaFSI electrolyte-based ES exhibited the lowest cell voltage due to the oxidation of Al current collector in this electrolyte.

To improve the performance of organic electrolyte-based ES, various surface treatment methods have been proposed for Al current collectors [48–51,54,57]. For example, Sato et al. [57] modified the Al foil surface with a thermocuring coating composed of glycol–chitosan, pyromellitic acid, and carbon powder. Compared with the unmodified Al foil current collector, the modified Al foil resulted in a much higher rate capability of the ES using the 1 M [DEME][BF_4]/propylene carbonate (PC) organic electrolyte. Lei et al. [49] found that the contact resistance of the carbon/Al in tetraethylammonium tetrafluoroborate ($TEABF_4$)/PC organic electrolyte greatly decreased by a carbon-modified Al current collector with Al_4C_3 nanowhiskers. Other methods, such as etching [48], sol–gel coating [48,54], and carbon modification [50]

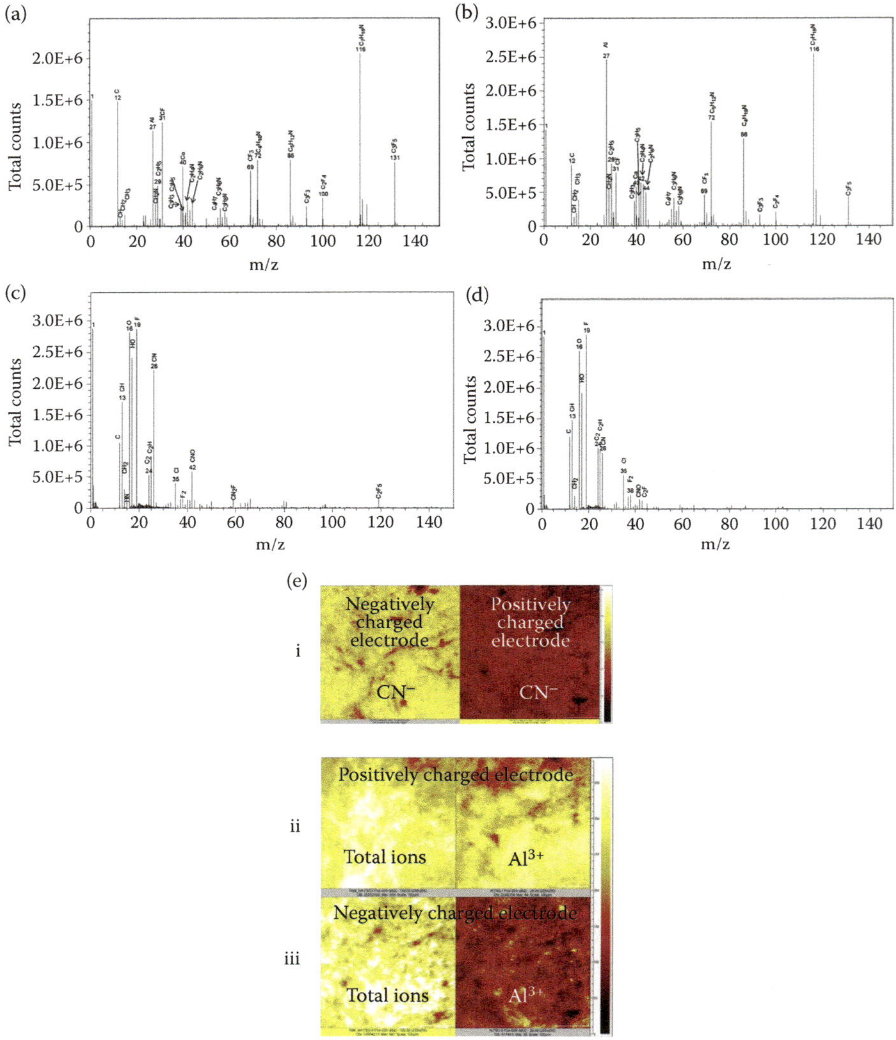

FIGURE 3.5 Time-of-flight–secondary ion mass spectrometry (TOF–SIMS) spectra for titanium carbide (TiC)–CDC supercapacitor for (a, c) positive and (b, d) negatively charged electrode for (c, d) negatively and (a, b) positively charged secondary ions after 40,000 charge/discharge cycles. (e) Positive and negative ion-specific images of (i) CN^- for negatively and positively charged TiC–CDC electrode; (ii) total ions and Al^{3+} ions for positively and (iii) negatively charged TiC–CDC electrode. Image sizes are $400 \times 400\ \mu m^2$. (Reproduced with permission from The Electrochemical Society from Jänes, A., J. Eskusson, R. Kanarbik, A. Saar, and E. Lust. 2012. *Journal of the Electrochemical Society* 159:A1141–A1147.)

for Al current collectors, have also been studied to decrease the contact resistance between the current collectors and the electrode materials. To decrease the interface resistance between the current collector and the electrode, hot pressing [58] and roll pressing [59] have been proposed for improving the contact quality between the current collector and the electrode.

For the ionic liquid (IL) electrolytes, Kühnel et al. [60,61] studied the anodic electrochemical stability of the Al current collector in three ILs with different anions: $[PYR_{14}][FSI]$, $[PYR_{14}][TFSI]$, and $[PYR_{14}][FTFSI]$. Figure 3.6 shows that after cyclic voltammetry (CV) tests, Al foil current collectors exhibit good passivation in $[PYR_{14}][TFSI]$ IL, while pitting corrosion of the Al current collector was observed in $[PYR_{14}][FSI]$ IL [61]. Interestingly, the authors also found that the anodic stability of Al current collectors in $[PYR_{14}][FSI]$ could be improved remarkably by controlling the chloride impurities level to below 1 ppm [61]. Regarding the $[PYR_{14}][FTFSI]$ IL electrolyte, the same group [60] found Al current collectors to exhibit excellent anodic corrosion resistance in this IL. The positive Al current collectors showed mild pitting corrosion only when electrical double-layer capacitors (EDLCs) were subjected to a constant cell potential of 3.2 or 3.5 V for 140 h. Brandt et al. [62] investigated the corrosion stability of Al foil current collectors in IL–organic electrolyte mixtures, that is, trimethyl-sulfonium bis([trifluoromethyl] sulfonyl)imide ($[Me_3S][TFSI]$)–PC. Two different electrolyte mixtures were studied: one with 3.8 M $[Me_3S][TFSI]$ in PC (86 wt% of $[Me_3S][TFSI]$) and another with 1.9 M $[Me_3S][TFSI]$ in PC (50 wt% of $[Me_3S][TFSI]$). As shown in Figure 3.7, CV measurements have shown that the oxidative current of the Al foil current collector in 3.8 M $[Me_3S][TFSI]$–PC was one order of magnitude lower than that obtained in 1.9 M $[Me_3S][TFSI]$–PC. After the CV tests, the corrosion of the Al current collector was clearly observed in 1.9 M $[Me_3S][TFSI]$–PC whereas the Al current collector exhibited much less corrosion in 3.8 M $[Me_3S][TFSI]$–PC (see inset to Figure 3.7). As a result, EDLCs using the 1.9 M $[Me_3S][TFSI]$–PC electrolyte had a lower operating cell voltage and poorer cycling performance compared with EDLCs with 3.8 M $[Me_3S][TFSI]$–PC electrolyte.

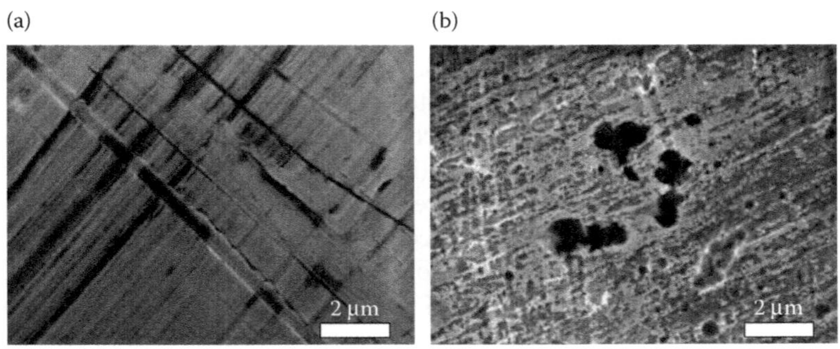

FIGURE 3.6 SEM images of Al electrodes taken after a CV experiment between 3.3 and 5 V in (a) $PYR_{14}TFSI$ and (b) $PYR_{14}FSI$. (Reprinted from Comparison of the anodic behavior of aluminum current collectors in imide-based ionic liquids and consequences on the stability of high voltage supercapacitors, 249, Kühnel, R. S., and A. Balducci, *Journal of Power Sources*, 163–171, Copyright 2014, with permission from Elsevier.)

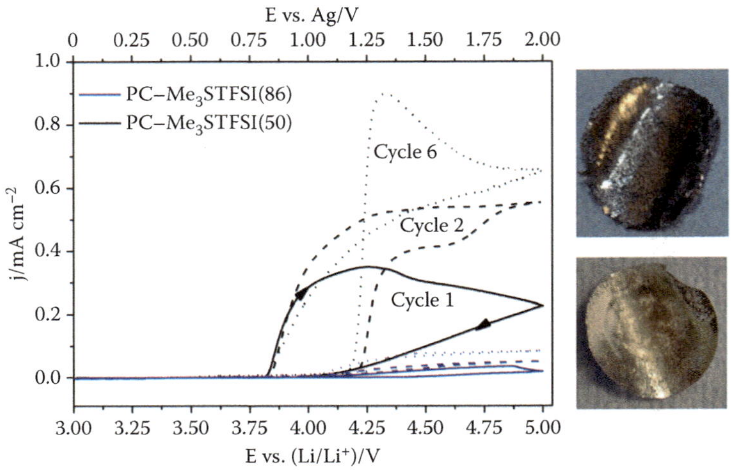

FIGURE 3.7 Cyclic voltammograms of the Al current collector in PC–Me_3STFSI(86) and PC–Me_3STFSI(50) at 20°C and at a scan rate of 0.5 mV s^{-1}. The images show the recovered Al current collectors after 100 cycles (top PC–Me_3STFSI(50), bottom PC–Me_3STFSI(86)). (Brandt, A. et al. 2013. An investigation about the use of mixtures of sulfonium-based ionic liquids and propylene carbonate as electrolytes for supercapacitors. *Journal of Materials Chemistry A* 1:12669–12678. Reproduced by permission of The Royal Society of Chemistry.)

With regard to the redox-active electrolyte, Lota and Frackowiak [63] found the cycling stability of the electrode in the 1 M KI electrolyte to be significantly affected by the type of the metallic current collector. After 10,000 cycles, a moderate decrease of the specific capacitance (<20%) was observed for the Au current collector, whereas an increase of specific capacitance from 235 to 300 F g^{-1} was found for the stainless-steel current collector. Meller et al. [64] also pointed out that it is not suitable to use an Au current collector in iodide electrolytes due to its reaction with iodides, while it is appropriate to use a stainless-steel current collector since it resulted in good long-term stability of ES.

In the case of solid- or quasi-solid-state electrolytes, there are other factors that need to be considered. For example, the choice of the current collectors should take into account the type of solid-state ES (i.e., whether it is flexible, stretchable, or transparent). For example, to meet the requirements of various solid-state ESs, a wide variety of carbon-based current collectors have been developed [65–67]. Notarianni et al. [68] found that for poly(vinyl alcohol) (PVA)/H_3PO_3 hydrogel electrolyte-based ES, the performance of the ES based on CNT film as the current collector was comparable to the Au current collector, and the CNT film also provides a stronger binding to the graphene film electrodes. Niu et al. [69] reported a highly compressible, all-solid-state ES based on polyaniline (PANI)–single-walled carbon nanotubes (SWCNTs)–sponge electrodes, using PVA/H_2SO_4 gel as the electrolyte. The performance of this all-solid-state ES could be largely maintained even when compressed under 60% strain (Figure 3.8).

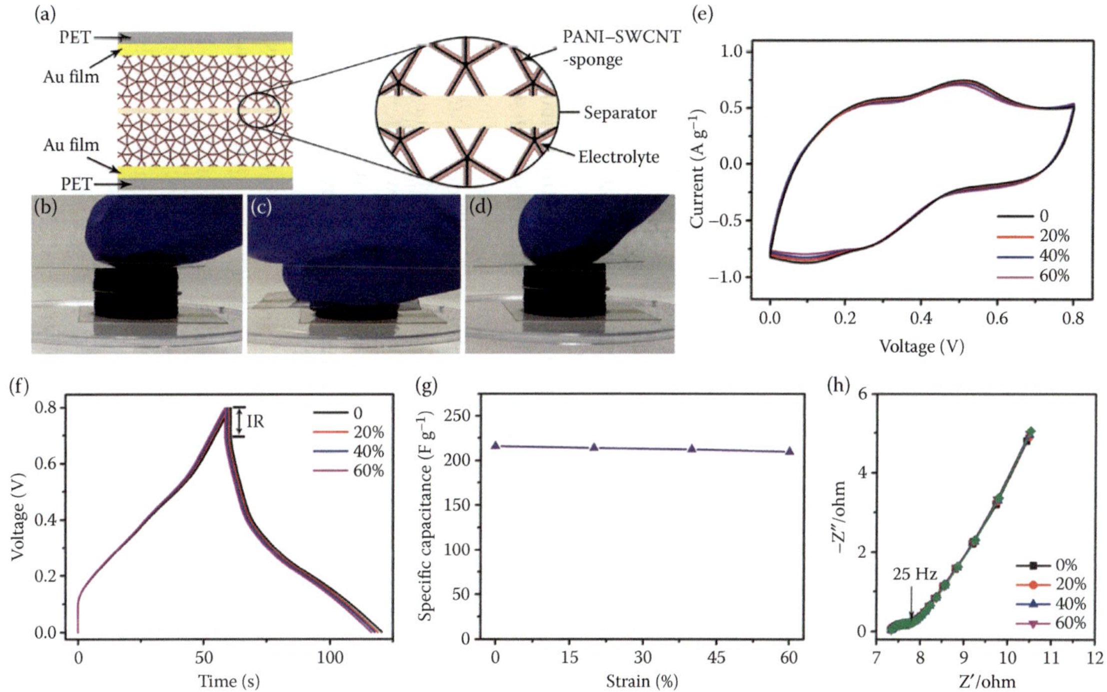

FIGURE 3.8 (a) Schematic diagram of an all-solid-state-integrated supercapacitor based on PANI–SWCNTs–sponge. (b–d) Real-time optical images of a supercapacitor showing the processes of compression and recovery, (e) CV curves, (f) galvanostatic charge/discharge curves, and (g) specific capacitances of a representative compressible supercapacitor at different strains. Scan rate: 10 mV s^{-1}, charge/discharge current: 0.64 A s^{-1}. (h) The Nyquist impedance plot of the compressible supercapacitor at different strains with frequency ranging from 10 kHz to 0.1 Hz. (Niu, Z. Q. et al.: Highly compressible and all-solid-state supercapacitors based on nanostructured composite sponge. 1–5. 2015. Copyright Wiley-VCH Verlag GmbH & Co. KGaA. Reproduced with permission.)

In addition, to simplify the fabrication process and also to avoid the potential problems associated with the use of additional current collectors, free-standing electrodes typically based on carbon-based materials have been developed [68–74]. ESs based on these free-standing electrodes have the advantage of avoiding the use of additional current collectors.

3.3 BINDERS

Electrode materials are generally composed of active particles or powders (e.g., AC powders) and binders. Binders can maintain the structural integrity of the active particles or powders and help electrode materials to adhere uniformly onto the current collector. To date, fluorinated polymeric binders, such as poly(vinylidene fluoride) (PVDF) and polytetrafluoroethylene (PTFE), have been extensively used for ESs.

The nature and the amount of the binders can also greatly influence the performance of the electrodes and the assembled ESs [75–78]. For instance, Tsay et al. [76] found that for Na_2SO_4 electrolyte-based ESs using carbon BP2000 electrodes, the maximum value of the specific capacitance was obtained when the PTFE binder content was 5 wt% (Figure 3.9). Decreasing or further increasing the content of PTFE decreased the specific capacitance. Since PTFE is hydrophobic, too large an amount of PTFE could inhibit the penetration of aqueous electrolyte ions into pores in the carbon electrode materials, resulting in a lower specific capacitance [76]. Paul et al. [79] added a small amount of polyvinylpyrrolidone (PVP) (3%) into the PTFE binder, and found it could effectively improve the wettability of the electrode materials in 1 M KCl aqueous electrolyte. The presence of binders was also found to change (usually decrease) the porosity of the porous electrodes, which affected the effective surface area and the ion diffusion in the pores [77]. Abbas et al. [77] demonstrated that such porosity change was also dependent on the nature of the carbon material and the property of the electrolyte. Timperman et al. [80] investigated two different binders for AC electrodes, that is, PVDF and carboxymethyl cellulose (CMC)–styrene–butadiene rubber (SBR). It was found that AC electrodes containing PVDF showed higher wettability in diisoproyl-ethyl-ammonium bis(trifluoromethanesulfonyl) imide ([DIPEA][TFSI]) IL compared with the electrodes containing CMC–SBR binders. However, AC electrodes containing PVDF or CMC–SBR binders exhibited similar wettability in pyrrolidinium bis(trifluoromethanesulfonyl)imide ([PYRR][TFSI]) IL. Owing to this different wettability behavior, the electrochemical performance such as the specific capacitance of the electrodes in certain IL electrolytes was also different.

Besides EDLCs, Liu and Pickup [81] found that adding a certain amount of Nafion® binder into the ruthenium (Ru) oxide powder-based electrodes was favorable for improving the power density of pseudocapacitors. As shown in Figure 3.10, the addition of an additional 5% Nafion binder led to a 39% increase in the average power density of the Ru oxide-based pseudocapacitor for full discharge at 1 A cm^{-2}. The high proton-conducting property could contribute to the improved power performance [81]. Cao et al. [78] compared the effect of PTFE and PVDF binders on the performance of lithium-ion capacitor (LIC) with the $LiPF_6$/ethylene

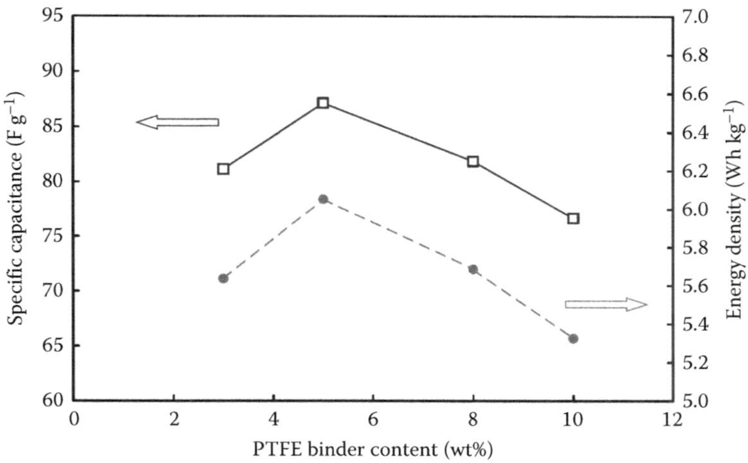

FIGURE 3.9 Specific capacitance and energy density of carbon BP2000 electrode material as a function of PTFE contents. (Reprinted from Effects of electrode layer composition/thickness and electrolyte concentration on both specific capacitance and energy density of supercapacitor, 60, Tsay, K. C., L. Zhang, and J. J. Zhang, *Electrochimica Acta*, 428–436, Copyright 2012, with permission from Elsevier.)

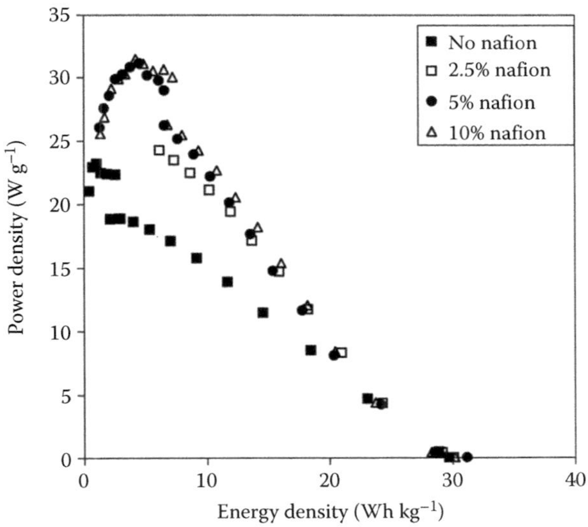

FIGURE 3.10 Ragone plots for supercapacitors with different amounts of Nafion binder. Ru oxide loadings were 9.51, 10.35, 10.34, and 10.14 mg for 0%, 2.5%, 5%, and 10% Nafion, respectively. The discontinuities at high powers are due to the use of a different potentiostat (Solartron 1286) with lower lead resistances for experiments at currents >1 A. (Reprinted from Ru oxide supercapacitors with high loadings and high power and energy densities, 176, Liu, X., and P. G. Pickup, *Journal of Power Sources*, 410–416, Copyright 2008, with permission from Elsevier.)

carbonate–dimethyl carbonate (EC–DMC) electrolyte. It was found that the negative electrode containing the PTFE binder resulted in higher energy and power densities of LIC compared with that containing the PVDF binder.

With the environmental concern about the selection of ES components, environmentally friendly binders such as fluorine-free binders have attracted growing attention. For example, some researchers investigated other binder materials, such as natural cellulose [82,83], PVP [84], and polyacrylic acid (PAA) [85]. These binders may be used as alternatives to fluorinated polymeric binders. Aslan et al. [84] found that in comparison with the PTFE or PVDF binder, the application of the PVP binder resulted in a significantly higher pore volume and larger specific surface area of the AC-based electrodes (normalized to the amount of AC). Lee et al. [85] reported that the PAA binders could promote a uniform distribution of the electrolyte throughout the active-electrode layer because of their ability to adsorb and swell the electrolytes. Shin et al. [86] found that deoxyribonucleic acid (DNA) was also able to act as a mechanical binder for SWCNT-based electrodes in the 2 M NaCl electrolyte. Some promising results have also been obtained for natural cellulose binders [82,83]. For example, Varzi et al. [82] found that cellulose-based EDLCs with the [PYR_{14}][TFSI] IL electrolyte showed a much better performance compared with the PVDF-based ones after float voltage tests at 3.7 V. In addition, some ECPs, such as polypyrrole (PPy) [87], poly(3,4-ethylenedioxythiophene) (PEDOT) [88,89], and PANI [90], have also been proposed as binders to be used in electrode materials of ESs. In contrast with traditional binders such as PTFE and PVDF, ECP-type binders are electrically conductive and can also have additional pseudocapacitance.

However, to avoid some negative influences associated with the binders (e.g., decrease in the effective surface area and increase in electrode resistance), binder-free electrode materials have been investigated for ESs [91–93].

3.4 SEPARATORS

As shown in Figure 1.3, the separator is placed between the negative and positive electrodes of the ES to avoid contact and the electron transfer between the two electrodes. There are several critical requirements for the separators of ESs [94]: (1) good conductivity for the transfer of electrolyte ions through the separator, (2) high electronic-insulating property to prevent short circuits, (3) high chemical and electrochemical stability in the electrolyte used, and (4) sufficient mechanical strength to contribute to ES durability. Thin and highly porous insulating films are generally used to make the separators of ESs to meet the above requirements. Several factors, such as the type of ES, electrode materials, electrolytes, working temperature, and operative cell voltage, should be considered for the selection or development of appropriate separators. Typically used materials for separators include cellulose, a porous polymer membrane, and glass fiber. Bittner et al. [55] found that the performance aging of EDLCs using a paper separator could be accelerated by the trace water in the $TEABF_4$/acetonitrile (ACN) organic electrolyte. When ILs are used as electrolytes for ESs, due to the viscous nature of ILs, the ion transfer resistance through the separator becomes highly important [95,96].

Furthermore, the physical (e.g., thickness, porosity, and pore size distribution) and chemical (e.g., composition) properties of the separator play an important role in a series of performances of EDLCs, such as an ideal polarizability limit, specific capacitance, characteristic time constant, ESR, energy, and power densities [95,97–100]. For example, several studies found that separators are thinner or those with a higher porosity and higher pore size led to a better power performance, due to the lower ESR [97,100]. Karabelli et al. [101] reported that a PVDF separator provided higher ionic conductivity for a $TEABF_4$/ACN electrolyte compared with commercial cellulose and Celgard® (based on polyethylene [PE]/polypropylene [PP] separators). Besides, it should be noted that the influence of the separator parameters on the performance of EDLCs also depends on the investigated systems. For example, some studies [98,102] found there is a weak dependence of the separator parameters on the limits of ideal polarizability and the specific energy density whereas other studies [95] have shown a noticeable dependence. To increase the wettability and thus to improve the performance of polymer-based separators (e.g., PP) in aqueous electrolytes, surface modifications of the separators have been proposed [103,104].

Besides EDLCs, the performance of pseudocapacitors and hybrid ESs is also dependent on the properties of separators. Liu and Pickup [105] compared the effect of four different separators on the performance of RuO_2-based pseudocapacitors using the 0.5-M H_2SO_4 electrolyte. As shown in Figure 3.11, the value of the energy density of ES using Nafion-115 was the highest (31.2 Wh kg^{-1} at 1 mA cm^{-2}), whereas the energy density value of ESs with Celgard separator was the lowest (23.4 Wh kg^{-1} at 1 mA cm^{-2}). The other Nafion separators led to intermediate energy densities.

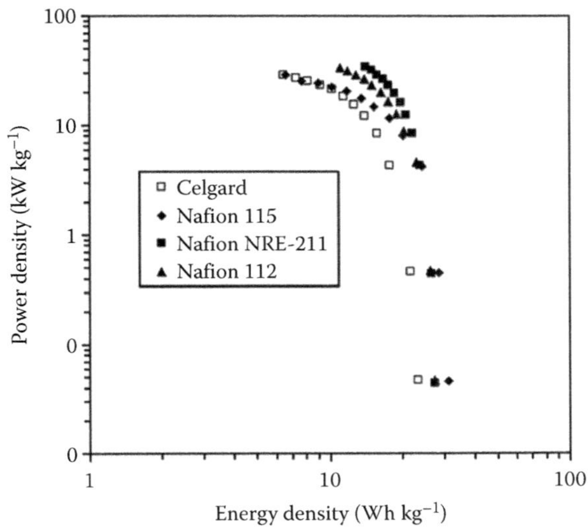

FIGURE 3.11 Ragone plots for Ru oxide (ca. 10 mg) supercapacitors with different separators. Measurements were made by constant current discharge from 1.0 V. (Liu, X. R., and P. G. Pickup. 2008. Performance and low temperature behavior of hydrous ruthenium oxide supercapacitors with improved power densities. *Energy and Environmental Science* 1:494–500. Reproduced by permission of The Royal Society of Chemistry.)

As mentioned in Section 2.6, the selection of a suitable separator was also beneficial for reducing the self-discharge of redox-active electrolyte-based ESs [106,107].

Recently, researchers have developed some new separator materials for ESs, such as graphene oxide film [108] and eggshell membrane [109]. For example, Yu et al. [109] found the use of eggshell membrane as the separator for ESs gave a lower resistance, higher specific capacitance, quicker charge/discharge ability (4.76 s), and good cyclic stability (92% capacitance retention after 10,000 cycles) compared with the PE separator.

3.5 SUMMARY

Since the components of ESs are strongly coupled with each other, it is highly desirable to consider the compatibility or possible interactions between the electrolytes and both the active-electrode materials and inactive components of ESs. Several important aspects were discussed in this chapter to enable readers to (1) understand the compatibility issues or possible interactions between electrolytes and inactive components (current collector, binder, and separator) of ESs; (2) recognize how the interactions between electrolytes and inactive components affect the performance of ESs; (3) understand the importance of the compatibility of inactive components of ESs with the electrolytes and electrode materials; (4) understand the importance of designing or choosing appropriate inactive components according to the electrolyte systems and type of ESs; and (5) obtain an overview of choices or design strategies of inactive components of ESs to improve the device performance with a given electrolyte.

REFERENCES

1. Pech, D., M. Brunet, T. M. Dinh, K. Armstrong, J. Gaudet, and D. Guay. 2013. Influence of the configuration in planar interdigitated electrochemical micro-capacitors. *Journal of Power Sources* 230:230–235.
2. Liu, C. C., D. S. Tsai, D. Susanti, W. C. Yeh, Y. S. Huang, and F. J. Liu. 2010. Planar ultracapacitors of miniature interdigital electrode loaded with hydrous RuO_2 and RuO_2 nanorods. *Electrochimica Acta* 55:5768–5774.
3. Ryu, I., M. Yang, H. Kwon et al. 2014. Coaxial RuO_2–ITO nanopillars for transparent supercapacitor application. *Langmuir* 30:1704–1709.
4. Sumboja, A., X. Wang, J. Yan, and P. S. Lee. 2012. Nanoarchitectured current collector for high rate capability of polyaniline based supercapacitor electrode. *Electrochimica Acta* 65:190–195.
5. Cho, H. W., L. R. Hepowit, H. S. Nam et al. 2012. Synthesis and supercapacitive properties of electrodeposited polyaniline composite electrode on acrylonitrile–butadiene rubber as a flexible current collector. *Synthetic Metals* 162:410–413.
6. Ahn, H. J., W. B. Kim, and T. Y. Seong. 2008. $Co(OH)_2$-combined carbon-nanotube array electrodes for high-performance micro-electrochemical capacitors. *Electrochemistry Communications* 10:1284–1287.
7. Quintero, R., D. Y. Kim, K. Hasegawa, Y. Yamada, A. Yamada, and S. Noda. 2014. Carbon nanotube 3D current collectors for lightweight, high performance and low cost supercapacitor electrodes. *RSC Advances* 4:8230–8237.

8. Wee, B. H., and J. D. Hong. 2014. Multilayered poly(p-phenylenevinylene)/reduced graphene oxide film: An efficient organic current collector in an all-plastic supercapacitor. *Langmuir* 30:5267–5275.
9. Folcher, G., H. Cachet, M. Froment, and J. Bruneaux. 1997. Anodic corrosion of indium tin oxide films induced by the electrochemical oxidation of chlorides. *Thin Solid Films* 301:242–248.
10. Yun, Y. S., M. E. Lee, M. J. Joo, and H. J. Jin. 2014. High-performance supercapacitors based on freestanding carbon-based composite paper electrodes. *Journal of Power Sources* 246:540–547.
11. Bo, Z., W. G. Zhu, W. Ma et al. 2013. Vertically oriented graphene bridging active-layer/current-collector interface for ultrahigh rate supercapacitors. *Advanced Materials* 25:5799–5806.
12. Gong, X. F., J. P. Cheng, F. Liu, L. Zhang, and X. B. Zhang. 2014. Nickel–cobalt hydroxide microspheres electrodepositioned on nickel cobaltite nanowires grown on Ni foam for high-performance pseudocapacitors. *Journal of Power Sources* 267:610–616.
13. Miller, J. R., R. A. Outlaw, and B. C. Holloway. 2010. Graphene double-layer capacitor with AC line-filtering performance. *Science* 329:1637–1639.
14. Gu, L., Y. W. Wang, R. Lu, L. Guan, X. S. Peng, and J. Sha. 2014. Anodic electrodeposition of a porous nickel oxide–hydroxide film on passivated nickel foam for supercapacitors. *Journal of Materials Chemistry A* 2:7161–7164.
15. Guo, D., H. Zhang, X. Yu et al. 2013. Facile synthesis and excellent electrochemical properties of $CoMoO_4$ nanoplate arrays as supercapacitors. *Journal of Materials Chemistry A* 1:7247–7254.
16. Huang, H. F., Y. M. Tang, L. Q. Xu, S. L. Tang, and Y. W. Du. 2014. Direct formation of reduced graphene oxide and 3D lightweight nickel network composite foam by hydrohalic acids and its application for high-performance supercapacitors. *ACS Applied Materials and Interfaces* 6:10248–10257.
17. Zhang, G. G., W. F. Li, K. Y. Xie, F. Yu, and H. T. Huang. 2013. A one-step and binder-free method to fabricate hierarchical nickel-based supercapacitor electrodes with excellent performance. *Advanced Functional Materials* 23:3675–3681.
18. Xing, W., S. Z. Qiao, X. Z. Wu et al. 2011. Exaggerated capacitance using electrochemically active nickel foam as current collector in electrochemical measurement. *Journal of Power Sources* 196:4123–4127.
19. Pan, S. J., Y. J. Shih, J. R. Chen, J. K. Chang, and W. T. Tsai. 2009. Selective micro-etching of duplex stainless steel for preparing manganese oxide supercapacitor electrode. *Journal of Power Sources* 187:261–267.
20. Xiang, F., J. Zhong, N. Y. Gu et al. 2014. Far-infrared reduced graphene oxide as high performance electrodes for supercapacitors. *Carbon* 75:201–208.
21. Shah, R., X. F. Zhang, and S. Talapatra. 2009. Electrochemical double layer capacitor electrodes using aligned carbon nanotubes grown directly on metals. *Nanotechnology* 20:395202.
22. Chen, L. Y., Y. Hou, J. L. Kang, A. Hirata, and M. W. Chen. 2014. Asymmetric metal oxide pseudocapacitors advanced by three-dimensional nanoporous metal electrodes. *Journal of Materials Chemistry A* 2:8448–8455.
23. Alhebshi, N. A., R. B. Rakhi, and H. N. Alshareef. 2013. Conformal coating of $Ni(OH)_2$ nanoflakes on carbon fibers by chemical bath deposition for efficient supercapacitor electrodes. *Journal of Materials Chemistry A* 1:14897–14903.
24. Huang, L., D. C. Chen, Y. Ding, Z. L. Wang, Z. Z. Zeng, and M. L. Liu. 2013. Hybrid composite $Ni(OH)_2$@$NiCo_2O_4$ grown on carbon fiber paper for high-performance supercapacitors. *ACS Applied Materials and Interfaces* 5:11159–11162.

25. Xu, J., Q. Wang, X. Wang et al. 2013. Flexible asymmetric supercapacitors based upon Co_9S_8 nanorod//Co_3O_4@RuO_2 nanosheet arrays on carbon cloth. *ACS Nano* 7:5453–5462.
26. Zhou, R. F., C. Z. Meng, F. Zhu et al. 2010. High-performance supercapacitors using a nanoporous current collector made from super-aligned carbon nanotubes. *Nanotechnology* 21:345701.
27. Ji, J., L. L. Zhang, H. Ji et al. 2013. Nanoporous $Ni(OH)_2$ thin film on 3D ultrathin-graphite foam for asymmetric supercapacitor. *ACS Nano* 7:6237–6243.
28. Shi, K. Y., and I. Zhitomirsky. 2013. Polypyrrole nanofiber–carbon nanotube electrodes for supercapacitors with high mass loading obtained using an organic dye as a co-dispersant. *Journal of Materials Chemistry A* 1:11614–11622.
29. Xiao, J. W., S. X. Yang, L. Wan, F. Xiao, and S. Wang. 2014. Electrodeposition of manganese oxide nanosheets on a continuous three-dimensional nickel porous scaffold for high performance electrochemical capacitors. *Journal of Power Sources* 245:1027–1034.
30. Liu, D. Q., Q. Wang, L. Qiao et al. 2012. Preparation of nano-networks of MnO_2 shell/Ni current collector core for high-performance supercapacitor electrodes. *Journal of Materials Chemistry* 22:483–487.
31. Su, Z. J., C. Yang, B. H. Xie et al. 2014. Scalable fabrication of MnO_2 nanostructure deposited on free-standing Ni nanocone arrays for ultrathin, flexible, high-performance micro-supercapacitor. *Energy and Environmental Science* 7:2652–2659.
32. Zheng, L., Y. Xu, D. Jin, and Y. Xie. 2010. Well-aligned molybdenum oxide nanorods on metal substrates: Solution-based synthesis and their electrochemical capacitor application. *Journal of Materials Chemistry* 20:7135–7143.
33. Wang, Z. Q., Z. S. Li, J. Y. Feng et al. 2014. MnO_2 nanolayers on highly conductive $TiO_{0.54}N_{0.46}$ nanotubes for supercapacitor electrodes with high power density and cyclic stability. *Physical Chemistry Chemical Physics* 16:8521–8528.
34. Su, Z. J., C. Yang, C. J. Xu et al. 2013. Co-electro-deposition of the MnO_2–PEDOT:PSS nanostructured composite for high areal mass, flexible asymmetric supercapacitor devices. *Journal of Materials Chemistry A* 1:12432–12440.
35. Ratajczak, P., K. Jurewicz, and F. Béguin. 2014. Factors contributing to ageing of high voltage carbon/carbon supercapacitors in salt aqueous electrolyte. *Journal of Applied Electrochemistry* 44:475–480.
36. Ratajczak, P., K. Jurewicz, P. Skowron, Q. Abbas, and F. Béguin. 2014. Effect of accelerated ageing on the performance of high voltage carbon/carbon electrochemical capacitors in salt aqueous electrolyte. *Electrochimica Acta* 130:344–350.
37. Ban, S., J. J. Zhang, L. Zhang, K. Tsay, D. T. Song, and X. T. Zou. 2013. Charging and discharging electrochemical supercapacitors in the presence of both parallel leakage process and electrochemical decomposition of solvent. *Electrochimica Acta* 90:542–549.
38. Brousse, T., P. L. Taberna, O. Crosnier et al. 2007. Long-term cycling behavior of asymmetric activated carbon/MnO_2 aqueous electrochemical supercapacitor. *Journal of Power Sources* 173:633–641.
39. Zhou, H., and Y. R. Zhang. 2014. Electrochemically self-doped TiO_2 nanotube arrays for supercapacitors. *Journal of Physical Chemistry C* 118:5626–5636.
40. Kim, J. S., S. S. Shin, H. S. Han et al. 2014. 1-D structured flexible supercapacitor electrodes with prominent electronic/ionic transport capabilities. *ACS Applied Materials and Interfaces* 6:268–274.
41. Liu, J., F. Mirri, M. Notarianni, M. Pasquali, and N. Motta. 2014. High performance all-carbon thin film supercapacitors. *Journal of Power Sources* 274:823–830.
42. Shen, J. M., A. D. Liu, Y. Tu et al. 2012. Asymmetric deposition of manganese oxide in single walled carbon nanotube films as electrodes for flexible high frequency response electrochemical capacitors. *Electrochimica Acta* 78:122–132.

43. Wang, Y. Q., A. B. Yuan, and X. L. Wang. 2008. Pseudocapacitive behaviors of nanostructured manganese dioxide/carbon nanotubes composite electrodes in mild aqueous electrolytes: Effects of electrolytes and current collectors. *Journal of Solid State Electrochemistry* 12:1101–1107.
44. Du, C. S., and N. Pan. 2006. High power density supercapacitor electrodes of carbon nanotube films by electrophoretic deposition. *Nanotechnology* 17:5314–5318.
45. Du, C. S., and N. Pan. 2006. Supercapacitors using carbon nanotubes films by electrophoretic deposition. *Journal of Power Sources* 160:1487–1494.
46. Miller, J. R., R. A. Outlaw, and B. C. Holloway. 2011. Graphene electric double layer capacitor with ultra-high-power performance. *Electrochimica Acta* 56:10443–10449.
47. Jiang, J., J. P. Liu, R. M. Ding et al. 2011. Large-scale uniform α-$Co(OH)_2$ long nanowire arrays grown on graphite as pseudocapacitor electrodes. *ACS Applied Materials and Interfaces* 3:99–103.
48. Portet, C., P. L. Taberna, P. Simon, and C. Laberty-Robert. 2004. Modification of Al current collector surface by sol–gel deposit for carbon–carbon supercapacitor applications. *Electrochimica Acta* 49:905–912.
49. Lei, C. H., F. Markoulidis, Z. Ashitaka, and C. Lekakou. 2013. Reduction of porous carbon/Al contact resistance for an electric double-layer capacitor (EDLC). *Electrochimica Acta* 92:183–187.
50. Portet, C., P. L. Taberna, P. Simon, and E. Flahaut. 2006. Modification of Al current collector/active material interface for power improvement of electrochemical capacitor electrodes. *Journal of the Electrochemical Society* 153:A649–A653.
51. Wu, H. C., Y. P. Lin, E. Lee et al. 2009. High-performance carbon-based supercapacitors using Al current-collector with conformal carbon coating. *Materials Chemistry and Physics* 117:294–300.
52. Väli, R., A. Laheäär, A. Jänes, and E. Lust. 2014. Characteristics of non-aqueous quaternary solvent mixture and Na-salts based supercapacitor electrolytes in a wide temperature range. *Electrochimica Acta* 121:294–300.
53. Zheng, J. P., and Z. N. Jiang. 2006. Resistance distribution in electrochemical capacitors with spiral-wound structure. *Journal of Power Sources* 156:748–754.
54. Portet, C., P. L. Taberna, P. Simon, E. Flahaut, and C. Laberty-Robert. 2005. High power density electrodes for carbon supercapacitor applications. *Electrochimica Acta* 50:4174–4181.
55. Bittner, A. M., M. Zhu, Y. Yang et al. 2012. Ageing of electrochemical double layer capacitors. *Journal of Power Sources* 203:262–273.
56. Jänes, A., J. Eskusson, R. Kanarbik, A. Saar, and E. Lust. 2012. Surface analysis of supercapacitor electrodes after long-lasting constant current tests in organic electrolyte. *Journal of the Electrochemical Society* 159:A1141–A1147.
57. Sato, T., S. Marukane, T. Morinaga, T. Uemura, K. Fukumoto, and S. Yamazaki. 2011. A thin layer including a carbon material improves the rate capability of an electric double layer capacitor. *Journal of Power Sources* 196:2835–2840.
58. Dsoke, S., X. Tian, C. Taubert, S. Schluter, and M. Wohlfahrt-Mehrens. 2013. Strategies to reduce the resistance sources on electrochemical double layer capacitor electrodes. *Journal of Power Sources* 238:422–429.
59. Kim, I. J., S. H. Yang, S. I. Moon, and H. S. Kim. 2007. Electrochemical characteristics of electric double layer capacitor with sheet type polarizable electrode. *Journal of Power Sources* 164:964–967.
60. Kühnel, R. S., J. Reiter, S. Jeong, S. Passerini, and A. Balducci. 2014. Anodic stability of aluminum current collectors in an ionic liquid based on the (fluorosulfonyl)(trifluoromethanesulfonyl)imide anion and its implication on high voltage supercapacitors. *Electrochemistry Communications* 38:117–119.

61. Kühnel, R. S., and A. Balducci. 2014. Comparison of the anodic behavior of aluminum current collectors in imide-based ionic liquids and consequences on the stability of high voltage supercapacitors. *Journal of Power Sources* 249:163–171.
62. Brandt, A., C. Ramirez-Castro, M. Anouti, and A. Balducci. 2013. An investigation about the use of mixtures of sulfonium-based ionic liquids and propylene carbonate as electrolytes for supercapacitors. *Journal of Materials Chemistry A* 1:12669–12678.
63. Lota, G., and E. Frackowiak. 2009. Striking capacitance of carbon/iodide interface. *Electrochemistry Communications* 11:87–90.
64. Meller, M., J. Menzel, K. Fic, D. Gastol, and E. Frackowiak. 2014. Electrochemical capacitors as attractive power sources. *Solid State Ionics* 264:61–67.
65. Chen, T., H. S. Peng, M. Durstock, and L. M. Dai. 2014. High-performance transparent and stretchable all-solid supercapacitors based on highly aligned carbon nanotube sheets. *Scientific Reports* 4:3612.
66. Chen, T., Y. H. Xue, A. K. Roy, and L. M. Dai. 2014. Transparent and stretchable high-performance supercapacitors based on wrinkled graphene electrodes. *ACS Nano* 8:1039–1046.
67. Jung, H. Y., M. B. Karimi, M. G. Hahm, P. M. Ajayan, and Y. J. Jung. 2012. Transparent, flexible supercapacitors from nano-engineered carbon films. *Scientific Reports* 2:773.
68. Notarianni, M., J. Liu, F. Mirri, M. Pasquali, and N. Motta. 2014. Graphene-based supercapacitor with carbon nanotube film as highly efficient current collector. *Nanotechnology* 25:435405.
69. Niu, Z. Q., W. Y. Zhou, X. D. Chen, J. Chen, and S. S. Xie. 2015. Highly compressible and all-solid-state supercapacitors based on nanostructured composite sponge. *Advanced Materials* 5:1–5. DOI: 10.1002/adma.201502263
70. Niu, Z. Q., W. Y. Zhou, J. Chen et al. 2011. Compact-designed supercapacitors using free-standing single-walled carbon nanotube films. *Energy and Environmental Science* 4:1440–1446.
71. Cheng, Y., H. Zhang, S. Lu, C. V. Varanasi, and J. Liu. 2013. Flexible asymmetric supercapacitors with high energy and high power density in aqueous electrolytes. *Nanoscale* 5:1067–1073.
72. Cheng, Y. W., S. T. Lu, H. B. Zhang, C. V. Varanasi, and J. Liu. 2012. Synergistic effects from graphene and carbon nanotubes enable flexible and robust electrodes for high-performance supercapacitors. *Nano Letters* 12:4206–4211.
73. He, Y. M., W. J. Chen, J. Y. Zhou et al. 2014. Constructed uninterrupted charge-transfer pathways in three-dimensional micro/nanointerconnected carbon-based electrodes for high energy-density ultralight flexible supercapacitors. *ACS Applied Materials and Interfaces* 6:210–218.
74. Niu, Z. Q., P. S. Luan, Q. Shao et al. 2012. A "skeleton/skin" strategy for preparing ultrathin free-standing single-walled carbon nanotube/polyaniline films for high performance supercapacitor electrodes. *Energy Environmental Science* 5:8726–8733.
75. Hsieh, Y. C., K. T. Lee, Y. P. Lin, N. L. Wu, and S. W. Donne. 2008. Investigation on capacity fading of aqueous $MnO_2 \cdot nH_2O$ electrochemical capacitor. *Journal of Power Sources* 177:660–664.
76. Tsay, K. C., L. Zhang, and J. J. Zhang. 2012. Effects of electrode layer composition/thickness and electrolyte concentration on both specific capacitance and energy density of supercapacitor. *Electrochimica Acta* 60:428–436.
77. Abbas, Q., D. Pajak, E. Frąckowiak, and F. Béguin. 2014. Effect of binder on the performance of carbon/carbon symmetric capacitors in salt aqueous electrolyte. *Electrochimica Acta* 140:132–138.
78. Cao, W. J., Y. X. Li, B. Fitch, J. Shih, T. Doung, and J. Zheng. 2014. Strategies to optimize lithium-ion supercapacitors achieving high-performance: Cathode configurations, lithium loadings on anode, and types of separator. *Journal of Power Sources* 268:841–847.

79. Paul, S., K. S. Choi, D. J. Lee, P. Sudhagar, and Y. S. Kang. 2012. Factors affecting the performance of supercapacitors assembled with polypyrrole/multi-walled carbon nanotube composite electrodes. *Electrochimica Acta* 78:649–655.
80. Timperman, L., F. Béguin, E. Frackowiak, and M. Anouti. 2014. Comparative study of two protic ionic liquids as electrolyte for electrical double-layer capacitors. *Journal of the Electrochemical Society* 161:A228–A238.
81. Liu, X., and P. G. Pickup. 2008. Ru oxide supercapacitors with high loadings and high power and energy densities. *Journal of Power Sources* 176:410–416.
82. Varzi, A., A. Balducci, and S. Passerini. 2014. Natural cellulose: A green alternative binder for high voltage electrochemical double layer capacitors containing ionic liquid-based electrolytes. *Journal of the Electrochemical Society* 161:A368–A375.
83. Böckenfeld, N., S. S. Jeong, M. Winter, S. Passerini, and A. Balducci. 2013. Natural, cheap and environmentally friendly binder for supercapacitors. *Journal of Power Sources* 221:14–20.
84. Aslan, M., D. Weingarth, N. Jäckel, J. S. Atchison, I. Grobelsek, and V. Presser. 2014. Polyvinylpyrrolidone as binder for castable supercapacitor electrodes with high electrochemical performance in organic electrolytes. *Journal of Power Sources* 266:374–383.
85. Lee, K. T., C. B. Tsai, W. H. Ho, and N. L. Wu. 2010. Superabsorbent polymer binder for achieving MnO_2 supercapacitors of greatly enhanced capacitance density. *Electrochemistry Communications* 12:886–889.
86. Shin, S. R., C. K. Lee, I. So et al. 2008. DNA-wrapped single-walled carbon nanotube hybrid fibers for supercapacitors and artificial muscles. *Advanced Materials* 20:466–470.
87. Lee, H., H. Kim, M. S. Cho, J. Choi, and Y. Lee. 2011. Fabrication of polypyrrole (PPy)/carbon nanotube (CNT) composite electrode on ceramic fabric for supercapacitor applications. *Electrochimica Acta* 56:7460–7466.
88. Lei, C., P. Wilson, and C. Lekakou. 2011. Effect of poly(3,4-ethylenedioxythiophene) (PEDOT) in carbon-based composite electrodes for electrochemical supercapacitors. *Journal of Power Sources* 196:7823–7827.
89. Luo, J. Y., V. C. Tung, A. R. Koltonow, H. D. Jang, and J. X. Huang. 2012. Graphene oxide based conductive glue as a binder for ultracapacitor electrodes. *Journal of Materials Chemistry* 22:12993–12996.
90. Kang, M., J. E. Lee, H. W. Shim, M. S. Jeong, W. B. Im, and H. Yoon. 2014. Intrinsically conductive polymer binders for electrochemical capacitor application. *RSC Advances* 4:27939–27945.
91. Mai, L. Q., A. Minhas-Khan, X. Tian et al. 2013. Synergistic interaction between redox-active electrolyte and binder-free functionalized carbon for ultrahigh supercapacitor performance. *Nature Communications* 4:1–7.
92. Pech, D., M. Brunet, H. Durou et al. 2010. Ultrahigh-power micrometre-sized supercapacitors based on onion-like carbon. *Nature Nanotechnology* 5:651–654.
93. Kang, J. L., A. Hirata, H. J. Qiu et al. 2014. Self-grown oxy-hydroxide@ nanoporous metal electrode for high-performance supercapacitors. *Advanced Materials* 26:269–272.
94. Zhong, C., Y. Deng, W. Hu, J. Qiao, L. Zhang, and J. Zhang. 2015. A review of electrolyte materials and compositions for electrochemical supercapacitors. *Chemical Society Reviews* 44:7484–7539.
95. Tõnurist, K., T. Thomberg, A. Jänes, I. Kink, and E. Lust. 2012. Specific performance of electrical double layer capacitors based on different separator materials in room temperature ionic liquid. *Electrochemistry Communications* 22:77–80.
96. Tõnurist, K., T. Thomberg, A. Jänes, T. Romann, V. Sammelselg, and E. Lust. 2013. Polymorphic behavior and morphology of electrospun poly(vinylidene fluoride) separator materials for non-aqueous electrolyte based electric double layer capacitors. *Batteries and Energy Technology (General Session)—222nd ECS Meeting/Prime 2012: In Honor of James Mcbreen* 50:49–58.

97. Laforgue, A., and L. Robitaille. 2012. Electrochemical testing of ultraporous membranes as separators in mild aqueous supercapacitors. *Journal of the Electrochemical Society* 159:A929–A936.
98. Tõnurist, K., T. Thomberg, A. Jänes, and E. Lust. 2013. Specific performance of supercapacitors at lower temperatures based on different separator materials. *Journal of the Electrochemical Society* 160:A449–A457.
99. Tõnurist, K., A. Jänes, T. Thomberg, H. Kurig, and E. Lust. 2009. Influence of mesoporous separator properties on the parameters of electrical double-layer capacitor single cells. *Journal of the Electrochemical Society* 156:A334–A342.
100. Jänes, A., K. Tõnurist, T. Thomberg, and E. Lust. 2009. Comparison of electrospun and commercially available separator materials for supercapacitors. *ECS Transactions* 19:23–32.
101. Karabelli, D., J. C. Lepretre, F. Alloin, and J. Y. Sanchez. 2011. Poly(vinylidene fluoride)-based macroporous separators for supercapacitors. *Electrochimica Acta* 57:98–103.
102. Tõnurist, K., T. Thomberg, A. Janes, T. Romann, V. Sammelselg, and E. Lust. 2013. Influence of separator properties on electrochemical performance of electrical double-layer capacitors. *Journal of Electroanalytical Chemistry* 689:8–20.
103. Stepniak, I., and A. Ciszewski. 2010. Grafting effect on the wetting and electrochemical performance of carbon cloth electrode and polypropylene separator in electric double layer capacitor. *Journal of Power Sources* 195:5130–5137.
104. Szubzda, B., A. Szmaja, M. Ozimek, and S. Mazurkiewicz. 2014. Polymer membranes as separators for supercapacitors. *Applied Physics A—Materials Science and Processing* 117:1801–1809.
105. Liu, X. R., and P. G. Pickup. 2008. Performance and low temperature behaviour of hydrous ruthenium oxide supercapacitors with improved power densities. *Energy and Environmental Science* 1:494–500.
106. Chen, L. B., Y. R. Chen, J. F. Wu, J. W. Wang, H. Bai, and L. Li. 2014. Electrochemical supercapacitor with polymeric active electrolyte. *Journal of Materials Chemistry A* 2:10526–10531.
107. Chen, L. B., H. Bai, Z. F. Huang, and L. Li. 2014. Mechanism investigation and suppression of self-discharge in active electrolyte enhanced supercapacitors. *Energy and Environmental Science* 7:1750–1759.
108. Shulga, Y. M., S. A. Baskakov, V. A. Smirnov, N. Y. Shulga, K. G. Belay, and G. L. Gutsev. 2014. Graphene oxide films as separators of polyaniline-based supercapacitors. *Journal of Power Sources* 245:33–36.
109. Yu, H. J., Q. Q. Tang, J. H. Wu et al. 2012. Using eggshell membrane as a separator in supercapacitor. *Journal of Power Sources* 206:463–468.

4 Electrolyte Performance Validation Using Electrochemical Supercapacitor Cells

Xiaopeng Han and Wenbin Hu

CONTENTS

4.1 INTRODUCTION

As discussed extensively in the above chapters, the electrolyte has been identified as one of the key components in determining the performance of assembled electrochemical supercapacitors (ESs). The accurate characterization of the electrolytes' performance is vital for optimizing, validating, and developing the electrolytes of ESs. In this chapter, we introduce the design and assembly of ESs, and the related cell configuration to meet the requirements of instrumental characterizations. The electrolyte preparation and purification processes are described. Much attention should be paid to the purification of electrolytes since this greatly affects the performance of electrolytes and the associated ESs. In addition, the testing protocols and procedures used to characterize the electrolyte performance are discussed. With the help of many electrochemical techniques such as cyclic voltammetry (CV), galvanostatic charging/discharging (GCD), and electrochemical impedance spectroscopy (EIS), a series of important parameters such as capacitance, equivalent series resistance (ESR), energy and power densities, and cycle life, can be measured so as to characterize the performance of the electrolyte and the related ESs. Some other physical instrumental analysis methods, especially the online or *in situ* methods, are also discussed. These instrumental methods are helpful in characterizing the properties of electrolytes or monitoring the changes or the adsorbing/desorbing behaviors of the electrolyte ions in real time. Finally, we demonstrate and analyze the electrolyte degradation diagnosis factors in practical ESs reported in the references. Much effort has been devoted to mitigating or overcoming the existing challenges with the aim of producing and developing high-performing and durable electrolytes for ESs.

4.2 SUPERCAPACITOR TEST CELL PREPARATION (DESIGN AND STRUCTURE)

To evaluate the performance of the electrolyte or the assembled ES with the investigated electrolyte, it was generally fabricated, constructed, and tested using a three-electrode electrochemical cell or a two-electrode system (Figure 4.1a and b) [1]. The three-electrode cell consists of the working electrode (WE), counter electrode (CE), and reference electrode. The WE is generally made of chemically inert material such as glassy carbon (GC), platinum (Pt), and/or gold (Au), which is deposited on, or coated by, the active electrode materials being tested. Usually, the binders (e.g., polytetrafluoroethylene [PTFE]) and conducting agents (e.g., carbon black) are incorporated into the active electrode materials to provide good electrical conductivity and ensure sufficient contact and mechanical durability to effectively reduce the connection resistances to avoid possible deterioration, loosening, and loss of the active materials from the electrodes upon electrochemical cycling [2]. The CE usually uses a high-surface-area electrode such as Pt foil, providing the transferred currents during the measurement. As for the reference electrode (e.g., saturated calomel electrode and silver chloride electrode (Ag/AgCl)), it is a stable electrode and possesses a well-known standard redox potential. The reference potentials can be converted to other standard potentials (e.g., normal hydrogen electrode and reversible hydrogen electrode) to allow a fair comparison. In a practical experiment of three electrodes, the three electrodes are put into the tested electrolyte to support the ionic transfer inside the electrolyte, while the electrons are transferred in the external circuit. In a given electrochemical testing system, the collected voltage potential data is the measured potential of the WE with respect to the reference electrode used. To avoid interference from the dissolved oxygen in the electrolyte during the electrochemical measurements, the electrolyte is generally deaerated by high-purity nitrogen gas. In the two-electrode cell, it has both positive and negative electrodes containing active materials immersed in the tested electrolyte. If the active materials on the surface of the two electrodes are the same, the investigated ES is known as a symmetric ES, otherwise it is an asymmetric one. The two electrodes are separated by the separator membrane. Good contact should be guaranteed between all components so as to minimize the ESR of the ES. For example, Tsay et al. [3] demonstrated an ES test cell using two metal end plates with three screws to form a three-point geometry to tighten the electrodes and separator together, thus providing a better pressure balance than a four-point geometry (Figure 4.2). There are metal sheets between the Teflon plates

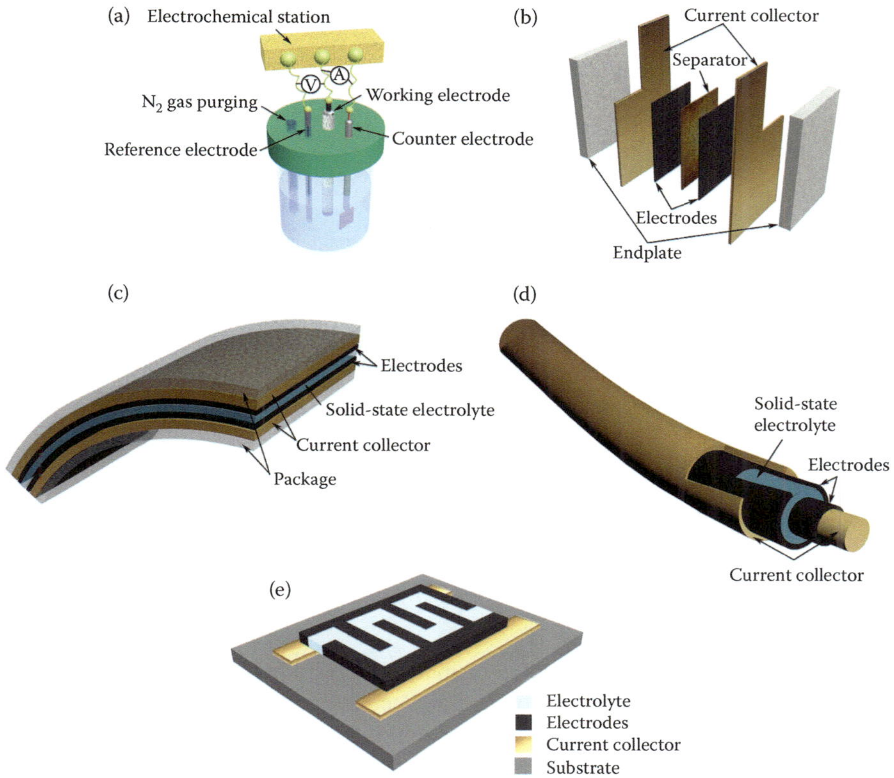

FIGURE 4.1 Schematic diagrams of (a) three-electrode cell, (b) two-electrode cell, (c) sandwiched flexible ES, (d) fiber ES, and (e) micro-ES. (Zhong, C. et al. 2015. A review of electrolyte materials and compositions for electrochemical supercapacitors. *Chemical Society Reviews* 44:7484–7539. Reproduced by permission of The Royal Society of Chemistry.)

and the electrodes, which not only serve as a current collector but also as holders to tighten the electrode separator, ensuring that the electrode–electrolyte assembly stays together. Benefiting from the back pressure coins on each cell Teflon plate, the pressure from the two end metal plates can be directly transferred to the electrode–electrolyte area to ensure good contact and resulting mechanical strength [3,4]. In contrast with the three-electrode cell, the electrode potential in the two-electrode cell is referenced from one electrode to the other. It should be noted that with the development of various types of ESs and electrolytes such as solid-state electrolytes, other types of two-electrode cell configurations have also been used. For example, Figure 4.1c–e shows the basic structure of sandwiched flexible ES, fiber ES, and micro-ES, respectively [1].

It also needs to be recognized that there is a primary difference between the three- and two-electrode cells in respect to characterizing electrolytes and ESs. Using the three-electrode system, the obtained information is only from the investigated WE

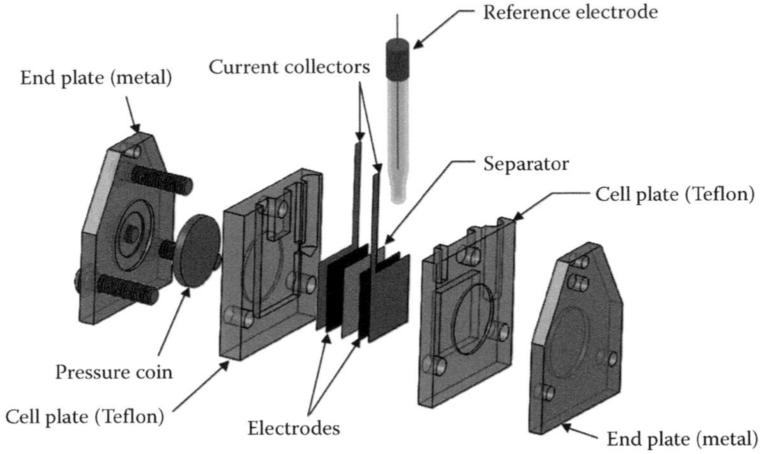

FIGURE 4.2 Schematic illustration of the two-electrode supercapacitor test cell. (Reprinted from *Electrochimica Acta*, 60, Tsay, K.-C., L. Zhang, and J. Zhang, Effects of electrode layer composition/thickness and electrolyte concentration on both specific capacitance and energy density of supercapacitor, 428–436, Copyright 2012, with permission from Elsevier.)

process while the collected signal contributed from both electrodes when using a two-electrode cell [4]. Compared to the two-electrode technique, the three-electrode technique is widely employed in rapid testing and characterizing of the electrode/electrolyte materials and the ESs, because it does not require the assembly of an entire ES device and thus it is simpler and more convenient. The three-electrode system has also been proven to be an efficient way to distinguish the individual contributions of each electrode in asymmetric ESs [4]. However, the results obtained by the three-electrode cell only make sense for material investigations and cannot directly represent the performance of practical ES devices [4,5]. Furthermore, in the case of the two-electrode cell, the total capacitance value of the ES can be regarded as the series connection of the positive and negative capacitances, which is quite different from the capacitance value obtained from a three-electrode cell. As a specific example, in the pseudocapacitive study, Khomenko et al. [5] reported the influence of the test cell construction on the measured capacitance values. 20 wt% of multiwalled carbon nanotubes (MWNTs) and 80 wt% of two conducting polymers (polyaniline [PANI] and polypyrrole [PPy]) composite electrodes were prepared and measured by both two- and three-electrode cells. The three-cell characterization (Figure 4.3a) shows the stable operation voltage range is from 0.2 to −0.4 V (vs. Hg/Hg_2SO_4) for PANI/MWNT electrodes. The upper and lower voltages are limited by the polymer degradation and switching to the charge isolation state of the conducting polymer, respectively. The collected results using the two-electrode cell are evidently different from the cell performance data above (Figure 4.3b). Capacitance values of 250–1100 F g^{-1} are found in the case of three-electrode measurements while for the same materials in the case of two-electrode cells, the different values are determined to be 190–360 F g^{-1}. The measured specific capacitances via two- and three-electrode

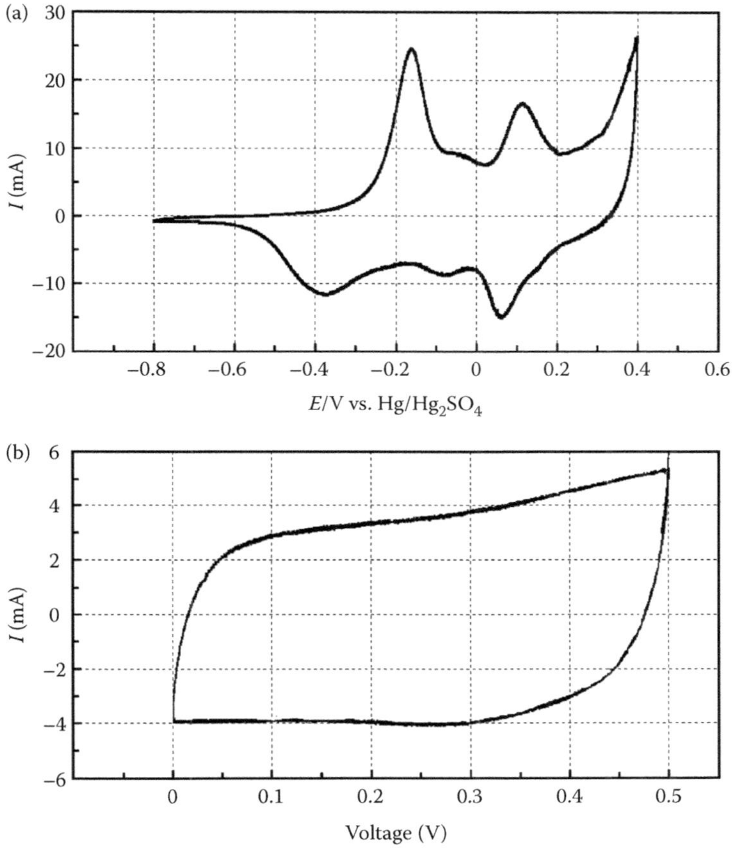

FIGURE 4.3 (a) Cyclic voltammogram (three-electrode cell) of a PANI/MWNTs pellet electrode; mass of composite: 5.4 mg; electrolyte: 1 M H_2SO_4 at a scan rate of 2 mV s^{-1}. (b) Cyclic voltammogram (2 mV s^{-1}) of a symmetric capacitor based on PANI/MWNTs composite electrodes (two-electrode cell); mass of each electrode: 10.4 mg. (Reprinted from *Electrochimica Acta*, 50, Khomenko, V., E. Frackowiak, and F. Béguin, Determination of the specific capacitance of conducting polymer/nanotubes composite electrodes using different cell configurations, 2499–2506, Copyright 2005, with permission from Elsevier.)

cell configurations for two different electrode materials are listed in Table 4.1. The authors point out the values of the specific capacitance obtained by a three- or two-electrode ES cell cannot simply be compared with each other [5]. Although a three-electrode cell is a powerful instrument to investigate the electrochemical behavior of a single electrode, the tested values of specific capacitance cannot be reached in practical ESs.

In addition, special capacitor cell structures are designed to meet the requirements of some instrumental measurements (especially *in situ* techniques) for electrolytes. Here, we describe three types of cells used for *in situ* Raman, nuclear magnetic resonance (NMR), and small-angle neutron scattering (SANS) characterizations.

TABLE 4.1
Values of Specific Capacitance (F g^{-1}) Measured via Different Cell Configurations

Conducting Polymers in the Electrode	Three-Electrode Cell CV	Three-Electrode Cell Galvanostatic Discharge	Two-Electrode Cell CV	Two-Electrode Cell Galvanostatic Discharge
PPy	506	495	192	200
PANI	670	650	344	360

Source: Khomenko, V., E. Frackowiak, and F. Béguin. 2005. *Electrochimica Acta* 50:2499–2506. With permission.

Together with other instrumental strategies, Raman spectroscopy acts as a complementary tool to characterize the physical and chemical composition changes of both the electrode and electrolyte materials [6–9]. An upgraded *in situ* Raman cell was designed by Hardwick et al. [9] to perform the Raman measurements (Figure 4.4). The employed cell consisted of three electrodes and the filled electrolyte was 1 M tetraethylammonium tetrafluoroborate ($TEABF_4$) in acetonitrile (ACN). A separator was added between two electrodes with the reference electrode located at the edge of the separator. The investigated graphite material was coated onto the upper electrode, acting as the WE. The used current collector was aluminum foil. An aluminum wire was pushed onto the Al current collector to provide a good electrical contact with the WE. From the reverse of this upper electrode, the laser beam lights in through the hole in the center of the metallic foil. Measurements are made in backscattering configuration with a Raman microscope and a laser with wavelength of 632.8 nm [9].

As for NMR experiments, the cells should also be specially designed. One critical requisite of the electrochemical cells for *in situ* NMR tests is the minimum metallic components [10]. Thus, a modified plastic bag cell was projected with two electrodes at opposite sides along the cell's long axis (Figure 4.5a) [11]. Using this cell construction, the acquired NMR signals come from both electrodes of the ES, and it is difficult to distinguish the resonances corresponding to each of the different electrodes. Aiming to overcome this issue, an optimized cell was developed in which the two electrodes were separated by a 4 mm gap (Figure 4.5b) [11]. This configuration is helpful in individually monitoring the NMR spectra of a single electrode by putting one studied electrode into the NMR coil. Although the plastic-type cell has many advantages including being light, easy to assemble, almost metal free, and flexible, it still suffers from other serious problems such as air and moisture contamination, poor sealing, electrolyte leakage, and possible loss of mechanical contacts between the electrodes owning to the weakening of extra external pressure [12]. In order to solve these problems to permit the application of standard NMR probes, a robust, home-made, cylindrical plastic (polypropylene), easily assembled cell structure was proposed by Poli et al. [10], as shown in Figure 4.6. The active powders filled inside

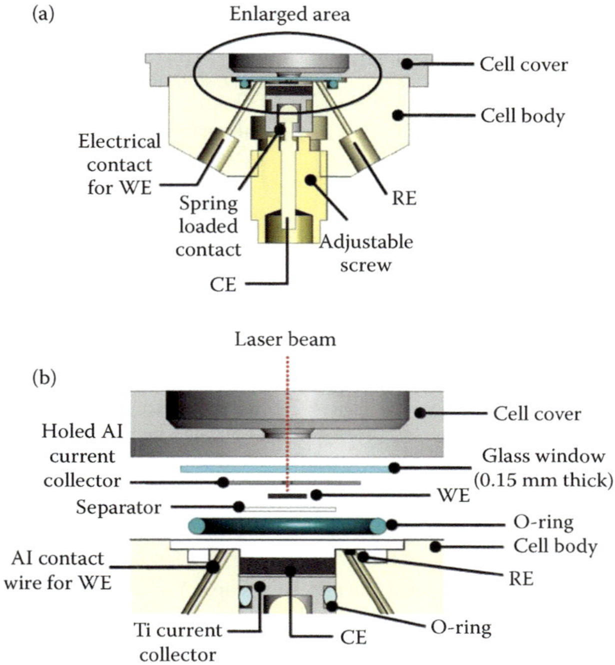

FIGURE 4.4 Three electrode *in situ* Raman cell in (a) the fully assembled state and (b) an enlarged area showing the separate components (expanded, not to scale). (Reprinted from *Electrochimica Acta*, 52, Hardwick, L. J. et al., An *in situ* Raman study of the intercalation of supercapacitor-type electrolyte into microcrystalline graphite, 675–680, Copyright 2006, with permission from Elsevier.)

the cell can be tailored to a large extent. A copper metal wire serves as a current collector here and passed through the bottom of the whole device. The cell is open on one side and the other side has a screw-like cap to hermetically seal the cell. The authors observed the cell performance decay and they attributed the reason to the reduced pressure of the total cell.

The SANS technique can not only provide the information of the molecule electroadsorption in different size pores, but also can directly visualize the influence of the potential on the electrolyte ion concentration changes. A simplified schematic of the *in situ* experimental setup for SANS measurements was recently demonstrated by Boukhalfa et al. [13], as displayed in Figure 4.7. An quartz optical cuvette of the tightly sealed chamber was used to accommodate the activated carbon fabric (ACF) electrodes [13]. The distance between the WEs and CEs is about 3–5 mm. The neutron beam should hit only on the surface of the WE, not the CE. The edge of the CE is at least 5 mm far away from the incident beam. The potential applied on the WE was controlled by the potentiostat. A two-dimensional neutron detector was used to collect the diffraction patterns passing through the WE. After correcting for the backgrounds and normalizing to the transmission, the

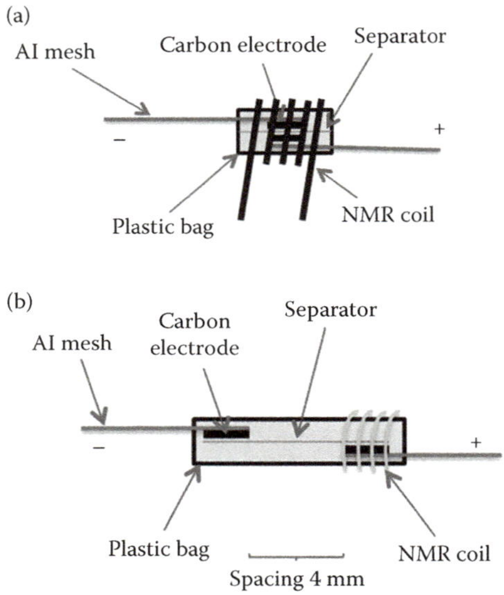

FIGURE 4.5 (a) Configuration of a normal supercapacitor cell design for *in situ* NMR study. (b) Configuration of a "long" supercapacitor with two separated electrodes. (Reprinted with permission from Wang, H. et al., Real-time NMR studies of electrochemical double-layer capacitors. *Journal of the American Chemical Society* 133, 19270–19273. Copyright 2011 American Chemical Society.)

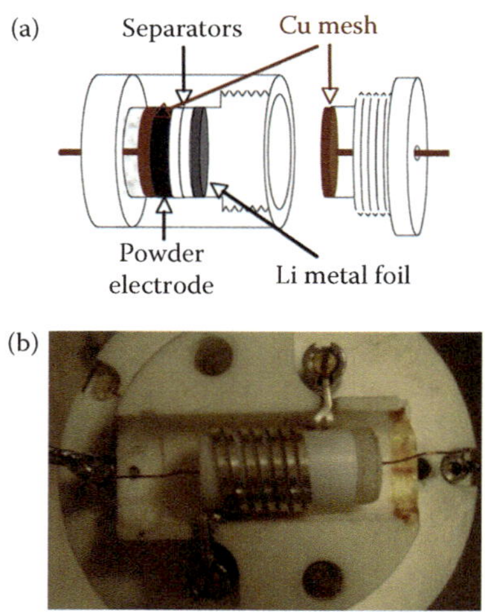

FIGURE 4.6 (a) Schematic drawing of the cylindrical cell design for *in situ* NMR measurements. (b) Top view of the NMR probe with the cylindrical cell inside a 10-mm solenoid. (Reprinted from *Electrochemistry Communications*, 13, Poli, F. et al., New cell design for *in-situ* NMR studies of lithium-ion batteries, 1293–1295, Copyright 2011, with permission from Elsevier.)

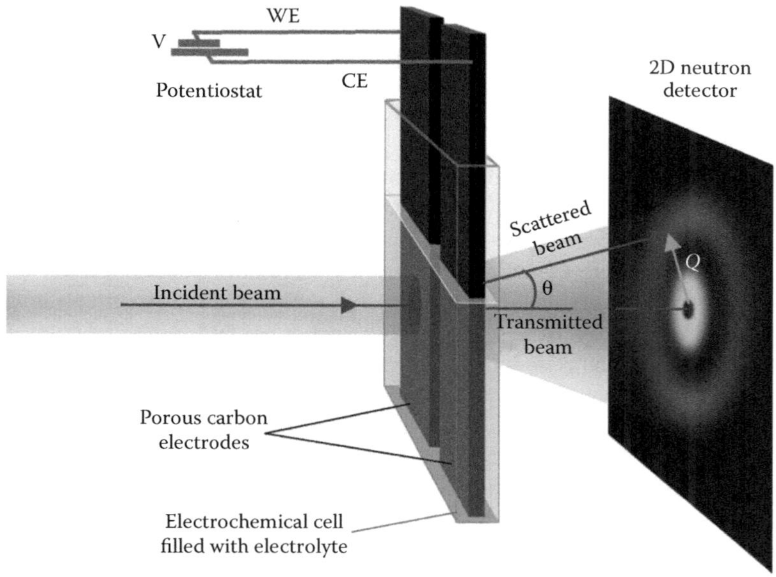

FIGURE 4.7 Experimental setup for *in situ* studies of ion adsorption on the surface of microporous carbon electrodes. (Reprinted with permission from Boukhalfa, S. et al. 2014. *In situ* small angle neutron scattering revealing ion sorption in microporous carbon electrical double layer capacitors. *ACS Nano* 8, 2495–2503. Copyright 2014 American Chemical Society.)

obtained data are analyzed on the basis of the three-phase model to monitor the adsorption behavior of gases and supercritical fluids in materials with nanoporous structure [14].

4.3 ELECTROLYTE PREPARATION

One of the most common problems of the electrolyte is impurities since the contamination in the electrolyte diffuses to the surface of electrode, adheres to it, and reacts there during the operation of the cells [15]. The side reaction competes with the examined electrode reaction, thus affecting the electrochemical properties of the ES [15]. Most impurities in the electrolyte can be decreased substantially through careful purification treatment processes of the solute and solvent materials. For an aqueous electrolyte, the solvent of water is purified by distillation or ultrafiltration procedures. The deionized water can be obtained by using a Millipore ion resin exchange deionizer system, as has been outlined in many articles [16–19]. As for the solute, high-purity salt compounds (mostly in analytical grade, 99.99 wt% pure) can be bought without extensive purification and handled under a controlled atmosphere. When this approach is not the case, standard cleaning procedures such as recrystallization and calcinations can be used to achieve the desired level of purification [15]. Acids can be further cleaned via normal or vacuum distillation [15]. The preparation of an aqueous electrolyte for ESs simply involves dissolving the corresponding solute materials in high-purity distilled water to obtain a specific concentration. The aqueous electrolytes for ESs investigated in the references are mainly divided into three conditions—acidic, neutral, and alkaline which conventionally correspond to H_2SO_4, Na_2SO_4, and KOH solutions, respectively.

Most nonaqueous electrolytes are more difficult to purify than aqueous solutions. Since residual impurities have a negative influence on the electrochemical stable potential window (ESPW) and possibly lead to an accelerated self-discharge process [1,20–22], it is essential to develop purification strategies to reduce the amount of impurities in nonaqueous electrolytes. Water is the common impurity in nonaqueous electrolytes. Butler proposed the water content in nonaqueous electrolytes should be less than 50 ppm for electrochemical applications [23]. The effect of impurities on the electrode reaction processes can be evaluated by detecting the background currents in the blank electrolyte. The smaller the background currents are the better and they should certainly be corrected so as to observe the signals from electroactive materials clearly. Therefore, using both high-purity solvent and solute is the critically important first step to obtain good results from the target electrolyte and ESs.

In the lithium-ion battery industry, considerable attention has been paid to the art of nonaqueous solvent purification together with the required technologies. The typical solvent includes propylene carbonate (PC), ethylene carbonate (EC), tetrahydrofuran (THF), dimethyl sulfoxide (DMSO), etc. The rapid purification of most organic solvents involves mixing and stirring with a strongly absorbed water substrate (e.g., CaH_2) overnight and then distilling the mixture at reduced pressure under a given temperature [24]. The required temperatures and the pressures are different for different organic compounds with different physicochemical properties. Collected anhydrous products should be stored over 4A molecular sieves in a glove box with an inert atmosphere. Electrolysis of the electrolyte is another common and effective purification method [15]. In practice, two electrodes are placed and immersed into the pretreated electrolyte. Then, an external potential is applied between the two electrodes to reduce or oxidize some of the impurities on the surface of the electrodes at a negative or positive voltage. The used electrodes act in the role of garbage disposal; thus they are called "sacrificial" electrodes [15]. To develop future nonaqueous electrolyte-based ESs, continuous efforts must be devoted to the purification processes of the electrolyte. As considerable experience has been gained in developing electrolyte systems for the battery industry, especially for Li-ion batteries, many of these technologies can be learned and transferred to the production of nonaqueous electrolytes for ESs.

Ionic liquids (ILs) also face the same purification problem. The traditional trace impurities are water, oxygen, and halide ions [25]. The less common impurities include the by-products of the hydrolysis of water-sensitive ions and also even the hydrogen fluoride [26]. The water in various ILs based on the bis(trifluoromethylsulfonyl) imide (TFSI) anion can be removed by high-temperature evaporation (e.g., 24 h at 130°C) [25]. Otherwise, special attention should be paid to ILs with hydrophobic anions, in which a small amount of water in the ILs may benefit the long-term durability of the electrolyte [1].

4.4 TESTING PROTOCOLS AND PROCEDURES

ESs have been extensively characterized, tested, and diagnosed by several important electrochemical characterization techniques such as CV, GCD, and EIS. Since the characterization methods for ESs' performance have already been reviewed by several other articles [27–30] and books [2,4,31,32], the following discussion will focus on some of the main tests and characterization methods used to investigate electrolytes. Based on these electrochemical characterization techniques, a series of key parameters of ESs, such as capacitance, ESR, energy and power densities, and cycle life, can be measured so as to record the performance of the investigated electrolyte. Meanwhile, several online or *in situ* instrumental technical strategies are also introduced in detail. These strategies can be employed to: (a) investigate electrolytes' properties; (b) monitor changes in electrolyte properties; or (c) observe ion adsorption/desorption behavior in the electrolytes in real time.

4.4.1 Cyclic Voltammetry

CV is a relatively swift and conventional potential-dynamic electrochemical technique to identify potential electrode materials and electrolytes for ESs. This method is achieved by sweeping potential cycles within anodic and cathodic defined voltages limited by the decomposition of the electrolyte and then measuring the resulting currents [2]. Normally, a three-electrode electrochemical cell, including WE, reference electrode, and CE, is used to perform the CV measurements, as described above. A potentiostat instrument is used to control the electrode potential. The applied potential on the WE is measured versus the reference electrode and linearly scans back and forth from the specified lower voltage limit to the upper voltage limit [4]. During sweeping the electrode potential, the current through the WE and CE can be recorded. The current passing across the WE and the corresponding electrode potential yield the CV curves. Apart from the three-electrode technique, CV can also be conducted using an alternative two-electrode cell configuration. In addition, from the CVs at different potential scan rates (v_s), the kinetics of the electrode reactions can be identified quantitatively and qualitatively by this scan rate dependence [4].

For an ideal capacitor, a constant sweep rate ($\pm v_s$) leads to a constant and equivalent current ($\pm I$) and the CV curve shows a rectangle-like shape (Figure 4.8) [2]. However, the ESs often deviate away from the ideal case in real operational conditions [2]. Other electrode materials may exhibit a faradaic characteristic with no constant capacitance, which is reflected by the responded current peaks owing to the reversible pseudocapacitive reactions, as shown in Figure 4.9 [2]. The area under the CV curves represents the total amount of the stored charge. The CV experimental data can be used to calculate the specific capacitance (C in F g^{-1}) through the integration of current response, according to the following equation [2,33]:

$$C = \int I \frac{dV}{\Delta V m v_s} \tag{4.1}$$

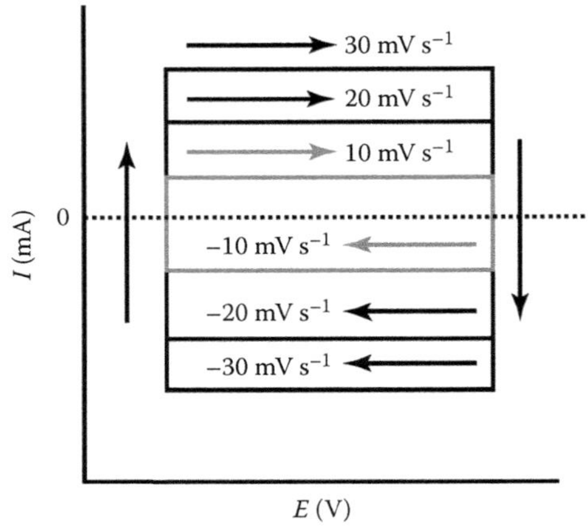

FIGURE 4.8 Cyclic voltammograms of an ideal capacitor. (Liu, R. et al.: *Electrochemical Technologies for Energy Storage and Conversion*. 2012. Copyright Wiley-VCH Verlag GmbH & Co. KGaA. Reproduced with permission.)

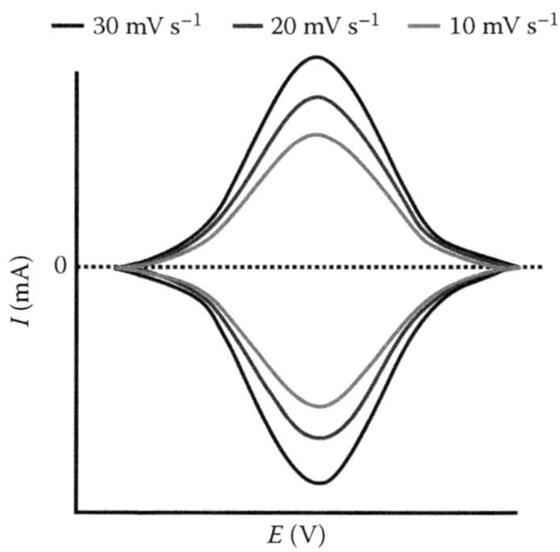

FIGURE 4.9 Cyclic voltammograms of a localized and reversible pseudocapacitive electrode reaction for an electrode thinly coated by a redox material at increasing scan rates. (Liu, R. et al.: *Electrochemical Technologies for Energy Storage and Conversion*. 2012. Copyright Wiley-VCH Verlag GmbH & Co. KGaA. Reproduced with permission.)

where I is the voltammetric current (A), ΔV the potential window (V), m the mass of the active material (g), and v_s the scan rate (V s^{-1}).

Palm et al. [34] employed the CV method to investigate the effect of the concentration of the addition of 1-butyl-3-methylimidazolium tetrafluoroborate ([BMIM][BF_4]) in 1-ethyl-3-methylimidazolium tetrafluoroborate ([EMIM][BF_4]) electrolyte on the electrochemical characteristics for electrical double-layer capacitors (EDLCs). They found the concentration of [BMIM][BF_4] has little influence on the electrochemical properties of the mixed electrolyte at the low scan rate of 1 mV s^{-1}. The distortion effect is visible at the higher scan rate of 10 mV s^{-1}, particularly with increased concentration of [BMIM][BF_4] in the electrolyte. They explained this as due to the lower conductivity of the mixture electrolyte with high content of [BMIM][BF_4] and the insufficient ion diffusion rate at larger scan rates.

Through the information of CV curves, especially a series of CV data at different scan rates and calculated capacitances, we can recognize the types of electrochemical reactions (faradaic or capacitive behaviors) involved in the charge/discharge processes of ESs. Tonisoo et al. [35] investigated the EDLC based on the Mo_2C-derived carbon electrodes and [EMIM][BF_4] electrolyte. The capacitance of the dry two-electrode cell was unclear while the capacitance increased remarkably when [EMIM][BF_4] was added. The CV profiles with potential ranging from 0.0 to 3.2 V are presented in Figure 4.10a and the shape is similar to the result obtained from [EMIM][BF_4] electrolyte at 1 mV s^{-1} published elsewhere [34]. The slow increase of charging currents measured at larger scanning rates (1, 5, 10, and 25 mV s^{-1}) between 0.5 and 2.5 V was due to the enhanced ions Gibbs adsorption [35]. When the potential

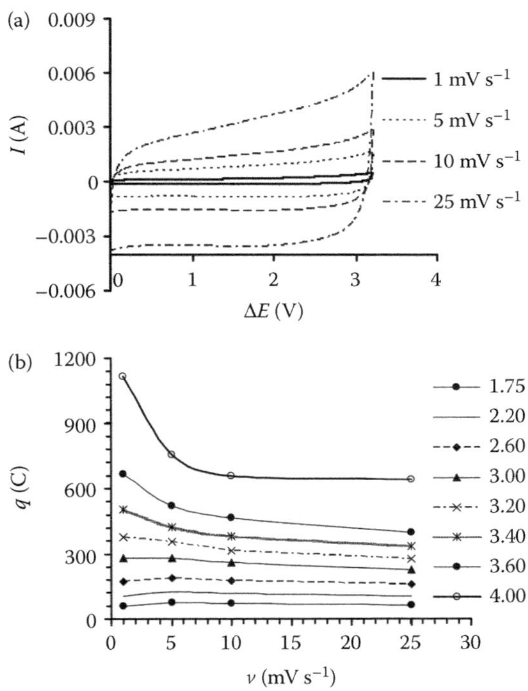

FIGURE 4.10 (a) Cyclic voltammograms measured for Mo_2C-derived carbon electrode-based EDLC with [EMIM][BF_4] electrolyte in a dry Ar 5.0 atmosphere in two-electrode cell for the potential range from 0.0 to 3.2 V at different potential sweep rates. (b) Dependence of total charge on the cell potential scan rate at different maximal cell potentials. (From Tonisoo, A. et al. 2013. *Journal of the Electrochemical Society*, 160, A1084–A1093. Reproduced by permission of The Electrochemical Society.)

window is over 3 V, faradaic reactions occur, which lead to the dimerization of imidazolium-based cations [36]. The result is also confirmed by the electrical charges (q) needed to charge the capacitor with different cell potential (ΔE) at various potential scanning rates, as displayed in Figure 4.10b [35]. When $\Delta E \leq 3.0$ V, the sweeping speed has no obvious effect on the values of q, demonstrating the electrode process is dominated by the electrical double layer (EDL) and there is no faradaic process. When $\Delta E > 3.0$ V, the characteristic faradaic reaction is observed in addition to the behavior of the EDLCs. When further increasing the ΔE above 3.2 V, more charge is consumed at a lower scan rate (Figure 4.10b) in comparison with that at larger scan rates, indicating a prominently increased proportion of faradaic process at the slower scan rate.

4.4.2 Galvanostatic Charging/Discharging

GCD is a traditional method to test the performance of the electrolyte and the ESs. It is performed by charging the electrochemical cell at constant current density and afterwards discharging it at constant current density [4]. GCD is a well-known measurement method to determine the capacitance of packaged ESs in industry. Through this method, a series of electrochemical performance parameters of the ESs such as capacitance, ESR, energy and power densities, and cycle stability can be characterized in detail. Like the CV characterization, both the conventional three- and two-electrode cells could be adopted for the GCD experiments, depending on practical requirements. The GCD technique is suitable for both the symmetric and the asymmetric capacitors. Apart from the constant current mode described above, the constant voltage mode can also be employed for characterizing the properties of the ESs, although it is seldom used [4].

The GCD method can be used to determine the capacitor's specific capacitance (F g^{-1}) according to the following equation [2,37]:

$$C = \frac{I}{m(dV/dt)} = I\frac{\Delta t}{m\Delta V} \tag{4.2}$$

where I is the constant discharge current (A), dV/dt the slope of the GCD discharge curve, and m the mass of active materials (g). Charging and discharging at different rates (current densities) can offer some information about the electrode mass transfer kinetics. The specific capacitance would be decreased with an increasing charging/discharging rate, which is a typical phenomenon for all types of ESs [3,38–42]. The reason for this is the limited mass transfer kinetics and the inaccessible pore portions in the porous electrode layer at high current rates [3]. The testing voltage range should be in consistence with the electrolyte's electrochemically stable window, generally from 0 to 1 V for aqueous media and from 0 to 2.5–2.7 V for organic electrolytes [30]. Maximum/minimum voltages for ESs depend on the oxidation/reduction of electrolyte materials and the decomposition of electrolytes. Figure 4.11a illustrates the GCD curves of an asymmetric ultracapacitor of manganese oxide/activated carbon (AC) electrodes at 100 mA g^{-1} in 2 M KNO_3 electrolyte with different maximum cell voltages [43]. When the maximum voltage reaches 2.2 V, a nonsymmetric

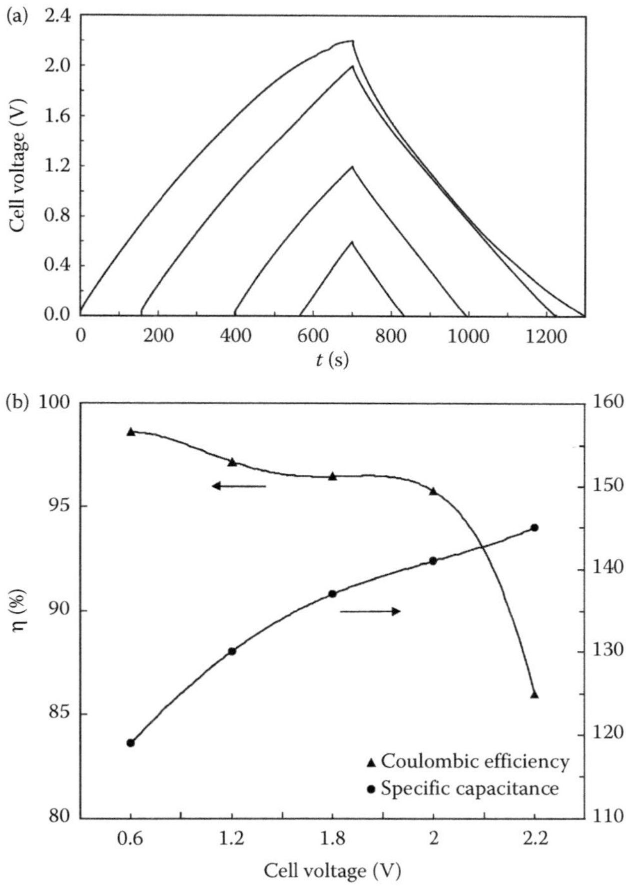

FIGURE 4.11 (a) Galvanostatic charge–discharge curves (100 mA g^{-1}) of an asymmetric manganese oxide/AC capacitor in 2 M KNO_3 electrolyte. (b) Coulombic efficiency and specific capacitance (F g^{-1}) of the asymmetric manganese oxide/AC supercapacitor in 2 M KNO_3 electrolyte versus the cell voltage. (Reprinted from *Journal of Power Sources*, 153, Khomenko, V., E. Raymundo-Piñero, and F. Béguin, Optimisation of an asymmetric manganese oxide/activated carbon capacitor working at 2 V in aqueous medium, 183–190, Copyright 2006, with permission from Elsevier.)

charge/discharge curve is obtained, suggesting the presence of noncapacitive behavior. To optimize the cell voltage and also to improve the energy utilization ratio, the coulombic efficiency (η) was also a concern [43]:

$$\eta = \frac{q_d}{q_c} \tag{4.3}$$

in which q_d and q_c are the discharging and charging quantity of charge of the capacitor obtained from the GCD experiments. The coulombic efficiencies and specific capacitance (F g^{-1}) based on the MnO_2 positive electrode and AC negative electrode material in an alkaline solution with a different maximum voltage range are presented in Figure 4.11b [43]. When the charging voltage is set higher than 2 V, the capacitance increases whereas the coulombic efficiency (η) obviously decreases, which is possibly ascribed to the hydrogen produced by the negative polarization not being completely adsorbed in the AC electrodes [44]. Actually, operating a cell above its true limited maximum voltage leads to irreversible reactions inside the cell, resulting in an overestimation of the specific capacitance, poor efficiency, and a short lifetime. Since the cell voltage plays a critical role in determining the capacitance, it is important to calculate the ES capacitance within the typical working voltage range. The typical operating potential range is from V_{max} to approximately $1/2V_{max}$ for most ESs and the recommended and accurate method to calculate the slope *(dv/dt)* is to use two data points of the discharge curve, that is Reference 30,

$$\frac{dV}{dt} = \frac{V_{max} - 1/2V_{max}}{T_2 - T_1} \tag{4.4}$$

T_2 and T_1 are the discharge time when the potential reaches maximum voltage (V_{max}) and half of the voltage ($1/2V_{max}$), respectively.

The input and output charge (*q*) and energy (*E*) of the capacitor can be calculated by discharging across a load resistor, see below [2]

$$q = \int i dt = \int \frac{V}{R_L} dt \tag{4.5}$$

$$E = \int iV dt = \int \frac{V^2}{R_L} dt \tag{4.6}$$

where *V* is the applied voltage across the resistor and R_L is the resistance value of a load resistor. By employing a square-wave current pulse procedure, the ESR can be measured via the instantaneous voltage decay. The ESR is typically characterized by the GCD tests using the IR drop at the beginning of the discharge curves (Figure 4.12) [45]. The internal resistance can be obtained from $\Delta V/\Delta I$, where ΔV is the change in voltage (*IR* drop) and ΔI is the total change in applied current [30,45]. It is advised that the current interruption or pulse (i.e., $I \rightarrow 0$ or $0 \rightarrow I$) is favorable for the measurements. The measured ESR of a tested cell or a full packaged capacitor device is dependent on all cell components like current collectors, electrodes, electrolyte, separator and the interface contacts among them, and so on [30]. Thus, the ESR can be affected by many factors including the cell size, electrolyte spacing, cell pressure, package type, etc. Actually, the contribution from the electrode material itself only accounts for a small portion. The ESR in an ES is a significantly important factor for the performance of the device and the value should be as low as possible. The magnitude of the internal resistance is closely related with the computed energy efficiency of the device (i.e., larger ESR leads

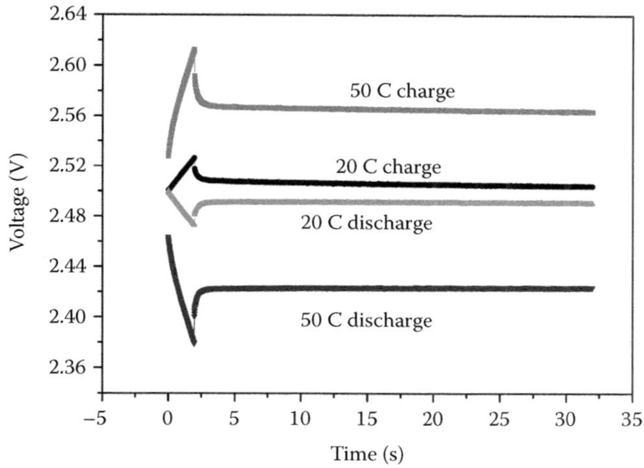

FIGURE 4.12 Voltage time response of a 350 F supercapacitor. The supercapacitor was first charged or discharged at 20 C (4.86 A) and 50 C (12.15 A) for 2 s, followed by current interruption to 0 A, and rested for 30 s. (Reprinted from *Electrochemistry Communications*, 12, Zhao, S. et al., A measurement method for determination of dc internal resistance of batteries and supercapacitors, 242–245, Copyright 2010, with permission from Elsevier.)

to lower energy efficiency). To find more details about the electrochemical testing methods, readers are referred to related references and previously published books [1,2,15,24,27,28,30–32,39,46–48].

The charging/discharging measurements can also be used to assess the cycle life and the degradation of electrolytes and ESs by observing the alterations in both specific capacitance and ESR over repeated charge and discharge cycles. It should be pointed out the two-electrode model is recommended for characterizing the cycle operational life when using the GCD characterization because it is closer to real operating conditions [4]. The cycle lives of ESs are normally much longer than batteries. EDLCs can operate as long as 100,000 cycles while commercial lithium-ion batteries only have a life for 400–1200 cycles [4]. With the prolonged charging/discharging cycles, the capacitance will decrease while the ESR increases, leading to gradually reduced energy and power densities of the ES.

4.4.3 Electrochemical Impedance Spectroscopy

EIS, also known as alternating current impedance spectroscopy, is a powerful technique for characterizing the electrode, electrolyte, and the dynamic properties of the solid–liquid interface [47]. This section focuses on the applications of the EIS in characterizing the ESs and the electrolyte materials.

The frequency-response behavior, specific capacitance, and ESR can also be evaluated by the EIS method. The EIS experiments are conducted at the open-circuit voltage (OCV) and an alternating potential with a small amplitude (5–10 mV) is applied over a wide range of frequencies from 10 mHz to 100 kHz. The measured impedance (Z) is composed of real (Z') and imaginary (Z'') parts. It can be defined as $Z = Z' + jZ''$. The specific capacitance can also be determined from an imaginary part (Z'') on the basis of the following equation [46]:

$$C = -\frac{1}{2\pi f Z'' m} \tag{4.7}$$

where f is the frequency (Hz) and m the mass of the electrode active materials (g).

Both the real (Z') and imaginary (Z'') parts of impedance are the function of angular frequency (ω, equal to $2\pi f$), which can be expressed as follows [39,49]:

$$Z(\omega) = Z'(\omega) + jZ''(\omega) = \frac{1}{j\omega C(\omega)} \tag{4.8}$$

When the capacitive behavior is ideally dominant, the capacitance can be calculated according to the following equation [39,46]:

$$C(\omega) = C'(\omega) - jC''(\omega) \tag{4.9}$$

$$C'(\omega) = \frac{-Z''(\omega)}{\omega|Z(\omega)|^2} \tag{4.10}$$

$$C''(\omega) = \frac{Z'(\omega)}{\omega|Z(\omega)|^2} \tag{4.11}$$

$C'(\omega)$ and $C''(\omega)$ are the real and imaginary parts of the capacitance, respectively. Figure 4.13a shows a typical curve of $C'(\omega)$ versus frequency [39]. With the increasing of frequency, the value of $C'(\omega)$ sharply decreases and then seems to be less frequency dependent. This result relates to the electrode/electrolyte structures and their interfaces [39]. As presented in Figure 4.13b, the imaginary of $C''(\omega)$ reaches a maximum value at frequency of f_0 (corresponding to a time constant of $t_0 = 1/f_0$). The t_0 is called the relaxation time constant which has also been introduced in Section 1.4 [50].

Plotting Z' versus Z'' to get a Nyquist curve is another widely used method to analyze the EIS data [51,52]. As shown in Figure 4.13c [52], in the Nyquist plot, a semicircle is observed in the high-frequency region, followed by a linear line in the low-frequency region. The semicircle is indicative of the electrical conductivity and the charge-transfer resistance whereas the line is an indicator of how close the device is to an ideal capacitor [39]. The more vertical the line the closer the capacitor is to an ideal capacitor, thereby demonstrating low resistance to ion diffusion and pure capacitive behavior [53]. The impedance spectra can be fitted using the electrical equivalent model (inset in Figure 4.13c) [52]. In this circuit, C_l and C_e are the limiting

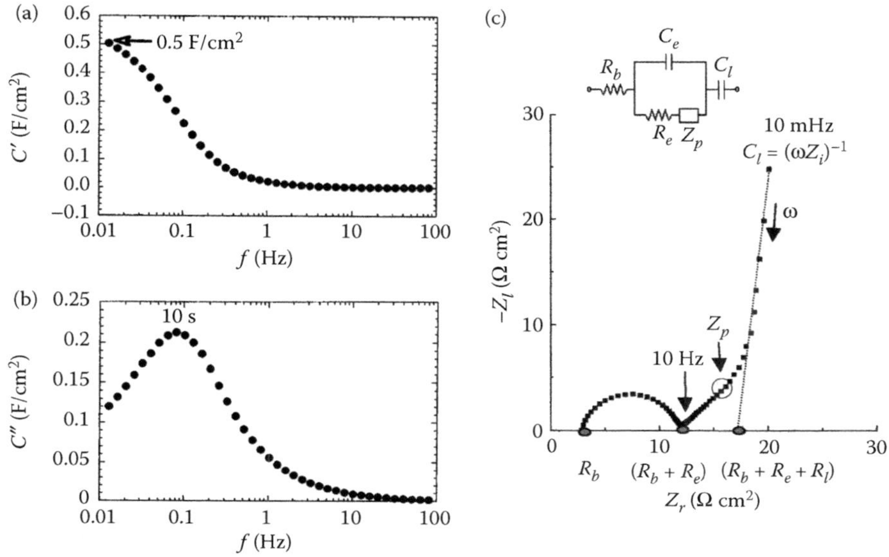

FIGURE 4.13 Evolution of the (a) real part and (b) imaginary capacitance versus frequency for a 4 cm^2 cell assembled with two electrodes containing 15 mg cm^{-2} of AC in the 1.5 M $TEABF_4$/ACN. (From Taberna, P. L., P. Simon, and J. F. Fauvarque. 2003. *Journal of the Electrochemical Society* 150:A292–A300. Reproduced by permission of The Electrochemical Society.) (c) Impedance spectrum (10 mHz–10 kHz) recorded in two-electrode mode of the charged activated carbon//activated carbon (ACAC) double-layer carbon supercapacitors and equivalent circuit for impedance analysis. (From Di Fabio, A. et al. 2001. *Journal of the Electrochemical Society* 148:A845–A850. Reproduced by permission of The Electrochemical Society.)

and double-layer capacitances, Z_p is the Warburg impedance owing to the ions diffusion/transport in the electrolyte [54], R_e the charge-transfer resistance, and R_b the combinational resistance of the substrate intrinsic resistance, the contact resistance at the interface between the active material and the current collector [46], respectively. The sum of R_e and R_b contributes mainly to the ESR, which in turn largely determines the specific power and energy densities of the ESs.

The ESR results from the Nyquist plots of the impedance spectra could agree well with the values obtained via the IR drop in GCD measurements. For instance, Stoller et al. [30] reported the Nyquist plots for three-type cells: packaged cell and two-electrode test cells with two different separators. The real component of impedance of the Nyquist plot (*X*-axis) described gradually increased series resistances that were observed from the packaged cell, the two-electrode cell with Celgard separator, and the two-electrode cell with separator from the package. These results are consistent with the values calculated by the IR drop from the GCD curves.

The EIS method can also be employed to reveal the properties of different electrolytes with the same electrode active material. Bettini et al. [55] reported the Nyquist impedance profiles of four different ILs (1-dodecyl-3-methylimidazolium bis(trifluoromethylsulfonyl)imide ([C_{12}MIM][NTf_2]), 1-butyl-1-methylpyrrolidinium bis(trifluoromethylsulfonyl)imide ([BPyr][NTf_2]), 1-butyl-3-methylimidazolium bis (trifluoromethylsulfonyl)imide ([BMIM][NTf_2]), and 1-ethyl-3-methylimidazolium bis(trifluoromethylsulfonyl)imide ([EMIM][NTf_2])) in the frequency of 0.1 MHz–10 mHz with nanostructured carbon thin-film electrodes (Figure 4.14). All the impedance spectra obtained exhibit similar trends with an arc at the high frequency and a straight line at the low frequency, as described above. The knee frequencies of [C_{12}MIM][NTf_2], [BPyr][NTf_2], [BMIM][NTf_2], and [EMIM][NTf_2] were found to be 1, 7, 10, and 32 Hz, respectively. The lower the knee frequency the larger the internal resistance. The fitted values of R_p were 3344, 667, 458, and 179 Ω for [C_{12}MIM] [NTf_2], [BPyr][NTf_2], [BMIM][NTf_2], and [EMIM][NTf_2] RTILs, respectively. The authors attributed the large differences in R_p to differences in ionic conductivity of the electrolytes employed [56]. Larger R_p values indicated lower ionic conductivity. Accordingly, the corresponding ionic conductivities were 0.6, 2.2, 4.0, and 9.2 mS cm^{-1}, following the trend: [C_{12}MIM][NTf_2] < [BPyr][NTf_2] < [BMIM][NTf_2] < [EMIM][NTf_2]. The change of cations in ILs resulted in different physical properties. Accordingly, the usage of the [C_{12}MIM][NTf_2] electrolyte leads to the lowest double-layer capacitance compared to the other capacitances of the other three ILs studied.

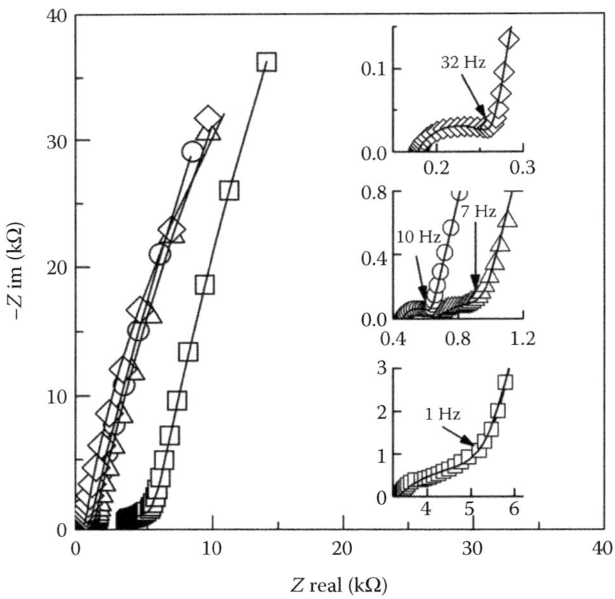

FIGURE 4.14 Nyquist plot of the complex impedance of a nanostructured carbon film electrode with 200 nm in thickness immersed in four different ILs (symbols: open squares, [C_{12}MIM][NTf_2]; open circles, [BMIM][NTf_2]; open triangles, [BPyr] [NTf_2]; and open rhombi, [EMIM][NTf_2]). (Reprinted from *Carbon*, 59, Bettini, L. G. et al., Planar thin film supercapacitor based on cluster-assembled nanostructured carbon and ionic liquid electrolyte, 212–220, Copyright 2013, with permission from Elsevier.)

4.4.4 Other Electrochemical Characterizations

Besides the cycling test, the constant voltage hold test has also been used to evaluate the stability of ESs. In most cases, cycling ability is assessed by repeated charge/discharge processes, which is rather time consuming as it typically lasts for thousands of cycles without significant capacitance loss for EDLCs. In order to save cycling time and more importantly, to more closely reflect the emergency conditions these devices are often employed in, the voltage hold experiment was developed as an alternative evaluation method [57,58]. This procedure involves holding the ES at a constant cell potential and determining the capacitance variation versus time. Weingarth et al. [58] described the detailed procedures in both cycling and voltage hold tests, and pointed out the advantages and disadvantages of each test. The solid lines stand for voltage and current of the common GCD cycle tests procedure while the dotted lines represent applied voltage and current of the constant voltage hold test (Figure 4.15). As can be seen, a constant voltage above the assumed stability limit is applied to the capacitor all the time when characterizing the stability using the hold test method (Figure 4.15, top), whereas the time when the voltage is above the stability limit is short in the cycling test. The current passing across the cell is also different for each of the two testing methods, see bottom in Figure 4.15. The current of the voltage hold test gradually decreases with prolonged time while the recorded

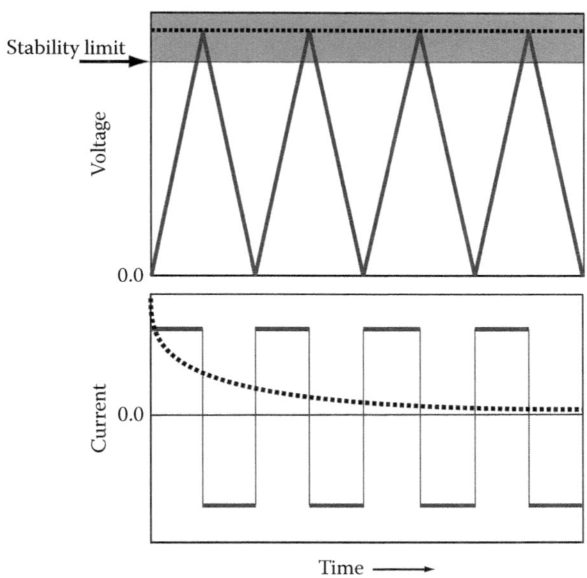

FIGURE 4.15 Curves of voltage (top) and current (bottom) during the cycle test and during the constant voltage hold test. Cycle test: full line, hold test: dashed lines. Gray area indicates voltage (and time) above the stability limit. (Reprinted from *Journal of Power Sources*, 225, Weingarth, D., A. Foelske-Schmitz, and R. Kotz, Cycle versus voltage hold—Which is the better stability test for electrochemical double layer capacitors? 84–88, Copyright 2013, with permission from Elsevier.)

current of the cycling test remains unchanged. The authors obtained different results from the cycling test and the voltage hold test with the applied voltage limit slightly higher than the accepted stability limit. They demonstrated that constant voltage hold tests are more demanding than cycling tests for AC-based EDLCs using either $TEABF_4$/ACN or [EMIM][BF_4] IL electrolytes.

The self-discharge is a meaningful performance metric for a capacitor device. The voltage of a capacitor in an open-circuit state will gradually decline during the self-discharge process. The driving force of the self-discharge process is the more positive free energy for a capacitor or battery in the charged conditions than in other states, such as discharged or partially charged states [32], as has been described in Chapter 1. Most of the ESs face the problem of self-discharge. Testing for self-discharge, investigating and exploring the mechanism, and further more proposing control strategies, are therefore of great importance for the evaluation and future improvement of such devices.

Through measuring the open-circuit potential of the charged cell with respect to time (from milliseconds to months), we can assess the self-discharge rate, see example, Figure 4.16, reported by Azaïs et al. [59]. They found that for fresh ESs, the decrease of OCV for OPTI (a typical synthetic AC prepared by steam activation

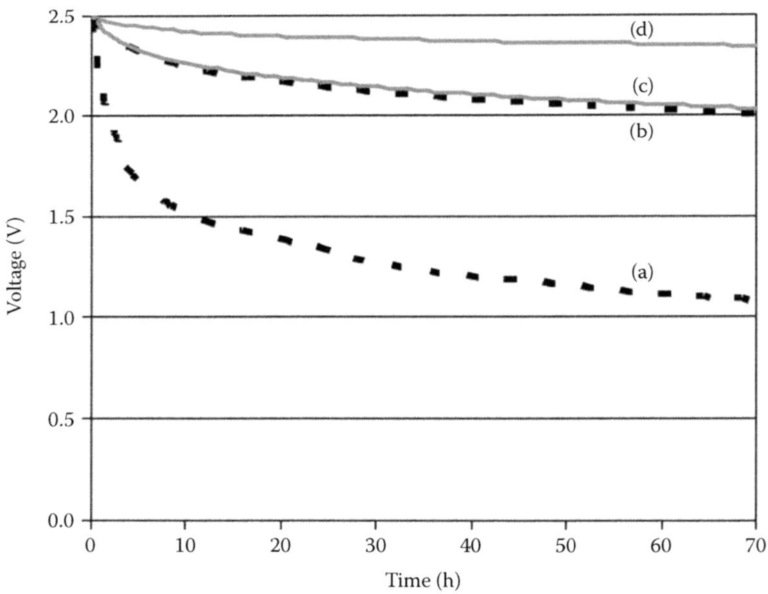

FIGURE 4.16 Time dependence of the OCV for supercapacitors based on OPTI and Maxsorb electrode. The supercapacitors were charged at 2.5 V during 30 min. (a) Supercapacitor based on Maxsorb electrodes (black dot-line); (b) the same as (a) after 500 h floating at 2.5 V (black dot-line); (c) supercapacitor based on OPTI electrodes (gray line); and (d) the same as (c) after 500 h floating at 2.5 V (gray line). (Reprinted from *Journal of Power Sources*, 171, Azaïs, P. et al., Causes of supercapacitors ageing in organic electrolyte, 1046–1053, Copyright 2007, with permission from Elsevier.)

of a phenolic resin char) is lower than that for Maxsorb (AC obtained after activation of a petroleum pitch carbon by KOH at 850°C) after 70 h using 1 M $TEABF_4$ in ACN electrolyte, suggesting that the self-discharge of the Maxsorb electrode is more serious than OPTI. This result can be ascribed to the different number of carboxylic groups in the Maxsorb and OPTI materials [60]. Another primary way to measure self-discharge is to test the time-dependent potentials using a three-electrode arrangement [32]. The potential of each electrode is recorded against a reference electrode and the experiment is carried out on an open circuit. In this case, the potential drops of each electrode can be measured separately. This method is particularly suitable when the mechanisms or rates of self-discharge at the positive and negative electrodes are different. In tests using the self-discharge measurement process, applying a float current to maintain a fully charged potential is another simple and convenient procedure [32]. The self-discharge rate is finely balanced by the float current to maintain a constant potential. This method, however, provides little information about what is taking place on the electrodes. It is noted that ES measuring systems should have a high input resistance, preferably at least $10^7\ \Omega$ or more, in order to make sure the measuring circuit itself does not draw a significant charge from the ES, thus resulting in an enhanced rate of self-discharge [32].

ESs self-discharge rates and their related mechanisms are principally the result of the type of electrolytes and their impurities and residual gases, has been discussed in detail in Chapter 2. The migration of electroactive components into the electrolyte can also contribute to accelerating the rate of the self-discharge process. Chen et al. [61] systematically investigated the processes of self-discharge and proposed the migration of the active electrolyte in the device (namely "the shuttle effect") is the main reason for accelerated self-discharge when soluble electroactive species are incorporated into a supercapacitor. Usage of an ion-exchange membrane separator (e.g., Nafion® film) and a novel electrolyte that convert into a solid species in the process of electrochemical reactions may be two of the strategies to explore to block the migration of active electrolytes [61].

It is worth recording that the testing conditions, such as: cell configuration; CV scan rate; GCD current density; and potential range window, can have a large influence on the measured results (e.g., specific capacitance) of the cell. For example, the specific capacitance of a two-electrode cell is contributed by two capacitive interfaces in series, and therefore it cannot be compared with that obtained from a three-electrode cell. Specific capacitances achieved at lower scan rates or current densities are commonly higher than those obtained at higher scan rates or current densities. Highly comparable electrode thicknesses are essential and too thin and/or too light electrode materials will lead to an overstatement of performance [30]. Electrochemical performance of the active electrode material should be much higher than that contributed from other ES components in order to get an ideal signal–noise ratio. Other factors related to the electrodes, such as the mass loading of the active material and the thickness of the electrodes themselves will also affect the measured outcomes [30,62].

Figure 4.17 displays the influence of the mass loading of the single-walled carbon nanotube (SWCNT) electrodes on the specific capacitances using a two-electrode cell [30,62]. With an increase in the mass loading of active materials, the specific capacitance

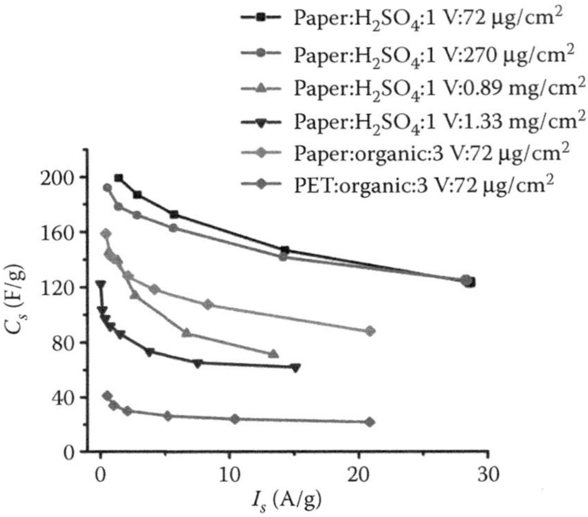

FIGURE 4.17 Mass loading dependence of gravimetric capacitance of carbon nanotube (CNT) on paper. The polyethylene terephthalate (PET) device with the lowest mass loading is plotted together for comparison. The data are calculated based on CNTs mass only. (Adapted from Hu, L. et al. 2009. *Proceedings of the National Academy of Sciences of the United States of America* 106:21490–21494 [in Supporting Information].)

decreased in a H_2SO_4 aqueous electrolyte. At 2 A g^{-1}, the measured capacitance decreased from 200 to 85 F g^{-1} when the mass of electrode increased from 0.072 to 1.33 mg cm^{-2}. Electrodes with different mass loadings also led to different electrode thickness. The electrode thickness of 1.33 mg cm^{-2} mass loading is 14 μm, which is about 18-fold larger than that of 72 μg cm^{-2} electrode (0.75 μm) [62]. Thus, it is essential to attach corresponding electrode mass and thickness to any meaningful reported specific capacitance and energy density. Therefore, to compare the performances of the studied electrolytes and their correspondingly assembled ESs from different investigators, special attention must be paid to the electrode factors and related measurement parameters.

In particular, the ESPW of the electrolyte is a crucial property that largely dominates maximum cell potential limits, and in turn, which determines two key performance metrics (i.e., energy density and power density). Thus, it is essential to develop accurate quantification methods to determine the ESPW of any given electrolyte. ESPW refers to the potential range where the electrolyte remains stable, and it is related to the resistance of the electrolyte components (e.g., solvent and dissolved salt) to the oxidative and reductive decomposition reactions at the charge electrodes. The conventional method to determine ESPW is generally based on the measurement of the cathodic reduction and anodic oxidation limits of the electrolyte by linear sweep voltammetry on a smooth WE, such as Pt and a GC [63–66]. As shown in Figure 4.18a, the oxidative and reductive decomposition of the electrolyte lead to an abrupt increase of the apparent current densities [63]. When the current density reaches a critical value, the applied positive and negative potentials are generally defined as the cathodic and anodic limits of the ESPW. Between the cathodic and

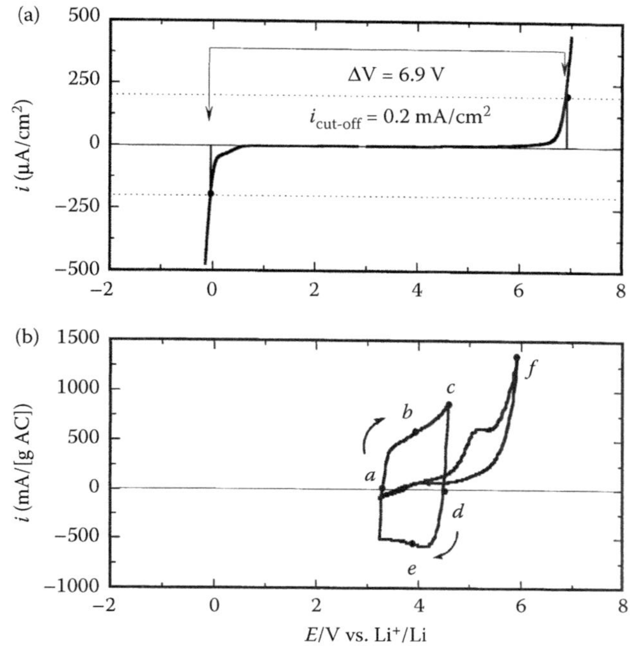

FIGURE 4.18 (a) CV and conventional determination of stability limits of a popular nonaqueous electrolyte for a double-layer capacitor: 1.0 M Et_3MeNPF_6 in EC/DMC (1:1 by weight) on nonporous WE (GC). (b) CV of the same electrolyte when porous composite based on M30 AC (95% with 5% PVdF) is used as WE. Scan rate: 5 mV s^{-1}; Li as reference and Pt as CEs. Successive scans were conducted with an interval of 0.25 V between their cutoff potential limits (only sixth and tenth scans are shown). (Reprinted from *Electrochimica Acta*, 46, Xu, K., M. S. Ding, and T. R. Jow, A better quantification of electrochemical stability limits for electrolytes in double layer capacitors, 1823–1827, Copyright 2001, with permission from Elsevier.)

anodic voltage limits, the electrolyte is recognized as an electrochemically stable media. As can be seen the definition of these limits is somewhat arbitrary, which brings about uncertainty for the comparison of the ESPW between different electrolytes. Olson et al. [67] proposed a general method for defining the ESPW limits that depends less on experimental parameters, which allowed more fair and meaningful comparison of data from different investigators. This method involved linear fits in the two approximately linear ranges of the current–potential curve. The ESPW limit was established by the intersection of the two linear fits. This method was found to be much less sensitive to variations in ion mobility and electrolyte concentration than the conventional measurement method. Furthermore, the actual electrodes exposed to the electrolyte in ESs are commonly highly porous. Consequently, this gives a much larger capacitive current response compared with that of nonporous electrodes (Figure 4.18b [63]) because the porous nanostructure provides more exposed active sites and benefits the immersion and transportation of the electrolyte. Several studies have been devoted to the development of a determination method of ESPW for porous electrodes used in ESs [63,68]. For example, Xu et al. [63] demonstrated an improved quantification measurement of ESPW for porous electrodes. In this study, the faradaic (correlated with electrolyte decomposition) and capacitive current were isolated. Through the proposed new method, reliable stability data can be obtained to describe the ESPW limits of nonaqueous electrolytes used in EDLCs. This new approach could also be employed to evaluate the interaction between electrolyte and electrode in more general systems, especially for electrodes with high surface area. In addition, the ESPW of organic liquid electrolytes are accurately predicted based on *ab initio* molecular orbital calculations by Maeshima et al. [69,70]. The authors calculated the reduction and oxidation values of different molecular species by considering the intermolecular interactions. This method offers a predictive methodology for the performance of a wide range of electrolyte materials.

4.4.5 Instrumental Characterizations

Apart from the electrochemical methods, other instrumental characterizations such as Fourier transform infrared (FTIR) spectrometer [71], NMR [12], SNAS [72,73], synchrotron X-ray diffraction (XRD) [74], etc. have also been employed to characterize the properties and electrochemical behaviors of electrolytes. For example, the *in situ* infrared spectroelectrochemical technique can quantitatively and directly characterize, on a molecular level and in real time, the ion dynamics (i.e., anions and cations adsorbing/desorbing or entering and exiting behaviors in nanoporous electrodes), thus helping to clarify and understand the deeper fundamental operating mechanisms of ESs [71]. In the EDLCs based on 1-ethyl-3-methylimidazolium bis-(trifluoromethylsulfonyl)imide ([EMIM][TFSI]) electrolyte, the *in situ* spectroelectrochemical results show the amounts of [EMIM] cations and [TFSI] anions both decreases as the voltage increases from 0 to 1 V due to the strong interaction between cations and anions [71] (similar to previously observed results by the same group [75]). When evaluating the potential higher than 1.0 V, however, more anions compared to cations positively enter into the charged pores of nanoporous carbon nanofibers (NCNFs), showing the phenomenon of cation–anion separation (Figure 4.19) [71]. Conversely, when charging the KOH-activated NCNF electrodes, the concentration of [EMIM] cation increases while the concentration of the TFSI anion decreases (Figure 4.20) [71]. As the cations exit the nanopores, the anions enter the nanopores of the NCNFs at the same time. These results indicate that through tailoring the surface chemistry of the nanofiber materials, the operating behavior of the [EMIM][TFSI] electrolyte can fundamentally change. This finding sheds light on the rational design of the electrolytes and electrodes and their interfaces in ESs for future device optimization. Such observation is impossible without the insights provided by *in operando* infrared spectroelectrochemistry.

Using *in situ* NMR investigation, susceptibility effects were exploited to gain a deeper understanding of the mechanisms of charge/discharge processes of ESs and key factors affecting performance [11]. Solid-state NMR is powerful in observing the dynamic behaviors of electrolyte ions, like detecting where they are (inside the carbon pores or adsorbed on the outer surface of carbon walls) [76]. The solid-state ^{11}B NMR revealed the behavior of the tetrafluoroborate (BF_4^-) anion of $TEABF_4$ in a

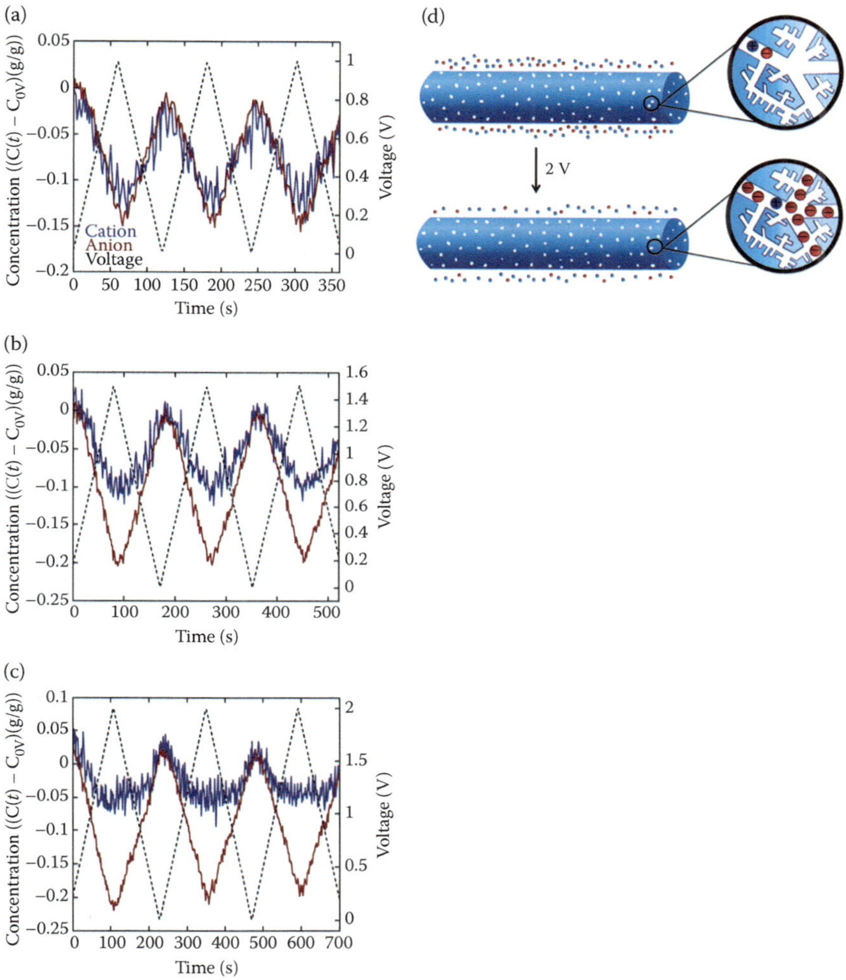

FIGURE 4.19 *In operando* infrared spectroelectrochemical results: initialized time-resolved concentration (concentration units g/g corresponds to $g_{[EMIM][TFSI]}/g_{electrode}$) of the EMIM cation and TFSI anion in NCNFs during three CV at 20 mV s^{-1} from (a) 0–1 V, (b) 0–1.5 V, and (c) 0–2 V. (d) Illustration showing ion adsorption within nanopores of NCNFs. (Reprinted with permission from Richey, F. W. et al., Ionic liquid dynamics in nanoporous carbon nanofibers in supercapacitors measured with in operando infrared spectroelectrochemistry. *The Journal of Physical Chemistry C* 118, 21846–21855. Copyright 2014 American Chemical Society.)

PC solvent electrolyte in an EDLC using AC as an electrode material [76]. Two peaks (sharp peak A and broad peak B) were distinguished by ^{11}B NMR upon the carbon in the states of impregnated, charged, and discharged (Figure 4.21), corresponding to the BF_4^- ions absorbed on the outer surface and inside the pores of the AC, respectively. More electrolyte ions were located on the outer surface, indicated by

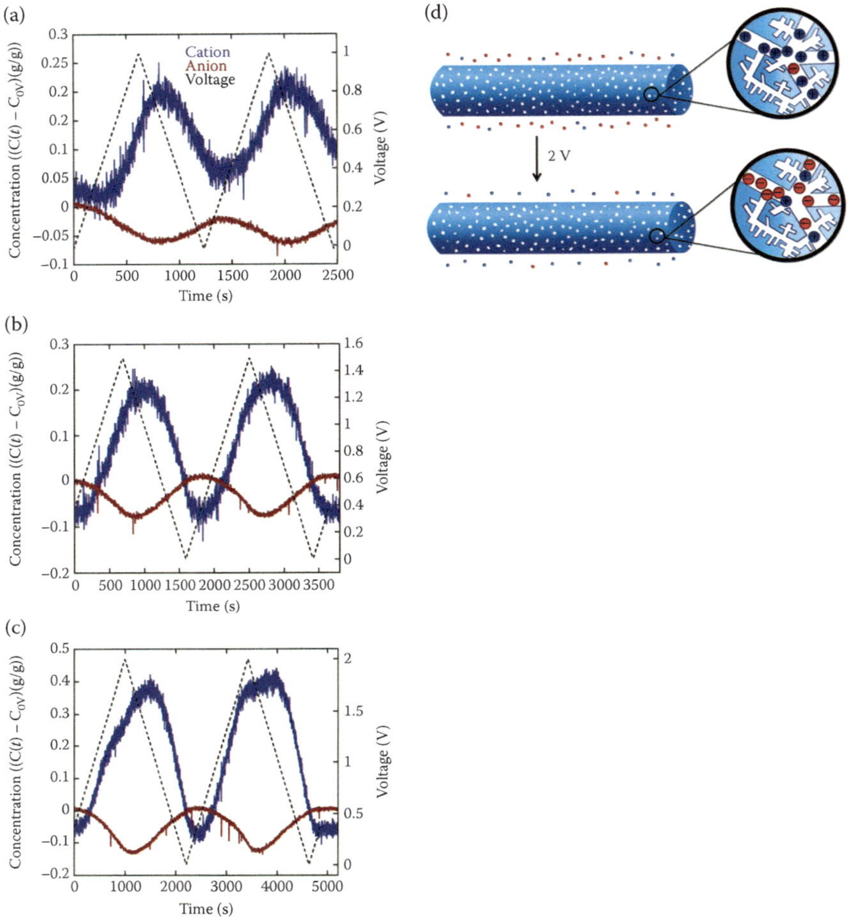

FIGURE 4.20 *In operando* infrared spectroelectrochemical results: initialized time-resolved concentration (concentration units g/g corresponds to $g_{[EMIM][TFSI]}/g_{electrode}$) of the EMIM cation and TFSI anion in KOH-activated NCNFs during three CV at 2 mV s^{-1} from (a) 0–1 V, (b) 0–1.5 V, and (c) 0–2 V. (d) Illustration showing ion adsorption/desorption in/out of nanopores of KOH-activated NCNF with a larger concentration of EMIM cations than TFSI anions in the nanopores of the NCNFs at OCV, followed by EMIM cations being replaced with TFSI anions during charging to 2 V. (Reprinted with permission from Richey, F. W. et al., Ionic liquid dynamics in nanoporous carbon nanofibers in supercapacitors measured with in operando infrared spectroelectrochemistry. *The Journal of Physical Chemistry C* 118, 21846–21855. Copyright 2014 American Chemical Society.)

the larger intensity of peak A. Notably, the anions confined in the AC pores showed significant ^{11}B diamagnetic up-shift and substantial line broadening compared to that of the free BF_4^- species [76]. Furthermore, Wang et al. [11] identified three major resonances when investigating the sorption of BF_4^- anions on the porous carbon. The three peaks were ascribed to strongly bound ions on the carbon surfaces, weakly

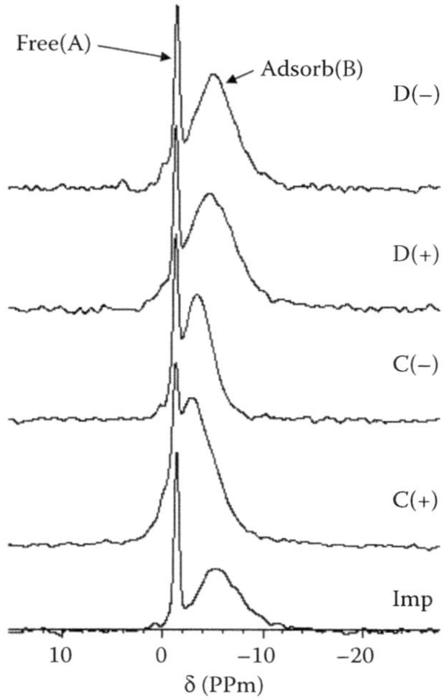

FIGURE 4.21 ^{11}B-NMR spectra of BF_4 in AC with specific surface area of 500 m^2 g^{-1} at impregnated, charged, and discharged states. A and B peaks were attributed to free and adsorbed BF_4 species, respectively. (Reprinted from *Carbon*, 44, Lee, S.-I. et al., ^{11}B NMR study of the anion in activated carbons at various stages of charge of EDLCs in organic electrolyte, 2578–2586, Copyright 2006, with permission from Elsevier.)

bound ions adsorption in the outer layers or within the larger mesopores, and free ions in the bulk electrolyte, respectively. On charging, the anions move positively to the charged electrode and the concentration changes. The *in situ* NMR experiments allowed us to identify the ionic coordination sites distinctly and also the ions' migration path during the operation of the ESs, NMR measurements on hierarchical porous carbons loaded with 1 M $TEABF_4$ in ACN, however, showed only one single NMR signal [77]. The authors attributed this to the rapid movement and exchange of BF_4^- anions between different pores systems of the utilized carbon with multi-pore size distribution [77]. They also observed (confirmed by the increase of the diamagnetic shift) that solvent removal reduced the average ion-wall distance in large pores due to the loss of their solvent shells, but did not have a large effect on that distance in smaller pores with diameter less than 1 nm [77]. This procedure provides experimental evidence of the absence of intact shells among electrolyte molecules in such small pores. The method is also promising for future investigations of the structure and dynamics of molecules and ions inside the ES systems and can be extended to study the solid–liquid interfaces and electrical double-layer constructions.

The transportation and electroadsorption of ions in the electrolyte are affected by many parameters of the electrolyte including wetting, dielectric constant, viscosity, conductivity, solvation energy, and shell size. Since these factors have been proven to greatly influence the performance of ESs [1,78], a better, in-depth understanding of their impacts is urgent. SANS has been developed to get additional information about material structure at the micrometer length level and supercritical fluid or ions adsorption behaviors in porous media [72]. Applying SANS technology to ESs, Boukhalfa et al. [73] recently reported they could elucidate the position where ion adsorption occurred through probing the distribution of the hydrogen atoms within the electrolyte under an external potential. In a H_2O-based electrolyte which wets the carbon surface well, proton adsorption is enhanced in subnanometer pores. However, both SANS and sorption measurement revealed incomplete access of D_2O to such pores and diminished proton adsorption. The SANS technique can also be utilized with organic and ILs systems. For example, Boukhalfa et al. [13] demonstrated the unique capability of *in situ* SANS to construct the relationship between the ion electroadsorption (concentration changes) and the applied potential or pore size in 1 M $TEABF_4$ in deuterated ACN (d-ACN) electrolyte. They found a positive potential led to a decreased intensity while a negative potential resulted in an increased intensity (Figure 4.22a). This result can be easily explained because the $N(C_2H_5)_4^+$ cations were absorbed into the pores under negative potential. Under a positive potential, the cations were expelled and/or replaced by BF_4^- anions and solvent molecules from the pores. Figure 4.22b shows the SANS profiles under voltage of 0, –2, and +2 after normalizing the intensities to the zero voltage SANS data. This approach enables the influence of pore size on the ion electroadsorption to be seen more clearly (Figure 4.22c). The higher scattering intensities for the smallest pores ($Q > 0.2$ Å^{-1}) indicated higher H concentration and lower B concentration at negative potentials. Opposite results were obtained using positive potentials for these pores. These results are in accordance with previously reported results and indicate enhanced ion adsorption in the subnanometer porous nanostructures [6,79].

In addition, synchrotron XRD was employed to investigate the structure and mechanism of nanoscale-confined electrolyte solutions. An example was recently reported by Ohba et al. [74], who used CNT as the electrode material and 1 M NaCl aqueous solution as the electrolyte. In this example, the terms "nanoelectrolyte" and "nanowater" respectively refer to the NaCl electrolyte and the water in the internal CNTs nanospaces. Identifying the Fourier transforms of the XRD patterns by subtracting the backgrounds tested *in vacuo*, the electron radial distribution functions (ERDFs) of the nanoelectrolyte, the nanowater, the bulk electrolyte solution, and the water were obtained, as shown in Figure 4.23. The nearest-neighbor distance was 0.34 for the nanoelectrolyte, which was about 0.03 nm longer than that in the bulk electrolyte solution. The first-neighbor distance in the nanoelectrolyte was ascribed to the intermolecular distances of water/water, Na^+/water, and Cl^-/water. The water hydrogen bonds were demonstrably stretched and broken at longer distances. The hydration shell of ions was formed instead of the hydrogen bonding. The interaction between water and ions favored the formation of first hydration shell. This insight provides a preliminary understanding about the mechanism of nanoscale aqueous electrolytes, although further work is needed to establish more detailed information like the hydration structures and hydration numbers.

Further characterization such as differential scanning calorimetry (DSC) using a thermal analyzer was performed to understand the structural and thermal properties

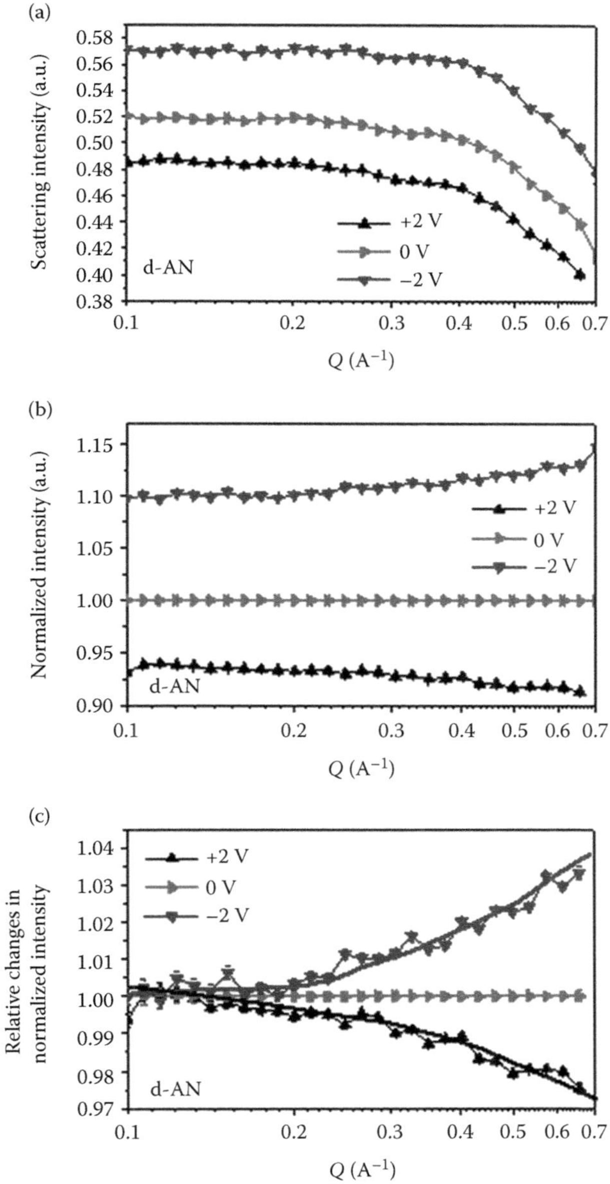

FIGURE 4.22 *In situ* neutron scattering experiments based on ACF electrodes immersed into 1 M $TEABF_4$/d-ACN electrolyte under a potential applied between the WE and CE: (a) SANS patterns and (b) SANS profiles normalized by the 0 V one, (c) relative changes in the intensity of the normalized SANS profiles. (Reprinted with permission from Boukhalfa, S. et al., *In situ* small angle neutron scattering revealing ion sorption in microporous carbon electrical double layer capacitors. *ACS Nano* 8, 2495–2503. Copyright 2014 American Chemical Society.)

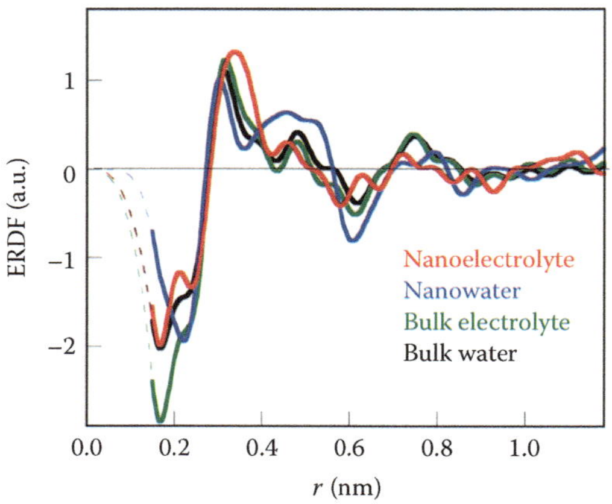

FIGURE 4.23 ERDFs for nanoelectrolyte, nanowater, bulk electrolyte, and bulk water. (Reprinted with permission from Ohba, T., K. Hata, and H. Kanoh. Significant hydration shell formation instead of hydrogen bonds in nanoconfined aqueous electrolyte solutions. *Journal of the American Chemical Society* 134, 17850–17853. Copyright 2012 American Chemical Society.)

of the investigated electrolytes, especially for solid electrolytes. For example, the solid electrolyte composed of poly(vinyl alcohol) (PVA) or Keggin-type heteropoly acid showed promising conductivities and stability and has been employed for solid-state ESs [80–82]. Gao et al. [83] using DSC thermograms, have provided the thermal behaviors of solid polymer electrolytes of phosphotungstic acid (PWA) and/or silicotungstic acid (SiWA) in a PVA matrix. Apart from this, the DSC signals also gave information about water contamination involved in the electrolyte. The temperature of endothermic peaks related to water departure revealed the type of water (e.g., adsorbed water, bonded water, and crystallized protonated water). The results showed the PVA–PWA and/or PVA–SiWA mixed polymer electrolytes retained some crystallized water. This may provide some insights into future water purification and electrolyte preparation.

4.5 ELECTROLYTE DEGRADATION DIAGNOSIS AND POSSIBLE MITIGATION

Aging and the related failure phenomenon of electrolytes and electrolyte-based ESs are important issues that need to be addressed. Understanding component aging mechanisms will be beneficial for the development of ESs with wider voltage windows and will also be necessary for guidance in the safe and reliable use of ESs. Performance degradation of ESs is mainly due to the following:

1. Wide ranging operation of cell potential leads to the oxidation of electrode materials and electrolyte decomposition. Higher operating cell voltage is pursued in order to maximize the energy density of the device. However,

taking organic electrolyte-based ESs as examples, once the cell voltage is larger than the typical value of 2.5–2.8 V, the undesirable oxidation reaction of electrode material and the decomposition reactions of the electrolyte are likely to occur. Increasing cell voltage has also been found to drastically decrease the lifetime of ESs [84]. For example, Kötz et al. [84,85] reported the acceleration factor for ESs degradation was about 2 for a voltage increase of 0.1 V. Therefore, a thorough understanding of component aging mechanisms will benefit the refinement of the voltage window in the further development of ESs.

2. Incompatibility of the electrolyte with the electrode materials, separators, and/or the current collectors. The cycle life of a pseudocapacitor with aqueous electrolytes is much shorter than that with an organic electrolyte whereas the mechanism is predominantly electrostatic charge separation [1,86]. With regard to current collectors, previous studies have demonstrated degradation of Al current collectors can also cause the aging of organic electrolyte-based ESs [87]. Besides, the electrochemical stability of an Al support in an aprotic electrolyte can be greatly affected by the type of conducting salt used [88]. The oxidation reaction of an Al component in a $NaN(SO_2F)_2$ (i.e., sodium bis(fluorosulfonyl)imide, NaFSI)-based electrolyte has been observed [88]. Separator materials can also cause performance degradation when used with incompatible electrolytes. For example, although cellulose separators can work well in organic electrolytes, they can suffer serious degradation in H_2SO_4 aqueous electrolyte, resulting in performance losses of related ESs [87].
3. Degradation of the electrode materials caused by electrolyte ion intercalation/deintercalation issues or negative electrochemical reactions with the electrolytes will lead to performance degradation of ESs [9,89].
4. Harsh operating conditions such as very high temperatures and too wide a working voltage range will also cause degradation of ESs' performance. For example, a high peak temperature was reported to decrease the cycle life of active carbon due to electrode material degradation caused by partial surface oxidation in an alkaline aqueous electrolyte [90]. Besides, the capacitance loss became more serious since the chemical reactions related to the fading processes were accelerated at high temperatures [90].
5. Residual impurities can lead to large performance degradation and fast self-discharge in ESs utilizing organic electrolytes. Common impurities consist of water, oxygen, and halide ions. Less common impurities may include the hydrolysis products of water-sensitive ions and even hydrogen fluoride [26]. For measurements of high-precision conductivity and voltammetry in ESs using organic electrolytes, water and oxygen must be completely removed.

From the above, it can be seen that most of the degradation factors in ESs are related to the electrolyte. Hahn et al. [91] used online mass spectrometry (differential electrochemical mass spectrometry, DEMS) to analyze the gas products evolved in an EDLC with AC electrodes in 1 M $TEABF_4$/PC during the charging/overcharging process. CO_2 ($m/z = 44$), propene ($m/z = 41$), and H_2($m/z = 2$) were

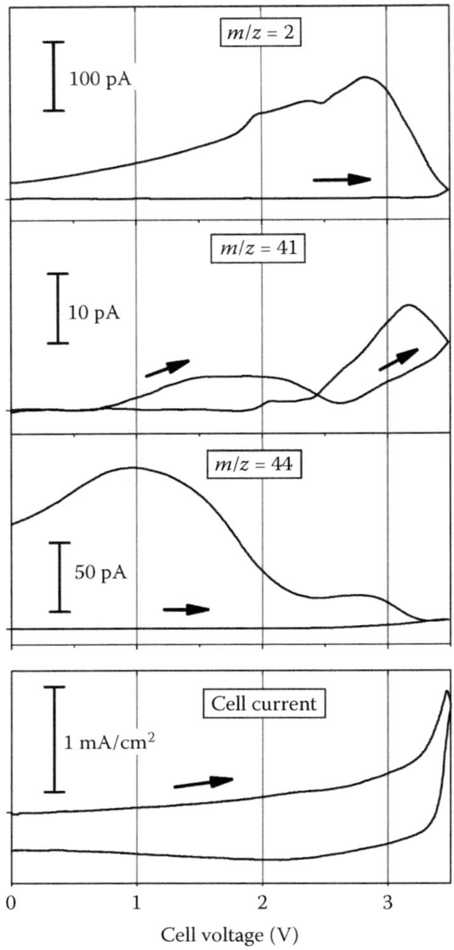

FIGURE 4.24 First cycle of CV (bottom) with scan rate of 0.5 mV s^{-1} and simultaneous gas evolution as detected by *on-line* mass spectroscopy. (Reprinted from *Electrochemistry Communications*, 7, Hahn, M. et al., Gas evolution in activated carbon/propylene carbonate based double-layer capacitors, 925–930, Copyright 2005, with permission from Elsevier.)

identified as the main gaseous decomposition products at cell voltages from 2.6 to 4 V (Figure 4.24). The formation of propene and H_2 were attributed to the solvent reduction at the negative electrode and CO_2 was believed to be formed by the solvent oxidation at the positive electrode. Notably, although the majority of the propene evolved above 2.6 V, a small amount of propene was already detected at cell voltages below 1 V. The gas evolution can lead to a significant pressure increase in the EDLCs, as reported by Hahn et al. [92] for an EDLC with 1 M $TEABF_4$/PC electrolyte at a cell voltage above 2.5 V (Figure 4.25). Kötz et al. [93] investigated the gas evolution and the resultant pressure formed in the EDLC

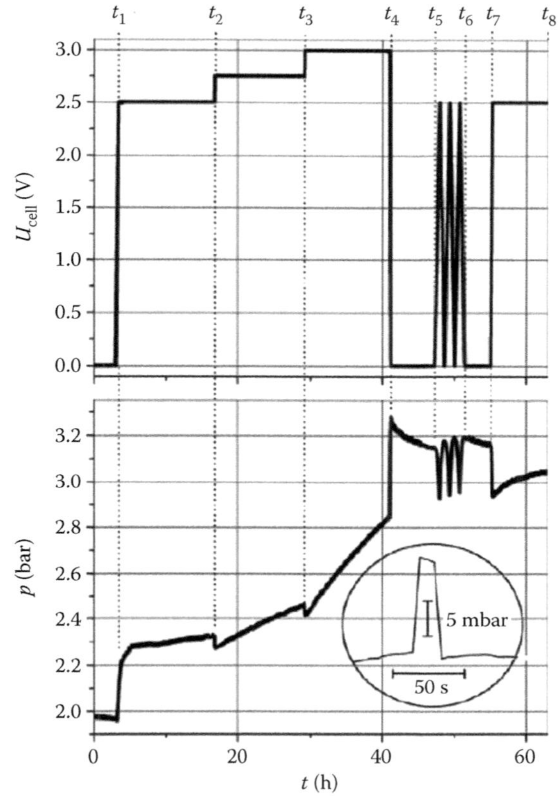

FIGURE 4.25 Applied cell voltage profile (top) and pressure response (bottom). The inset shows the pressure hump with injection/removal of 15 μL of liquid into the cell. The used electrolyte is 1 M $TEABF_4$ in PC. (Reprinted from *Electrochimica Acta*, 52, Hahn, M. et al. Pressure evolution in propylene carbonate based electrochemical double layer capacitors, 1709–1712, Copyright 2006, with permission from Elsevier.)

with 1 M $TEABF_4$ in three different organic solvents, including ACN, PC, and γ-butyrolactone (GBL), respectively (Figure 4.26). At the investigated cell voltages (2.5–3.25 V), the GBL-based electrolyte resulted in a significantly higher gas evolution compared to the PC and ACN-based electrolytes. The gas evolution in the PC-based electrolyte was 10 times higher than that in the ACN-based electrolyte at a cell voltage of 3.25 V. The authors thus concluded ACN-based electrolyte was favorable for high-voltage ESs. However, it should be noted the toxicity of the ACN-based electrolyte is much higher than that of the PC-based electrolyte.

The above work concerning the formation of gas products in ESs during high voltage operation was also confirmed by Ishimoto et al. [89,94], who separately analyzed the gases evolved from the positive and negative compartments in an H-type cell after float-tests (Figure 4.27). They found that for an EDLC with 1 M $TEABF_4$/PC electrolyte, charging beyond 2.7 V caused serious decomposition of the

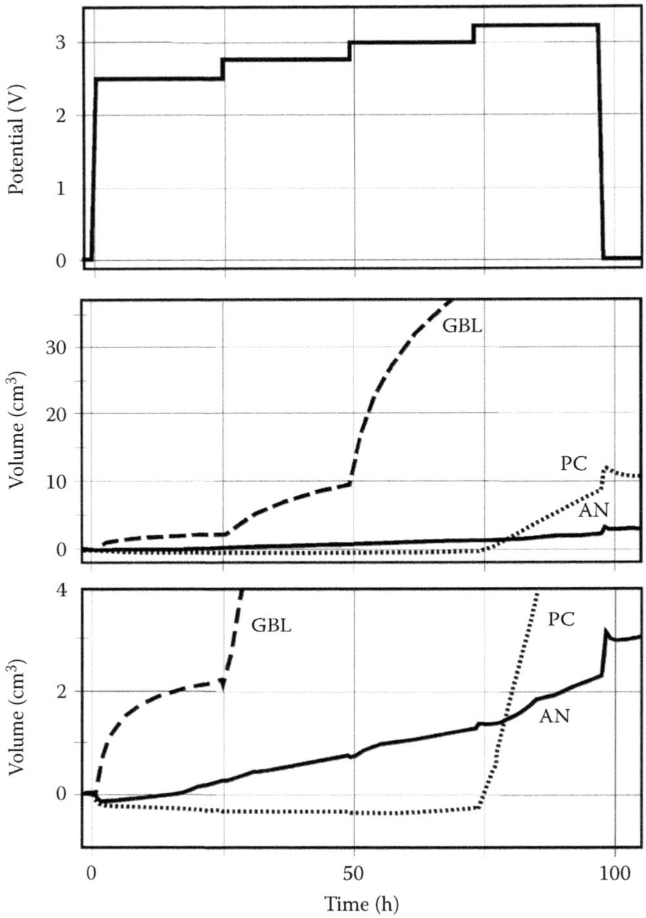

FIGURE 4.26 Applied cell voltage (top) and evolved gas volume (at 1 bar) calculated from pressure rise and compressibility. Bottom graph shows an enlarged region. 1 M $TEABF_4$ in PC (dotted line), acetonitrile (AN) (full line), and GBL (dashed line). (Reprinted from *Electrochemistry Communications*, 10, Kötz, R. et al. Comparison of pressure evolution in supercapacitor devices using different aprotic solvents, 359–362, Copyright 2008, with permission from Elsevier.)

electrolyte leading to gas evolution. As a result, the capacitance decreased and the internal resistance increased with an increase in the applied voltage. When the cell voltage ranged from 3.0 to 4.0 V, two main gaseous products (CO_2 and CO) were found in the positive compartment, due to the electrochemical oxidation of PC and the surface functional groups on AC [89,91,94]. Other functional groups, such as the phenol and ketone, produce CO gas at voltages higher than 3.3 V [89]. In the negative compartment, the adsorbed water was reduced to H_2 gas and OH^-. Because of the formation of OH^-, the hydrolysis of PC occurred in the negative compartment.

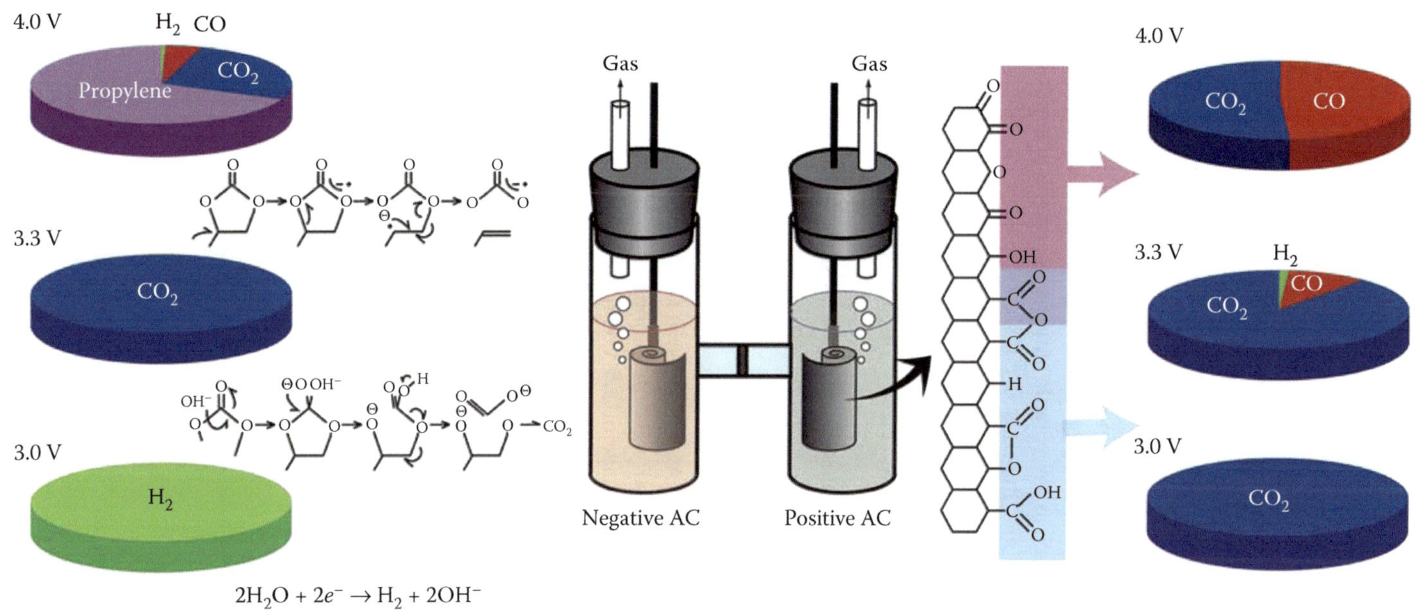

FIGURE 4.27 Gas evolution from an EDLC cell upon over-voltage application. (Naoi, K. et al. 2012. Second generation "nanohybrid supercapacitor": Evolution of capacitive energy storage devices. *Energy & Environmental Science* 5:9363–9373. Reproduced by permission of The Royal Society of Chemistry.)

Other gases, such as propylene, CO_2, ethylene, and CO, were detected at voltages higher than 4.0 V, indicating the direct electrochemical oxidation of PC [89]. The accumulation of solid electrolyte interface films was observed at higher cell voltage (e.g., 3.7–4.0 V), which was caused by the electrochemical oxidation and the reduction of the PC on the electrode surface.

The decomposition of the organic electrolyte at high cell voltage was also reported by Azaïs et al. [59], who characterized the AC electrodes by X-ray photoelectron spectroscopy (XPS) and NMR in the 1 M $TEABF_4$/ACN before and after prolonged floating (4000–7000 h) at a cell voltage of 2.5 V. It was found the aging of the ES was strongly related to the decomposition of the organic electrolyte. The decomposition products blocked the pores of the ACs, resulting in the decrease of the capacitance and increase of ESR (Figure 4.28). A series of decomposition products of 1 M $TEABF_4$/ACN, such as the volatile amines, intermediate amides, carboxylate anions of organic acids, and fluorinated compounds, were identified by Kurzweil et al. [95] for an EDLC after testing at cell voltages above 4 V (Figure 4.29). Taking into account the power- and energy density of a supercapacitor is proportional to the square of the cell potential, the attempts to develop novel electrolytes to increase the working potential range of the electrolyte and the corresponding ESs are very welcome.

Different aging degrees at the negative and positive electrodes have also been noticed [96]. For example, it was found the positive electrode aged much more rapidly than the negative electrode for an EDLC with $TEABF_4$/ACN electrolyte [57,96]. Furthermore, the aging rate of the negative and positive electrodes also depends on both the electrode polarity and the solvents. Ruch et al. [97] found the aging process was much faster on the positive AC electrode in the ACN solvent with 1 M $TEABF_4$ whereas the aging was much stronger on the negative electrode in the PC solvent with the same salt at a cell voltage of 3.5 V (Figure 4.30). The capacitance loss is

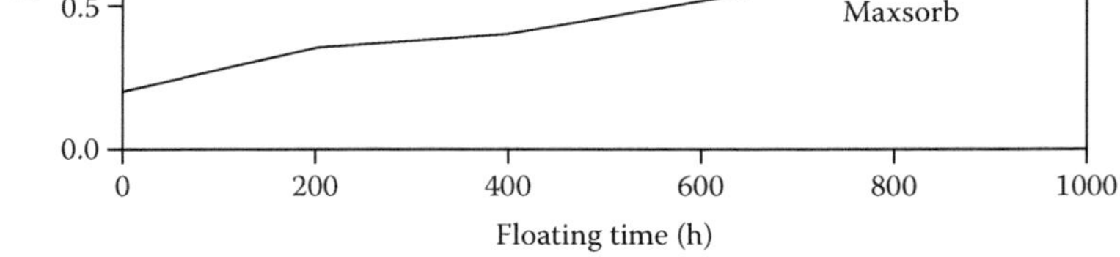

FIGURE 4.28 Evolution of the ESR at 0.01 Hz of Maxsorb and OPTI with the floating time at 2.5 V. Maxsorb and OPTI are the two ACs synthesized through two different methods. (Reprinted from *Journal of Power Sources*, 171, Azaïs, P. et al., Causes of supercapacitors ageing in organic electrolyte, 1046–1053, Copyright 2007, with permission from Elsevier.)

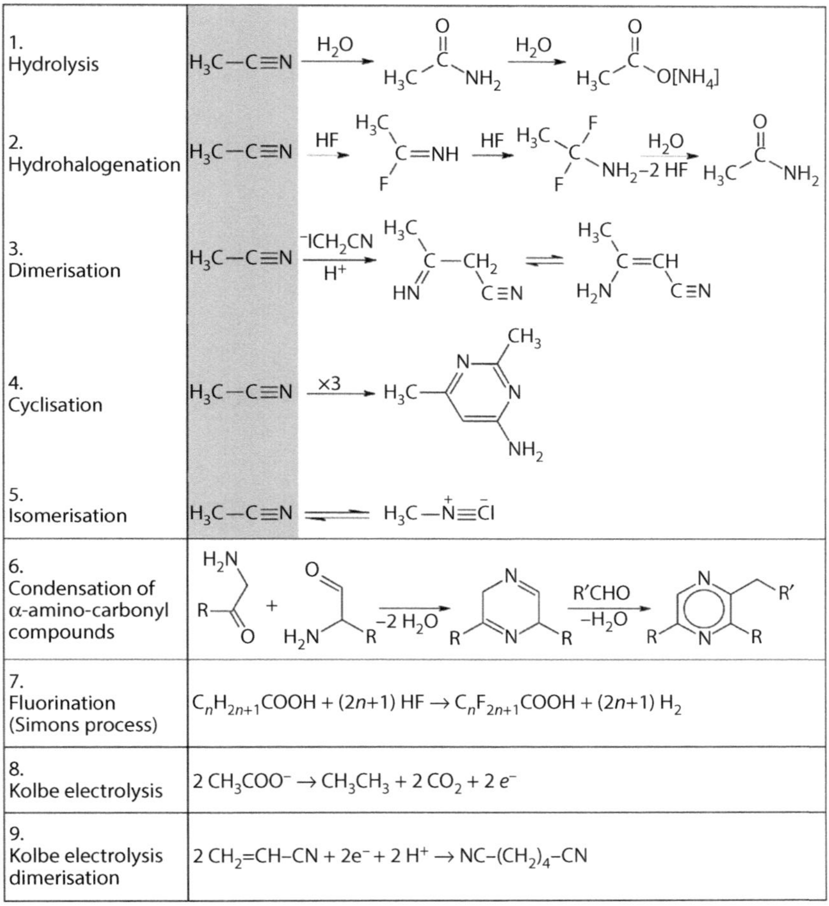

FIGURE 4.29 Plausible reactions of acetonitrile and its oxidation products which are consistent to the analytical findings. (Reprinted from *Journal of Power Sources*, 176, Kurzweil, P., and M. Chwistek, Electrochemical stability of organic electrolytes in supercapacitors: Spectroscopy and gas analysis of decomposition products, 555–567, Copyright 2008, with permission from Elsevier.)

ascribed to the increased electrode resistance during the aging period caused by the increased contact resistances between the electrode and the current collectors or the individual grains of the electrodes, or an increase in the distributed resistance within the porous microstructure. The above results suggest an asymmetric design of the carbon-based ESs may be beneficial for increasing the lifetime.

The degradation rate of the electrolyte can be mitigated through optimizing the electrode mass ratio in carbon-based ESs. For instance, Cericola et al. [98] reported that for carbon-based ESs with 1 M $TEABF_4$ in ACN or in PC, the degradation rate can be decreased through optimizing the carbon mass of the positive electrode. Figure 4.31 shows the cell capacitances of the AC electrodes-based supercapacitors

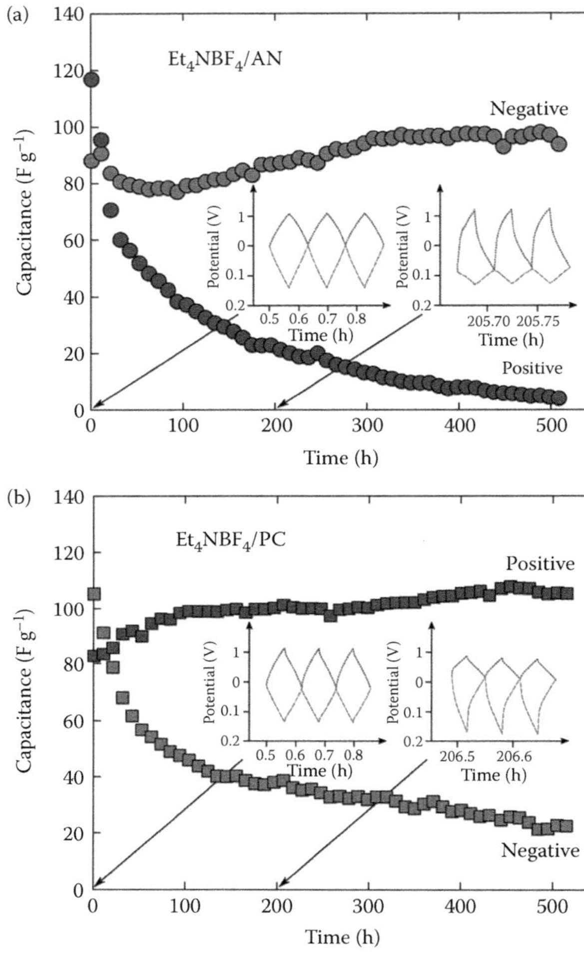

FIGURE 4.30 Measured capacitances of single electrode with the (a) ACN-based and (b) PC-based electrolytes. The capacitances are all normalized to the active mass of single electrodes. The inset graphs represent the potential versus time curves of the negative (dashed lines) and positive (dotted lines) electrodes of the full cell during galvanostatic cycling at the indicated times. (Reprinted from *Electrochimica Acta,* 55, Ruch, P. W. et al., A comparison of the aging of electrochemical double layer capacitors with acetonitrile and propylene carbonate-based electrolytes at elevated voltages, 2352–2357, Copyright 2010, with permission from Elsevier.)

with different mass ratios under accelerated aging tests in $TEABF_4$ in ACN for different aging times. The optimized mass ratio was 0.65, which exhibited the best stability. After aging at 3.5 V for 400 h, the capacitance of the electrode with mass ratio of 0.65 maintained half of the original value (from 21.5 to 10.5 F) while that with mass ratio of 0.50 decreased to 15% of its initial value (from 25 to 3.5 F). Moreover, it can be seen that the increased mass of the positive side greatly reduced

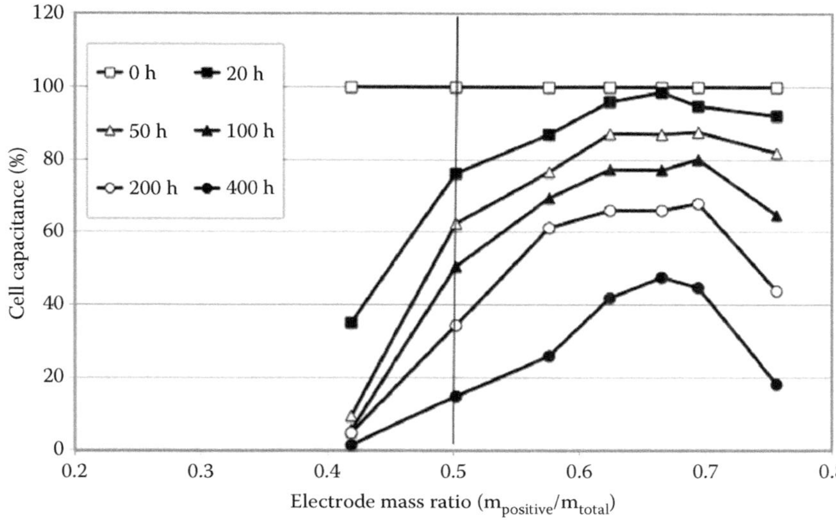

FIGURE 4.31 Cell capacitance as a function of electrode mass ratio (r) after various times of accelerated constant voltage test at a cell voltage of 3.5 V, electrolyte: 1 M $TEABF_4$ in ACN. (Reprinted from *Journal of Power Sources*, 196, Cericola, D., R. Kötz, and A. Wokaun, Effect of electrode mass ratio on aging of activated carbon based supercapacitors utilizing organic electrolytes, 3114–3118, Copyright 2011, with permission from Elsevier.)

the degradation process while the degradation of the negative side was accelerated in the same electrolyte (Figure 4.32).

The aged electrodes and the electrolyte decomposition were investigated by XPS spectra. Decomposition products were found to form mainly on the surface of the electrodes after aging. Azaïs et al. [59] analyzed the electrolyte (1 M $TEABF_4$/ACN) and the carbon electrodes before and after prolonged floating tests using XPS technique. Detailed results are listed in Table 4.2. Sodium and a high amount of oxygen were detected in the fresh electrodes. Both elements originated from the binder (carboxymethylcellulose). The compositions of elements on the aged positive and negative electrodes are quantitatively different. The positive electrode contains more nitrogen while the negative electrode contains more fluorine. Both the nitrogen and fluorine elements observed on the positive and negative electrodes are evidence of the decomposition of Et_4N^+ and BF_4^- ions in the electrolyte. Furthermore, a small amount of boron was found in the aged negative electrodes, again suggesting the decomposition of BF_4^-. The concentrations of Na, N, and F on the surface were much larger than those inside the aged electrodes, indicating the decomposition reactions happened mainly on the surface. Another unusual result was the decreasing amount of the oxygen elements on the electrodes' surface after aging, which was caused by the partial dissolution of the binder into the electrolyte. For different AC electrodes, the aging mechanism of the electrodes was the same, as a result of the similar electrochemical reactions during the electrolyte decomposition. The NMR signals further confirmed the XPS analysis. The formation of insulating products blocked the

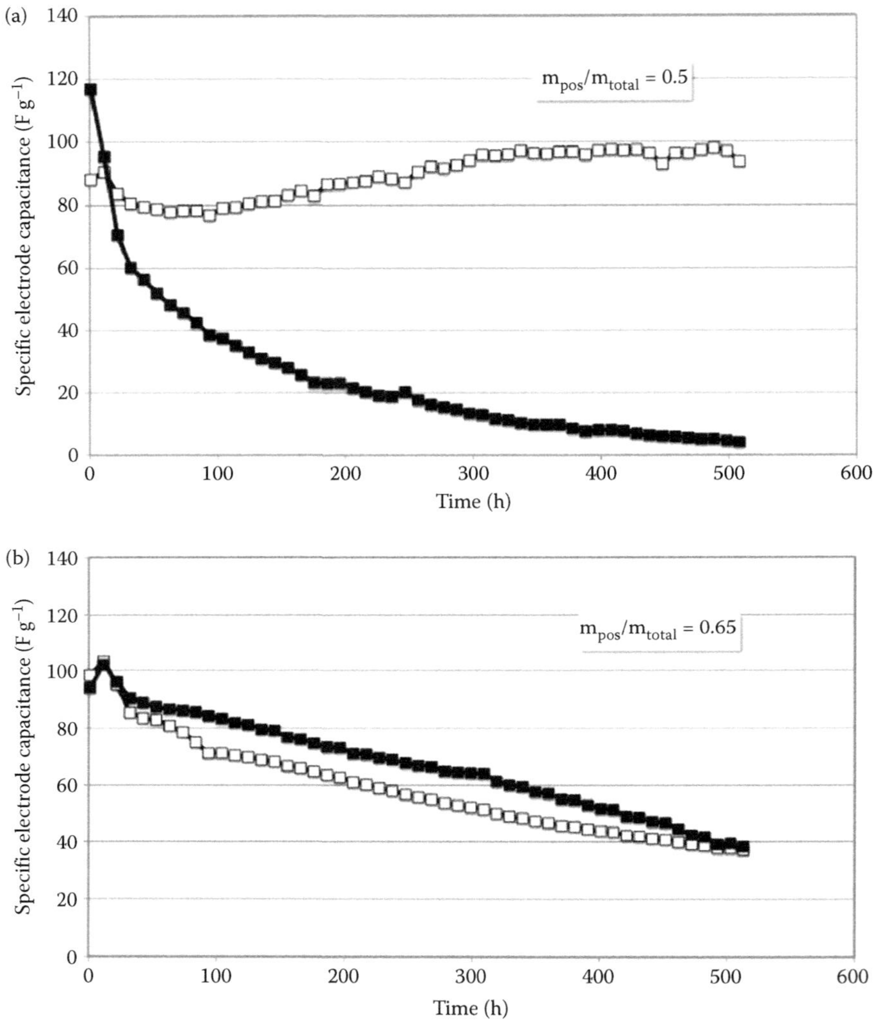

FIGURE 4.32 Specific capacitance of the single electrodes during accelerated aging tests in 1 M $TEABF_4$ in ACN with mass ratio of (a) 0.5 and (b) 0.65. The full and open squares represent positive electrode and negative electrode, respectively. (Reprinted from *Journal of Power Sources*, 196, Cericola, D., R. Kötz, and A. Wokaun, Effect of electrode mass ratio on aging of activated carbon based supercapacitors utilizing organic electrolytes, 3114–3118, Copyright 2011, with permission from Elsevier.)

pores and hindered the ions diffusion, thus leading to the increase of resistance [59]. Note that the decreasing of electrolyte concentration caused by its decomposition may also affect electrolyte conductivity and capacitance [99].

Water contamination in the organic electrolyte may dramatically deteriorate ES performance. Wang et al. [100] demonstrated the water contamination in the

TABLE 4.2
Used XPS to Determine the Elements Forming the Composition of Maxsorb Carbon Materials in 1 M $TEABF_4$ in ACN Electrolyte before and after Aging by Floating for 4000 h at 2.5 V

Element	As-Received Powder	Fresh Electrode	Aged Electrode (Negative)		Aged Electrode (Positive)	
			Surface	Inside	Surface	Inside
B	0	0	0	0.3	0	0
Na	0	1.1	0.8	0.3	0	0
F	0	0	2.2	0.6	0.7	1
N	0	0	0.9	0	12.5	4
O	4	17.7	9.2	19.8	13.2	18.6

Source: Azaïs, P. et al. 2007. *Journal of Power Sources* 171:1046–1053. With permission.
Note: Analysis of the aged electrodes was carried out on the surface after removing a few layers.

electrolyte (1.5 M $TEMAPF_6$/PC, TEMA stands for triethylmethylammonium) led to a significant initial irreversible capacitance of AC/graphite-based ES. The fading trend of the discharge capacity with prolonged cycles became more serious as the amount of water increased in the AC/graphite ES. The authors also found the cathodic electrolyte decomposition (negative side) became more obvious in the water-containing electrolyte than in the water-free electrolyte, while the anodic decomposition (positive side) seemed to be milder. Cericola et al. [101] also investigated the influence of water contamination on the capacitance aging of an AC electrode-based EDLC using constant voltage load tests in 1 M $TEABF_4$ in ACN or PC electrolytes. An initial loss of capacitance was also observed for a cell voltage of 2.75 V with both electrolytes. However, when the cell voltage was elevated to 3.5 V, the introduced water contamination had no effect on the capacitance loss for both the cell and the single electrode. This is in contradiction to results from Ishimoto et al. [89], who demonstrated that the side reactions induced by water impurity may led to the degradation processes of the electrolyte at higher voltage. Water seems to be stable at the positive electrode while possibly consumed at the negative electrode. Water reduced to H_2 gas as well as the formation of an equal amount of OH^- at the negative electrode may be the primary reason for a capacitance loss of EDLCs in water-polluted electrolytes. It is therefore urgent to search for more pure and electrochemically more stable electrolyte systems.

In order to understand the aging or failure mechanisms of the IL electrolyte-based EDLCs, the electrochemical decomposition of the ILs in EDLCs beyond the ESPW has been investigated [35,102,103]. Using *in situ* infrared and electrochemical spectroscopy methods, Romann et al. [102] reported that for [EMIM][BF_4] IL electrolyte, imidazolium cation dimerized and led to the formation of 1,5-diethyl-4,8-dimethyl-1,4,5,8-tetraazafulvalene at the electrode potential below –2 V (vs. Ag/AgCl wire in the same IL), while the formation of BF_3 complexes with BF_4^- anion and fluorination of the cations occurred at 2 V (Figure 4.33). The fluorination of the IL limited the cell voltage of symmetric [EMIM][BF_4]-based EDLCs below 2.8 V. *In situ* XPS have also been used to investigate the possible electrochemical reactions taking place on the WE in ILs [35,103]. For example, Kruusma et al. [103] found the oxidative polymerization of imidazolium cation started at the positively polarized Mo_2C-derived carbon (C(Mo_2C))|[EMIM][BF_4] interface in the symmetric EDLCs at the cell potential above 3.4 V (Figure 4.34). Meanwhile, the color of the [EMIM][BF_4] was observed to change from colorless to brown at both electrodes, suggesting the possible presence of new $EMIM^+$ cation polymeric compounds or the cathodic reduction or anodic oxidation of the Al current collectors. Besides the electrochemical decomposition of ILs, electrolyte ion intercalation induced degradation of the CNT electrodes in [EMIM][TFSI] ILs were also observed by using Raman spectroscopy [104]. Thus, the voltage operation window (ΔU) mainly determines the degree of capacity fade in prolonged cycling testing. When the cell operated at $\Delta U \leq 2.8$ V, the two-electrode potentials both located within the limits, therefore, the system is highly reversible and the capacitance showed little or no decay with increasing cycle number. When $\Delta U = 3.4$ V, the positive electrode was beyond the limit but the negative electrode was not. The cell degradation was moderate, because the intercalation only occurred on the one side. When $\Delta U = 3.8$ V, both electrodes went over their limits, and the destructions were exhaustive. Hence, the cycle number was much shorter than the cases of ESs with $\Delta U \leq 2.8$ V.

Another critical topic is the hydrolysis of ILs, especially in the presence of some quantities of water. The results from Freire et al. [105] obtained by combination of

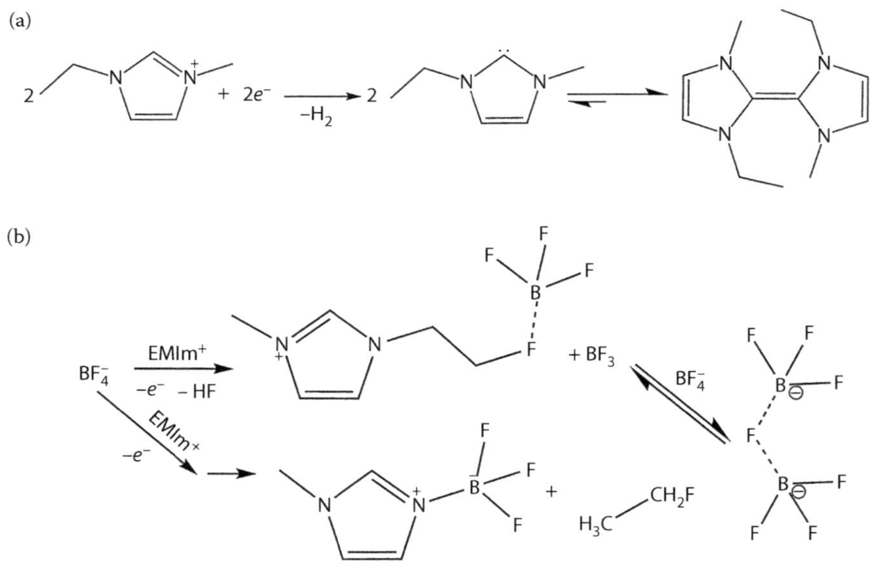

FIGURE 4.33 Reaction schemes at (a) −2.4 V and (b) at $E > 2$ V. (Reprinted from *Electrochimica Acta*, 125, Romann, T. et al., Surface chemistry of carbon electrodes in 1-ethyl-3-methylimidazolium tetrafluoroborate ionic liquid—An *in situ* infrared study, 183–190, Copyright 2014, with permission from Elsevier.)

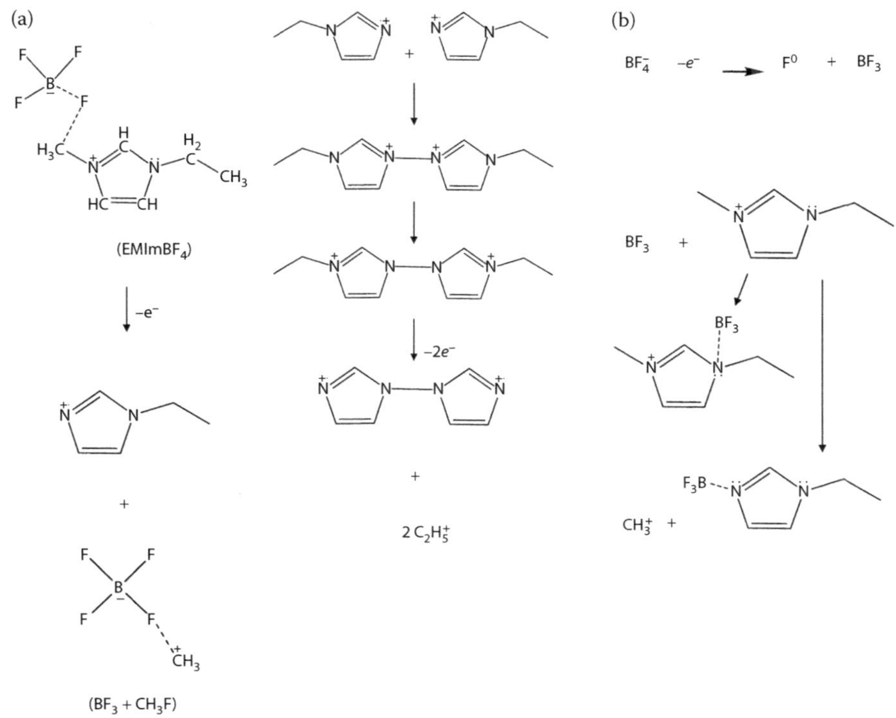

FIGURE 4.34 Proposed electrochemical processes occurring on the Mo_2C-derived carbon electrode at high positive polarizations in [EMIM][BF_4] electrolyte: (a) oxidative dimerization of $EMIM^+$ cations via N–N bond formation; (b) possible complexation of *in situ* electrochemically formed BF_3 with a free electron pair of a nitrogen atom in the $EMIM^+$ cation and/or the substitution of an alkyl chain to the $EMIM^+$ cation. (From Kruusma, J. et al. 2014. In situ XPS studies of electrochemically positively polarized molybdenum carbide derived carbon double layer capacitor electrode. *Journal of the Electrochemical Society*, 161:A1266–A1277. Reproduced by permission of The Electrochemical Society.)

NMR spectroscopy, electrospray ionization mass spectrometry (ESI-MS), and PH measurements clearly suggested the hydrolysis of [C_4MIM][BF_4], [C_8MIM][BF_4], [C_4MIM][PF_6], and [C_8MIM][PF_6] ILs was a reaction process only correlated with $[PF_6]^-$ and $[BF_4]^-$ and the cation of imidazolium remained remarkably stable. The $[BF_4]^-$-based ILs hydrolyzed at all conditions even at room temperature (Table 4.3) while $[PF_6]^-$-based ILs were still electrochemically stable at moderate conditions. The decomposition of $[PF_6]^-$ appeared to be significant under high temperature (343 and 373 K) or acidic conditions (PH = 3). Moreover, the comparison of hydrolysis behavior between the [C_4MIM][PF_6] and [C_8MIM][PF_6] indicated the longer the alkyl chain, the larger the extent of the anion hydrolysis (Table 4.3), which might be due to the related cation–anion interaction strength.

To summarize, the appropriate choice of electrolytes, together with suitable electrode materials, current collectors, and also separators is important in minimizing the degradation of the ESs, thus improving the cycling stability. To mitigate

TABLE 4.3

Relative Abundance (%)[a] of the Ions Determined by the ESI-MS Spectra of an Aqueous Solution of $[C_4MIM][BF_4]$ and of the Aqueous-Rich Phase of $[C_8MIM][BF_4]$

IL	Equilibrium Conditions	*m/z* 51 $[CH_3OH \cdot F]^-$	*m/z* 87 $[^{11}BF_4]^-$	*m/z* 99[b] $[^{11}BF_3OCH_3]^-$	*m/z* 111[b] $[^{11}BF_2(OCH_3)_2]^-$
$[C_4MIM][BF_4]$	24 h at 298 K	1.24	100	2.43	0.61
	24 h at 318 K	0.19	100	4.41	0.93
	24 h at 298 K; PH 3	0.69	100	1.96	0.50
	24 h at 298 K; PH 12	0.72	100	0.04	<0.1
	30 min at 343 K	0.22	100	3.49	0.64
	30 min at 373 K	0.55	100	1.83	0.56
$[C_8MIM][BF_4]$	24 h at 298 K	2.17	100	5.10	1.76
	24 h at 318 K	3.41	100	12.16	3.31
	24 h at 298 K; PH 3	1.46	100	2.4	0.86
	24 h at 298 K; PH 12	3.16	100	0.27	0.14
	30 min at 343 K	1.71	100	17.71	4.65
	30 min at 373 K	1.13	100	25.61	6.68

Source: From Freire, M. G. et al. 2010. *The Journal of Physical Chemistry A* 114:3744–3749 2010. With permission.

[a] Relative to the base peak taken as 100%.

[b] Ions corresponding to ^{10}B were observed with an abundance corresponding to the isotopic pattern of boron.

electrolyte degradation, it is definitely required that degradation mechanisms and failure modes be understood. More attention should especially be paid to the electrolyte reaction and decomposition correlated with the higher cell voltage operation. This can be achieved by using both experimental characterization methods and theoretical model approaches in molecular and electronic levels. Because of the very thin electrode/electrolyte interface and the small amount of the products formed on the surface of the WE, new highly sensitive surface-analysis methods should be employed. Apart from FTIR, XPS, and NMR spectra technologies used in the current literature, a variety of other instrumental analysis methods, especially the online and *in situ* characterizations (such as DEMS, Raman, Auger electron spectroscopy, Rutherford backscattering, and secondary ion mass spectrometry), should be developed in the near future to monitor the changes or detect the formed compounds in the electrolytes during lifetime tests. In addition, increasing the purity of the electrolyte and the working temperature range would be helpful to improve the performance of the electrolyte and the resultant ESs.

4.6 SUMMARY AND POSSIBLE RESEARCH DIRECTIONS

In the past few decades, great progress has been achieved in the field of characterization and validation of electrolytes' performance. Various types of cell configuration have been designed and used to test the performance of electrolytes and related ESs. Several important electrochemical characterization methods together with various instrumental technical strategies have been widely employed to characterize, test, and diagnose ESs with the investigated electrolyte. A series of important parameters of ESs, such as capacitance, ESR, energy and power densities, and cycle life have been presented so as to identify the performance and the stability of the related electrolyte. With the increasing number of investigations in this area, it is urgent to develop suitable and standardized methods to measure, evaluate, and also compare the performance of various electrolytes and related ESs. Faster screening and higher output methods are necessary for testing the performance of electrolytes to accelerate the development of novel electrolytes. In addition, some electrolytes are still very high cost and it is also highly urgent to develop purification procedures to decrease the impurities in organic solvents. Special attention should be paid to the matching between electrolytes (especially the newly developed ones) and electrode materials and also the inactive components in the ESs. Furthermore, much effort should be devoted to mitigating and overcoming the challenges related to the degradation and failure of electrolytes. This requires both experimental and theoretical model approaches to get a deep, fundamental understanding about the degradation mechanisms and failure modes in more realistic and complex electrodes.

REFERENCES

1. Zhong, C., Y. Deng, W. Hu, J. Qiao, L. Zhang, and J. Zhang. 2015. A review of electrolyte materials and compositions for electrochemical supercapacitors. *Chemical Society Reviews* 44:7484–7539.

2. Liu, R., L. Zhang, X. Sun, H. Liu, and J. Zhang. 2012. *Electrochemical Technologies for Energy Storage and Conversion*. Weinheim, Germany: Wiley-VCH Verlag GmbH & Co. KGaA.
3. Tsay, K.-C., L. Zhang, and J. Zhang. 2012. Effects of electrode layer composition/thickness and electrolyte concentration on both specific capacitance and energy density of supercapacitor. *Electrochimica Acta* 60:428–436.
4. Yu, A., V. Chabot, and J. Zhang. 2013. *Electrochemical Supercapacitors for Energy Storage and Delivery: Fundamentals and Applications*. Boca Raton, FL: CRC Press.
5. Khomenko, V., E. Frackowiak, and F. Béguin. 2005. Determination of the specific capacitance of conducting polymer/nanotubes composite electrodes using different cell configurations. *Electrochimica Acta* 50:2499–2506.
6. Chmiola, J., G. Yushin, Y. Gogotsi, C. Portet, P. Simon, and P. L. Taberna. 2006. Anomalous increase in carbon capacitance at pore sizes less than 1 nanometer. *Science* 313:1760–1763.
7. Nam, K. W., and K. B. Kim. 2002. A study of the preparation of NiO_x electrode via electrochemical route for supercapacitor applications and their charge storage mechanism. *Journal of the Electrochemical Society* 149:A346–A354.
8. Vivekchand, S. R. C., C. S. Rout, K. S. Subrahmanyam, A. Govindaraj, and C. N. R. Rao. 2008. Graphene-based electrochemical supercapacitors. *Journal of Chemical Sciences* 120:9–13.
9. Hardwick, L. J., M. Hahn, P. Ruch et al. 2006. An *in situ* Raman study of the intercalation of supercapacitor-type electrolyte into microcrystalline graphite. *Electrochimica Acta* 52:675–680.
10. Poli, F., J. S. Kshetrimayum, L. Monconduit, and M. Letellier. 2011. New cell design for *in-situ* NMR studies of lithium-ion batteries. *Electrochemistry Communications* 13:1293–1295.
11. Wang, H., T. K. J. Koster, N. M. Trease et al. 2011. Real-time NMR studies of electrochemical double-layer capacitors. *Journal of the American Chemical Society* 133:19270–19273.
12. Blanc, F., M. Leskes, and C. P. Grey. 2013. *In situ* solid-state NMR spectroscopy of electrochemical cells: Batteries, supercapacitors, and fuel cells. *Accounts of Chemical Research* 46:1952–1963.
13. Boukhalfa, S., D. Gordon, L. L. He et al. 2014. *In situ* small angle neutron scattering revealing ion sorption in microporous carbon electrical double layer capacitors. *ACS Nano* 8:2495–2503.
14. Melnichenko, Y. B., H. Mayama, G. Cheng, and T. Blach. 2010. Monitoring phase behavior of sub- and supercritical CO_2 confined in porous fractal silica with 85% porosity. *Langmuir* 26:6374–6379.
15. Bockris, J. O. M., A. K. N. Reddy, and M. Gamboa-Aldeco. 2002. *Modern Electrochemistry 2A: Fundamentals of Electrodics*. New York: Kluwer Academic.
16. Li, S., L. Qi, L. Lu, and H. Wang. 2012. Facile preparation and performance of mesoporous manganese oxide for supercapacitors utilizing neutral aqueous electrolytes. *RSC Advances* 2:3298–3308.
17. Zhang, X. J., W. H. Shi, J. X. Zhu et al. 2010. Synthesis of porous NiO nanocrystals with controllable surface area and their application as supercapacitor electrodes. *Nano Research* 3:643–652.
18. Wang, W., W. Lei, T. Yao et al. 2013. One-pot synthesis of graphene/SnO_2/PEDOT ternary electrode material for supercapacitors. *Electrochimica Acta* 108:118–126.
19. Peng, X. Y., X. X. Liu, D. Diamond, and K. T. Lau. 2011. Synthesis of electrochemically-reduced graphene oxide film with controllable size and thickness and its use in supercapacitor. *Carbon* 49:3488–3496.
20. Joseph, J., R. Rajagopalan, S. S. Anoop et al. 2014. Shape tailored $Ni_3(NO_3)_2(OH)_4$ nano-flakes simulating 3-D bouquet-like structures for supercapacitors: Exploring the effect of electrolytes on stability and performance. *RSC Advances* 4:39378–39385.

21. Andreas, H. A., K. Lussier, and A. M. Oickle. 2009. Effect of Fe-contamination on rate of self-discharge in carbon-based aqueous electrochemical capacitors. *Journal of Power Sources* 187:275–283.
22. Oickle, A. M., and H. A. Andreas. 2011. Examination of water electrolysis and oxygen reduction as self-discharge mechanisms for carbon-based, aqueous electrolyte electrochemical capacitors. *The Journal of Physical Chemistry C* 115:4283–4288.
23. Butler, J. N. 1970. *Advances in Electro-Chemistry and Electrochemical Engineering*, Vol. 7. New York: CRC Press, Interscience.
24. Cheng, N. 2015. *Solvent Handbook*, 5th edn. Beijing, China: Chemical Industry Press.
25. Fletcher, S., V. J. Black, I. Kirkpatrick, and T. S. Varley. 2013. Quantum design of ionic liquids for extreme chemical inertness and a new theory of the glass transition. *Journal of Solid State Electrochemistry* 17:327–337.
26. Endres, F. 2010. Physical chemistry of ionic liquids. *Physical Chemistry Chemical Physics* 12:1648–1648.
27. Devillers, N., S. Jemei, M. C. Pera, D. Bienaime, and F. Gustin. 2014. Review of characterization methods for supercapacitor modelling. *Journal of Power Sources* 246:596–608.
28. Burke, A., and M. Miller. 2010. Testing of electrochemical capacitors: Capacitance, resistance, energy density, and power capability. *Electrochimica Acta* 55:7538–7548.
29. Zhang, J. T., and X. S. Zhao. 2012. On the configuration of supercapacitors for maximizing electrochemical performance. *ChemSusChem* 5:818–841.
30. Stoller, M. D., and R. S. Ruoff. 2010. Best practice methods for determining an electrode material's performance for ultracapacitors. *Energy & Environmental Science* 3:1294–1301.
31. Lu, M., F. Beguin, and E. Frackowiak. 2013. *Supercapacitors: Materials, Systems and Applications*. Weinheim, Germany: Wiley-VCH Verlag GmbH & Co. KGaA.
32. Conway, B. E. 1999. *Electrochemical Supercapacitors: Scientific Fundamentals and Technological Applications*. New York: Kluwer Academic/Plenum.
33. Zhang, L., T. Wei, W. Wang, and X. Zhao. 2009. Manganese oxide–carbon composite as supercapacitor electrode materials. *Microporous and Mesoporous Materials* 123:260–267.
34. Palm, R., H. Kurig, K. Tonurist, A. Janes, and E. Lust. 2012. Is the mixture of 1-ethyl-3-methylimidazolium tetrafluoroborate and 1-butyl-3-methylimidazolium tetrafluoroborate applicable as electrolyte in electrical double layer capacitors? *Electrochemistry Communications* 22:203–206.
35. Tonisoo, A., J. Kruusma, R. Parna et al. 2013. *In situ* XPS studies of electrochemically negatively polarized molybdenum carbide derived carbon double layer capacitor electrode. *Journal of the Electrochemical Society* 160:A1084–A1093.
36. Romann, T., O. Oll, P. Pikma, and E. Lust. 2012. Abnormal infrared effects on bismuth thin film–$EMImBF_4$ ionic liquid interface. *Electrochemistry Communications* 23:118–121.
37. Wang, H., Q. Hao, X. Yang, L. Lu, and X. Wang. 2009. Graphene oxide doped polyaniline for supercapacitors. *Electrochemistry Communications* 11:1158–1161.
38. Bisquert, J. 2000. Influence of the boundaries in the impedance of porous film electrodes. *Physical Chemistry Chemical Physics* 2:4185–4192.
39. Taberna, P. L., P. Simon, and J. F. Fauvarque. 2003. Electrochemical characteristics and impedance spectroscopy studies of carbon–carbon supercapacitors. *Journal of the Electrochemical Society* 150:A292–A300.
40. Barcia, O. E., E. D'Elia, I. Frateur, O. R. Mattos, N. Pébère, and B. Tribollet. 2002. Application of the impedance model of de Levie for the characterization of porous electrodes. *Electrochimica Acta* 47:2109–2116.
41. Niu, Z., W. Zhou, X. Chen, J. Chen, and S. Xie. 2015. Highly compressible and all-solid-state supercapacitors based on nanostructured composite sponge. *Advanced Materials* 27:6002–6008.

42. Peng, S., L. Li, Y. Hu et al. 2015. Fabrication of spinel one-dimensional architectures by single-spinneret electrospinning for energy storage applications. *ACS Nano* 9:1945–1954.
43. Khomenko, V., E. Raymundo-Piñero, and F. Béguin. 2006. Optimisation of an asymmetric manganese oxide/activated carbon capacitor working at 2 V in aqueous medium. *Journal of Power Sources* 153:183–190.
44. Jurewicz, K., E. Frackowiak, and F. Béguin. 2004. Towards the mechanism of electrochemical hydrogen storage in nanostructured carbon materials. *Applied Physics A* 78:981–987.
45. Zhao, S., F. Wu, L. Yang, L. Gao, and A. F. Burke. 2010. A measurement method for determination of dc internal resistance of batteries and supercapacitors. *Electrochemistry Communications* 12:242–245.
46. Zhang, J., and X. Zhao. 2012. On the configuration of supercapacitors for maximizing electrochemical performance. *ChemSusChem* 5:818–841.
47. Yuan, X., C. Song, H. Wang, and J. Zhang. 2010. *Electrochemical Impedance Spectroscopy in PEM Fuel Cells*. New York: Springer.
48. Cheng, F. Y., J. Liang, Z. L. Tao, and J. Chen. 2011. Functional materials for rechargeable batteries. *Advanced Materials* 23:1695–1715.
49. Portet, C., P. L. Taberna, P. Simon, and E. Flahaut. 2005. Influence of carbon nanotubes addition on carbon–carbon supercapacitor performances in organic electrolyte. *Journal of Power Sources* 139:371–378.
50. Cole, K. S., and R. H. Cole. 1941. Dispersion and absorption in dielectrics I. alternating current characteristics. *The Journal of Chemical Physics* 9:341–351.
51. Cheng, F., H. Wang, Z. Zhu et al. 2011. Porous $LiMn_2O_4$ nanorods with durable high-rate capability for rechargeable Li-ion batteries. *Energy & Environmental Science* 4:3668–3675.
52. Di Fabio, A., A. Giorgi, M. Mastragostino, and F. Soavi. 2001. Carbon-poly(3-methylthiophene) hybrid supercapacitors. *Journal of the Electrochemical Society* 148:A845–A850.
53. Yan, J., T. Wei, B. Shao et al. 2010. Electrochemical properties of graphene nanosheet/carbon black composites as electrodes for supercapacitors. *Carbon* 48:1731–1737.
54. Stoller, M. D., S. Park, Y. Zhu, J. An, and R. S. Ruoff. 2008. Graphene-based ultracapacitors. *Nano Letters* 8:3498–3502.
55. Bettini, L. G., M. Galluzzi, A. Podestà, P. Milani, and P. Piseri. 2013. Planar thin film supercapacitor based on cluster-assembled nanostructured carbon and ionic liquid electrolyte. *Carbon* 59:212–220.
56. Largeot, C., P. L. Taberna, Y. Gogotsi, and P. Simon. 2011. Microporous carbon-based electrical double layer capacitor operating at high temperature in ionic liquid electrolyte. *Electrochemical and Solid State Letters* 14:A174–A176.
57. Ruch, P. W., D. Cericola, A. Foelske-Schmitz, R. Kotz, and A. Wokaun. 2010. Aging of electrochemical double layer capacitors with acetonitrile-based electrolyte at elevated voltages. *Electrochimica Acta* 55:4412–4420.
58. Weingarth, D., A. Foelske-Schmitz, and R. Kotz. 2013. Cycle versus voltage hold—Which is the better stability test for electrochemical double layer capacitors? *Journal of Power Sources* 225:84–88.
59. Azaïs, P., L. Duclaux, P. Florian et al. 2007. Causes of supercapacitors ageing in organic electrolyte. *Journal of Power Sources* 171:1046–1053.
60. Yoshida, A., I. Tanahashi, and A. Nishino. 1990. Effect of concentration of surface acidic functional groups on electric double-layer properties of activated carbon fibers. *Carbon* 28:611–615.
61. Chen, L. B., H. Bai, Z. F. Huang, and L. Li. 2014. Mechanism investigation and suppression of self-discharge in active electrolyte enhanced supercapacitors. *Energy & Environmental Science* 7:1750–1759.

62. Hu, L., J. W. Choi, Y. Yang et al. 2009. Highly conductive paper for energy-storage devices. *Proceedings of the National Academy of Sciences of the United States of America* 106:21490–21494.
63. Xu, K., M. S. Ding, and T. R. Jow. 2001. A better quantification of electrochemical stability limits for electrolytes in double layer capacitors. *Electrochimica Acta* 46:1823–1827.
64. Ue, M., K. Ida, and S. Mori. 1994. Electrochemical properties of organic liquid electrolytes based on quaternary onium salts for electrical double-layer capacitors. *Journal of the Electrochemical Society* 141:2989–2996.
65. Ue, M., M. Takeda, M. Takehara, and S. Mori. 1997. Electrochemical properties of quaternary ammonium salts for electrochemical capacitors. *Journal of the Electrochemical Society* 144:2684–2688.
66. McEwen, A. B., H. L. Ngo, K. LeCompte, and J. L. Goldman. 1999. Electrochemical properties of imidazolium salt electrolytes for electrochemical capacitor applications. *Journal of the Electrochemical Society* 146:1687–1695.
67. Olson, E. J., and P. Buhlmann. 2012. Unbiased assessment of electrochemical windows: Minimizing mass transfer effects on the evaluation of anodic and cathodic limits. *Journal of the Electrochemical Society* 160:A320–A323.
68. Weingarth, D., H. Noh, A. Foelske-Schmitz, A. Wokaun, and R. Kotz. 2013. A reliable determination method of stability limits for electrochemical double layer capacitors. *Electrochimica Acta* 103:119–124.
69. Maeshima, H., H. Moriwake, A. Kuwabara, C. A. J. Fisher, and I. Tanaka. 2014. An improved method for quantitatively predicting the electrochemical stabilities of organic liquid electrolytes using *ab initio* calculations. *Journal of the Electrochemical Society* 161:G7–G14.
70. Maeshima, H., H. Moriwake, A. Kuwabara, and C. A. J. Fisher. 2010. Quantitative evaluation of electrochemical potential windows of electrolytes for electric double-layer capacitors using *ab initio* calculations. *Journal of the Electrochemical Society* 157:A696–A701.
71. Richey, F. W., C. Tran, V. Kalra, and Y. A. Elabd. 2014. Ionic liquid dynamics in nanoporous carbon nanofibers in supercapacitors measured with in operando infrared spectroelectrochemistry. *The Journal of Physical Chemistry C* 118:21846–21855.
72. Melnichenko, Y. B., and G. D. Wignall. 2007. Small-angle neutron scattering in materials science: Recent practical applications. *Journal of Applied Physics* 102:021101.
73. Boukhalfa, S., L. He, Y. B. Melnichenko, and G. Yushin. 2013. Small-angle neutron scattering for *in situ* probing of ion adsorption inside micropores. *Angewandte Chemie-International Edition* 52:4618–4622.
74. Ohba, T., K. Hata, and H. Kanoh. 2012. Significant hydration shell formation instead of hydrogen bonds in nanoconfined aqueous electrolyte solutions. *Journal of the American Chemical Society* 134:17850–17853.
75. Richey, F. W., B. Dyatkin, Y. Gogotsi, and Y. A. Elabd. 2013. Ion dynamics in porous carbon electrodes in supercapacitors using *in situ* infrared spectroelectrochemistry. *Journal of the American Chemical Society* 135:12818–12826.
76. Lee, S.-I., K. Saito, K. Kanehashi et al. 2006. ^{11}B NMR study of the anion in activated carbons at various stages of charge of EDLCs in organic electrolyte. *Carbon* 44:2578–2586.
77. Borchardt, L., M. Oschatz, S. Paasch, S. Kaskel, and E. Brunner. 2013. Interaction of electrolyte molecules with carbon materials of well-defined porosity: Characterization by solid-state NMR spectroscopy. *Physical Chemistry Chemical Physics* 15:15177–15184.
78. Gu, W., and G. Yushin. 2014. Review of nanostructured carbon materials for electrochemical capacitor applications: Advantages and limitations of activated carbon, carbide-derived carbon, zeolite-templated carbon, carbon aerogels, carbon nanotubes, onion-like carbon, and graphene. *Wiley Interdisciplinary Reviews: Energy and Environment* 3:424–473.

79. Raymundo-Piñero, E., K. Kierzek, J. Machnikowski, and F. Béguin. 2006. Relationship between the nanoporous texture of activated carbons and their capacitance properties in different electrolytes. *Carbon* 44:2498–2507.
80. Gao, H., Q. Tian, and K. Lian. 2010. Polyvinyl alcohol-heteropoly acid polymer electrolytes and their applications in electrochemical capacitors. *Solid State Ionics* 181:874–876.
81. Lian, K., and Q. Tian. 2010. Solid asymmetric electrochemical capacitors using proton-conducting polymer electrolytes. *Electrochemistry Communications* 12:517–519.
82. Lian, K., and C. Li. 2009. Solid polymer electrochemical capacitors using heteropoly acid electrolytes. *Electrochemistry Communications* 11:22–24.
83. Gao, H., and K. Lian. 2010. Characterizations of proton conducting polymer electrolytes for electrochemical capacitors. *Electrochimica Acta* 56:122–127.
84. Kötz, R., M. Hahn, and R. Gallay. 2006. Temperature behavior and impedance fundamentals of supercapacitors. *Journal of Power Sources* 154:550–555.
85. Kötz, R., P. W. Ruch, and D. Cericola. 2010. Aging and failure mode of electrochemical double layer capacitors during accelerated constant load tests. *Journal of Power Sources* 195:923–928.
86. Ruiz, V., C. Blanco, M. Granda, and R. Santamaria. 2008. Enhanced life-cycle supercapacitors by thermal treatment of mesophase-derived activated carbons. *Electrochimica Acta* 54:305–310.
87. Bittner, A. M., M. Zhu, Y. Yang et al. 2012. Ageing of electrochemical double layer capacitors. *Journal of Power Sources* 203:262–273.
88. Vali, R., A. Laheaar, A. Janes, and E. Lust. 2014. Characteristics of non-aqueous quaternary solvent mixture and Na-salts based supercapacitor electrolytes in a wide temperature range. *Electrochimica Acta* 121:294–300.
89. Ishimoto, S., Y. Asakawa, M. Shinya, and K. Naoi. 2009. Degradation responses of activated-carbon-based EDLCs for higher voltage operation and their factors. *Journal of the Electrochemical Society* 156:A563–A571.
90. Tian, Y., J. Yan, R. Xue, and B. Yi. 2011. Influence of electrolyte concentration and temperature on the capacitance of activated carbon. *Acta Physico—Chimica Sinica* 27:479–485.
91. Hahn, M., A. Wursig, R. Gallay, P. Novak, and R. Kötz. 2005. Gas evolution in activated carbon/propylene carbonate based double-layer capacitors. *Electrochemistry Communications* 7:925–930.
92. Hahn, M., R. Kötz, R. Gallay, and A. Siggel. 2006. Pressure evolution in propylene carbonate based electrochemical double layer capacitors. *Electrochimica Acta* 52:1709–1712.
93. Kötz, R., M. Hahn, P. Ruch, and R. Gallay. 2008. Comparison of pressure evolution in supercapacitor devices using different aprotic solvents. *Electrochemistry Communications* 10:359–362.
94. Naoi, K., S. Ishimoto, J. I. Miyamoto, and W. Naoi. 2012. Second generation "nanohybrid supercapacitor": Evolution of capacitive energy storage devices. *Energy & Environmental Science* 5:9363–9373.
95. Kurzweil, P., and M. Chwistek. 2008. Electrochemical stability of organic electrolytes in supercapacitors: Spectroscopy and gas analysis of decomposition products. *Journal of Power Sources* 176:555–567.
96. Zhu, M., C. J. Weber, Y. Yang et al. 2008. Chemical and electrochemical ageing of carbon materials used in supercapacitor electrodes. *Carbon* 46:1829–1840.
97. Ruch, P. W., D. Cericola, A. Foelske, R. Kotz, and A. Wokaun. 2010. A comparison of the aging of electrochemical double layer capacitors with acetonitrile and propylene carbonate-based electrolytes at elevated voltages. *Electrochimica Acta* 55:2352–2357.

98. Cericola, D., R. Kötz, and A. Wokaun. 2011. Effect of electrode mass ratio on aging of activated carbon based supercapacitors utilizing organic electrolytes. *Journal of Power Sources* 196:3114–3118.
99. Zheng, J. P., and T. R. Jow. 1997. The effect of salt concentration in electrolytes on the maximum energy storage for double layer capacitors. *Journal of the Electrochemical Society* 144:2417–2420.
100. Wang, H., and M. Yoshio. 2010. Effect of water contamination in the organic electrolyte on the performance of activated carbon/graphite capacitors. *Journal of Power Sources* 195:389–392.
101. Cericola, D., P. W. Ruch, A. Foelske-Schmitz, D. Weingarth, and R. Kotz. 2011. Effect of water on the aging of activated carbon based electrochemical double layer capacitors during constant voltage load tests. *International Journal of Electrochemical Science* 6:988–996.
102. Romann, T., O. Oll, P. Pikma, H. Tamme, and E. Lust. 2014. Surface chemistry of carbon electrodes in 1-ethyl-3-methylimidazolium tetrafluoroborate ionic liquid—An *in situ* infrared study. *Electrochimica Acta* 125:183–190.
103. Kruusma, J., A. Tonisoo, R. Parna, E. Nommiste, and E. Lust. 2014. *In situ* XPS studies of electrochemically positively polarized molybdenum carbide derived carbon double layer capacitor electrode. *Journal of the Electrochemical Society* 161:A1266–A1277.
104. Tsai, D. S., C. H. Chang, W. W. Chiang, K. Y. Lee, and Y. S. Huang. 2014. Effects of ion insertion on cycling performance of miniaturized electrochemical capacitor of carbon nanotubes array. *Nanotechnology* 25:425401.
105. Freire, M. G., C. M. S. S. Neves, I. M. Marrucho, J. A. P. Coutinho, and A. M. Fernandes. 2010. Hydrolysis of tetrafluoroborate and hexafluorophosphate counter ions in imidazolium-based ionic liquids. *The Journal of Physical Chemistry A* 114:3744–3749.

5 Challenges and Perspectives of Electrochemical Supercapacitors

Cheng Zhong, Yida Deng, Wenbin Hu, Jinli Qiao, and Jiujun Zhang

CONTENTS

5.1 INTRODUCTION

To date, significant achievements have been made in understanding electrolyte effects on the performance of electrochemical supercapacitors (ESs). Novel electrolytes have also been created to improve performance and expand the commercialization of ES products.

This chapter provides a comprehensive review of developments over the last 20 years and identifies recent research trends regarding electrolytes for ESs. To improve the overall ES performance, a wide variety of electrolytes, such as aqueous electrolytes; organic electrolytes; ionic liquids (ILs); solid-state or quasi-solid-state electrolytes; and redox-active electrolytes, have been thoroughly investigated in the literature and summarized in this book. Extensive laboratory studies have focused on revealing the influence of electrolyte properties (e.g., electrochemical stability, ionic conductivity, and viscosity) on the primary performance characteristics of ESs (e.g., capacitance, energy and power densities, equivalent series resistance [ESR], rate capacity, cycling stability, and temperature stability). A major research trend in this area has been efforts to develop novel electrolytes for ESs with high-energy density without sacrificing other advantages such as high-power density and excellent cycling stability. Although organic electrolytes currently dominate the commercial ES market, aqueous electrolytes are the most extensively used electrolytes in academic research since they are inexpensive, easy to handle, and do not require complicated drying and purification processes compared to those used in the synthesis of organic electrolytes. Traditional aqueous electrolyte-based ESs generally have low cell voltages up to about 1.0 V, which severely limited their energy density. However, with recent advances such as neutral electrolytes and the development of asymmetric and hybrid ESs, the operative cell voltage of aqueous electrolyte-based ESs can now effectively be increased to more than 2.0 V. These developments result in a much higher energy density.

To further increase the operative cell voltage, IL electrolytes have attracted a great deal of attention due to their high electrochemical stability. Besides increasing cell voltage, redox-active electrolytes have also been investigated to increase the energy density of ESs through the pseudocapacitance contribution from the electrolytes themselves. Interestingly, the creation of solid-state electrolytes greatly facilitates the development of novel ESs with diverse shapes, structures, and applications.

Since the components of ESs are strongly coupled with each other, it is important to consider not only the interplay between electrolytes and electrode materials, but also possible interactions between the electrolytes and inactive components including current collectors, binders, and separators.

Despite remarkable improvements in recent years in ES electrolytes, several significant challenges still exist. In this chapter, the major challenges of ES electrolyte development will be discussed, followed by a discussion of some perspectives and future potential research directions.

5.2 CHALLENGES

Several major challenges in the field of electrolytes for ESs still exist.

1. Usually trade-offs exist between important properties of the electrolytes for ESs. In some cases, such trade-offs are acceptable whereas in other cases, they may have a strong negative effect on the overall performance of ESs. One important trade-off exists between electrochemical stability and ionic conductivity. Electrolytes with a higher electrochemical stability allow increases in the operating cell voltage of the ESs, which significantly improves the energy density. However, higher electrochemical stability often causes deterioration to other properties of the electrolytes.

   As can be clearly seen in Table 2.1, although ESs using organic electrolytes can achieve a much higher operative cell voltage compared to aqueous electrolyte-based ESs, the ionic conductivity of the organic electrolyte is generally about one magnitude lower than the aqueous electrolyte. This differential greatly increases the ESR and limits the power density of the ES. The application of IL electrolytes for ESs can further increase the working cell voltage up to 4 V, but the ionic conductivity of IL is much lower even than that of the organic electrolytes. A similar trade-off also exists between the electrochemical stability and viscosity of the electrolytes. As a result, the application of ES's electrolytes with higher electrochemical stability often increases ESR and decreases the rate and power performance, unless the power density loss resulting from the increased ESR can be compensated by an increase in cell voltage. Besides, electrolytes with better thermal stability such as ILs usually have lower ionic conductivity. Regarding specific capacitance, organic or IL electrolyte-based ESs commonly exhibit a smaller specific capacitance value compared to aqueous electrolyte ESs. It is therefore, rather a challenge to develop ESs with high-energy density while maintaining other advantages (e.g., high-power density and excellent cycling stability) of ESs.

2. An in-depth, fundamental understanding about the charge storage mechanism of ESs especially with newly developed electrolytes. Recently, the fundamental understanding of the effect of an electrolyte on an ES's performance has been greatly advanced by both theoretical and *in situ* experimental methods. Nevertheless, with further development of new electrolytes (e.g., ILs, redox-active electrolytes, and solid-state electrolytes) and new electrode materials (e.g., lithium-ion capacitor electrodes), a challenge still exists in the understanding of charge storage and release mechanisms for these novel systems. For example, conflicting results have been reported regarding the relationship between the pore size and the capacitance normalized by surface area [1]. Several studies found the specific capacitance increased significantly if the average pore size of the carbon electrode was similar to the electrolyte ion size of $TEABF_4$/ACN organic electrolyte (desolvated ion size) and IL [2,3]. However, Centeno et al. [4] reported the specific capacitance did not change obviously for 28 porous carbons with the pore size ranging from 0.7 to 15 nm (Figure 5.1). On the basis of density functional theory (DFT) studies, Jiang et al. [5] found the relationship between the capacitance and the pore size of the nanoporous electrode in the IL electrolyte exhibited the oscillation pattern (Figure 2.57). This finding based on DFT calculations needs to be confirmed by further direct experimental work. Similarly, regarding IL electrolytes, some different results have also been reported regarding the effects of the organic solvent [6,7] and temperature [8] on the electrical double-layer (EDL) structure and the corresponding capacitance at the interface between the electrode and IL. Besides, compared to the extensively studied planar or carbon powder-based electrodes, the behavior of electrolyte ions in some complex structures, such as a porous hierarchical nanostructure, is less well understood. Furthermore, although extensive fundamental studies have been done regarding the EDL charge storage mechanisms, there have been limited studies on the theoretical and modeling work associated with pseudocapacitive materials-based electrodes [9–13].
3. Standardized characterization methods to evaluate the performance of electrolytes for ESs. As can be seen in Table 2.4, it is not easy to compare the performance data from the different literature since the experimental data are usually obtained under different testing conditions or parameters. As a result, it is relatively difficult to identify candidate electrolytes from different work [14]. In addition, the reported values of the specific capacitance, energy, and power densities of ESs are mostly normalized by the mass of electrode materials. However, the total mass and cost of ESs are also strongly dependent on the amount of electrolyte used, especially when

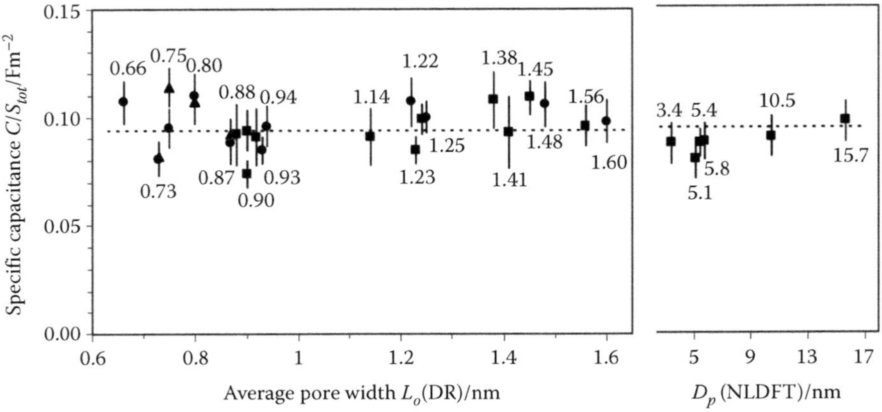

FIGURE 5.1 Specific capacitances of porous carbons in the TEABF$_4$/ACN electrolyte. Left: C/S_{tot} versus the average pore width L_o(DR) for microporous carbons. The data show no anomalous behavior below 1 nm and in the range of 0.66–1.60 nm, the average capacitance is 0.094 ± 0.011 F m^{-2}. Right: mesoporous carbons with pore diameters D_p between 3.4 and 15.7 nm, confirming the absence of a trend. S_{tot} and L_o(DR) are obtained from CCl_4 isotherms (•) supplemented by norbornadiene (▲), or only by small molecules (■) (mainly nitrogen and phenol), where the enthalpies of immersion indicate full accessibility of the pore system to the $(C_2H_5)_4N^+$ ion. (Centeno, T. A., O. Sereda, and F. Stoeckli. 2011. Capacitance in carbon pores of 0.7 to 15 nm: A regular pattern. *Physical Chemistry Chemical Physics* 13:12403–12406. Reproduced by permission of The Royal Society of Chemistry.)

electrode materials with low-packing density are involved. To date, there has been limited work taking into account the quantity of electrolytes utilized when evaluating the performance of ESs.

4. High cost of some electrolytes. The cost of state-of-the-art ESs is \$10–20 Wh^{-1}, which is much higher than that of lithium-ion batteries (\$1–2) [15]. The high cost of ESs could greatly limit their commercialization. Aqueous electrolytes are inexpensive; however, their low electrochemical stability limits the operative cell voltage and thus the energy density of ESs (especially symmetric ESs). Although the use of organic and IL electrolytes can effectively increase the operative cell voltage, they can also dramatically increase the cost of ESs.

5.3 FUTURE PERSPECTIVES AND POSSIBLE RESEARCH DIRECTIONS

To overcome the above-mentioned challenges, several possible research directions are suggested including:

1. Improvement of the energy density of the ESs while maintaining other advantages by exploring innovative electrolytes or optimizing electrolytes of interest.

Index

E

F

G

H

T

U

V

W

X